Limiting Greenhouse Effects
Controlling Carbon Dioxide Emissions

Goal of this Dahlem Workshop:

to provide a rational basis for national and
international efforts to reduce the rate of
accumulation of carbon dioxide in the atmosphere

Environmental Sciences Research Report ES 10

Held and published on behalf of the
Freie Universität Berlin

Sponsored by:
Stiftungsfonds Unilever
im Stifterverband für die Deutsche Wissenschaft

Limiting Greenhouse Effects
Controlling Carbon Dioxide Emissions

Edited by

G.I. PEARMAN

Report of the Dahlem Workshop on
Limiting the Greenhouse Effect:
Options for Controlling
Atmospheric Carbon Dioxide Accumulation

Berlin 1990, December 9–14

Program Advisory Committee:
G.I. Pearman, Chairperson
P.J. Crutzen, J.A. Edmonds, J. Jäger
K.M. Meyer-Abich, I. Mintzer

JOHN WILEY & SONS
Chichester · New York · Brisbane · Toronto · Singapore

Copyright © 1992 by John Wiley & Sons Ltd,
Baffins Lane, Chichester,
West Sussex, PO19 1UD, England

All rights reserved.

No part of this book may be reproduced by any means,
or transmitted, or translated into a machine language
without the written permission of the publisher.

Library of Congress Cataloging-in-Publication Data

Dahlem Workshop on Limiting the Greenhouse Effect : Options for
  Controlling Atmospheric Carbon Dioxide Accumulation (1990 : Berlin,
  Germany)
    Limiting greenhouse effects : controlling carbon dioxide emissions
  : report of the Dahlem Workshop on Limiting the Greenhouse Effect:
  Options for Controlling Atmospheric Carbon Dioxide Accumulation,
  Berlin 1990, December 9–14 / edited by G.I. Pearman.
      p. cm. — (Environmental sciences research
  reports : 10) (Dahlem workshop reports)
    Includes bibliographical references and index.
    ISBN 0-471-92945-X
    1. Atmospheric carbon dioxide—Environmental aspects—Congresses.
  2. Power resources—Environmental aspects—Congresses.
  3. Greenhouse effect, Atmospheric—Congresses. 4. Air quality
  management—Congresses. I. Pearman, G.I. II. Title.
  III. Series. IV. Series : Dahlem workshop reports.
  TD885.5.C3D34 1990
  628.5′3—dc20   91-38136
                                                                    CIP

British Library Catologuing-in-Publication Data

A catalogue record for this book is available from the British Library

ISBN 0-471-92945-X

Dahlem Editorial Staff: J. Lupp, C. Rued-Engel, A. Cezeaux
Typeset in 10/12 pt Times by Photo-graphics, Honiton, Devon
Printed and bound in Great Britain by Biddles Ltd, Guildford, Surrey

Contents

The Dahlem Konferenzen	ix
List of Participants with Fields of Research	xi

INTRODUCTION

1 Global Climate Change and the Energy Community
   G.I. Pearman   1

GENERAL BACKGROUND PAPERS

2 Greenhouse Gases: What Is Their Role in Climate Change?
   J.A. Edmonds, D. Wuebbles, and W.U. Chandler   13

3 What Are the Current Characteristics of the Global Energy Systems?
   I.A. Bashmakov   59

4 Insurance Against the Heat Trap: Estimating the Costs of Reducing the Risks
   I.M. Mintzer   83

OPTIONS FOR REDUCING GREENHOUSE POTENTIAL PER UNIT OF ENERGY SERVICE

5 Carbon Dioxide Emissions Control and Global Environmental Change
   J. Reilly   111

6 To What Extent Can Renewable Energy Systems Replace Carbon-based Fuels in the Next 15, 50, and 100 Years?
   C.-J. Winter   135

7 The Potential Role of Nuclear Power in Controlling CO_2 Emissions
   W. Fulkerson, J.E. Jones, J.G. Delene, A.M. Perry, and R.A. Cantor   165

8 The Likely Roles of Fossil Fuels in the Next 15, 50, and 100 Years, With or Without Active Controls on Greenhouse-Gas Emissions
   R.L. Kane and D.W. South   189

vi Contents

9 Group Report: What Are the Economic Costs, Benefits, and Technical Feasibility of Various Options Available to Reduce Greenhouse Potential per Unit of Energy Service?
S.T. Boyle, Rapporteur
W. Fulkerson, R. Klingholz, I.M. Mintzer, G.I. Pearman, G. Pinchera, J. Reilly, F. Staiß, R.J. Swart, C.-J. Winter   229

OPTIONS FOR REDUCING ENERGY USE PER UNIT OF GNP

10 Prospects for Efficiency Improvements in the Electricity Sector
K. Yamaji   261

11 Reducing Greenhouse-Gas Emissions from the Transportation Sector
L.L. Jenney   283

12 Global Options for Reducing Greenhouse-Gas Emissions from Residential and Commercial Buildings
W.B. Ashton, S.C. McDonald, and A.K. Nicholls   303

13 Potentials to Reduce Greenhouse-Gas Emissions by Rational Energy Use and Structural Changes
E.K. Jochem   327

14 Least-Cost Climatic Stabilization
A.B. Lovins and L.H. Lovins   351

15 Energy Saving in the U.S. and Other Wealthy Countries: Can the Momentum Be Maintained?
L. Schipper   443

16 Group Report: What Are the Options Available for Reducing Energy Use per Unit of GDP?
W.U. Chandler, Rapporteur
W.B. Ashton, I.A. Bashmakov, S. Büttner, L.L. Jenney, E.K. Jochem, A.B. Lovins, R.G. Richels, L. Schipper, K. Yamaji   463

SOCIAL AND INSTITUTIONAL BARRIERS TO CO_2 EMISSION REDUCTIONS

17 Regimes for Reducing Greenhouse-Gas Emissions
K. von Moltke   475

18 Reduction of Greenhouse-Gas Emissions: Barriers and Opportunities in Developing Countries
M. Munasinghe   487

Contents

19 Group Report: Social and Institutional Barriers to Reducing CO_2 Emissions
J. Ausubel, Rapporteur
E. Arrhenius, R.E. Benedick, O. Davidson, A.K. Jain,
J.-H. Kim, S. Rayner, B. Schlomann, R. Ueberhorst,
K. von Moltke   513

LINKAGES BETWEEN STRATEGIES FOR CO_2 EMISSION REDUCTION AND OTHER NATIONAL AND INTERNATIONAL GOALS

20 Implications for Greenhouse-Gas Emissions of Strategies Designed to Ameliorate Other Social and Environmental Problems
D.A. Tirpak and D.R. Ahuja   535

21 How Will Climatic Changes and Strategies for the Control of Greenhouse-Gas Emissions Influence International Peace and Global Security?
P.H. Gleick   561

22 Winners and Losers in Climatic Change: How Will Greenhouse-Gas Emissions and Control Strategies Influence International and Intergenerational Equity?
K.M. Meyer-Abich   573

23 Group Report: What Is the Significance of the Linkage between Strategies to Reduce CO_2 Emissions and Other National and International Goals?
M. Lönnroth, Rapporteur
C.L. Caccia, U. Fritsche, P.H. Gleick, H. Hilse,
K.M. Meyer-Abich, T. Peter, U.E. Simonis, D.A. Tirpak,
I.J. Walker   603

Author Index   617

Subject Index   619

The Dahlem Konferenzen

The purpose of the Dahlem Konferenzen is to promote international interdisciplinary exchange of scientific information and ideas and to stimulate international cooperation in research. This is achieved by arranging discussion workshops, mainly in the life sciences and environmental sciences, organized according to a model developed and tested by the Dahlem Konferenzen.

The Dahlem Konferenzen was founded in 1974 by the Stifterverband für Deutsche Wissenschaft[1] in collaboration with the Deutsche Forschungsgemeinschaft[2] to promote more effective communication between scientists. It was named after the Dahlem district of Berlin, long a home of the sciences and arts. In January 1990, it became incorporated into the Freie Universität Berlin. Financial support comes from the Senate of the Land Berlin, the Deutsche Forschungsgemeinschaft and private foundations.

As scientific research has become increasingly interdisciplinary, a growing need for specialists in one field to understand the problems and work with the concepts of related fields has emerged. New insights can be gained when a problem is approached from the standpoint of another discipline, and because no existing form of scientific meeting provided the forum necessary for such exchanges, Dahlem Konferenzen created a new concept, which has been tested and refined over the years. Now internationally recognized as the Dahlem Workshop Model, it provides a framework for coherent interdisciplinary discussion of a topic of contemporary international interest in five working days, culminating in the draft manuscript for a book.

Dahlem Workshops provide a unique opportunity for posing questions to colleagues from different disciplines and for soliciting alternative opinions on contentious issues. The aim is not to solve problems or to reach a consensus but to identify gaps in our knowledge and possible new ways of approaching stubborn issues, and to define priorities for research. This approach is well summed up in the instruction to participants to state what they do not know rather than what they do.

Workshop topics are proposed by leading scientists and are approved by an independent scientific board advised by qualified referees. Approximately one year before the workshop, an advisory committee meets to determine the scientific program and select participants, who are invited according to

---

[1] The Donors Association for the Promotion of Sciences and Humanities, a foundation created in 1921 in Berlin and supported by German trade and industry to fund basic research in the sciences
[2] German Science Foundation

their scientific reputations, with the exception of a number of places reserved for junior German scientists.

The discussions at each workshop are organized around four key questions, each tackled by a group of about twelve participants with a range of expertise. There are no lectures; instead, prior to the workshop, selected participants write background papers that review particular facets of the topic and serve as the basis for the discussion. These papers are distributed to all participants before the meeting and selected participants act as referees. During the workshop, each of the four discussion groups prepares a report reflecting the ideas, opinions, and controversies that have emerged from their discussion as well as identifying problem areas and directions for future research.

The revised background papers and group reports are published as the Dahlem Workshop Reports. Each volume is edited by the chairperson(s) of the workshop and staff of the Dahlem Konferenzen. The Reports provide multidisciplinary surveys by an international group of distinguished scientists, based on discussions of advanced concepts, techniques and models. The Dahlem Workshop Reports are published in two series: Life Sciences and Environmental Sciences (formerly Physical, Chemical, and Earth Sciences).

Dr. Jennifer Altman, Scientific Director
Dahlem Konferenzen der Freien Universität Berlin
Tiergartenstr. 24–27, W-1000 Berlin 30, F.R. Germany

List of Participants with Fields of Research

E. ARRHENIUS   The World Bank, Room S-5045, 1818 H Street NW, Washington, DC 20433, U.S.A.

Implementation of scientific and technological knowledge in international financing

W.B. ASHTON   Battelle, Pacific Northwest Laboratory, 901 D Street SW, Suite 900, Washington, DC 20024-2115, U.S.A.

Technology responses to global climate change; energy conservation; strategic technology planning methods

J.H. AUSUBEL   The Rockefeller University, Box 234, 1230 York Avenue, New York, NY 10021, U.S.A.

Technology and environment

I.A. BASHMAKOV   Dept. of World Energy, Institute for Energy Research AS/USSR, 44/2 Vavilov Street, 117333 Moscow, U.S.S.R.

Future world and the U.S.S.R. energy development with minimum greenhouse gases emissions

R.E. BENEDICK   World Wildlife Fund and The Conservation Foundation, 1250 24th Street NW, Washington, DC 20037, U.S.A.

International environmental policy; ozone layer protection; climate change; multilateral negotiations and institutions.

S.T. BOYLE   Greenpeace, Canonbury Villas, London N1 2PN, U.K.

Previous affiliation: Association for the Conservation of Energy, 9 Sherlock Mews, London W2M 3RH, U.K.

The role of energy efficiency in reducing greenhouse-gas emissions; the development and assessment of energy/CO_2 scenarios

List of Participants with Fields of Research

S. BÜTTNER   Institut für Ökologische Wirtschaftsforschung, Giesebrechtstr. 13, 1000 Berlin 12, F.R. Germany

Instruments of ecological economics, ecological balancing, eco-controlling; solar hydrogen economy

C.L. CACCIA   House of Commons, Room 353S, Centre Block, Ottawa K1A 0A6, Canada

Parliamentary Center for environmentally sustainable development

W.U. CHANDLER   Battelle, Pacific Northwest Laboratory, 901 D Street, Suite 900, Washington, DC 20024-2115 U.S.A.

Global climate change/energy efficiency

O. DAVIDSON   University Research & Development Services Bureau, University of Sierra Leone, Private Mail Bag, Freetown, Sierra Leone

Energy use and global climate change in African less-developed countries

U.R. FRITSCHE   Öko-Institut, Büro Darmstadt, Prinz-Christians-Weg 7, 6100 Darmstadt, F.R. Germany

Energy/environment systems analysis

W. FULKERSON   Oak Ridge National Laboratory, P.O. Box 2008, Oak Ridge, TN 37831-6247, U.S.A.

Energy technology

P.H. GLEICK   Pacific Institute for S.i.D.E.S., 1681 Shattuck Avenue, Suite H, Berkeley, CA 94709, U.S.A.

Impacts of global climate change; relationship between climate change and international security; water resources

H. HILSE   Technische Universität Dresden, Institut für Energietechnik, Mommsenstr. 13, 8027 Dresden, F.R. Germany

Thermal hydraulics in nuclear power plants safety analysis

A.K. JAIN   Center for Applied Climatology and Environmental Studies (ACE), Climate and Energy Research Unit, Institute for Geography, University of Münster, Robert-Koch-Str. 26, 4400 Münster, F.R. Germany

Global climate change; development of climate models to test the efficiency of different measures to reduce the additional greenhouse warming due to CO_2, CFCs, CH_4, and N_2O emissions.

List of Participants with Fields of Research xiii

L.L. JENNY Transportation Research Board, 2101 Constitution Ave. NW, Washington, DC 20418, U.S.A.

Transportation

E.K. JOCHEM Fraunhofer Institut für Systemtechnik und Innovationsforschung, Breslauer Str. 48, 7500 Karlsruhe 1, F.R. Germany

Rational energy use; technical, economic and policy aspects; social costs of energy use; technology assessment

J.-H. KIM Korea Energy Economics Institute, C.P.O. Box 4311, Seoul, Korea

Long-term projection of energy demand in Korea

M. LÖNNROTH The Prime Minister's Office, 103 33 Stockholm, Sweden

Energy, environment, health, policy

A.B. LOVINS Rocky Mountain Institute, 1739 Snowmass Creek Road, Snowmass, CO 81654-9199, U.S.A.

Efficient end-use of energy, especially electricity (technology, implementation methods, policy, climatic implications)

K.M. MEYER-ABICH Kulturwissenschaftliches Institut im Wissenschaftszentrum NRW, Hagmanngarten 5, 4300 Essen 15, F.R. Germany

Philosophy of nature, energy systems, implications of climate change

I.M. MINTZER Center for Global Change, University of Maryland, The Executive Building, Suite 401, 7100 Baltimore Avenue, College Park, MD 20740, U.S.A.

Risk communication; role of national energy strategies in the problems of climate change, ozone depletion, acid rain; analysis of investments by electric utilities and problems of implementation

G.I. PEARMAN CSIRO, Division of Atmospheric Research, PMB 1, Mordialloc, Victoria 3195, Australia

Global atmospheric chemistry and climate change

T. PETER Abt. Atmosphärische Chemie, Max-Planck-Institut für Chemie, Postfach 3060, 6500 Mainz, F.R. Germany

Stratospheric ozone

List of Participants with Fields of Research

G.C. PINCHERA ENEA, Area Energia, Ambiente e Salute, Viale Regina Margherita 125, Roma 00198, Italy

Energy and environment: assessment studies of technological and strategic options for limiting the greenhouse effect

S. RAYNER Global Environmental Studies Center, Oak Ridge National Laboratory, P.O. Box 2008, Oak Ridge, TN 37831-6206, U.S.A.

Policy, energy, and human systems analysis for global environmental change

J.M. REILLY U.S. Dept. of Agriculture, Economic Research Service, 1301 New York Avenue NW, Rm.528, Washington, DC 20005-4788, U.S.A.

Agricultural economics, resource economics, climate change

R.G. RICHELS Electric Power Research Institute, 3412 Hillview Avenue, Palo Alto, CA 94304, U.S.A.

The economics of emissions abatement

L. SCHIPPER Lawrence Berkeley Laboratory, Room 4046, Bldg. 90, 1 Cyclotron Road, Berkeley, CA 94720, U.S.A.

Energy demand

B. SCHLOMANN Fraunhofer-Institut für Systemtechnik und Innovationsforschung, Breslauer Str. 48, 7500 Karlsruhe 1, F.R. Germany

Rational energy use, external costs of energy systems, long-term projections of energy demand

U.E. SIMONIS Wissenschaftszentrum Berlin, Reichpietschufer 50, 1000 Berlin 30, F.R. Germany

Environmental policy

F. STAIß Zentrum für Sonnenenergie- und Wasserstoff-Forschung, Pfaffenwaldring 38–40, 7000 Stuttgart 80, F.R. Germany

Integration of renewable energies in the electricity supply

R.J. SWART RIVM, Global Change Division, P.O. Box 1, 3720 BA Bilthoven, The Netherlands

Greenhouse-gas emissions and preventive options

List of Participants with Fields of Research

D.A. TIRPAK Environmental Protection Agency, Room 3220WT-PM221, 401 M Street SW, Washington, DC 20460, U.S.A.

Global policy analysis

R. UEBERHORST Beratungsbüro für diskursive Projektarbeiten und Planungsstudien, Marktstr. 18, 2200 Elmshorn, F.R. Germany

Alternative futures, conflict resolution, political discourse

K. VON MOLTKE Environmental Studies Program, 324 Murdough Center, Hanover, NH 03755, U.S.A.

International environmental relations

I.J. WALKER Energy Programs Division, Dept. of Primary Industries and Energy, GPO Box 858, Canberra City 2601, Australia

Energy research

C.-J. WINTER c/o DLR, Pfaffenwaldring 38–40, 7000 Stuttgart 80, F.R. Germany

Renewable energies, hydrogen technologies and systems development, energy efficiency

K. YAMAJI Central Research Institute of Electric Power Industry (CRIEPI), 1-6-1 Otemachi, Chiyoda-ku, Tokyo 100, Japan

Energy systems analysis, technology assessment

1
Global Climate Change and the Energy Community

G.I. PEARMAN
CSIRO Division of Atmospheric Research, Mordialloc, 3195, Australia

ABSTRACT

There exists irrefutable evidence that the composition of the global atmosphere is changing as a result of human activities. There are strong scientific grounds to believe that this will bring about general global warming in the next few decades with concomitant changes to regional patterns of climate and sea level, and resultant impacts on societal, economic, and environmental aspects of planetary existence. It is too soon to know precisely the patterns of these changes and their impacts, or to accurately assess their costs. Yet there is sufficient indication that these costs may be so considerable that action to avert or at least slow down the changes is warranted now.

The use of energy to service an ever-growing global population, with steadily increasing energy and nutritional demands, is the major reason for the changing atmospheric composition and, in particular, increasing atmospheric concentrations of the greenhouse gas, carbon dioxide. Growing energy demands are at the heart of the aspirations of those who seek to achieve more and those who simply need the basic elements of food, water, and shelter.

So we have a major dilemma on our hands. The energy goose that lays that golden egg of prosperity, that provides a key ingredient for global equity and a better future, turns out to be a serious hindrance to the long-term sustainability of our planetary systems.

This paper provides background to the atmospheric science of the global warming issue and describes how an integral part of the total greenhouse debate is the assessment by the global energy community of how it can contribute to this debate. It describes the objectives of the workshop, of which this volume is the proceedings. This conference was a small (compared with the magnitude of the issue) but determined attempt to look afresh at our global energy systems with the object of discovering what options exist for the reduction of emissions of carbon dioxide, what are the barriers to implementing those reductions, and how these barriers can be overcome.

Limiting the Greenhouse Effect: Options for Controlling Atmospheric CO_2 Accumulation
Edited by G.I. Pearman © 1992 John Wiley & Sons Ltd

INTRODUCTION

By the mid-1980s, the atmospheric science community throughout the world had prepared two major sets of reviews concerning the likelihood of general global warming as a result of the accumulation of certain trace gases in the atmosphere. The first of these (Bolin et al. 1986) resulted from a conference held in Villach, Austria. That conference released a statement which had a significant impact throughout the world (WMO 1986, 1988). The thrust of the statement was that the science community, although well aware of the imperfections of the science on global warming, concluded that there was a high probability of significant planetary warming within the next 30 years. Further, it was argued that, given this confidence, it was time for the wider community to assess what this warming might mean in terms of impacts and the possible need for adaptive or abatement responses. The second set of reviews was prepared at about the same time by the U.S. Department of Energy (DOE 1986a–e) and these complemented the Villach report.

Following the Villach Conference there was a rapid growth in interest in the topic throughout the wider community including the energy sector. In 1988 at the conference on "The Changing Atmosphere: Implications for Global Security" held in Toronto (Pearman et al. 1989; WMO 1989), there was the first widely publicized reference to global targets for the reduction of greenhouse-gas emissions. The Toronto target referred only to carbon dioxide (CO_2) emissions and suggested a 20% reduction from 1988 levels by the year 2005. It was not based on an exhaustive analysis of what really is desirable in order to slow the rate of increase of CO_2 in the atmosphere significantly, nor was it based on what is achievable economically, sociologically, and politically. However, this target has markedly influenced the thinking of governments internationally. For example, in Australia, several state governments, and more recently the Australian Federal Government, have set reduction targets similar to those of the Toronto target. Currently more than 20 nations have adopted targets for greenhouse-gas emission reduction.

STATE OF THE SCIENCE OF GLOBAL WARMING

Considerably more research has occurred since these earlier reviews. During 1989–1990, the United Nations Intergovernmental Panel on Climate Change (IPCC) was established and worked on a further review of the climate change issue. The scientific basis of the issue was reported by Panel 1 of the IPCC (Houghton et al. 1990), while the second and third Panels produced reports which dealt with the questions of the impacts of climate change and the possible response options, respectively (Tegart et al. 1990; IPCC 1991).

The results of deliberations of all three Panels of the IPCC were reported at the Second World Climate Conference in Geneva (SWCC 1990a). The following is a brief overview of the results of the science review. A more comprehensive consideration of the science is given by Edmonds et al. (this volume).

Changing Composition of the Atmosphere

Developments over the last one to two decades have led to a rather precise picture of the changing composition of the global atmosphere, at least with respect to the more important greenhouse gases, CO_2, methane (CH_4), nitrous oxide (N_2O), and the major chlorofluorocarbon (CFC) gases. This picture results from the number of high quality observational programs initiated during the 1970s and 1980s, and the new techniques for the retrieval of air trapped in bubbles in polar ice at definable previous times. Table 1.1 gives a summary of the changes in greenhouse-gas atmospheric concentrations.

Climate Forcing

The assessment of the relative importance of these various greenhouse gases in causing global warming is in itself a complex and, until recently, poorly understood problem. The difficulty arises because several factors need to be taken into consideration when comparing the gases. These include:

- the different radiative properties of the gases, including the strength of the absorption of infrared radiation and the wavelength of that absorption with respect to other gases in the atmosphere;
- the lifetime of the gas in the atmosphere. How quickly does it get removed by surface uptake or chemical reaction to become alternative species? Clearly, long-lived gases are more important than those which decay rapidly. This is a particularly difficult concept in the case of CO_2, given that its loss from the atmosphere results from exchange with carbon reservoirs on a wide range of timescales. Decision-makers are faced with first deciding over what timescale they wish to make their comparisons before they can assess the relative importance of CO_2 with respect to other greenhouse gases; and finally
- the secondary effects that some greenhouse gases have on other gases. For example, CH_4 reacts to produce water vapor in the stratosphere or competes with ozone (O_3) for OH radicals in the troposphere. As both O_3 and H_2O are themselves greenhouse gases, then secondary interactions need to be evaluated.

Table 1.1 Historical and current atmospheric concentrations of greenhouse gases (based on Houghton et al. 1990)

Greenhouse Gas[1]	Preindustrial atmospheric concentration (1750–1800)	Current atmospheric concentration (1990)[2]	Current rate of annual atmospheric accumulation	Atmospheric lifetime (years)[3]
Carbon dioxide	280 ppmv[4]	353 ppmv	1.8 ppmv (0.5%)	(50–200)
Methane	0.8 ppmv	1.72 ppmv	0.015 ppmv (0.9%)	10
CFC-11	0	280 pptv[4]	9.5 pptv (4%)	65
CFC-12	0	484 pptv	17 pptv (4%)	130
Nitrous oxide	288 ppbv[4]	310 ppbv	0.8 ppbv (0.25%)	150

[1]Ozone has not been included in the table because of lack of precise data.
[2]The current (1990) concentrations have been estimated based upon an extrapolation of measurements reported for earlier years, assuming that the recent trends remained approximately constant.
[3]For each gas in the table, except CO_2, the "lifetime" is defined here as the ratio of the atmospheric content to the total rate of removal. This time scale also characterizes the rate of adjustment of the atmospheric concentrations if the emission rates are changed abruptly. CO_2 is a special case since it has no real sinks, but is merely circulated between various reservoirs (atmosphere, ocean, biota). The "lifetime" of CO_2 given in the table is a rough indication of the time it would take for the CO_2 concentration to adjust to changes in the emissions.
[4]ppmv = parts per million by volume; ppbv = parts per billion by volume; pptv = parts per trillion by volume.

This has led to the development of the concept of the Global Warming Potential (GWP), which is the warming effect of a unit mass of a gas released into the atmosphere, divided by the warming effect of the same mass of CO_2. Table 1.2 shows the GWPs for key greenhouse gases.

It is on the basis of these GWPs that a comparison can be made between the warming effects of past, current, or future emissions of the gases. Table 1.2 shows such comparisons for the 1990 emissions and the estimated warming expected by 2025 in the "business-as-usual" scenario of the IPCC. These comparisons show that CO_2, primarily released from the activities of the energy sector, has and will in the near future continue to contribute about half of all global warming.

Climate Change Predictions

It was the reasonable level of agreement between climate modelers that led to the Villach, and more recently, the IPCC consensus concerning probable

Table 1.2 Global warming potentials of key greenhouse gases for three time horizons, the current emissions of those gases, and, given future emissions (business-as-usual scenario), the anticipated relative contribution to warming for a 100-year time horizon (based on Houghton et al. 1990)

Greenhouse Gas	Time Horizon 20 yr	Time Horizon 100 yr	Time Horizon 500 yr	Emissions 1990 (Tg)	Relative contribution over 100 yr (%)	1765–2025 Relative contrib. over 100 yr[1] (%)
Carbon dioxide	1	1	1	26000[2]	61	63
Methane[3]	63	21	9	300	15	16
Nitrous oxide	270	290	190	6	4	5
CFC-11	4500	3500	1500	0.9	11	7
CFC-12	7100	7300	4500			
HCFC-22	4100	1500	510	0.1	0.5	4
Others					8.5	

[1] Percentage of the effect considering only the 5 species and based on expected emissions with a business-as-usual scenario to 2025.
[2] 26000 Tg (teragrams) of carbon dioxide = 7100 Tg (= 7 Gt) of carbon
[3] These values include the indirect effect of these emissions on other greenhouse gases via chemical reactions in the atmosphere. Such estimates are highly model-dependent and should be considered preliminary and subject to change. The estimated effect of ozone is included under "others." The gases included under "others" are given in the source document.

warming. This is not to say that these scientists are 100% certain of what will happen nor that they agree in all respects with the current methodologies for representing the climate system in computer models. Disagreements do exist, despite which there is an overwhelming belief that global warming is likely. Much needs to be done to improve the confidence in the predictions of the way in which climate is likely to change. Given the expected growth of greenhouse-gas concentrations in the atmosphere over the next few decades, the best estimate of the climate sensitivity to these gases, and the time delay due to oceanic warming, the IPCC Report predicts a warming of 0.3°C per decade (see Table 1.3). Associated with this warming are anticipated changes to global average sea level and precipitation.

The confidence that exists is confined almost entirely to the gross global warming response and not to the impact this will have at the regional level. In any case, should climatologists agree on the precise level of global warming expected for a particular scenario of gas release, this would be of little value to the policymaker. Policy substantially relates to the impacts on particular communities and thus must respond to regional or local impacts of change. Such regional climate changes are not agreed upon by the modeling community. What the models do indicate at this stage is that

Table 1.3 Best estimates of the realized climate changes expected as a result of "business-as-usual" societal changes through the next century (based on Houghton et al. 1990)

TEMPERATURE	
Global mean	Rate 0.3°C (0.2–0.5) per decade
	1°C above the present (1990) by 2025
	3°C above the present (1990) by 2100
Regional changes	variable
RAINFALL	
Large-scale	5–10% increase on average in middle and high latitude continents (33–55°N) by 2030
Regional	highly variable
SEA LEVEL	
Global mean	6 cm (3–10) per decade
	20 cm above current (1990) levels by 2030
	65 cm above current (1990) levels by 2100
Regional changes	variable

regionally responses to global warming are expected to vary in intensity from the global average.

Therefore climate modelers have ahead of them the task of improving confidence in the prediction of climate warming, particularly at the regional level. This is important in determining the real impacts on hydrology, water resources, soil water content and thus vegetation, both natural and agricultural. Currently, the state of knowledge is such that we cannot make these predictions and consequently we cannot really assess exactly how important climate changes are really going to be. Ideally such an assessment is needed to evaluate the costs of letting the warming happen unabated, for comparison against the cost of abatement (Table 1.4).

It should not be underestimated just how difficult a task it is to develop regional predictions of climate change. Huge efforts are required before we are likely to be successful, and indeed, with regard to some aspects it is likely that we will be able to detect the changes before predictions can be made with confidence.

TOWARDS AN INTERNATIONAL RESPONSE TO GLOBAL WARMING

The potential importance of global warming is now widely acknowledged. For example, the Heads of State and Government of the seven major

Table 1.4 Simplified relationships between the science, impact assessment, and policy response aspects of global warming

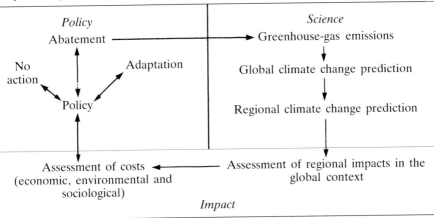

industrial democracies and the President of the Commission of the European Communities recently made a clear statement on the importance of climate change and their commitment to undertake efforts to limit emissions of greenhouse gases (see Table 1.5).

Further, the Ministers representing 137 countries and those from the European Communities at the Second World Climate Conference made clear statements concerning the state of the science, the potential importance of climate change, and the need for strong and prompt attention to international agreements on emission limitation (see Table 1.6).

While many nations have already announced strategies designed to slow the rate of growth of emissions of greenhouse gases, few will be prepared to do so unilaterally. In an economically competitive world, nations will be reluctant to undertake emission controls if such controls jeopardize their competitiveness. Enormous differences exist between factors such as the per capita contribution to certain greenhouse-gas emissions, the accuracy with which certain emissions are known, the importance of wealth-generation versus survival activities in contributing to emissions, and economic capacity to mitigate. There can be little doubt that the development of precise and effective international agreements of this kind by the time of the United Nations Conference on Environment and Development (June 1992) will be very difficult.

AIM OF THE DAHLEM CONFERENCE

As CO_2 is the primary greenhouse gas, and its main source is from the energy sector, the energy community needs to consider what options are

Table 1.5 Excerpts from The Houston Economic Declaration signed by the Heads of State and Government of the seven major industrial democracies and the President of the Commission of the European Communities, July 1990

> "... One of our most important responsibilities is to pass on to future generations an environment whose health, beauty, and economic potential are not threatened. Environmental challenges such as climate change, ozone depletion, deforestation, marine pollution, and loss of biological diversity require closer and more effective international cooperation and concrete action. We as industrialized countries have an obligation to be leaders in meeting these challenges. We agree that, in the face of threats of irreversible environmental change, lack of full scientific certainty is no excuse to postpone actions which are justified in their own right . . . .
>
> Climate change is of key importance. We are committed to undertake common efforts to limit emissions of greenhouse gases, such as carbon dioxide. We strongly support the work of the Intergovernmental Panel on Climate Change (IPCC) and look forward to the release of its full report in August . . . .
>
> We welcome the amendment of the Montreal Protocol to phase out the use of chlorofluorocarbons (CFCs) by the year 2000 and to extend coverage of the Protocol to other ozone depleting substances . . . .
>
> To cope with energy-related environmental damage, priority must be given to improvements in energy efficiency and to the development of alternative energy sources . . . .
>
> Cooperation between developed and developing countries is essential to the resolution of global environmental problems . . . ."

available for reducing CO_2 (and CH_4, N_2O) emissions. Options clearly include conservation of energy, efficiency improvements, development of nonfossil alternative sources, fuel switching, and societal changes. But quantitatively, are these really options, given the implications of economic, societal, and environmental costs? Again, there will be no simple answer. The one thing that is clear is that policymakers need information about the existing energy sector and the future options available.

On the one hand, nations need inventories of releases of greenhouse gases. Clearly each nation needs to know which are the most significant emissions before it can best plan for a reduction strategy. Emission profiles will vary between nations and even within regions of larger nations. On the other hand, nations need to know what options exist through conservation, nonfossil and renewable fuels, etc. to reduce emissions and the associated costs.

The specific aim of the Dahlem Workshop on "Limiting the Greenhouse

Table 1.6 Excerpts from the Ministerial Declaration of the Second World Climate Conference, Geneva, 29 October–7 November 1990 (SWCC 1990b)

> "... we commit ourselves and intend to take active and constructive steps in a global response, without prejudice to sovereignty of States. ...
> Recognizing that climate change is a global problem of unique character and taking into account the remaining uncertainties in the field of science, economics and response options, we consider that a global response, while ensuring sustainable development of all countries, must be decided and implemented without further delay ....
> We welcome the decisions and commitments undertaken by the European Community with its Member States, Australia, Austria, Canada, Finland, Iceland, Japan, New Zealand, Norway, Sweden, Switzerland, and other developed countries to take action aimed at stabilizing their emissions of CO_2, or CO_2 and other greenhouse gases not controlled by the Montreal Protocol, by the year 2000 in general at 1990 level, ...
> Financial resources channelled to developing countries should *inter alia*, be directed to:
> (i) Promoting efficient use of energy, development of lower and non-greenhouse gas emitting energy technologies and paying special attention to safe and clean new and renewable sources of energy ....
> We note that energy production and use account for nearly half of the enhanced radiative forcing resulting from human activities ... We recognize the promotion of energy efficiency as the most cost-effective immediate measure, in many countries, for reducing energy-related emissions of carbon dioxide, methane, nitrous oxide and other greenhouse gases ....
> We call for negotiation on a framework convention on climate change to begin without delay ... we urge all countries and regional economic integration organizations join in these negotiations and recognize that it is highly desirable that an effective framework convention on climate change, containing appropriate commitments, and any related instruments as might be agreed upon on the basis of consensus, be signed in Rio de Janeiro during the United National Conference on Environment and Development."

Effect: Options for Controlling Atmospheric CO_2 Accumulation" was, therefore, to provide a rational basis for national and international efforts to reduce the rate of accumulation of CO_2 in the atmosphere. Several comments need to be made about this aim.

First, while it is appreciated that CO_2 is not the only greenhouse gas increasing in the atmosphere, and indeed that the cycles and interactions between various greenhouse gases will sometimes be quite important, we chose to consider only CO_2 in order to keep the scope of the meeting reasonably constrained.

Second, it was decided that while we were ultimately interested in providing material and advice for policymaking, establishing a set of policy recommendations was not the primary aim of this meeting. The primary aim was to provide background information and to investigate the kinds of research and questions that need to be addressed to provide the rational basis for decision-making.

Third, it was appreciated that the aims as defined would not be achievable at one Dahlem Workshop. The task is enormous. It involves bringing together a huge existing knowledge base and applying it to this objective. It involves defining gaps in that knowledge base and stimulating research to fill those gaps. It involves the transmission of this information to those who have to synthesize the data into policy.

Fourth, we are not the only group attempting to do this. For example, in my country we have already held a national meeting on the subject (Swaine 1990). Other meetings are being held almost weekly around the world. But, the Dahlem Workshop did bring together an impressive group of experts on the topic. Our authors have produced papers describing background information. The discussions of the meeting have assisted the authors in refining their papers and allowed the rapporteurs to write group reports to highlight the research needed to complete the task.

These were the charges for the meeting and in attending to these charges, several themes were kept in mind:

- It was appreciated that the options available would depend strongly on the timescales for consideration. Thus, where possible, we tried to distinguish between the options in timescales of a few years to 15 years; intermediate (50 years); and longer (100 years).
- We also acknowledged that options available would depend on individual countries and their economies. We attempted to keep our discussions broad where necessary, to look at the relative merits of actions in both developed and developing countries.
- The report attempts to indicate not only what is technically feasible, but what is acceptable economically and socially, and has insignificant alternative environmental consequences.
- Even though it was decided to concentrate on options for the control of CO_2 emissions, it was appreciated that in doing so, we may have tended to underestimate the value of options that contribute to reductions of greenhouse gases in addition to CO_2.

ACKNOWLEDGEMENTS

Planning and conducting an international meeting such as a Dahlem Conference is a major undertaking and one that requires the dedication and skills of a number of people. The details of the meeting were developed by the program advisory committee members, Paul Crutzen, Jae Edmonds, Jill Jäger, Klaus Meyer-Abich, and Irving Mintzer. Their wisdom and enthusiasm for the meeting were key elements in bringing together a most interesting mixture of expertise. I wish to thank the authors for being remarkably prompt in the preparation of their papers and responding to the suggestions of the referees and my editorial queries. The referees, who will remain anonymous, played an important part in assuring a high publication standard.

The meeting itself was conducted with the usual efficiency of the Dahlem Konferenzen office. This was my third Dahlem conference and each has been conducted with the same professionalism and enthusiasm. The Konferenzen office is a remarkable team and they all deserve much credit. My special thanks to Gloria Custance for her patience and support, Julia Lupp for her efforts in the long editorial process, and Jennifer Altman for her support at the time of the meeting. I thank the moderators of the working groups, Brad Ashton, Charles Caccia, Irving Mintzer, and Steve Rayner and the rapporteurs Jesse Ausubel, Stewart Boyle, Bill Chandler, and Måns Lönnroth. Finally, a very special thanks to Silke Bernhard for her friendship and her encouragement in the early stages of this project.

REFERENCES

Barnola, J.M., D. Raynaud, Y.S. Korotkevich, and C. Lorius. 1987. Vostock ice cores provides 160,000 year record of atmospheric CO_2. *Nature* **329**:408–414.

Bolin, B., B.R. Döös, J. Jäger, and R.A. Warrick. 1987. The Greenhouse Effect, Climate Change and Ecosystems, pp. 541. New York: Wiley.

DOE (U.S. Dept. of Energy). 1985a. Glaciers, ice sheets and sea level: Effects of a CO_2-induced climatic change. Report of a workshop held in Seattle, September 1984, DOE/ER-60235-1. Springfield, VA: U.S. Information Service.

DOE. 1985b. Characterization of information requirements for studies of CO_2 effects: Water resources, agriculture, fisheries, forests and human health, ed. M.R. White, DOE/ER-0236. Springfield, VA: U.S. Information Service.

DOE. 1985c. Detecting the climatic effects of increasing carbon dioxide, ed. M.C. McCracken and F. M. Luther, DOE/ER-0237. Springfield, VA: U.S. Information Service.

DOE. 1985d. Direct effects of increasing carbon dioxide on vegetation, ed. B.R. Strain and J.D. Cure, DOE/ER-0238. Springfield, VA: U.S. Information Service.

DOE. 1985e. Atmospheric carbon dioxide and the global carbon cycle, ed. J.R. Trabalka, DOE/ER-0239. Springfield, VA: U.S. Information Service.

Houghton, J.T., G.J. Jenkins, and J.J. Ephraums, eds. 1990. Climate Change. The IPCC Scientific Assessment, 365 pp. Cambridge: Cambridge Univ. Press.

IPCC (Intergovernmental Panel on Climate Change). 1991. Climate Change. The IPCC Response Strategies, 273 pp. Washington, D.C.: Island Press.

Pearman, G.I. 1988. Greenhouse gases: Evidence for atmospheric changes and anthropogenic causes. In: Greenhouse: Planning for Climate Change, ed. G. I. Pearman, pp. 3–21. Melbourne: CSIRO.

Pearman, G.I., N.J. Quinn, and J.W. Zillman. 1989. The changing atmosphere. *Search* **20**:59–65.

Swaine, D.J., ed. 1990. Greenhouse and Energy, 482 pp. Australia: CSIRO.

SWCC (Second World Climate Conference). 1990a. Abstracts of Scientific/Technical Papers, 87 pp. Geneva: World Meteorological Organization.

SWCC 1990b. Ministerial Declaration of the Second World Climate Conference, 9 pp. Geneva: World Meteorological Organization.

Tegart, W.J., G.W. Sheldon, and D.C. Griffiths, eds. 1990. Climate Change. The IPCC Impacts Assessment. Canberra: Australian Gov. Publishing Service.

WMO (World Meteorological Organization). 1986. Conference statement from the UNEP/WMO/ICSU international assessment of the role of carbon dioxide and other greenhouse gases in climate variations and associated impacts. *WMO Bulletin* **35**:129–134.

WMO. 1988. Developing policies for responding to climatic change. A summary of the discussions and recommendations of the workshops held at Villach (28th September–2nd October 1987) and Bellagio (9–13th November 1987) under the auspices of the Beijer Institute, Stockholm. WCIP, WMO TD-No. 225, 53pp. Geneva: World Meteorological Organization.

WMO. 1989. The changing atmosphere—implications for global security. *WMO Bulletin* **38**:41–42.

2
Greenhouse Gases: What Is Their Role in Climate Change?

J.A. EDMONDS[1], D. WUEBBLES[2], and W.U. CHANDLER[1]
[1]Pacific Northwest Laboratory, Battelle, Washington, D.C., 20024-2115, U.S.A.
[2]Lawrence Livermore National Laboratory, Livermore, CA 94550, U.S.A.

ABSTRACT

This chapter summarizes information relevant to understanding the role of greenhouse gases in the atmosphere. We address the issue of the Earth's radiation budget and the greenhouse effect itself in the first section. In the second section we examine trends in atmospheric concentration of greenhouse gases, emissions sources, and natural processes which regulate the concentration of greenhouse gases. In the third section we examine the natural carbon cycle and its role in determining the atmospheric residence time of carbon dioxide (CO_2). In the fourth section, the role atmospheric chemistry plays in determining the concentrations greenhouse gases is examined. In the final section we discuss the global warming potential concept.

This chapter is not intended to be an exhaustive treatment of these issues. Exhaustive treatments can be found in other volumes, many of which are cited herein. Rather, this chapter is intended to summarize some of the major findings, unknowns, and uncertainties associated with the current state of knowledge regarding the role of greenhouse gases in the atmosphere. Other, somewhat more detailed summaries which might be of interest include: Trabalka (1985), MacCracken and Luther (1985a, 1985b), Bolin et al. (1986), Wuebbles and Edmonds (1988), MacCracken et al. (1990), Houghton et al. (1990), and Wuebbles and Edmonds (1991).

THE GREENHOUSE EFFECT

A suite of naturally occurring gases, including water vapor (H_2O), carbon dioxide (CO_2), methane (CH_4), ozone (O_3), nitrous oxide (N_2O), and gases of anthropogenic origin, such as the chlorofluorocarbons (CFCs) and CFC substitute chemicals, have the property that they are transparent to incoming solar radiation but absorb infrared radiation reemitted by the Earth. Such gases are referred to as greenhouse gases because of this energy "trapping" property. Ironically, the major heat-trapping mechanism for a greenhouse is quite different than for a greenhouse gas. For the greenhouse, glass blocks the convective rise of warm air. For a greenhouse gas, warming of the Earth's surface and cooling of the stratosphere are achieved via molecular absorption and reradiation in the infrared spectrum.

Figure 2.1 illustrates the relationship between solar radiation, the Earth's reemission of energy, and the greenhouse gases. The Sun emits most of its energy between 0.2 and 4.0 μm, primarily in the ultraviolet, visible, and near-infrared wavelength regions. A very small fraction of this energy is

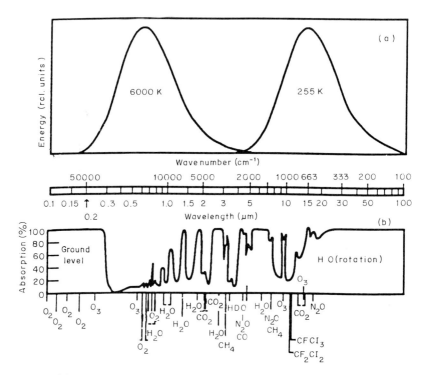

Figure 2.1 (a) Black-body curves showing the variation of emitted energy with wavelength for temperatures typical of the Sun and the Earth, respectively, and (b) percentage of atmospheric absorption for radiation passing from the top of the atmosphere to the surface. Note the comparatively weak absorption of the solar spectrum, and the region of weak absorption from 7 to 13 μm in the longwave spectrum, referred to as the "window region" (based on MacCracken and Luther 1985b)

intercepted by the Earth as it orbits the Sun. The atmosphere absorbs approximately 23% (MacCracken and Luther 1985b) of the incoming solar radiation, principally by O_3 in the ultraviolet and visible, and by H_2O in the near-infrared.

The Earth reemits the energy it absorbs back to space, thus maintaining an energy balance. Satellite measurements indicate that the incoming and outgoing radiation of the atmosphere is in balance (Shine 1989). Because the Earth is much colder than the Sun, the bulk of this emission takes place at longer wavelengths than those for incoming solar radiation. Most of this radiation is emitted in the wavelength range from 4 to 100 μm, which is the region generally referred to as longwave or infrared radiation (Fig. 2.1). While H_2O, CO_2, and other greenhouse gases are relatively inefficient absorbers of solar radiation, these gases are strong absorbers of longwave radiation. Clouds also play an important role in determining the energy balance (Fig. 2.2); one of the largest uncertainties in determining the climate change expected from greenhouse gases is the current limitation in understanding cloud processes (Ramanathan 1988; Houghton et al. 1990).

The greenhouse gases (and clouds) reemit the absorbed longwave radiation at a wavelength and intensity dependent on their local atmosphere temperature, which tends to be cooler than the Earth's surface temperature. Some of this radiation reaches space. Some of the radiation, however, is transmitted downward, leading to a net trapping of longwave radiation and a warming of the surface. As the concentrations of greenhouse gases increase, this net trapping of infrared radiation is enhanced. The net result is that surface temperatures rise until the amount of radiative energy being emitted to space balances the incoming solar radiation.

The Direct Radiative Influence

The contribution of a gas to the greenhouse effect depends on the wavelength at which the gas absorbs infrared radiation, the concentration of the gas, the strength of the absorption per molecule (line strength), and whether or not other gases absorb at the same wavelengths (e.g., see Ramanathan et al. 1985, 1987; Wang et al. 1985, 1986; Mitchell 1989). Gases absorb and emit radiation at wavelengths that correspond to transitions between discrete energy levels. Absorption at infrared (greenhouse) wavelengths occurs for triatomic or larger molecules, where vibrational and rotational energy transitions occur at appropriate wavelengths. Although each transition is associated with a discrete wavelength, the interval over which absorption occurs is "broadened" by addition or removal of energy due to molecular collision (pressure broadening) or the Doppler frequency shift due to the random velocities of molecules (Doppler broadening). If absorption is

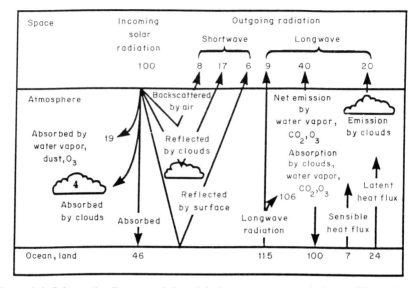

Figure 2.2 Schematic diagram of the global average energy balance. The units are percent of incoming solar radiation. Solar fluxes are shown on the left-hand side; longwave (IR) fluxes are on the right-hand side. Of the 100 units of incoming solar radiation (each equivalent to about 3.4 W/m²), about 23 are absorbed in the atmosphere, 46 are absorbed by the surface, and 31 are reflected to space. The planetary energy balance is achieved by the emission of 69 units to space as longwave radiation. The solar energy absorbed by the surface is used in part to heat the atmosphere directly (sensible heat flux) and to evaporate moisture (latent heat flux). Atmospheric emission of longwave radiation downward to the surface is about equal to the solar radiation reaching the top of the atmosphere and more than twice as large as the amount of solar radiation absorbed at the surface. This greenhouse energy permits the surface to warm significantly more than would be permitted by the solar radiation alone (from MacCracken and Luther 1985b)

strong, there may be complete absorption (saturation) around the central wavelength of the spectral line.

As shown in Figure 2.1, at a number of wavelengths the high concentrations of H_2O and CO_2 almost completely absorb the radiation emitted at the surface before it can be lost to space. Increases in the concentrations of these species, therefore, lead only to increased absorption in the wings of the absorption lines, with the results that the net trapping of infrared radiation due to these gases increases logarithmically rather than linearly with concentration. Because atmospheric temperature changes with altitude, additional concentrations of these gases also change the effective altitude of emission, thereby changing the infrared flux and further enhancing the greenhouse effect of these gases. Gases absorbing at similar wavelengths as CO_2 and H_2O will contribute little to the greenhouse effect, unless they have comparable concentrations. However, Figure 2.1 shows that there is a region from about 7 to 13 μm where absorption by CO_2 and H_2O is weak; this is referred to as the "window region." Nearly 80% of the radiation emitted by the surface in the window region escapes to space (Ramanathan 1988). Most of the non-CO_2 gases with the potential to lead to climate change, including CH_4, N_2O, O_3, and the CFCs, have absorption lines in the "window" region. Some of these gases, such as CH_4 and N_2O have absorption lines that lead to saturation of the line cores and emission from the pressure-broadened Lorentz line wings; the net result is that their absorption increased approximately as the square root of their concentration. Gases with little overlap, such as CFC-11 and CFC-12, exhibit absorption that increases linearly with concentration.

The net result of these varying radiative characteristics is that comparable increases in concentrations of different greenhouse gases have vastly different effects on radiative forcing, as shown in Table 2.1. The instantaneous

Table 2.1 Radiative forcing (ΔF) relative to CO_2 per unit molecule change and per unit mass change in the atmosphere for various greenhouse gases for present-day concentrations (based on Houghton et al. 1990)

TRACE GAS	ΔF per molecule relative to CO_2	ΔF per unit mass relative to CO_2
CO_2	1	1
CH_4	21	58
N_2O	206	208
CFC-11	12,400	3,970
CFC-12	15,800	5,750
CFC-113	15,800	3,710
CFC-114	18,300	4,710
CFC-115	14,500	4,130
HCFC-22	10,700	5,440
CCl_4	5,720	1,640
CH_3CCl_3	2,730	900
CF_3Br	16,000	4,730
Possible CFC substitutes:		
HCFC-123	9,940	2,860
HCFC-124	10,800	3,480
HFC-125	13,400	4,920
HFC-134a	9,570	4,130
HCFC-141b	7,710	2,900
HCFC-142b	10,200	4,470
HFC-143a	7,830	4,100
HFC-152a	6,590	4,390

radiative forcing of the surface–troposphere system is the change in the net radiative flux at the tropopause; equivalently, it can be defined as the surface temperature change with zero climate feedbacks. The addition of one molecule of CH_4 has about 21 times the effect on climate as the addition of one CO_2 molecule; one CFC-11 molecule is about 12,400 times as effective as one molecule of CO_2. These differences occur because CH_4 and CFC-11 absorb in the window region, whereas the CO_2 molecule competes not only with H_2O but also with many other CO_2 molecules (i.e., the 15-μm region of CO_2 absorption is saturated). Although added CO_2 has the least effect on radiative forcing of the gases considered on a per molecule basis, it is still the primary gas of concern for climate change due to the larger absolute change in concentration.

Ozone plays an important dual role in affecting climate. Although the climate effect of CO_2 and the other trace constituents, considered above, depends primarily on their concentration in the troposphere, the climatic effect of O_3 depends on its distribution throughout the troposphere and stratosphere. Ozone and molecular oxygen are the primary absorbers of the ultraviolet and visible solar radiation in the atmosphere; absorption of solar radiation by O_3 is responsible for the increase in temperature with altitude in the stratosphere. Ozone is also an important absorber of infrared radiation. It is the balance between these radiative processes and the local changes in O_3 with altitude (see Lacis et al. 1990) that determines the net effect of O_3 on climate. Increases in O_3 above about 30 km tend to decrease surface temperature; increases in O_3 below 30 km tend to increase surface temperature.

Several research studies using global atmospheric models have attempted to examine the relative effects of other trace gases in affecting radiative forcing and climate as compared with the effects from the increasing CO_2 concentrations (Lacis et al. 1981; Ramanathan et al. 1985, 1987; Wang et al. 1986; Wigley 1987; Hansen et al. 1988). Each of these studies indicates that the combined effects of the other trace gases are comparable to that for CO_2 emissions, both for recent observed changes in species abundances and for reasonable assumptions about future changes in their concentrations. For example, model calculations by Lacis et al. (1981) and Hansen et al. (1988, 1989) suggest that observed changes in concentrations of CH_4, CFC-11, CFC-12, N_2O, O_3, and other gases provided about half of the total change in greenhouse forcing during recent decades. Results of the Hansen et al. (1988) calculations are shown in Figure 2.3; note that the radiative forcing is only the direct radiative effect on climate and does not account for the additional temperature change expected from the climate feedback effects of clouds, water vapor, sea ice, etc. These results do not account for the delay of the expected temperature change resulting from ocean–atmosphere interactions. These calculations indicate that by the 1980s,

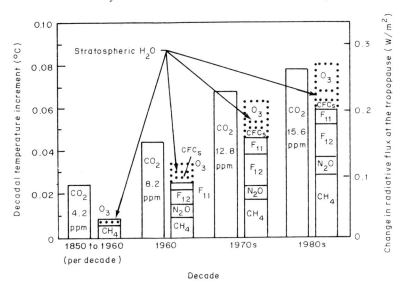

Figure 2.3 Decadal additions to global mean greenhouse radiative forcing of the climate system based on studies using a one-dimensional radiative–convective model of Hansen et al. (1988). The computed commitment to future surface temperature change (°C) once equilibrium is reestablished ($t \to \infty$) is estimated for decadal increases in trace gas abundances, without including the amplification that would be induced by climate feedbacks. The change in net radiative forcing of the troposphere-surface system is indicated on the right ordinate

the effects of the other greenhouse gases on radiative forcing have together become as large as the effect from the increasing CO_2 concentration.

Evaluations of future scenarios, such as those by Ramanathan et al. (1985, 1987), Wang et al. (1985, 1986), and Hansen et al. (1988), suggest that these gases could effectively double, or more than double, the climate effect from the increasing CO_2 concentration alone. It is not surprising, given the significant uncertainties concerning future trace constituent concentrations, that published scenarios for future changes in radiative forcing extend over a wide range of values.

The Greenhouse Gases

A number of greenhouse gases and other gases of climatic importance or potential importance have been mentioned in the prior discussion. Table 2.2 contains a synopsis of basic information on these gases. For these chemicals, Table 2.3 indicates the direct and indirect means by which they affect climate.

Table 2.2 Synopsis of basic information on greenhouse gases

Chemical symbol	Are source/sink rates well known?	Are major sources energy related?	Current concentration ca. late 1980s (ppmv)	Trend in concentration	Radiatively interactive?	Chemically interactive?
CO_2	yes/no	yes	350	increasing	yes	no
O_3	no/no	no	0.02–10	troposphere: increasing stratosphere: decreasing	yes	yes
CH_4	no/no	yes	1.7	increasing	yes	yes
N_2O	no/yes	yes	0.31	increasing	yes	yes
$CFCl_3$	yes/yes	yes	2.8×10^{-4}	increasing	yes	yes
CF_2Cl_2	yes/yes	yes	4.8×10^{-4}	increasing	yes	yes
CFC-113	yes/yes	no	60×10^{-6}	increasing	yes	yes
CFC-115	yes/yes	yes	5×10^{-6}	increasing	yes	yes
HCFC-22	yes/no	yes	120×10^{-6}	increasing	yes	yes
CCl_4	yes/yes	no	146×10^{-6}	increasing	yes	yes
CH_3CCl_3	yes/no	no	158×10^{-6}	increasing	yes	yes
OH	no/no	no	$1.5 \times 10^{-7} - 3 \times 10^{-4}$	?	no	yes
CO	no/no	yes	0.12 (N.H.), 0.06 (S.H.)	increasing	yes (weak)	yes
NO_x (=NO + NO_2)	no/no	yes	$1 \times 10^5 - 0.02$	?	yes (weak)	yes
CF_2ClBr	no/no	no	2×10^{-6}	increasing	yes	yes
CF_3Br	no/no	no	2×10^{-6}	increasing	yes	yes
SO_2	no/no	yes	$(0.1–2) \times 10^{-4}$	?	yes	yes
COS	no/no	yes	5×10^{-4}	increasing	yes	yes

N.H. = northern hemisphere; S.H. = southern hemisphere

Table 2.3 Greenhouse gases and their importance to climate

Trace Constituent	Common Name	Importance for Climate
CO_2	Carbon dioxide	Absorbs infrared radiation; affects stratospheric O_3
O_3	Ozone	Absorbs UV and infrared radiation
CH_4	Methane	Absorbs infrared radiation; affects tropospheric O_3 and OH; affects stratospheric O_3 and H_2O; produces CO_2
N_2O	Nitrous oxide	Absorbs infrared radiation; affects stratospheric O_3
$CFCl_3$	CFC-11	Absorbs infrared radiation; affects stratospheric O_3
CF_2Cl_2	CFC-12	Absorbs infrared radiation; affects stratospheric O_3
$C_2F_3Cl_3$	CFC-113	Absorbs infrared radiation; affects stratospheric O_3
C_2F_5Cl	CFC-115	Absorbs infrared radiation; affects stratospheric O_3
CHF_2Cl	HCFC-22	Absorbs infrared radiation; affects stratospheric O_3
CCl_4	Carbon tetrachloride	Absorbs infrared radiation; affects stratospheric O_3
CH_3CCl_3	Methyl chloroform	Absorbs infrared radiation; affects stratospheric O_3
OH	Hydroxyl	Scavenger for many atmospheric pollutants, including CH_4, CO, CH_3CCl_3, and CHF_2Cl
CO	Carbon monoxide	Affects tropospheric O_3 and OH cycles; produces CO_2
NO_x	Nitrogen oxide	Affects O_3 and OH cycles; precursor of acidic nitrates
CF_2ClBr	Ha-1211	Absorbs infrared radiation; affects stratospheric O_3
CF_3Br	Ha-1301	Absorbs infrared radiation; affects stratospheric O_3
SO_2	Sulfur dioxide	Forms aerosols, which scatter solar radiation
$(CH_3)_2S$	Dimethyl sulfide	Produce cloud condensation nuclei, affecting cloudiness and albedo
C_2H_2, etc.	NMHC	Absorbs infrared radiation; affects tropospheric O_3 and OH
COS	Carbonyl sulfide	Forms aerosol in stratosphere which alters albedo

The Climate Response

It is relatively easy to determine the direct forcing effect on surface temperature due to increases in greenhouse gases because the radiative properties of the greenhouse gases are reasonably understood. In contrast, there are many uncertainties remaining about the extent of processes acting to amplify (through positive feedbacks) or reduce (through negative feedbacks) the expected warming. Some of the most important feedbacks that have been identified include:

H_2O Greenhouse Feedback: As the lower atmosphere (the troposphere) warms, it can hold more H_2O vapor. The enhanced H_2O vapor traps more infrared radiation and amplifies the greenhouse effect. Ramanathan (1988) indicates that, based on studies with one-dimensional climate models, this feedback amplifies the air temperature by a factor of approximately 1.5 and the surface warming by a factor of approximately 3. The recent Intergovernmental Panel on Climate Change (IPCC) report (Houghton et al. 1990) determined a surface temperature amplification factor of 1.6 for H_2O vapor feedback.

Ice–Albedo Feedbacks: Global warming induced by warming melts sea ice and snow cover. Whether it is ocean or land, underlying surface is much darker (i.e., it has a lower albedo) than the ice or snow such that it absorbs more solar radiation, thus amplifying the initial warming. According to Ramanathan et al. (1988), ice–albedo feedback amplifies the global warming by 10–20%, with larger effects near sea-ice margins and in polar oceans.

Cloud Feedback: Cloud feedback mechanisms are extremely complex and are still poorly understood. Changes in cloud type, cloud amount, cloud altitude, and cloud water content can all affect the extent of the climate feedback. The sign of the feedback is also not understood, although current climate models generally find this to be a positive feedback. (This result is consistent with the findings of Ramanathan et al. [1989], that the overall effect of clouds in the current atmosphere results in net cooling.) Since clouds are still treated rather crudely in existing three-dimensional climate models (general circulation models or GCMs), it is premature to make reliable conclusions on the magnitude of cloud feedback processes, but they cannot actually change the sign of the effect.

Ocean–Atmosphere Interactions: Oceans influence the climate in two fundamentally important ways (Ramanathan 1988). First, because the importance of the H_2O greenhouse feedback, the air and land temperature response is affected by the warming of the ocean surface. If the oceans do not respond to the greenhouse heating, the H_2O feedback would be turned off since increased evaporation from the warmer ocean is the primary source of increasing atmospheric H_2O. Second, oceans can sequester the radiative heating into the deeper layers, which, because of their enormous heat capacity, can significantly delay the overall global warming effect over land surfaces. Current climate models suggest this delaying effect may cause a lag in the expected temperature response of at least a few decades, with the best estimate being about 50 years (Schlesinger and Jiang 1990).

Other feedbacks exist, such as changes in albedo related to changes in land features and biomass, but many of these feedbacks are not well understood. Figure 2.4 gives a schematic illustration of the many components of the coupled atmosphere, ocean–ice–land, climatic system. It is the combined effect of the many uncertainties in feedback processes that has resulted in the factor of 3 (1.5° to 4.5°C) difference in the amount of

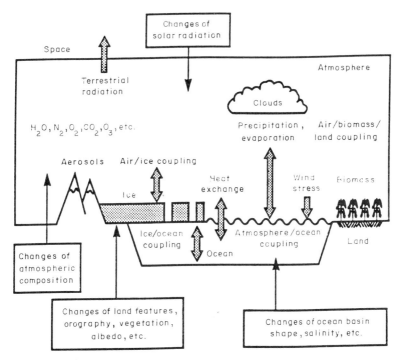

Figure 2.4 Schematic illustration of the components of the coupled atmosphere–ocean–ice–land climatic system. Full arrows represent external processes; the ocean arrows represent internal processes in climatic change (taken from Houghton et al. 1990)

equilibrium warming estimated for a radiative equivalent doubling of CO_2 concentrations that was mentioned earlier.

ATMOSPHERIC CONCENTRATIONS AND EMISSIONS SOURCES OF GREENHOUSE GASES

The quality and detail of the state of knowledge of historical emissions of greenhouse gases vary considerably among gases and sources. The general relationship between gaseous emissions and human activities is shown in Table 2.4. We will discuss the atmospheric concentration and emissions sources of the major emitted gases associated with the greenhouse issue.

Carbon Dioxide

Accurate observations of the concentration of greenhouse gases began in 1958, with measurements at the Mauna Loa Observatory in Hawaii. Figure 2.5 shows that the average annual concentration of CO_2 in the atmosphere

Table 2.4 Primary anthropogenic sources of greenhouse gases

Gas	Primary Anthropogenic Source
CO_2	Fossil-fuel burning; land-use conversion
CH_4	Ruminant animals; rice paddies; biomass burning; gas and mining leaks
CO	Energy use; agriculture; biomass burning
N_2O	Cultivation and fertilization of soils; combustion
$NO_x (= NO + NO_2)$	Fossil-fuel burning; biomass burning
$CFCl_3$	Chemical industry
CF_2Cl_2	Chemical industry
$C_2Cl_3F_3$	Chemical industry
CH_3CCl_3	Chemical industry
CF_2ClBr	Fire extinguishing
CF_3Br	Fire extinguishing
SO_2	Coal and petroleum burning
COS	Biomass burning; fossil-fuel burning
DMS	Primarily natural
NMHC	Incomplete combustion; agriculture
O_3	Not directly emitted, created by reactions involving NO_x in troposphere and by photolysis of O_2 in the stratosphere
OH	Not directly emitted, created naturally
H_2O	Anthropogenic emissions are small compared with natural evaporation

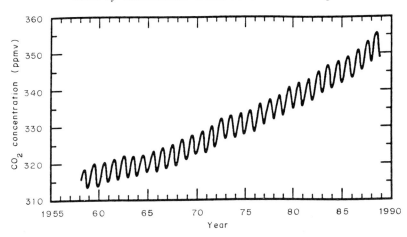

Figure 2.5 Average CO_2 annual atmospheric concentration in Mauna Loa, Hawaii. Data obtained by C.D. Keeling (Scripps Institution of Oceanography, University of CA, La Jolla, CA) and were obtained from files in the Carbon Dioxide Information Analysis Center, Oak Ridge National Laboratory, Oak Ridge, TN

has risen in the period from 1958 from approximately 315 ppmv to approximately 350 ppmv in 1988. The average annual rate of growth of the CO_2 concentration is approximately 0.4% y^{-1}. There is a clear annual cycle in the Mauna Loa data which corresponds to the annual cycle of plant growth and decay in the northern hemisphere. Carbon dioxide concentrations increase during the fall and winter and decline during spring and summer seasons. This cycle is reversed and of smaller amplitude in the southern hemisphere. The smaller amplitude in the southern hemisphere reflects the smaller extent of landmass in that hemisphere relative to the North. Estimates of the preindustrial concentration of CO_2 have been made by sampling air trapped in ice cores. These estimates indicate that the preindustrial concentration of CO_2 was approximately 280 ppmv. These estimates are shown in Figure 2.6 along with more recent observations.

Anthropogenic emissions of CO_2 are released principally from two human activities: fossil-fuel use, approximately 5.9 GtC y^{-1} in 1988 (Boden et al. 1990), and land-use changes (deforestation), approximately 1.5 GtC y^{-1} in 1988 MacCracken et al. (1990), with lesser amounts released by industrial processes such as cement manufacture, 0.15 GtC y^{-1} in 1988 (Boden et al. 1990). Estimates of fossil-fuel CO_2 emissions are known with relatively good confidence. Confidence bounds for global fossil-fuel CO_2 emissions of plus or minus 10% are frequently cited. Even if these bounds are overly optimistic, they are considerably narrower than the bounds that surround most other radiatively important emissions estimates, with the exception of

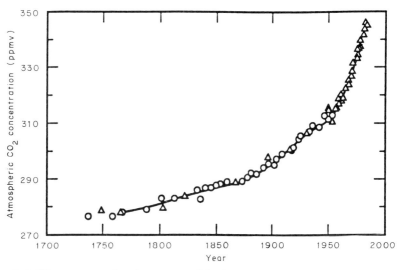

Figure 2.6 CO_2 concentration (parts per million) based on Lashof and Tirpak (1989)

CFCs. Fossil-fuel CO_2 emissions are known with much greater certainty than CO_2 emissions from land-use change. Human-made emissions of CO_2 take place against a background of natural exchanges between the atmosphere and oceans and the atmosphere and the terrestrial biosphere that are enormous. Each of these fluxes is estimated to be in the range of 100 GtC y^{-1} or more (Trabalka 1985). Sources of net CO_2 emissions and associated data quality are given in Table 2.5.

More is known about fossil-fuel CO_2 emissions than of any other gas-emission source. Since 1860, global annual emissions of fossil-fuel CO_2 have increased from 0.1 (Trabalka 1985; Marland et al. 1989) to approximately 5.9 GtC y^{-1} in 1988. During the period 1945 through 1979 the rate of CO_2

Table 2.5 Historical CO_2 emissions

Sources	Quality of emissions data	Uncertainty	Notes
Fossil-fuel use	Good	± 10–20%	1988; 5.9 GtC y^{-1} Available 1860–1988 by country, by fuel
Cement manufacture	Good	± 10–20%	Relatively minor importance
Land-use change deforestation	Fair/poor	± 50–100%	1980; 0.4–2.6 GtC y^{-1} Primarily tropical

emissions from fossil-fuel use grew at 4.5% y^{-1} (Marland et al. 1989). Emissions declined from 1979 until 1983, but emissions have risen subsequently (Marland et al. 1989; Boden et al. 1990). The U.S., USSR, and P.R. China account for half of the world's fossil-fuel CO_2 emissions (Marland et al. 1989). U.S. fossil-fuel CO_2 emissions accounted for more than 40% of global emissions in 1950 (Trabalka 1985; Marland et al. 1989). This share has steadily declined to less than 25% in 1988 (Boden et al. 1990). U.S. CO_2 emissions peaked in 1973 (1.27 GtC y^{-1}) and again in 1979 (1.30 GtC y^{-1}) and remained below that level until 1988 (1.31 GtC y^{-1}) (Boden et al. 1990). Global average per capita emissions of fossil-fuel CO_2 to the atmosphere were approximately 1.2 tC y^{-1} per capita in 1988 (Boden et al. 1990). U.S. emissions were approximately 5.3 tC y^{-1} per capita in 1988 (Boden et al. 1990).

Carbon content per unit energy varies by fuel. Of the fossil fuels, the average value of carbon per unit energy for natural gas is lowest, approximately 13.7 MtC/EJ^{-1}; coal is highest, 23.8 MtC/EJ^{-1}; and oil falls between the two, 19.2 MtC/EJ^{-1} (Edmonds and Reilly 1985). The mining of oil shales in carbonate rock formations would add an additional stream of CO_2 to the atmosphere whose magnitude depends on the grade of the resource and the technology employed to extract it. The transformation of primary fossil-fuel energy, as for example from coal to electricity or from coal to synoil or syngas, releases carbon in the conversion process. Energy technologies such as hydroelectric power, nuclear power, solar energy, and conservation (including energy efficiency improvements) emit no CO_2 to the atmosphere in their direct application. Some traditional biomass fuels, such as crop residues and dung, release CO_2 to the atmosphere in a balanced cycle of absorption and respiration whose time frame is short. The use of other biomass fuels such as firewood may provide either a net annual source or sink for carbon depending upon whether the underlying biomass stock is growing or being exhausted. Improvements in the efficiency of energy conversion technologies reduce the rate of emission of greenhouse gases per unit energy service provided.

There are approximately 600 GtC in the form of terrestrial biomass, principally stored in forests, with Trabalka (1985) estimating the value at 560 GtC and Bolin (1986) giving 640 GtC. This is estimated to be about 15–20% ($\approx$ 120 GtC) less than was present in the mid-nineteenth century (Trabalka 1985). On a global basis, this is estimated to vary less than about 10 GtC through the seasons as leaves and grasses grow and die. The phase pattern of northern and southern hemispheric cycles is, of course, reversed (Trabalka 1985; Bolin 1986; Boden et al. 1990).

Knowledge of the net annual emissions of carbon from land-use changes is far less certain than emissions estimates for fossil-fuel use. Emissions of net annual CO_2 release from land-use changes have been estimated for the year 1980 by various researchers. Net release is calculated as the difference between annual gross harvests of biomass, plus releases of carbon from soils, less biomass carbon whose oxidation is long delayed (e.g., stored in forest products such as telephone poles, furniture, and housing) and additions to the stock of standing biomass. Houghton et al. (1985) estimate 1980 emissions to be between 0.5 and 4.5 GtC y^{-1}. This range is narrowed in Houghton et al. (1987) to between 1.0 to 2.6 GtC y^{-1}. Detwiler and Hall (1988) estimate 1980 emissions from the tropics to be in the range 0.4 to 1.6 GtC y^{-1}. Net emissions from land-use change are dominated by tropical deforestation. Houghton et al. (1987) estimate that all but 0.1 GtC y^{-1} of net release are from tropical forests.

Estimates of deforestation in 1980 are greatest for Brazil, Colombia, the Ivory Coast, Indonesia, Laos, and Thailand (Houghton et al. 1987). Estimates of net CO_2 emissions from land-use change indicate increases in recent decades. Prior to 1950, significant deforestation is estimated to have occurred in the temperate latitudes as well as in the tropics (Trabalka 1985).

Conventional estimates of net CO_2 release from land-use change do not take the possibility of a CO_2 fertilization effect into account. While a matter of heated debate, it has been suggested that increases in the atmospheric concentration of CO_2 could act to accelerate the rate at which the terrestrial biosphere stores carbon. The conventional wisdom, see for example Bolin (1986), is that the CO_2 fertilization effect, if it exists at all, must be small and that the terrestrial biosphere is a net source of carbon release to the atmosphere. Recent papers including Tans et al. (1990), Goudriaan (1991, 1989), Goudriaan and Ketner (1984), and Esser (1991) have estimated that the terrestrial biosphere may be a large net sink for carbon. Tans et al. (1990) imputed a missing sink of carbon, most likely located in the northern midlatitudes, from a detailed analysis of atmospheric and oceanographic records. Goudriaan (1991) estimates that the CO_2 fertilization effect could more than compensate for carbon releases from land-use changes and still provide a net sink for carbon in the range of 0.5 to 1.5 GtC y^{-1}. Esser (1991) finds that the CO_2 fertilization effect may have resulted in a net carbon uptake of 70 Pg over the period 1860–1980.

Atmospheric chemistry is also a small net source of atmospheric CO_2. All the carbon species, in particular CO and CH_4, eventually oxidize to CO_2. Total annual oxidation of CO and CH_4 is estimated to be less than 1 GtC y^{-1}, most of this in the form of CO (WMO 1985; MacCracken and Luther 1985b; Houghton et al. 1990). Most of the carbon released in one year is oxidized to CO_2 during that same year. In addition, the convention by which CO_2 emissions are calculated assumes complete oxidation of all the carbon from fossil-fuel combustion byproducts and deforestation. This convention does not always reflect reality, but is a useful first approximation and reflects the eventual disposition of all of the carbon release.

Methane

Ice-core data going back 160,000 years indicate that concentrations of CH_4 in the preindustrial atmosphere were less than half the present concentration of 1700 ppbv (Pearman et al. 1986; Khalil and Rasmussen 1987; Raynaud et al. 1988; Figure 2.7). Methane concentrations were about 700 ppbv until the beginning of the 19th century. Concentrations during glacial periods were even smaller, as low as 350 ppbv (Raynaud et al. 1988). Current concentrations have been growing at approximately 0.9% y^{-1} (Steele et al. 1987; Blake and Rowland 1988; Houghton et al. 1990; Figure 2.8).

Annual observations of emissions sources are not generally available for CH_4. Source strength uncertainties are so high that emissions budgets are typically referenced to by decade rather than by individual year. Whereas emissions of fossil-fuel and land-use change CO_2 are developed for specific years based on data bases for a small number of human activities, the sources and sinks for CH_4 are developed using an observed globally averaged atmospheric burden of CH_4 (approximately 4800 Tg), the annual average rate of increase, and an atmospheric lifetime (approximately 12.5 years), derived from an atmospheric chemistry model to calculate a global emissions budget constraint of approximately 540 (400 to 640) Tg y^{-1}. Information

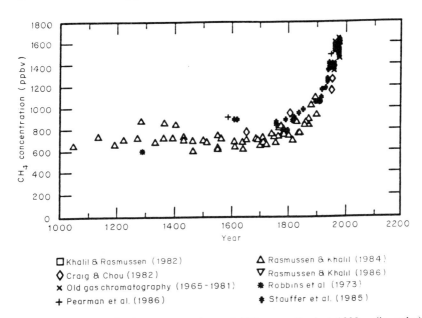

Figure 2.7 The atmospheric concentrations of CH_4 over the last 1000 y (in ppbv). The figure shows the complementary relationship between different studies, the time when CH_4 concentrations began to increase rapidly, and the doubling of CH_4 over the last several hundred years (Khalil and Rasmussen 1987)

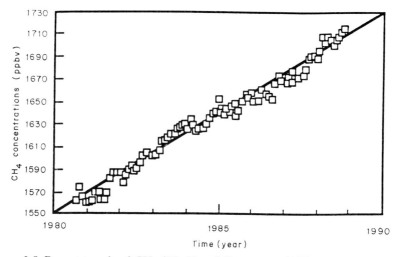

Figure 2.8 Recent trends of CH_4 (Khalil and Rasmussen 1988)

about the changing isotopic ratio of CH_4 (i.e., $^{14}CH_4/^{12}CH_4$) is used to infer the contribution of fossil fuels to total emissions. A summary of the best current understanding of the sources and sinks of CH_4 are given in Table 2.6.

At this point it is not clear whether all major sources of CH_4 have been identified and whether the emissions rates of those that have been identified are subject to significant uncertainty. Anthropogenic activities are currently thought to contribute at least half of all CH_4 emissions to the atmosphere. The three principal human activities that have been identified as emissions sources are cattle raising, rice production, and energy production and use. While human activities have been identified as major sources of atmospheric emissions, there remains great uncertainty surrounding emissions source estimates and the time profile of those emissions.

Roughly a quarter of the total atmospheric CH_4 emissions are attributable to the production, transfer, conversion, and consumption of energy (Barns and Edmonds 1990). These include the mining of coal as well as the gathering, transmission, distribution, venting, and flaring of natural gas. Landfill material representing the residue of the consumption process is a rich source of CH_4 that is only very slightly exploited as a source of energy at the present time. Burning of biomass can occur naturally, as in forest fires, or can be initiated by human activity such as in the clearing of land for agriculture. Some fraction of the human contribution is for direct energy consumption such as the burning of fuel wood and incomplete combustion in automobiles, and in inefficient residential and commercial heating systems. Finally, each of the combustion processes associated with the conversion of

Table 2.6 Historical CH_4 emissions

Sources and sinks	Quality of emissions data	Uncertainty	Notes
Sources			
Natural sources	Poor	Large	Total global emissions from natural and anthropogenic sources are estimated to range from 400 to 640 $TgCH_4\ y^{-1}$. Natural sources are approximately 50% of global emissions. It is not clear that all sources and sinks for CH_4 have been identified. Total emissions are derived from atmospheric observations and calculations from atmospheric chemistry models. Current source estimates are (in $TgCH_4\ y^{-1}$): enteric fermentation (wild animals), 4 (1–7); wetland, 110 (60–160); lakes, 4 (2–6); tundra, 3 (1–5); oceans, 10 (0–20); termites and other insects, 12 (5–45); methane hydrates, 5 (0–100); other 40 (0–80).
Agriculture	Fair/poor	± 40–50%	25–40% of global emissions. Current source estimates are (in $TgCH_4\ y^{-1}$): rice cultivation 70 (40–110); ruminant digestive systems of domesticated animals (cattle), 77 (40–110); slash-and-burn agriculture/land cleaning, 55 (25–85).
Energy	Fair/poor	± 40–50%	10–30% of global emissions. Current source estimates are (in $TgCH_4\ y^{-1}$): deep coal mining 25 (10–35); natural gas production, transport and distribution, 40 (20–60); incomplete combustion (e.g., automotive exhaust, fuel wood use) 15 (5–15); landfills, 30 (10–60).
Sinks			
Atmospheric chemistry	NA	± 25–50%	Current sink estimates are (in $TgCH_4\ y^{-1}$): reactions with tropospheric OH, 400 (300–500), transport to and reaction with OH, Cl, or O in stratosphere, 50 (30–70); uptake by soil micro-organisms, 32 (15–45); accumulation in the atmosphere, 43 (40–46).

fossil fuel to thermal energy may be attended by the emission of some quantity of CH_4, depending upon the constituents of the fuel, the temperature of combustion, and the efficiency of the process. Emissions from natural gas production, coal mining, and landfills currently appear to be more important sources of CH_4 than combustion process byproducts. The U.S., USSR, and the P.R. China are the largest sources of fossil-fuel CH_4 emissions (Barns and Edmonds 1990).

As a result of climatic warming, carbon emissions could significantly increase through the release of carbon now stored in frozen soils in Arctic regions and as peat and humus in other regions. Estimates of potential releases are highly speculative but potentially of an order of magnitude equivalent to current fossil-fuel CO_2 emissions. This carbon may be released in the form of CH_4. The attendant effect on the CH_4 budget would be significantly greater than on the CO_2 budget. Another relatively short-term climatic feedback may occur from increases in the decomposition of vegetation resulting from increased mortality due to rapid climatic change.

Nitrous Oxide

The atmospheric concentration of N_2O is increasing at the rate of about 0.3% y^{-1} (Watson et al. 1988; Khalil and Rasmussen 1988; Houghton et al. 1990; Figure 2.9). Current tropospheric concentrations are about 310 ppbv (Houghton et al. 1990). Ice core data indicate that the concentration of N_2O was stable for approximately 2000 to 3000 years at about 285 ppbv, though there is some doubt about the absolute calibration while concentrations began to increase beginning about 200–300 years before the present (Houghton et al. 1990).

The sources of N_2O emissions are poorly documented, as indicated in Table 2.7. Emissions rates are small relative to atmospheric stocks. While the atmospheric burden and annual rate of increase are known with some confidence, the atmospheric lifetime is uncertain within the range 100 to 175 years. This leads to significant uncertainties in the estimates of sources and sinks, which can be derived from atmospheric chemistry models. Individual source terms are subject to even greater uncertainty. Total emissions are estimated to be between 7 and 21 TgN y^{-1}. Emissions studies are inconsistent with regard to their categorization of emission-producing activities. The chief sources of emissions are presently thought to be biogeochemical activities in soils and combustion activities. The bio-geochemical activities include N_2O releases from both cultivated and uncultivated soils, and fertilized and unfertilized soils. Combustion activities include savanna burning, forest clearing, fuelwood use, and fossil-fuel combustion. Other sources of emissions include oceans and contaminated aquifers.

The dominant human activities associated with N_2O emissions are

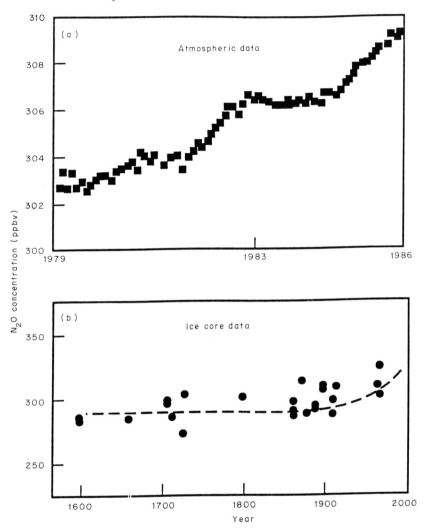

Figure 2.9 Nitrous oxide concentrations (parts per billion) (Lashof, D.A. and Tripak, D.A. 1989)

agricultural (savanna burning, soil cultivation, and fertilizer application) and energy (wood burning and fossil-fuel use). Until recently, the dominant human-made emissions source was thought to be fossil-fuel combustion. This conclusion was based on flask samples taken from combustion experiments, which recently have been shown to be subject to a sampling artifact, producing N_2O in the flask between the time the sample was taken and the time the flask was analyzed (Muzio and Kramlich 1987; Linak et

Table 2.7 Historical N_2O emissions

Sources and sinks	Quality of emissions data	Uncertainty	Notes
Sources			
Natural sources	Poor	Great	Total global emission from natural and anthropogenic sources are estimated to range from 10 to 19 TgN y^{-1}. Natural sources are approximately 45–65% of global emissions. Natural sources include (units = TgN y^{-1}): oceans and estuaries, 2 (1–3); natural soils, 6.5 (3–10); aquifers, wildfires, lightning and volcanos, 0.8 (0.5–1.1). It is not clear that all sources and sinks for N_2O have been identified. While a highly stable gas in the atmosphere, the average lifetime uncertainty ranges from 120 to 150 years, leading to large uncertainty in the global emission rate.
Nonenergy	Poor	± 50–100%	35–55% of global emissions. Current source estimates are (TgN y^{-1}): natural soil cultivation, 1.5 (1–2); nitrogen fertilizer applications, 1.1 (0.12–2.2); slash-and-burn agriculture/land clearing, 0.2 (0.1–0.9).
Energy	Poor	Great	3–15% of global emissions. Fossil-fuel combustion estimated to produce 0–3 TgN y^{-1}. Combustor studies originally showed high rates of release of N_2O from fossil-fuel use in high-temperature combustion. Later analysis revealed a sampling artifact which, when removed, greatly reduces the emission coefficient.
Sinks			
Atmospheric chemistry	N/A	± 15–40%	Current sink estimates are (in TgN y^{-1}): photolysis and reaction with $O(^1D)$ in stratosphere 10 (7–13); accumulation, 3.5 (3–4).

al. 1989). It is possible that fossil-fuel emissions are a relatively minor source of N_2O emissions, but this is by no means certain. It is also possible that the chemistry that occurred in the flask may also occur in nature.

Chlorofluorocarbons

The term chlorofluorocarbons refers to a family of compounds derived from the CH_4 or ethane (C_2H_6) molecules. A CFC is formed by replacing all hydrogen molecules with the halogens, chlorine (Cl), or fluorine (F). When the bromine (Br) atom is also used as a replacement, the compounds are referred to as halons.

The CFCs receiving the most attention, primarily because of their larger concentrations and potentially significant effects on stratospheric O_3, are $CFCl_3$ (referred to as CFC-11) and CF_2Cl_2 (referred to as CFC-12). Other similar compounds, including CCl_4, CH_3CCl_3, and the halons, are currently exerting less, but still significant, influence; their effects could become quite large if present emission trends continue, but this is most unlikely given the wide international support for reductions of emissions of these gases.

Chlorofluorocarbons CFC-11 and CFC-12 have the highest concentrations of the human-made CFCs: 0.280 and 0.484 ppbv (measured at Mauna Loa), respectively, in 1990 (Houghton et al. 1990). These concentrations are far more dilute than is the case for other radiatively important gases. The surface air concentrations of these two gases are currently increasing at a rate of over 4% y^{-1} (Houghton et al. 1990; Figure 2.10). These gases have relatively long-lifetimes: 75 years for CFC-11, 110 years for CFC-12. In addition, they are strong infrared absorbers in the "atmospheric window." A molecule of CFC-12 has several thousand times the radiative forcing impact of a molecule of CO_2 (Houghton et al. 1990; Ramanathan et al. 1987). CFC-11 and CFC-12 are thought to have contributed about one-quarter of the radiative forcing of gases other than CO_2 during the 1980s (Ramanathan et al. 1985; Hansen et al. 1989).

Other important CFCs include CFC-113 ($CF_2ClCFCl_2$), HCFC-22 (CHF_2Cl), and methyl chloroform (CH_3CCl_3). The atmospheric concentration of CFC-113 is increasing by about 10% y^{-1}, with the present surface air concentration being about 0.06 ppbv in 1990 (Houghton et al. 1990). CFC-22 and CH_3CCl_3 are used primarily as solvents, while CFC-22 is used in air conditioning and refrigeration equipment. Carbon tetrachloride is used primarily as a feedstock in the production of CFCs by the chemical industry. Halons are used extensively in fire extinguishing applications (EPA 1989).

All of these synthesized chlorocarbons have relatively long atmospheric lifetimes, that is relatively slow removal rates. Methyl chloroform has the shortest chemical lifetime, about six to seven years, while CFC-13 has the longest lifetime, about 400 years (Houghton et al. 1990). The long lifetimes

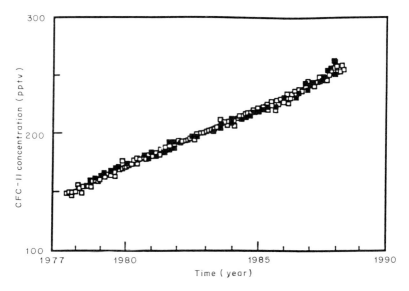

Figure 2.10 Measurements of CFC-11 at Mauna Loa Observatory since mid-1977, based on NOAA GMCC data (J. Peterson, pers. comm. 1989)

contribute to the rapidly increasing concentrations of these gases. The fully halogenated chlorocarbons, particularly CFC-11 and CFC-12, are of primary concern in affecting stratospheric O_3 concentrations because they are destroyed in the stratosphere, by photolysis, which releases all their chlorine atoms to act as catalysts for O_3 destruction. Other chlorinated halocarbons also release chlorine to the stratosphere, whereas the destruction of halons releases bromine. The chlorine and bromine atoms can then react catalytically to destroy O_3.

Total worldwide production of these two compounds is unknown. However, most CFC-producing companies in the OECD and developing world (but not eastern European nations, the USSR, or P.R. China) report production to the Chemical Manufacturing Association (CMA). CMA (1988) reports that for 1987 the total cumulative production of CFC-11 was 7.1 Tg y^{-1} while CFC-12 was 9.4 Tg y^{-1}. Actual production in 1987 was 0.4 Tg y^{-1} of CFC-11 and 0.4 Tg y^{-1} of CFC-12. The U.S. accounts for about one-quarter of the estimates of global production of CFC-11 and about one-third of the production of CFC-12.

Production does not equate to emission. Of the total cumulative production of 16.4 Tg of CFC-11 and CFC-12 by 1987, 15.0 Tg (or about 90%) were thought to have been emitted to the atmosphere (CMA 1988). The remaining CFCs are "banked" in products such as air conditioners and refrigerators (as working fluids), spray cans (as aerosols), and structural and flexible foam products (as blowing agents). The rate of emission of these banked CFCs depends upon the quantity held in relatively slow-release usages, such as home refrigerators, as compared with short-term release products such as spray cans. Nearly all aerosol uses have been banned in the U.S. and a few other nations since 1978.

Carbon Monoxide

Carbon monoxide is not important as a radiatively active gas, but it is extremely important because it reacts with the hydroxyl radical (OH). Peak concentrations of about 200 ppbv occur at higher northern latitudes, with minimum concentrations of about 50 ppbv found throughout the southern hemisphere. Concentrations are increasing by about 1.1% y^{-1} (globally averaged) although there is little evidence of an increase occurring in the southern hemisphere (Khalil and Rasmussen 1988; Cicerone 1988). Carbon monoxide has a relatively short lifetime before being transformed to CO_2; because of the relatively high efficiency of combustion processes, this source of CO_2 is relatively minor.

Even greater uncertainty surrounds the atmospheric CO budget than the CH_4 budget. As indicated in Table 2.8, CO is generated by incomplete combustion processes (complete combustion yields CO_2 rather than CO), oxidation of anthropogenic hydrocarbons, the decomposition of CH_4, and other minor sources. Because CO is highly reactive, it has a relatively short atmospheric lifetime (0.4 years) and is poorly mixed in the global atmosphere (Houghton et al. 1990; Wuebbles and Edmonds 1988). Annual emissions estimates range from 600 to 1700 TgC y^{-1} (Wuebbles and Edmond 1988), with the Houghton et al. (1990) estimating it at 1030 TgC y^{-1}. Concentrations of CO are significantly higher in the northern hemisphere than in the southern hemisphere, indicating higher emissions in those latitudes (Houghton et al. 1990). This is consistent with the pattern of combustion activities. U.S. emissions of CO from fossil-fuel use were estimated to be between 24 and 29 TgC y^{-1} in 1985 (NAPAP 1989; EPA 1990).

THE CARBON CYCLE

Over the period 1860 to 1980, approximately 160 GtC were introduced into the atmosphere by fossil-fuel use (Marland et al. 1989). Estimates of the emissions of carbon from the terrestrial ecosystems vary. From 1860 to 1980, Peng et al. (1983) estimate that 265 GtC were released from forests and soils. Emanuel et al. (1984) and Stuiver et al. (1984) derive estimates of 230 GtC and 150 GtC, respectively. Trabalka (1985) estimates the net carbon flux from land biosphere since 1800 to be only 120 GtC, with a

Table 2.8 Historical CO emissions

Sources and sinks	Quality of emissions data	Uncertainty	Notes
Sources			
Natural Sources	Poor	Great	Total global emission from natural and anthropogenic sources are estimated to range from 600 to 1700 TgC y^{-1}. Natural sources are approximately 50% of global emissions. Current source estimates are (in TgC y^{-1}): plants, 55 (20–90); wildfires, 10 (5 to 20); oceans, 20 (10 to 40); oxidation of natural hydrocarbons, 250 (50 to 500); and oxidation of CH_4, 260 (75 to 450). It is not clear that all sources and sinks for CO have been identified. CO is highly reactive and therefore global source/sink relationships are highly uncertain even in aggregate.
Nonenergy	Poor	± 50–100%	40–60% of global emissions. Current source estimates are (in TgC y^{-1}): slash-and-burn agriculture/land clearing, 270 (120–520).
Energy	Fair/Poor	±40–50%	40–60% of global emissions. Current source estimates are (in TgC y^{-1}: incomplete combustion (e.g., automotive exhaust, fuel wood use), 200 (150–250); oxidation of anthropogenic hydrocarbons, 40 (0–80).
Sinks			
Atmospheric chemistry	N/A	± 25–50%	Current sink estimates are (in TgC y^{-1}): 900 (500–1200).
Soil uptake	N/A	± 25–50%	Current sink estimates are (in TgC y^{-1}): 110 (80–140).

range of 90–180 GtC. Net deforestation was the principal source of carbon release into the atmosphere in the earlier part of the period, while fossil-fuel use dominated releases in the latter portion. The average carbon content of the atmosphere increased by approximately 125 GtC over the period 1960 to 1986. Thus, the increase in atmospheric carbon corresponds to between 30 and 45% of the carbon released over the period from 1860 to 1980. Bolin (1986) obtains a range of 32 to 60% by including a range of estimates for preindustrial atmospheric carbon in addition to a range of estimates for carbon released from the terrestrial biosphere. This phenomenon can be observed over shorter time periods as well. Approximately 3 GtC y^{-1} accumulate in the atmosphere. The annual net injection from anthropogenic activities is thought to be ranged from 6 to 8 GtC y^{-1}. Atmospheric retention is therefore approximately half of anthropogenic emissions.

These observations raise two questions. First, are the accumulation of CO_2 in the atmosphere and the net emission to the atmosphere of carbon by human activities related? And second, if the two are related, what is the nature of that relationship? Why is the buildup of carbon in the atmosphere smaller than the anthropogenic emission?

Background on the Carbon Cycle

Attempts to answer the two questions posed above have led researchers to study the flows of carbon throughout the global system. Estimates of the stocks (figures found within boxes) and the flows (figures associated with arrows between boxes) of carbon are found given in Figure 2.11. Estimates of the stocks and fluxes between various reservoirs of carbon are subject to great uncertainty. Only the atmospheric concentration of CO_2 and the fossil-

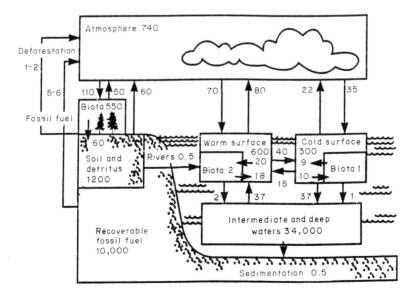

Figure 2.11 The global carbon cycle

fuel CO_2 injection are known with relatively good precision. The stock of carbon in the atmosphere in 1988 was approximately 742 GtC (Boden et al. 1990). This compares with the approximately 5.9 GtC y^{-1} that was injected by fossil-fuel use (Boden et al. 1990) and the 0 to 3 GtC y^{-1} which were estimated to be added by deforestation. Trabalka (1985) gives a range of 0 to 2.6 GtC y^{-1}, while Houghton et al. (1990) gives a range of 0.6 to 2.6 GtC y^{-1}. As noted earlier, the estimates of net deforestation make no allowance for either a CO_2 fertilization carbon uptake or other, significant, unconventional terrestrial carbon uptake mechanism.

Flows of carbon into and out of the atmosphere are very large. Annual fluxes into and out of the atmosphere are estimated to be on the order of 200 GtC y^{-1} (Trabalka 1985; Bolin et al. 1986; Houghton et al. 1990).

CO_2 is a stable form of carbon in the atmosphere. It is therefore not an active atmospheric constituent and there are no significant sinks for CO_2 in the atmosphere. The only generally agreed upon permanent sink for carbon is deep ocean and soils. The net buildup of carbon in soils is presently thought to proceed at a slower pace than the removal of carbon to oceans (Trabalka 1985; Bolin 1986). At present, soils may be a net source of carbon to the atmosphere as a result of land-use changes. Ocean models can account for only part of the 3 to 5 GtC y^{-1} removal of carbon from the atmosphere associated with human activities. The disposition of the so-called "missing carbon" cannot yet be adequately explained by present research models. Houghton et al. (1990) presents the accounts the following budget:

	GtC y^{-1}
Emissions from fossil fuels into the atmosphere	5.4 ± 0.5
Net emission from deforestation and land use	1.6 ± 1.0
Total Net Emissions	7.0 ± 1.5
Accumulation in the atmosphere	3.4 ± 0.2
Uptake by the ocean	2.0 ± 0.8
Total Disposition	5.4 ± 1.0
Net imbalance ("missing carbon")	1.6 ± 1.4

Evidence of the Relationship between Anthropogenic Net Carbon Emissions and the Accumulation of CO_2 in the Atmosphere

If the fluxes of carbon into and out of the atmosphere generated by natural sources are so large relative to those generated by human activities, it is natural to question whether the observed buildup of CO_2 in the atmosphere is simply the outgrowth of natural processes and merely coincident with the emission by human activities. After all, total annual emissions from fossil-fuel use and land-use changes amount to only 3% of the annual natural flux, well within the bounds of uncertainty of the natural flux rates. Nevertheless, circumstantial evidence provides a strong case that human activities are principally responsible for the observed increase in atmospheric CO_2.

First, ice core records show concentrations of atmospheric CO_2 remaining between 270 and 290 ppmv over the past 1000 years. Concentrations have risen steadily to 350 ppmv since 1860 (Neftel et al. 1985; Raynaud and Barnola 1985; Friedli et al. 1986; Pearman et al. 1986). Despite the exchange of carbon to the atmosphere from terrestrial and oceanic sources, and great uncertainty surrounding the actual rates, the atmosphere, whose carbon content is known with relative certainty, shows a slow but steady increase, which closely parallels the release of CO_2 into the atmosphere from fossil-fuel use. The relationship has been particularly strong in the period since 1958, when precise measurements of atmospheric CO_2 began at Mauna Loa (Trabalka 1985; Houghton et al. 1990). The ratio of accumulation in the atmosphere to fossil-fuel CO_2 release has been a remarkably stable 0.58. The rate of growth of atmospheric CO_2 also shows breaks in 1973 (corresponding to the OPEC oil embargo) and 1979 (corresponding to the Iran–Iraq War price spike).

Another important piece of evidence is the observed buildup of geologically older carbon in the atmosphere. The ratio of $^{12}CO_2$ to $^{14}CO_2$ has been decreasing in the atmosphere, signaling the increasing proportion of radiocarbon-free (and hence stored longer than 40,000 years) CO_2 in the atmosphere. This is consistent with the introduction of a large quantity of fossil-fuel CO_2 into the atmosphere (Trabalka 1985; Houghton et al. 1990).

Furthermore, the differential between the concentration of CO_2 in the northern and southern hemispheres has grown from approximately 1 ppmv (1960) to 3 ppmv (1985). This is consistent with the larger release of CO_2 in the northern hemisphere than in the southern hemisphere (Trabalka et al. 1985; Houghton et al. 1990).

Finally, the current rate of change in the concentration of CO_2 in the atmosphere (1.4 ppmv y^{-1}) is unprecedented. The most rapid rate of change in concentration observed in ice core records extending over the preceding 160,000 years is significantly less (0.5 ppmv y^{-1}). The present concentration of 350 ppmv is also unprecedented in the preceding 160,000 years. The previous maximum concentration associated with an interglacial period was 320 ppmv, observed 120,000 years before the present.

Alternative Models of the Atmospheric Retention of Fossil-Fuel CO_2

Several models have been used to describe the relationship between the net emission of anthropogenic CO_2 into the atmosphere and atmospheric concentrations of CO_2. These models range from very simple to extremely complex. The two most popular, simple models will be referred to as the airborne fraction model (AFM) (WMO 1981; Nordhaus and Yohe 1983; Edmonds and Reilly 1983; Bolin 1986), and the atmospheric decay model (ADM) (Nordhaus and Yohe 1983). These models are reduced-form, not realistic models. They make no attempt to trace the movement of carbon through the Earth system. The question is, are they useful predictive tools? Their usefulness as predictive tools depends upon the specific issue being addressed.

The AFM assumes that the change in the carbon content of the atmosphere is proportional to net carbon emission. Two variants of this model exist. The first computes the ratio of accumulation of carbon in the atmosphere to the release of fossil-fuel CO_2. This ratio is relatively stable in the post-1958 period at 0.58. The second computes the ratio of accumulation of carbon to the atmosphere to the net anthropogenic release of CO_2 (the sum of fossil-fuel CO_2 release plus release from deforestation). The second ratio is less certain because the net release rate from deforestation and other land-use changes is less certain. As noted earlier, estimates of the release rate for land-use changes vary between 0 and 3 GtC y^{-1}. Assuming a net release rate for land-use changes of approximately 1 GtC y^{-1} yields an airborne fraction of approximately 0.5; however, the ratio can vary between 0.6 (no net release from land-use change) and 0.4 (3 GtC y^{-1} release rate for land-use change; see Bolin 1986 for further discussion). These calculations are consistent with those obtained using Houghton et al.'s (1990) values for carbon balance cited earlier.

The strength of this model is that it is statistically tied to recent experience. Its principal weakness is that it is not based on a realistic model of the carbon cycle and therefore is an uncertain guide to atmospheric accumulation under emissions scenarios outside the bounds of historical experience. One of the implications of the model is that CO_2 will continue to accumulate in the atmosphere as long as any net emissions continue. That is, fossil-fuel releases of CO_2 plus net releases from land-use change would have to be cut to zero to stabilize the concentration of CO_2 in the atmosphere. This result contradicts the findings of more sophisticated carbon cycle models (Bolin 1986; Trabalka 1985; Houghton et al. 1990).

The ADM is a different model of CO_2 accumulation in the atmosphere. The ADM is based on the assumption that the average molecule of CO_2 has a fixed lifetime in the atmosphere. A fixed fraction of the stock of carbon in the atmosphere is removed annually, presumably into the oceans. Such models can also replicate the post-1958 experience reasonably well, using for example the Mauna Loa atmospheric CO_2 record and fossil-fuel carbon release estimates reported in Boden et al. (1990), on the assumption that the average lifetime of a molecule of carbon is 250 years and that approximately 1 GtC y^{-1} is released net from the biosphere. The implied instantaneous removal rate, without additional injections of carbon, for this model is approximately 3 GtC y^{-1}. Because the removal rate depends on the stock of carbon in the atmosphere, the rate of carbon removal varies only slowly, as long as the rate of emission to the atmosphere is small relative to the stock. As a first approximation, a secession of emissions would result in approximately 3 GtC y^{-1} continuing to be removed by the oceans annually. The model predicts that the atmospheric concentration of CO_2 would stabilize if emissions were cut to 3 GtC y^{-1}.

The strength of this model is that it can track the post-1958 experience reasonably well. Also, it can be loosely tied to more sophisticated carbon cycle processes, specifically ocean carbon uptake. The major problem with this model is that it cannot explain why there was any CO_2 in the atmosphere at all in 1860, i.e., if the removal rate for carbon from the total atmosphere were 1/250 per year, then at the end of a millennium there would be no atmospheric carbon left. However, the atmosphere was clearly stable over the preceding millennium.

Research models of the carbon cycle have striven to provide detailed descriptions of the flows of carbon in the Earth system. These models include interactions between living biomass, soils and detritus, rivers, and oceans including segmented ocean elements, ocean biota, and ocean sediments. No model of the Earth's carbon system can hope to approximate reality unless it can adequately explain the behavior of oceans in the removal and release of carbon. Initial models of oceans segmented the waters into a set of average layers differentiated by depth, and carbon diffused from the atmosphere into the surface layer and then from the surface layer into ever deeper layers. Such models are termed box-diffusion models. Diffusion coefficients were established so as to agree with observations of diffusion of radiocarbon and other tracers.

The earliest models of ocean carbon processes were two-box models containing simple reservoirs for carbon in surface and deep-water layers. Such models were explored by, among others, Craig (1957), Bolin and Eriksson (1959), and Keeling (1973). The course of research made it increasingly clear that simple two-dimensional models of the ocean's carbon system were inadequate for the purposes of either explaining the system or forecasting potential future ocean carbon uptake. Oeschger et al. (1975) proposed a box model which disaggregated the ocean into two parts: a well-mixed surface layer approximately 75 m in depth and a deep ocean. Carbon transport within the deep ocean was accomplished by eddy-diffusion. Work by, for example, Björkstrom (1979), Hoffert et al. (1981), Killough and Emanuel (1981), Enting and Pearman (1982, 1985), Bolin et al. (1983), Peng et al. (1983), and Siegenthaler (1983) developed a range of alternative models. More recent models have included three-dimensional transport and chemistry (e.g., Maier-Reimer and Hasselmann 1987). These models include improved understanding of ocean chemistry, deep water formation, upwelling, biological sequestering, and three-dimensional transport and chemistry. Interestingly, the most sophisticated models of ocean carbon cycle yield net results for carbon uptake similar to the simpler box-diffusion models.

Oceans and the terrestrial ecosystems are weakly coupled, except through their interchanges with the atmosphere, and are therefore treated as if they were two separate systems. The terrestrial system is generally thought to be in rough equilibrium with the atmosphere. Carbon dioxide uptake by plants, 110 GtC y^{-1}, is just balanced by releases from the biosphere to the atmosphere, 50 GtC y^{-1}, and from soils and detritus to the atmosphere, 60 GtC y^{-1} (Figure 2.11). Carbon leaving the biota to soils, 60 GtC y^{-1}, leaves both the biota and soils as neutral carbon sources (sinks). Uncertainty surrounds this conventional view of the terrestrial system. Because CO_2 is a fertilizer for plants, the net increase in atmospheric CO_2 should, in principle, stimulate plant growth and add to the net carbon stock in the biota. This is referred to as the CO_2 fertilization effect. While the CO_2 fertilization effect is a well-established phenomenon, the extent to which it affects the carbon cycle is a matter of debate. As noted earlier, estimates of up to 2.5 GtC y^{-1} net removal of carbon from the atmosphere by some terrestrial system mechanism have been estimated either by inference or by direct calculation (Enting and Mansbridge 1989; Goudriaan 1989; Esser 1991; Tans et al. 1990). Empirical evidence provides an inconclusive basis for resolving the question at present.

Holistic models of the global carbon cycle that include both oceans and terrestrial ecosystems have been developed by various researchers including Peng et al. (1983), Emanuel et al. (1984), Goudriaan and Ketner (1984), and Goudriaan (1989). These models allow the direct computation of atmospheric concentrations, given external information on emissions from fossil-fuel use and land-use change.

Research carbon cycle models, run to determine the amount of net anthropogenic CO_2 emissions that would be consistent with a stable concentration of atmospheric CO_2, indicate that net emissions would have to be reduced to approximately 1 GtC y^{-1} (EPA 1989).

The Lifetime of CO_2 in the Atmosphere

Unlike other greenhouse gases, the concept of an atmospheric lifetime for CO_2 is troublesome. As we shall see, the concept of a lifetime of carbon in the atmosphere is inherently different from the concept as it has been developed for gases that are chemically active and which have permanent sinks in the atmosphere. For gases other than CO_2, a simple exponential decay model is applied to a stock of the gas resident in the atmosphere to estimate the removal rate of the gas from the atmosphere.

This model, however, begins to break down for CO_2. The first problem that we encounter with CO_2 is the problem of gross versus net fluxes. The large gross annual fluxes between the atmosphere and oceans, and the atmosphere and terrestrial systems imply that carbon atoms leave the atmosphere and enter oceans or terrestrial systems, but are replaced by other atoms leaving oceans or terrestrial systems and entering the atmosphere. If all we were interested in was the time that an average CO_2 atom resided in the atmosphere as a CO_2 atom, that time would be rather short, on the order of 4 (2 to 6) years, depending upon the actual gross flux rates with the atmosphere. But large bi-directional gross flux rates can exist without any change in the stock of CO_2 in the atmosphere.

If a simple exponential decay function is fitted to the entire atmospheric stock of CO_2, a lifetime of 250 years is reasonable. This is the reciprocal of the net removal rate of carbon from the atmosphere. As we have seen, there is some problem reconciling an ADM that is consistent with recent experience with the presence of carbon in the atmosphere in the year 1860.

Alternatively, an ADM could be applied only to that fraction of CO_2 exceeding the preindustrial, and presumably steady-state, concentration of 280 ppmv. The application of that model to the post-1958 period yields a lifetime for CO_2 of approximately 40 years. The principal problem with this model is, how do oceans tell the difference between carbon that is eligible for removal and carbon that is ineligible for removal.

To circumvent the above problems, coupled ocean–atmosphere models have been run in a simulation mode to examine their response to small changes in the initial stock of carbon in the atmosphere over time (Maier-Reimer and Hasselmann 1987; Siegenthaler and Oeschger 1987). These models are dynamic equilibrium models and are tuned to reproduce preindustrial concentrations at initial conditions. They can be perturbed by adding carbon to the system and observing the fraction of the additional carbon remaining in the atmosphere at any future point in time. If the output from the Maier-Reimer and Hasselmann (1987) pulse-of-carbon experiment is used to fit a simple exponential decay model, an interesting result is observed. The lifetime of the incremental carbon changes depending upon the period over which the ADM is fit. If one year is allowed to pass, the computed lifetime of CO_2 is 16 years. That is, incremental carbon is disappearing from the atmosphere at a rate such that if it persisted over a 16-year period, $1/e$ would be left. But, the longer the period allowed to pass before the exponential decay model is applied, the longer the calculated lifetime. Table 2.9 displays some lifetimes of incremental carbon and the number of years after initial introduction of the carbon.

We note that for the Maier-Reimer and Hasselmann calculation, $1/e$ of

Table 2.9 The relationship between the computed lifetime of incremental carbon in Maier-Reimer and Hasselmann (1987) and the period of the experiment (units = years)

Period of the experiment	Lifetime of incremental carbon
1	16
5	26
10	34
25	50
50	69
10	100
40	250
50	294
100	516

the incremental carbon remains at the end of 100 years. This experiment presumes a preindustrial equilibrium and therefore does not necessarily yield the same result as an experiment in which an additional kilogram of carbon is released into the present, disequilibrium atmospheric.

ATMOSPHERIC PROCESSES

The actual atmospheric composition of greenhouse gases depends not only on surface emissions but also on any atmospheric chemical processes affecting their concentrations and distributions. Table 2.10 indicates that with the exception of CO_2, atmospheric chemical processes largely determine the rate of removal of greenhouse gases from the atmosphere. These processes are described in more detail in Ramanathan et al. (1987), Penner et al. (1988), and Wuebbles et al. (1989). The primary removal mechanism for some species, such as CO and CH_4, is reaction with OH in the troposphere. Other greenhouse gases such as N_2O and the CFCs are destroyed primarily by photodissociation (dissociation via reaction with a solar photon). In addition, there are indirect effects on climate forcing as a result of chemical interactions affecting the atmospheric concentrations of O_3 (in both the troposphere and stratosphere) or H_2O vapor (in the stratosphere; tropospheric H_2O vapor concentrations are largely unaffected by chemical processes). Table 2.11 indicates the ways in which chemical processes affect climate.

Table 2.10 Concentration and lifetimes of important atmospheric trace-gas constituents

Gas	Tropospheric concentration[a] (ppmv)	Trend in atmospheric concentration[a] (% per year)	Atmospheric lifetime[b] (years)	Primary removal process
SO_2	$1-20 \times 1^{-5}$	unknown	~ 0.02	Dry deposition and conversion to sulfate that is removed by precipitation
COS	5×10^4	< 3	2–2.5	Conversion is sulfate and removal by precipitation
DMS	$5-20 \times 10^{-5}$	unknown	~ 0.01	Chemical reaction leading to removal by precipitation
NMHC	< 1	unknown	< 0.02	Reaction with O_3 and OH
O_3	0.02–0.1	troposphere stratosphere	< 0.1 ≦ 0.1 to 1	Chemical reaction or removal at surface
OH	$4-100 \times 10^{-9}$	unknown	< 0.01	Chemical reactions to H_2O
H_2O	$1-20 \times 10^3$	unknown	~ 0.03	Precipitation or chemical reaction

[a] Corresponds to mid- to late-1980 values.
[b] These estimates of atmospheric lifetimes are about 10–20% uncertain for many of these species; a mid-range value is given.

Table 2.11 The role of chemical processes in affecting the atmospheric concentrations of greenhouse gases

Gas	Are concentrations effected by atmospheric chemistry?	Indirect effects on Tropospheric chemistry?	Indirect effects on Stratospheric chemistry?
CO_2	no	no	yes, affects O_3
CH_4	yes, reacts with OH	yes, affects OH, O_3	yes, affects O_3, H_2O
CO	yes, reacts with OH	yes, affects OH, O_3	not significantly
N_2O	yes, photolyzes in stratosphere	no	yes, affects O_3
NO_x	yes, reacts with OH, O_3	no	yes, affects O_3
$CFCl_3$	yes, photolyzes in stratosphere	no	yes, affects O_3
CF_2Cl_2	yes, photolyzes in stratosphere	no	yes, affects O_3
$C_2Cl_3F_3$	yes, photolyzes in stratosphere	no	yes, affects O_3
CH_3CCl_3	yes, reacts with OH	no	affects O_3
CF_2ClBr	yes, photolyzes	no	affects O_3
CF_3Br	yes, photolyzes	no	affects O_3
SO_2	yes, reacts with OH	yes, but not significant to climate	yes, affects aerosols
COS	yes, reacts with OH	no	yes, affects aerosols
DMS	yes, reacts with OH	yes, through cloud interaction	no
NMHC	yes, reacts with OH	yes, affects OH, O_3	no
O_3	yes, produced and destroyed via chemical reaction	yes, affects OH	yes
OH	yes, produced and destroyed via chemical reaction	yes, affects O_3, CH_4, CO, CH_3, CCl_3, etc.	yes, affects O_3, H_2O
H_2O	yes, in stratosphere, via oxidation of CH_4	yes, affects OH, O_3	yes

The Importance of OH in Climate Change

The hydroxyl radical itself is not a greenhouse gas with a direct effect on climate, but is an important chemical scavenger of many trace gases in the atmosphere that are greenhouse gases. It is the primary tropospheric scavenger of CH_4, CO, CH_3CCl_3, CH_3Cl, CH_3Br, H_2S, SO_2, DMS, and other hydrocarbons and hydrogen-containing halocarbons. The atmospheric concentrations of OH, therefore, affect the atmospheric lifetime of these species and thereby affects their abundances and ultimately the effect these species have on climate.

The production of OH depends on the reaction of excited oxygen atoms (produced via photolysis of O_3 with H_2O. Therefore, increases in the concentrations of O_3 or H_2O in the troposphere would increase the amount of OH. Global increases in temperature, driven by climate change, are expected to lead to an increase in the tropospheric H_2O concentration.

The primary sinks for hydroxyl are through the reactions of OH with CO and CH_4. These interactions suggest that the increases in CO and CH_4 now occurring should lead to a decrease in tropospheric OH concentrations, with a subsequent positive feedback on the lifetime and abundance of CH_4, CO, and other gases scavenged by OH. This feedback could play an important role in the future concentration trends of CH_4 and other greenhouse gases. Because of its small atmospheric concentrations and short lifetime, existing atmospheric measurements are insufficient to determine a trend in tropospheric OH concentrations. It is not known whether the concentrations of OH are increasing or decreasing.

THE CONCEPT OF GLOBAL WARMING POTENTIAL

The previous sections have dealt with greenhouse-gas emissions only in terms of gross emissions. However, for the purposes of policy formulation it would be useful to be able to compare the warming effect of the various gases which differ in their direct and indirect contributions to radiative forcing and time profiles of decay, as well as direct influence on atmospheric composition. For example, on a per molecule basis, CH_4 is about thirty times more effective than CO_2 as an infrared absorber; however, its lifetime is a small fraction of that of CO_2. But then, at the end of its lifetime, virtually all atmospheric CH_4 is oxidized to CO_2, so that each central atom of carbon continues to contribute to warming in a new form. Hence, if a comprehensive approach is to be applied, a common metric is required to enable direct comparison to be made between different policy instruments that affect both the magnitude and timing of greenhouse-gas emissions.

One such metric, developed by the IPCC, is referred to as the global warming potential (GWP). Initial thinking about the concept began in 1989 and is reflected in a series of draft papers that were reviewed by Wuebbles et al. (1989). Formal papers began to appear in 1990 and include Lashof and Ahuja (1990), Derwent (1990), Nordhaus (1990), and Houghton et al. (1990). All of these early papers adopted a change in radiative forcing as its common unit of impact. Models of the atmosphere and radiation are used to simulate the effect on radiative forcing of the instantaneous release of a kilogram of CO_2, in that year and subsequent years, over a period of time into the future, relative to a standard reference scenario. The change in radiative forcing is then added over a specific period of time, e.g., 20 years, 100 years, 500 years, or infinity. The experiment is then run for other gases. The cumulative effect of each gas on radiative forcing is then compared to the cumulative effect of CO_2 on radiative forcing.

While all the above cited formulations are similar in construct, each varies somewhat in the details of its implementation. Differences include the way different gases are treated with regard to their residence time in the atmosphere after emission, their solar and infrared radiative absorption properties, and the indirect climatic effects resulting from chemical interactions with other greenhouse gases. In addition, different studies treat the time profile of the change in radiative forcing differently. Some studies simply add changes on an equal basis over different numbers of years. Others apply discount rates to each year's change in radiative forcing before summation. From the perspective of its natural science, the construct developed by Derwent, Rodhe, and Wuebbles for IPCC (Houghton et al. 1990) is the most complete to date (values of IPCC GWP are shown in Table 2.12). All these studies define the GWP for a greenhouse gas relative to CO_2, the greenhouse gas of current primary concern to climate change.

While natural scientists continue to refine the construct of GWP coefficients for different gaseous emissions, social scientists have also explored the concept. Reilly (1990) points out that the whole point in constructing a GWP index in the first place is to obtain a relative measure of damage. Implicit in the standard formulation of a GWP is the assumption that the effect of a one degree change in radiative forcing at any two points in time is the same. This need not be the case. Marginal damages are likely to vary over time. Reilly (1990, p. 2) also points out that gases have nonclimate-related economic effects that differ among gases, and these should be counted as credits (e.g., direct CO_2 fertilization of crops) or debits (e.g., CFCs as contributors to O_3 depletion). While the introduction of a calculation of full damage and benefit from a pulse emission of a gas adds considerable complexity to the problem, Reilly (1990) shows that the successful introduction of such a concept could significantly change GWP measures. Reilly (1990) developed coefficients using alternative damage functions for the effects of a change in radiative forcing on a steady-state economy and alternative allowances for a CO_2 fertilization effect on agriculture. He found that the application of these varying assumptions, in conjunction with a simple model of the atmosphere, yielded a range of values for coefficients at least as large as the range of values associated with range of integration times investigated by the IPCC. It is important to note that these figures are intended to be illustrative only. The results of this experiment are reported in the Table 2.13.

A direct implication of Reilly's work is that GWP measures are not directly comparable from year to year. That is, the base atmosphere will change over time as a result of the accumulation of greenhouse gases from

Table 2.12 Values of global warming potential as used by the IPCC (Houghton et al. 1990)

	Global Warming Potential		
	Integration Time Horizon (yrs)		
Trace-Gas	20	100	500
Direct Effects			
CO_2	1	1	1
CH_4 (incl. Indirect Effects)	63	21	9
N_2O	270	290	190
CFC-11	4500	3500	1500
CFC-12	7100	7300	4500
HCFC-22	4100	1500	510
CFC-113	4500	4200	2100
CFC-114	6000	6900	5500
CFC-115	5500	6900	7400
HCFC-123	310	85	29
HCFC-124	1500	430	150
HFC-125	4700	2500	860
HFC-134a	3200	1200	420
HCFC-141b	1500	440	150
HCFC-142b	3700	1600	540
HFC-143a	4500	2900	1000
HFC-152a	510	140	47
CCl_4	1900	1300	460
CH_3CCl_3	350	100	34
CF_3Br	5800	5800	3200
Indirect Effects			
CH_4 effect on tropospheric O_3	4	8	3
CH_4 as CO_2	3	3	3
CH_4 as stratospheric H_2O	10	4	1
CO effect on tropospheric O_3	5	1	0
CO as CO_2	2	2	2
NO_x effect on tropospheric O_3	150	40	14
NMH as tropospheric O_3	28	8	3
NMH as CO_2	3	3	3

Note: NMH = non-methane hydrocarbon

previous emissions. This will not only change the chemical and radiative interactions of the gases; the damage associated with each degree change in radiative forcing will be affected by the scale of previously inflicted damage.

It is also important to note that GWP measures are marginal, i.e., they measure the consequences of a one time, one kilogram release of an individual gas into a standard atmosphere. The consequences of teragram-

Table 2.13 Comparisons of trace-gas indices

	Economic-Based			Physical Effects Only		
	Climate + CO_2 fert (quad.)	Climate Effects (quad.)	Climate Effects (linear)	Radiative Forcing	IPCC-GWP	
					20 y	500 y
CO_2	1	1	1	1	1	1
CO	3.7	2.9	0.9	2.3	7	2
CH_4	92	74	21	58	63	9
N_2O	260	208	201	206	270	190
HCFC-22	8950	7174	1427	5440	4100	510
CFC-11	6343	5085	1389	3970	4500	1500
CFC-12	9119	7309	2140	5750	7100	4500
CFC-113	5917	4743	1319	3710	4500	2100

Source: Reilly (1990), p. 16, Table 3.
Notes: CO_2 fert = CO_2 fertilization effect on plant productivity and water-use efficiency taken into account in addition to climate effects; quad. = economic damage function associated with a change in radiative forcing is assumed to be quadratic in form; linear = economic damage function associated with a change in radiative forcing is assumed to be linear in form.

and petagram-scale releases of gases into the atmosphere are not taken into account, neither are the consequences of simultaneous releases of gases. As these gases interact both chemically and radiatively, scale effects are potentially nontrivial.

CONCLUSIONS

The presence of greenhouse gases in the atmosphere has a significant influence on the energy balance of Earth and plays an important role in regulating surface temperature. Concentrations of greenhouse gases have been increasing during the industrial era. Most of the important greenhouse gases, such as H_2O, CO_2, CH_4, N_2O, and O_3, occur naturally. Some, such as CFCs and CFC substitutes, are of strictly human origin. Increases in the concentrations of greenhouse gases have been associated with byproduct emissions of human activities.

It is known that human activities release greenhouse gases into the atmosphere in quantities that are significant relative to the global scale natural system. Furthermore, the relationship between historic fossil-fuel use and CO_2 emissions is reasonably well understood. Similarly, the relationship between CFC manufacture and release into the atmosphere are also reasonably well known. The relationship between human activities and byproduct emissions of other gases is known less precisely.

While the qualitative relationship between emissions of greenhouse gases (and greenhouse-related gases, such as CO) and concentrations of greenhouse gases in the atmosphere are well known, significant work remains before precise relationships between potential future emissions and atmospheric concentrations can be established. For example, uncertainties regarding mechanisms of ocean uptake of carbon require further work. Important uncertainties surround the role of terrestrial ecosystems in the carbon cycle. In addition, atmospheric chemistry models need further refinement.

That the greenhouse effect is real is indisputable. There is no question that the presence of greenhouse gases in the atmosphere is a major factor determining the energy distribution of the planet and therefore a major factor determining climate. Furthermore, there is no question that large changes in atmospheric concentrations result in large changes in surface temperature and planetary climate. Much further work remains to be done in reducing uncertainty surrounding the detailed relationship between atmospheric concentrations and climate dynamics. Even if future atmospheric concentrations were known precisely, there still remains a factor of three between upper and lower estimates of mean global surface warming for a doubling of atmospheric CO_2. The issue of the role of clouds as a climate feedback mechanism is of particular importance.

ACKNOWLEDGEMENTS

We wish to express appreciation to Sean McDonald, who assisted in the accumulation of information on CFCs, and to Bruce Kinzey, Graeme Pearman, and Rob Swart for helpful comments on earlier drafts of this paper. This paper is based on material forthcoming in "A Primer on Greenhouse Gases", by D. Wuebbles and J. Edmonds, Lewis Publishers, Chelsea, MI, U.S.A.

REFERENCES

Barns, D.W., and J.A. Edmonds. 1990. An evaluation of the relationship between the production and use of energy and atmospheric methane emissions. TR047. DOE/NBB-0088P. Springfield, VA: Natl. Technical Info. Service, U.S. Dept. of Commerce.

Björkstrom, A. 1979. A model of CO_2 interaction between atmosphere, oceans and land biota. In: The Global Carbon Cycle, ed. B. Bolin, E.T. Degens, S. Kempe, and P. Ketner. SCOPE 13, 491 pp. New York: Wiley.

Blake, D.R. and F.S. Rowland. 1988. Continuing worldwide increase in tropospheric methane, 1978 to 1987. *Science* **239**:1129–1131.

Boden, T.A., F. Kanciruk, and M.P. Farrell. 1990. Trends '90: a compendium of data on global change. ORNL/CDIAC-36, 257 pp. Oak Ridge, TN: Oak Ridge Natl. Laboratory.

Bolin, B. 1986. How much CO_2 will remain in the atmosphere? In: The Greenhouse

Effect, Climatic Change, and Ecosystem, ed. B. Bolin, B.R. Doos, J. Jaeger, and R.A. Warrick, SCOPE 29. New York: Wiley.

Bolin, B., A. Björkstrom, K. Holmen, and B. Moore. 1983. The simultaneous use of tracers for ocean circulation studies. *Tellus* **35B**:206–236.

Bolin, B., B.R. Doos, J. Jaeger, and R.A. Warrick, eds. 1986. The Greenhouse Effect, Climatic Change, and Ecosystem, SCOPE 29, 541 pp. New York: Wiley.

Bolin, B., and E. Eriksson. 1959. Changes of the carbon dioxide content of the atmosphere and sea due to fossil fuel combustion. In: Atmosphere and Sea in Motion, ed. B. Bolin, 509 pp. New York: Rockefeller Inst. Press.

Cicerone, R. 1988. How has the atmospheric concentration of CO_2 changed? In: The Changing Atmosphere, ed. E.S. Rowland and I.A.S. Isaksen. Dahlem Workshop Report PC 7, 282 pp. Chichester: Wiley.

CMA. 1988. World Production and Sales of Chlorofluorocarbons CFC-11 and CFC-12. Chemical Manufacturers Association, Fluorocarbon Program Panel.

Craig, H. 1957. The natural distribution of radiocarbon and the exchange time of carbon dioxide between atmosphere and sea. *Tellus* **9**:1–17.

Derwent, R.G. 1990. Trace gases and their relative contribution to the greenhouse effect. AERE-R 13716. Harwell, U.K.: Atomic Energy Authority Harwell Lab.

Derwent, R., H. Rodhe, and D. Wuebbles. 1990. Global warming potential of greenhouse gases. Special report for the UN Env. Prog.

Detwiler, R.P., and C.A.S. Hall. 1988. Tropical forests and the global carbon cycle. *Science* **239**:4274.

Edmonds, J., and J. Reilly. 1985. Global energy: assessing the future. New York: Oxford Univ. Press.

Emanuel, W.R., G.G. Killough, W.M. Post, and H.H. Shugart. 1984. Modeling terrestrial ecosystems in the global carbon cycle with shifts in carbon storage capacity by land-use change. *Ecology* **65**:970–983.

Enting, I.G., and J.V. Mansbridge. 1989. Seasonal sources and sinks of atmosphere CO_2 direct inversion of filtered data. *Tellus* **41B**:111–126.

Enting, I.G., and G.I. Pearman. 1982. Description of a one-dimensional global carbon cycle model (Tech. Paper No. 42), 96 pp. Div. of Atmospheric Research, CSIRO, Melbourne, Australia.

Enting, I.G., and G.I. Pearman. 1985. Description of a one-dimensional carbon cycle model calibrated using techniques of constrained inversion. *Tellus* **39B**:459–476.

EPA. 1989. Policy options for stabilizing global climate. Draft Report to Congress. Washington, D.C.: U.S. Environmental Protection Agency, Office of Policy, Planning and Evaluation.

EPA. 1990. National air pollutant emission estimates 1940–1988. EPA-450/4-90-001. Office of Air Quality Planning and Standards, Technical Support Division, Natl. Air Data Branch, Research Triangle Park, NC.

Esser, G. 1991. Uncertainties in the dynamics of biosphere developments with the accent on deforestation. In: Proc. of the workshop on Assessment of Uncertainties in the Projected Concentrations of CO_2, organized by IUPAC and ECN, July 3–6, 1990, Petten, NE.

Friedli, H., H. Loetscher, H. Oeschger, U. Siegenthaler, and B. Stauffer. 1986. Ice core record of the $^{13}C/^{12}C$ record of atmospheric CO_2 in the past two centuries. *Nature* **324**:237–238.

Goudriaan, J. 1989. Modeling biospheric control of carbon fluxes between atmosphere, ocean and land in view of climatic change. In: Climate and Geo-Sciences, ed. A. Berger, S. Schneider, and J.C. Duplessy, NATO-ASI Series C, vol. 285. Dordrecht, NE: Kluwer Academic Publishers.

Goudriaan, J. 1991. Uncertainties in biosphere/atmosphere exchanges, CO_2 enhanced growth. In: Proc. of the workshop on Assessment of Uncertainties in the Projected Concentrations of CO_2, organized by IUPAC and ECN, July 3–6, 1990, Petten, NE.

Goudriaan, J., and F. Ketner. 1984. A simulation study for the global carbon cycles including man's impact on the biosphere. *Climatic Change* **6**:167–192.

Hansen, J., I. Fung, A. Lacis, D. Rind, S. Lebedeff, R. Ruedy, G. Russell, and P. Stone. 1988. Global climate changes as forecast by Goddard Institute for Space Studies three-dimensional model. *J. Geophys. Res.* **93**:9341–9364.

Hansen, J., A. Lacis, and M. Prather. 1989. Greenhouse effect of chlorofluorocarbons and other trace gases. *J. Geophys. Res.* **94**:16412–16421.

Hoffert, M.I., A.J. Callegari, and C.T. Hsieh. 1981. A box-diffusion carbon cycle model with upwelling, polar bottom water formation and a marine biosphere. In: Carbon Cycle Modeling, ed. B. Bolin, pp. 287–305. SCOPE 16, 390 pp. New York: Wiley.

Houghton, J.T., G.J. Jenkins, and J.J. Ephraums, ed. 1990. Climate Change. The IPCC Scientific Assessment, 365 pp. Cambridge: Cambridge Univ. Press.

Houghton, R.A., R.D. Boone, J.M. Melillo, C.A. Palm, G.M. Woodwell, N. Myers, B. Moore III, and D.L. Skole. 1985. Net flux of carbon dioxide from tropical forests in 1980. *Nature* **316**:617–620.

Houghton, R.A., B.D. Boone, J.R. Fruci, J.E. Hobbie, J.M. Melillo, C.A. Palm, B.J. Peterson, G.R. Shaver, and G.M. Woodwell. 1987. The flux of carbon from terrestrial ecosystems to the atmosphere in 1980 due to changes in land use: geographic distribution of the global flux. *Tellus* **39b**:122–139.

Keeling, C.D. 1973. The carbon dioxide cycle: reservoir models to depict the exchange of atmospheric carbon dioxide with the oceans and land plants. In: Chemistry of the Lower Atmosphere, ed. S.I. Rasool, 335 pp. New York: Plenum Press.

Khalil, M.A.K., and R.A. Rasmussen. 1987. Atmospheric methane: trends over the last 10,000 years. *Atmos. Env.* **21**:2445.

Khalil, M.A.K., and R.A. Rasmussen. 1988. Nitrous oxide: trends and global mass balance over the last 3000 years. *Ann. Glaciology* **73**.

Killough, G.G., and W.R. Emanuel. 1981. A comparison of several models of carbon turnover in the oceans with respect to their distributions of transit time and age and responses to atmospheric CO_2 and ^{14}C. *Tellus* **33**:274–290.

Lacis, A.A., J. Hansen, F. Lee, T. Mitchell, and S. Lebedeff. 1981. Greenhouse effect of trace gases 1920–1980. *Geophys. Res. Lett.* **8**:1035–1038.

Lacis, A.A., D.J. Wuebbles, and J.A. Logan. 1990. Radiative forcing of global climate changes in the vertical distribution of ozone. *J. Geophys. Res.* in press.

Lashof, D.A., and D.R. Ahuja. 1990. Relative global warming potentials of greenhouse gas emissions. *Nature* **344**:529–531.

Lashof, D.A., and D.A. Tirpak. 1989. Policy options for stabilizing global climate. Draft Report to Congress. Washington, D.C.: U.S. Environmental Agency, Office of Policy, Planning, and Evaluation.

Linak, W.P., J.A. McSorley, E. Hall, J. Ryan, R.K. Srivastava, J.O.L. Wendt, and J.B. Mereb. 1989. N_2O emissions from fossil fuel combustion. Air and Waste Management Association paper 89–4.6.

MacCracken, M. 1985. Carbon dioxide and climate: background and overview. In: Projecting the Climatic Effects of Increasing Carbon Dixoide, ed. M. MacCracken and F. Luther, DOE/ER-0237, 381 pp. Springfield, VA: Natl. Technical Info. Service, U.S. Dept. of Commerce.

MacCracken, M.C., and E.M. Luther. 1985a. Detecting the climatic effects of increasing carbon dioxide, DOE/ER-0235, 198 pp. Springfield, VA: Natl. Tech. Info. Service, U.S. Dept. of Commerce.

MacCracken, M.C., and E.M. Luther. 1985b. Projecting the climate effects of increasing carbon dioxide, DOE/ER-0237, 381 pp. Springfield, VA: Natl. Technical Information Service, U.S. Department of Commerce.

MacCracken, M.C. et al. 1990. Energy and Climate Change: Report of the DOE Multi-Laboratory Climate Change Committee. Chelsea, MI: Lewis Publishers.

Maier-Reimer, E., and K. Hasselmann. 1987. Transport and storage of CO_2 in the ocean: an inorganic ocean-circulation carbon cycle model. *Climate Dynamics* **2**:63–90.

Marland, G., T.A. Boden, R.C. Griffin, S.E. Huang, F. Kanciruk, T.R. Nelson. 1989. Estimates of CO_2 emissions from fossil fuel burning and cement manufacturing, based on the UN Energy Statistics and the U.S. Bureau of Mines Cement Manufacturing Data. ORNL/CDIAC-25 NDP-030. Oak Ridge, TN: Carbon Dioxide Information Analysis Center, Oak Ridge Natl. Laboratory.

Mitchell, J.F.B. 1989. The greenhouse effect and climate change. *Rev. Geophys.* **27**:115–139.

Moore, B., III. 1988. Presentation to the U.S. Dept. of Energy, Global Warming Round Table, Nov. 22, 1988, Washington, D.C.

Muzio, L.J., and J.C. Kramlich. 1987. An artifact in the measurement of N_2O from combustion sources. *Geophys. Res. Lett.* **15**:1369–1372.

NAPAP. 1989. The 1985 NAPAP emissions inventory (version 2): Development of the Annual Data and Modelers' Tapes. Prepared for the Natl. Acid Precipitation Assessment Program, EPA-600/789-012a, prepared by Air and Energy Engineering Research Laboratory. Research Triangle Park, NC: EPA.

Neftel, A., E. Moor, H. Oeschger, and B. Stauffer. 1985. Evidence from polar ice cores for the increase in atmospheric CO_2 in the past two centuries. *Nature* **315**:45–57.

Nordhaus, W.D. 1990. Contribution of different greenhouse gases to global warming: A new technique for measuring impact. New Haven, CT: Yale Univ. Dept. of Economics.

Nordhaus, W.D., and G.W. Yohe. 1983. Future CO_2 emissions from fossil fuels. In: Changing Climate, pp. 87–153. Washington D.C.: Natl. Academy Press.

Oeschger, H., U. Siegenthaler, U. Schotterer, and A. Gugelmann. 1975. A box diffusion model to study the carbon dioxide exchange in nature. *Tellus* **27**:168–192.

Oeschger, H., and B. Stauffer. 1986. Review of the history of atmospheric CO_2 records in ice cores. In: The Changing Carbon Cycle, A Global Analysis, ed. J.R. Trabalka and D.E. Reichle, pp. 89–108. New York: Springer-Verlag.

Pearman, G.I., D. Etheridge, E. DeSilva, and P.J. Fraser. 1986. Evidence of changing concentrations of atmospheric CO_2, N_2O, and CH_4 from air bubbles in Antarctic ice. *Nature* **320**:248–250.

Pearman, G.I., and P. Hyson. 1986. Global transport and inter-reservoir exchange of carbon dioxide with particular reference to stable isotope distributions. *J. Geophys. Res.* **86**:9839–9843.

Peng, T-H., W.S. Broecker, H.D. Freyer, and S. Trumbore. 1983. A deconvolution of the tree-ringbased ^{13}C record. *J. Geophys. Res.* **88**:3609–3620.

Penner, J.E., F.S. Connell, D.J. Wuebbles, and C.C. Covey. 1988. Climate change and its interactions with air chemistry: Perspectives and research needs. Lawrence Livermore Natl. Laboratory report UCRL 2111. Washington, D.C.: EPA.

Ramanathan, V. 1988. The greenhouse theory of climate change: a test by an inadvertent global experiment. *Science* **242**:293–299.

Ramanathan, V., R. Callis, R. Cess, J. Hansen, I. Isaksen, W. Kuhn, A. Lacis, E. Luther, J. Mahlman, R. Reck, and M.S. Schlesinger. 1987. Climate change interactions and effects of changing atmospheric trace gases. *Rev. Geophys.* **25**:1441–1482.

Ramanathan, V., R. Cess, E.E. Harrison, P. Minnis, B.R. Barkstrom, E. Ahmad, and D. Hartmann. 1989. Cloud-radiative forcing and climate: results from the Earth radiation budget experiment. *Science* **243**:57–63.

Ramanathan, V., R.J. Cicerone, H.B. Singh, and J.T. Kiehl. 1985. Trace gas trends and their potential role in climate change. *J. Geophys. Res.* **90**:5547–5566.

Raynaud, D., and J.M. Barnola. 1985. An Antarctic ice core reveals atmospheric CO_2 variations over the past few centuries. *Nature* **315**:309–311.

Raynaud, D.J., J. Chappellaz, J.M. Barnola, Y.S. Korotkevich, and C. Lorius. 1988. Climatic and CH_4 cycle implications of glacial-interglacial CH_4 change in the Vostok ice core. *Nature* **333**:655–659.

Reilly, J.M. 1990. Climate change damage and the trace gas index issues. Washington, D.C.: U.S. Dept. of Agriculture, Economic Research Service, Room 528.

Schlesinger, M.E., and X. Jiang. 1990. The timescale of climate change induced by increased carbon dioxide. *J. Climate* **3**:1297–1315.

Shine, K.P. 1989. The greenhouse effect. In: Ozone Depletion: Health and Environmental Consequences, ed. R. Russell Jones and T. Wigley. New York: Wiley.

Siegenthaler, U. 1983. Uptake of excess CO_2 by an outcrop-diffusion model of the ocean. *J. Geophys. Res.* **88**:3599–3608.

Siegenthaler, U., and H. Oeschger. 1987. Biospheric CO_2 emissions during the past 200 years reconstructed by deconvolution of ice core data. *Tellus* **39B**:140–154.

Steele, L.F., F.J. Fraser, R.A. Rasmussen, M.A.K. Khalil, T.J. Conway, A.J. Crawford, R.H. Gammon, K.A. Masarie, and K.W. Thoning. 1987. The global distribution of methane in the troposphere. *J. Atmos Chem.* **5**:125–171.

Stuiver, M., R.L. Burk, and P.D. Quay. 1984. $^{13}C/^{12}C$ ratios in tree-rings and the transfer of biospheric carbon to the atmosphere. *J. Geophys. Res.* **89(D7)**:11, 731–11,748.

Tans, F.P., I.E. Fung, and T. Takahashi. 1990. Observational constraints on the global atmospheric CO_2 budget. *Science* **24T**:1431–1438.

Trabalka, J., ed. 1985. Atmospheric carbon dioxide and the global carbon cycle. DOE/ER-0239, 316 pp. Springfield, VA: Natl. Tech. Info. Service, U.S. Dept. of Commerce.

Wang, W., D.J. Wuebbles, and W.M. Washington. 1985. Potential climatic effects of perturbations other than carbon dioxide. In: Projecting the Climate Effects of Increasing Carbon Dioxide, ed. M.C. MacCracken and E.M. Luther. DOE/ER-0237. Springfield, VA: Natl. Tech. Info. Service, U.S. Dept. of Commerce.

Wang, W., D.J. Wuebbles, W.M. Washington, R.G. Issacs, and G. Molnar. 1986. Trace gases and other potential perturbations to global climate. *Rev. Geophys.* **24(1)**:110–140.

Watson and Ozone Trends Panel, M.J. Prather, the Ad Hoc Theory Panel, M.J. Kurylo, and the NASA Panel for Data Evaluation. 1988. Present state of knowledge of the upper atmosphere 1988: an assessment report. Washington D.C.: NASA Ref. Pub. 1208.

Wigley, T.M.L. 1987. Relative contributions of different trace gases to the greenhouse effect. *Climate Monitor* **16**:14–29.

WMO (World Meteorological Organization). 1981. On the assessment of the role of CO_2 on climate variations and their impact. Geneva: WMO.

WMO (World Meteorological Organization). 1985. Atmospheric ozone 1985: global ozone research and monitoring project, Report No. 16. U.S. Natl. Aeronautics and Space Administration, Earth Science and Applications Division, Code EE, Washington D.C.

Wuebbles, D. 1989. Beyond CO_2—The other greenhouse gases. UCRL-99883. Livermore, CA: Lawrence Livermore Natl. Lab. (also paper 89-119.4, Pittsburgh, PA: Air and Waste Management Ass.).

Wuebbles, D., and J. Edmonds. 1988. A primer on greenhouse gases, DOE/NBB0083. Springfield, VA 22161: Natl. Tech. Info. Service, U.S. Dept. of Commerce.

Wuebbles, D. and J. Edmonds. 1991. A Primer on Greenhouse Gases. Chelsea, MI: Lewis Publishers, in press.

Wuebbles, D., K.E. Grant, F.S. Connell, and J.E. Penner. 1989. The role of atmospheric chemistry in climate change. *JAPCA* **39**:22–28.

3
What Are the Current Characteristics of the Global Energy Systems?

I.A. BASHMAKOV
Energy Research Institute, Academy of Sciences of the USSR, Moscow, USSR
Visiting fellow, Battelle Memorial Institute,
Pacific Northwest Laboratory, Washington, D.C. 20024-2115 U.S.A.

ABSTRACT

This paper describes the current global energy system and its long-term evolution. A general view of the trends in energy conservation is described. The structure of the 1990 world energy budget in terms of fuel type, economic sectors, and regional dimensions is also a subject of analysis. Finally, an estimate is presented of energy-related carbon dioxide and methane emissions by world region.

VIEW FROM THE PAST

Forty years ago Putnam (1953) published one of the first long-term energy-development forecasts. Most of the current issues connected with this subject were presented in his book, including the analysis of population growth, energy resources estimates, the analysis of fuel transformation technologies, input–output efficiency of fuel use and even, amazingly, the carbon dioxide (CO_2) emission problem. At that time the world faced an energy dilemma: properly used energy provided the basis for further growth of well-being, but continuation of the current trend could lead to the exhaustion of energy resources and environmental degradation.

Putnam concluded that if present trends continued in population growth, in per capita energy demand, and in preferences for liquid fuels and electricity, then growth in the real cost of coal would cause a strong demand for new energy sources sooner than many realized. If one replaces the word

---

Limiting the Greenhouse Effect: Options for Controlling Atmospheric CO_2 Accumulation
Edited by G.I. Pearman © 1992 John Wiley & Sons Ltd

coal by oil in the last sentence, one obtains the same main conclusions as the majority of relatively recent long-term predictions of world energy development.

Even if we use complex models, the limits of our knowledge about complex interconnections between the economy, energy, and the ecology usually force us to extrapolate our present experience to the future (Figure 3.1). That is why our way to the future is studded by the ruins of prior forecasts. What is also interesting is that predictions of the 1990s made in the 1950s and 1960s may be closer to reality than those made in the 1970s. This irony is explained more by the fact that present patterns correspond better to simple extrapolations made in the 1950s and 1960s than to any superiority in forecast capabilities. For example, Putnam's forecast is very close to recent predictions, but the details of his analysis vary greatly from reality. He thought that world population in the year 2000 would reach 3.5 billion, but the current population of the planet Earth is 5.3 billion people; he thought that coal prices would permanently increase, but they were more or less stable, with some variations during the period 1950–1990; he thought that recoverable resources of world oil were no more than 85 Gt of oil, but during the period 1950–1990 the world consumed 83 Gt of oil with more than 120 Gt of proved reserves remaining for future production.

To prevent us from automatically transferring our limited experience of the last several decades to the future, it seems necessary to observe the current situation against the background of its long-term prehistory. We chose the middle of the last century as a starting point because statistics

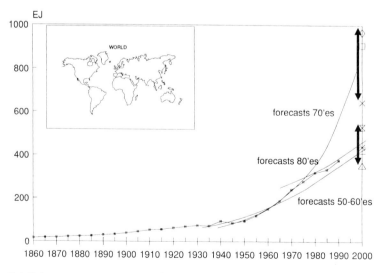

Figure 3.1 Primary energy consumption: history and forecasts

concerning earlier years are poor. Economic progress of humankind relied heavily on increases in energy use. Overall primary energy consumption grew by 23 times and commercial energy consumption by 80 times. This growth was uneven and the overall experience can be described by a set of S-shaped curves. One can distinguish three phases of relatively low rates (2% y^{-1} in 1860–1895; 1.2% y^{-1} during the period 1920–1945; 1.8% y^{-1} for 1975–1990) as well as two phases of comparatively high rates (2.7% y^{-1} for 1895–1920 and 4% y^{-1} during 1945–1975). Right now we are in the slow phase of the third cycle.

What has been the mechanism behind these variations? The fundamental mechanism can be described as follows. The economy cannot develop forever with the same technological structure. Sooner or later in a world of limited resources, limited not necessarily in physical terms but at least in technological and economical terms, growth will eventually deplete these resources. Prices and costs must eventually rise and the average rate of profit must decrease and slow the rate of economic activity. This in turn provides a favorable environment for the introduction of accumulated earlier innovations, which before had been unprofitable. This complex and expensive process of technological adaptation to the new cost proportions restores the possibility of further economic growth. This phase is characterized by lower rates of economic growth and a high share of investment to increase the productivity of scarce factors. Menshikov and Klimenko (1989) argue that these share fluctuations are responsible for long-term economic cycles. Simultaneous growth of productivity of several factors has to increase energy productivity as well. And it was the case.

For the period 1850–1990, there was a general historical trend of decreasing energy intensity (Figure 3.2), although this trend was punctuated by periods of increase. The energy-consumption data in Figure 3.2 include commercial energy and wood. When animal power is added, this trend becomes even clearer. The further inclusion of the consumption of water and wind energy would most likely further strengthen this thesis. Therefore, a decrease of energy intensity is not a new trend. It is at least 130 years old. What is also interesting to observe is that the rate of decrease of energy intensity was limited. To decrease energy intensity by a factor of two it took from 50 to 70 years or an average annual rate of decrease 1.0–1.5%. Finally, historical data show that the higher the rate of economic growth, the lower the rate of decrease of energy intensity.

Historical analysis does not provide us with any experiences in which long-term per capita energy consumption declined in conjunction with sustained economic growth. It is held that population growth is the main determinant of the future energy consumption growth. But in the last 100 years the increase of world population, from 1.5 to 5.3 billion, was responsible for only approximately 2% y^{-1} of energy consumption growth.

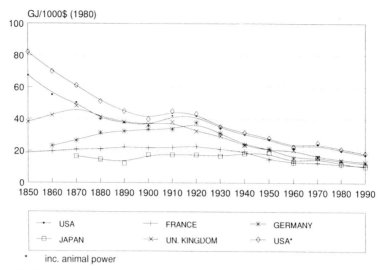

Figure 3.2 Energy/gross national product ratio: commercial energy and wood

In the U.S., energy consumption per capita grew from 130 GJ to approximately 363 GJ between 1860 and 1990. At the same time the U.K. demonstrated a more or less stable energy consumption per capita (130–140 GJ) and economic development for 100 years. If per capita energy consumption had plateaued at this level in the U.S. and the USSR alone, present global energy consumption would have been 19% lower.

History also shows that each long wave in the evolution of energy consumption was fueled by a new, higher-quality energy resource (measured in energy content per unit mass or hydrogen/carbon ratio) and finally, what is more important, this new resource provided the possibility of overall increases of productivity. It is not only the physical limits of energy resources which engender the introduction of new fuels. Each fuel, even after it has lost the leading position, continues to be produced in the same or even larger volumes for a long period of time (Figure 3.3). But lower-quality fuels fail to maintain exponential growth in primary energy consumption. The loss in relative importance is mainly due to the fact that additional capital costs to supply lower-quality fuels with previous annual rates of growth detract from well-being by nearly the same amounts as they contributed to well-being. Therefore to maintain economic growth there arises a need for new, cheap, and abundant energy resources. Until now, humans have successfully found successive generations of such resources. During the period 1860–1990 humans consumed 13,920 EJ of primary energy, including 2528 EJ of biomass, 5100 EJ of coal, 3899 EJ of oil, 1616 EJ of natural gas, 189 EJ of hydropower, and 5 EJ of nuclear power.

Analysis of historical data not only provides some explanations of the past, but gives rise to several important questions. Will we have a new wave of increased energy consumption in the future? If so, what resource will "fuel" that wave? Will energy conservation play this role? Can we at least maintain the present level of global per capita consumption? Is it possible to have both this new wave and a healthy environment? If not, is humankind, whose activities have become comparable in scale to the effects of long-term geological forces, able to stop the increased consumption and at the same time have sustainable economic growth?

ENERGY AND ECONOMIC GROWTH

At present, humankind consumes 370 EJ of primary energy (including biomass) to provide energy for 5.3 billion people, occupying around one billion dwellings, driving 500 million motor vehicles, producing a variety of agricultural and industrial products as well as services (Davis 1990). Energy serves as an important instrument for realizing further improvements in well-being across the planet. But at the same time, it is energy use that pollutes the air, water, and soil at the local, regional, and global scale. This raises the question of whether economic growth is sustainable. We need a better understanding of the relationship between economic growth and energy development to answer this question properly.

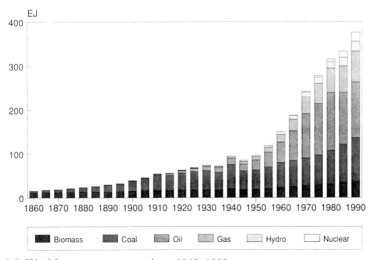

Figure 3.3 World energy consumption: 1860–1990

Energy consumption is distributed very unevenly over the face of our planet (Figure 3.4). To examine this we divided the world into eleven regions: the USSR; the U.S.; western Europe; eastern Europe; Japan; Canada, Australia and New Zealand (CANZ); Africa; Latin America; centrally planned Asia (CPA); Middle East; Southeast Asia. Numbers provided are preliminary estimates for 1990. Any cross-regional analysis of energy–economy relationships must proceed in the face of problems associated with international comparability of energy and the gross national product (GNP) statistics. In this analysis we employ the GNP reported in the International Comparison Project (Summers and Heston 1988) and energy budgets structure as reported in Bashmakov et al. (1990).

The main factor explaining differences in per capita energy consumption between regions is the level of economic development. Southern Asia, which has 14 times less GNP per capita than the U.S., consumes 23 times less energy per capita. But at the same time it is obvious that differences in the level of economic development are not the only explanatory factor. Low energy prices in the U.S., CANZ, the USSR, and eastern Europe are partly responsible for relatively high rates of energy consumption in these regions. Some differences are also due to varying regional climates. For the USSR and eastern Europe, inefficiencies associated with centrally planned economies are in large part responsible for their relatively high levels of per capita energy consumption.

For a long time the majority of countries have increased per capita income by increasing per capita energy consumption at rates corresponding to the

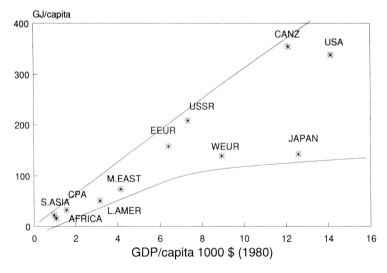

Figure 3.4 Primary energy consumption per capita, 1980

upper bound of scatter of points shown in Figure 3.4. But some regions recently have followed patterns traced by the lower bound of that scatter. It is very important to develop energy policies which will enable regions, now displayed in the upper part of the figure, to shift toward the lower bound, especially for developing countries. Goldemberg et al. (1988) believe that paths of extremely low energy intensity development are available. Whether energy patterns below the "lower bound" of Figure 3.4 are available or not is a matter of heated debate. Regardless, developing nations must avoid repeating the history of the industrialized world.

Limited potential exists for reduction in energy intensity for any given time period. Among the main factors that determine rates of energy intensity decrease are: energy efficiency of new technologies; rates of their introduction to market places; growth of energy efficiency for old technologies and level of their utilization. The absolute and relative importance of each of these factors varies. In the short term, rates of energy-intensity decrease appear to be limited to 3–4% y^{-1}, but long-term rates (for 20 years or more) are lower, 1.5–2.0% y^{-1}.

The question about the existence of these limits is a subject of heated debate. Some of the apparent differences of opinions between participants in these debates can be traced to differences in the definitions of terms. Of particular importance is the distinction between technical, economic, and market energy conservation potentials. Figure 3.5 shows energy conservation potentials for the USSR for each of these three notions.

Technological potential is the amount of energy conservation possible with the immediate application of the best existing technologies. The

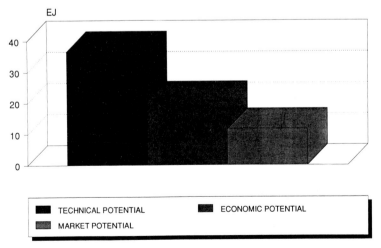

Figure 3.5 USSR: Energy conservation potentials, 2010

economic potential is the maximum conservation possible, satisfying the criterion of social cost minimization, for a given level of energy services. It is always less than technical potential. Finally, market potential is energy conservation obtained when private economic viability is satisfied. Market potential must always be less than economic potential.

Most recent research has been devoted to the evaluation of technological and economic potentials. It uses what might be referred to as the "mouse" approach, which seeks every opportunity for conservation possible in every "corner" of energy consumption. But there is also an "elephant" approach, which estimates market conservation potential using econometric models and postulates of neoclassic economic theory. Research using the latter approach usually obtains much lower estimates of energy conservation potential than those using the "mouse" approach. It is also very interesting to note that, according to neoclassic economic theory, economies cannot have rapid increases of energy conservation over a long period of time without decreasing overall productivity. That is, the pure market economy is not the best institutional setting for progress in long-term energy conservation. Similarly, the centrally planned economy is equally unable to maximize potential for energy conservation. The present situation in the U.S. and the USSR, where energy intensity has remained nearly stable over the period 1986 to 1990, illustrates this well.

Hence, there is a need for more government intervention in market economies to provide more conservation incentives. But at the same time, this means switching some resources from the present most profitable application, to obtain more long-term benefits. The clever balance between market forces and government regulations can help decrease differences between market and economic potential and produce real rates of decrease of energy intensity closer to the upper mark of 2% y^{-1}. This would be a very important achievement compared with historical 1.5% y^{-1} upper level. In other words, energy intensity would decrease by a factor of two every 40 years. Therefore, with rates of population growth close to 1% y^{-1} (now 1.8%) globally, it is possible to have 3% y^{-1} economic growth and constant per capita energy use indefinitely with strong government policies in the field of energy conservation.

Is economic growth of 3% y^{-1} enough for global population? These rates were achieved by many nations in the 1980s, a decade which was, however, a very difficult decade for many parts of the planet. Can we limit the growth of well-being? Generally speaking, my answer is no. To do that, sooner or later would require the introduction of an administrative-command system, which would consume the entire natural resources base very inefficiently. From another perspective the answer may be yes, if we speak about economic growth and the possibility to decrease military spending and conserve materials and energy for creatively solving other urgent problems.

Characteristics of the Global Energy Systems

This is especially true for the problem of creating a rational style of living, especially for regions which now have relatively low standards of living. If rates of economic growth increase, especially in developing countries, what will happen to rates of energy intensity improvement?

The question about interconnections between rates of economic growth and rates of energy conservation has not received enough attention. It is interesting to note, however, that for developed countries, these two variables are negatively correlated (Figure 3.6). To increase rates of economic growth it is necessary to increase the share of fixed investment in GNP, which in turn stimulates the growth of basic industries and, as a consequence, decreases the contribution of structural factors to energy conservation. For rates of economic growth in the range of 2–3% y^{-1}, which are anticipated for the majority of developed countries, this is not an important influence. But if an upward trend in a new economic long wave appears, then growth of rates of economic activity may lead to a decrease of energy productivity improvements.

For developing countries the picture looks very different: the higher the rate of economic growth, the lower the rate of increase of commercial energy intensity (Figure 3.7). The explanation for this is as follows. Developing countries have a very large share of GNP in the primary sectors (agriculture and mining). These sectors rely on traditional, noncommercial energy sources. Overall economic growth is driven by the relatively rapidly developing manufacturing and construction sectors, whose energy intensity is much higher than in traditional sectors. When the primary sector develops

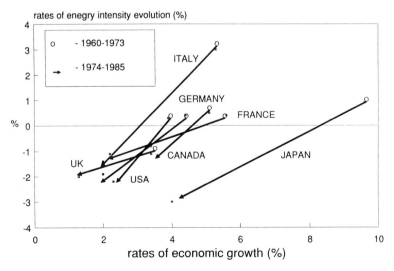

Figure 3.6 Relationship between rates of economic growth and energy intensity

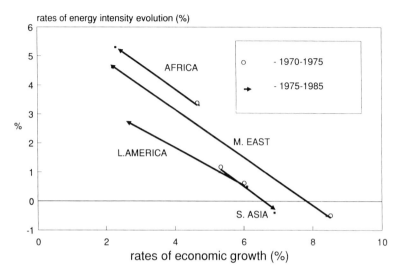

Figure 3.7 Relationship between rates of economic growth and energy intensity: developing countries

very slowly or experiences production declines, rates of growth of GNP become lower than rates of growth of energy consumption. In addition, the elasticity of residential energy consumption is relatively low in these countries.

Therefore, increasing the rate of economic growth in developing countries, say by 1% y^{-1}, would be accompanied by increased growth in commercial energy consumption of only 0.1–0.4% y^{-1}. This is a surprising result, but if a developing economy is going to follow an energy–economy trajectory near the lower bound shown in Figure 3.4, it will need higher rates of economic growth. If noncommercial sources are taken into account, then according to our estimates the general trend of decreasing energy intensity ("law of energy conservation") can be seen for developing countries as well as for developed.

STRUCTURE OF WORLD ENERGY BUDGET

The world energy budget has several dimensions. As a result we have a very sophisticated picture in multidimensional space. Certainly it is impossible to draw this picture in detail, but to understand the present state of the world energy system better we have to draw some crude projections of this picture. A two-dimensional picture of the world energy budget is presented

Characteristics of the Global Energy Systems

in Table 3.1. The data presented here represents an aggregation of the budgets for eleven regions.

Total global energy consumption in 1990, according to this estimate, was equal to 368 EJ. Fossil fuels predominated, contributing 89% of primary energy consumption with regional variation from 80% in western Europe to 97% in CPA. Oil had the leading position providing 35.2%, followed by coal (26%), natural gas (19.6%), biomass and other solid fuels (8.4%), hydropower and other renewable (6.2%), and finally nuclear power (5.2%).

If we add to this picture a regional dimension, then we see that 66% of oil consumption was accounted for by four regions (western Europe, the U.S., the USSR, and Japan); 82% of coal consumption was in just five regions: CPA, the U.S., the USSR, eastern and western Europe; 60% of natural gas was consumed in the USSR and the U.S.; 82% of biomass was consumed in developing regions; 69% of hydropower and renewable was produced in four regions (CANZ, Latin America, western Europe, and the U.S.); 89% of nuclear power generation originated in four regions (western Europe, the U.S., Japan, and the USSR).

Differences in the structure of primary energy consumption, to a large degree, result from inequalities in the structure of energy resource availability for different regions. But the further we go in the direction of final energy use, the more we find common features in structures of energy budgets.

From the very beginning it is important to note that only 72% of primary resources consumed reach final users. Part of these resources is lost in processes of electricity generation (19.6%), while part is consumed by the energy sector in processes of energy production, transformation, transportation, and distribution. One-third of the world's primary energy consumption goes to the electricity-generation sector to produce electricity and heat in cogeneration plants. Part of this stream is thermodynamic losses and another part is transferred to electricity (11.5%) and heat (2.9%). More than half of energy resources going to this sector are attributable to just three regions: the U.S., the USSR, and western Europe.

The energy sources' dimension of the picture is as follows: coal, 38.6%; hydropower and other renewable, 18%; nuclear power, 15.3%; natural gas, 15%; oil, 11.8%; biomass, 1.2%. There are sharp differences for different regions. The share of coal varies from 3% in Latin America to 72% in eastern Europe; the share of oil varies from 2% in CANZ to 47% in the Middle East; the share of natural gas varies from 1% in CPA to 40% in the USSR; the share of hydropower varies from 5% in the Middle East to 69% in Latin America; the share of nuclear power varies from zero in CPA and the Middle East to 35% in western Europe. The structure of energy resources consumption for the electricity-generation sector mainly coincides with the structure of the energy resource base for these regions.

As mentioned before, the energy sector accounted for 8.6% of global

Table 3.1 The world energy budget for 1990 (EJ). The data contained in the tables of this paper are from numerous and diverse sources including publications of the United Nations, OECD, WEC, BP, DOE, Asian Development Bank, OLADE and a wide range of other regional and national data books and publications

Energy Sectors	Energy Sources							
	coal	biomass & solid	oil	gas	hydro & renew.	nuclear	electricity	heat	total
Primary energy consumption	95.8	30.8	129.5	70.4	22.7	19.1	0.1		368.4
Electricity generation	48.3	1.5	14.9	18.8	22.7	19.1	−42.5	−10.8	72.1
Energy sector	4.7	0.5	8.2	9.2			7.6	1.5	31.5
Final energy consumption	42.8	28.8	106.5	42.5			35.1	9.3	264.8
Industry	27.8	5.8	20.8	21.2			17.7	6.6	99.8
Transport	1.1	0.2	54.1	0.1			0.7	0.1	26.3
Residential & commercial	13.7	22.7	16.9	17.4			16.6	2.6	89.8
Non-energy use	0.2	0.2	14.7	3.8					18.9

energy consumption. On average, for each EJ of coal consumed by final users, 0.11 EJ is consumed by the energy sector itself. For oil this proportion is 0.08 EJ; for natural gas, 0.22 EJ; for electricity 0.22 EJ; and for district heat from cogeneration utilities, 0.38 EJ. These numbers can also be viewed as indicators of how much energy we can save in the energy sector in addition to each EJ conserved by final consumers.

The share of final energy consumption in the industrial sector, which incorporates mining, manufacturing, construction and agriculture, is equal to 27% of primary energy consumption. If we include also 5% of primary energy for nonenergy consumption, then the industrial sector accounts for 40% of final energy consumption. The main contribution to this aggregate comes from planned and post-planned economies (47%). This is much more than their share of world industrial output. Therefore, economic efficiency is a very important factor in the evolution of industrial energy consumption.

Rates and proportions of overall energy consumption, to a large extent, are affected by the industrial sector. An analysis of the relationship between the level of economic development and industrial energy consumption per capita in various countries, as well as the dynamic aspect of this picture (Figure 3.8), leads to the following conclusions: a positive relationship exists between these two indexes; the higher the level of economic development, the greater the contribution from other factors; it was possible in some developed countries to have simultaneously increasing growth of well-being with stable or even decreasing levels of industrial energy consumption per capita; a negative relationship occurred in the period 1980–1985 under

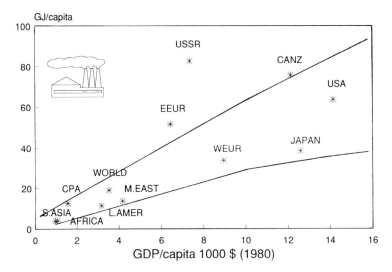

Figure 3.8 Industrial energy consumption per capita (1990)

conditions of low rates of economic growth and high energy prices; the growth of economic activity in 1985–1990 restored the trend of per capita energy increases. Therefore, it is possible to assert that for countries with GNP per capita higher than U.S. $7000, industrial energy consumption per capita saturates with deviations along this trend induced by rates of economic activity and energy price variations. For countries with GNP per capita lower than U.S. $7000, positive dependence is displayed very clearly.

In the industrial energy budget, coal predominates (27.9%), followed by natural gas (21.3%), oil (20.9%), electricity (17.8%), heat (6.6%), and biomass (5.8%). While there are substantial regional deviations from this average world picture, there are also some common patterns: the higher the level of economic development, the lower the share that coal has as a fuel with a tendency for stabilization; the share of biomass and other solid fuels tends to decrease with income growth from low to medium levels, but then, with utilization of modern technologies and more concern about waste utilization, contribution of these fuels grows; for oil the opposite picture is strictly observed—there is growth of the contribution of oil when countries are on the way to medium levels of economic development and decrease thereafter; the share of natural gas tends to increase with a growth of economic levels, with some saturation close to 30%; there is a positive relationship between GNP per capita and electricity contribution. There is no sign of saturation; district heat consumption from utilities, cogeneration are used mainly in planned economies. For these regions there is also a positive correlation between the heat share and GNP per capita level.

Transport is responsible for 15% of global primary energy consumption (56.3 EJ). Only two regions, the U.S. and western Europe, accounted for 55% of this consumption. Per capita transport energy consumption in the U.S. is 68 times as much as in CPA countries. Analysis of the relationship between GNP per capita and per capita transport energy consumption (Figure 3.9) leads one to conclude: there is a very strong positive correlation between these two indexes; there is no sign of saturation in terms of per capita energy consumption; on average, growth of GNP per capita by U.S. $1000 leads to an increase in the per capita transport energy consumption by 2.2 GJ.

In the fuel structure of world transport energy consumption, oil obviously dominates (96%). The second important fuel is coal (1.9%), followed by electricity (1.3%), biomass (0.3%), and natural gas (0.2%). The main coal contribution comes from CPA and southern Asia. More than two-thirds of electricity consumption is provided by planned economies. These countries have more diverse transport energy budgets. Only one clear trend exists in terms of fuel substitution along with economic growth and that is the substitution of coal by oil.

The residential and commercial sector share in the global primary energy

Characteristics of the Global Energy Systems 73

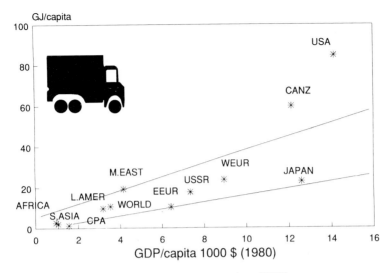

Figure 3.9 Transport per capita energy consumption (1990)

consumption in 1990, estimated at 24.4%. Per capita energy consumption for these sectors is distributed more evenly compared with other sectors. The highest level for the U.S. (69.4 GJ) is 13 times more than the lowest level for southern Asia (5.3 GJ). In general terms, the higher the level of economic development, the higher the per capita energy consumption in this sector (Figure 3.10).

Adding time to this picture, it is possible to draw several conclusions. First, the uncertainty of future consumption is determined to a large extent by differences in climate conditions and energy prices. Further, slopes of the curves for different countries are very closed. A stabilization or even a decrease of the residential and commercial energy consumption per capita has occurred mainly in countries with a high level of economic development. In some developing countries and regions, energy consumption in this sector increased, even when GDP per capita decreased. Hence in situations with low rates of economic growth, this factor contributes to a growth of energy intensity.

Globally averaged, the residential and commercial sectors derive much of their energy from biomass and other solid fuels (25.2% the global average, with variations from zero in Japan and the Middle East to 67% in Latin America). The second most important source of energy is natural gas (19.3% of the global average and variations in the range from 0.4% in CPA to 43% in the U.S.). Third place is occupied by oil (18.8% the world average, with variations from 4.8% in CPA to 67% in the Middle East). Fourth place is occupied by electricity (18.5% the global average, with variations in range

Table 3.2 Structure of 1990 world energy budget (percent). Calculations are based on data from Table 3.1.

Energy Sectors	Energy Sources							
	coal	biomass & solid	oil	gas	hydro & renew.	nuclear	electricity	heat	total
Primary energy consumption	26.0	8.4	35.2	19.1	6.2	5.2			100.0
Electricity generation	13.1	0.4	4.0	5.1	6.2	5.2	-11.5	-2.9	19.6
Energy sector	1.3	0.1	2.2	2.5			2.1	0.4	8.6
Final energy consumption	11.6	7.8	28.9	11.5			9.5	2.5	71.9
Industry	7.5	1.6	5.7	5.8			4.8	1.8	27.0
Transport	0.3	0.1	14.7	0.0			0.2	0.0	15.3
Residential & commercial	3.7	6.2	4.6	4.7			4.5	0.7	24.4
Non-energy use	0.1	0.0	4.0	1.0					5.1

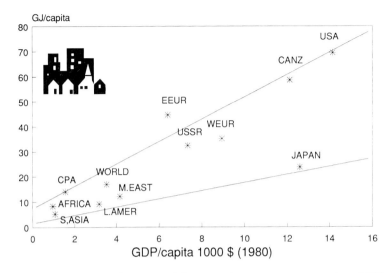

Figure 3.10 Residential and commercial per capita energy consumption: 1990

from 2.4% in CPA to 38.3% in CANZ). Next is coal (15.3% average and limits from 0.8% in Japan, CANZ, and Latin America to 41% in eastern Europe). Finally there is district heat (2.9% global average, with zero contribution for most regions and 15.4% share in eastern Europe).

As the level of economic development grows, so too does the balance of the energy budget of this sector. Biomass, with an efficiency of heating in the range from 5 to 30%, disappears from the energy budget. First of all it is substituted by coal with higher efficiency (60–70%), then by oil and natural gas (70–80%), and finally by electricity and district heat (100%). Thus trends in the shares of energy sources for this sector are closely related with the level of development.

The final component of the world energy budget is nonenergy use of energy resources. The importance of this item, which mainly includes energy resources which are consumed as feedstock in different industrial and construction processes, also depends on the stage of development (Figure 3.11). It accounts for 5.1% of global primary energy consumption. The main contribution comes from oil (77.8%) and natural gas (20.3%) and in regional terms, from the U.S. (23.6%), the USSR (22.2%), and western Europe (16.9%).

The uneven distribution of the world's fossil-fuel reserves necessitates a worldwide trade between regional energy systems. Now some 44% of oil, 14% of natural gas, and 11% of coal is traded internationally (Davis 1990). Therefore rates and structure of the evolution of the budgets of world and regional energy are heavily dependent on fluctuations of world energy prices.

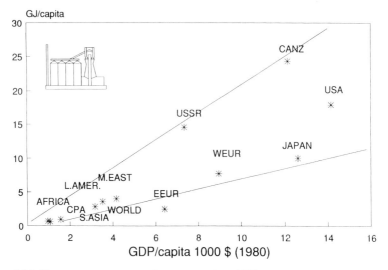

Figure 3.11 Non-energy consumption per capita: 1990

ENERGY PRICING

The variations of oil prices over the last 100 years are a manifestation of the uneven development of the world energy systems (Figure 3.12). The ups and downs of the curves attract the attention of many businessmen, experts, and consumers. Most of them would pay dearly for a device that could warn of such fluctuations. But alas, such a device is not available. The price shock caused by Iraq's invasion of Kuwait, unexpected by many experts, resulted in rises of oil prices of above U.S. $30 per barrel, or nearly a doubling of price. The situation of the 1980s and the beginning of the 1990s disproved the opinion that the real price of energy resources would inevitably increase. It is clear now that, as in the past, the future tendency of fuel prices to go up as fuel resources deplete may be punctuated by periods when prices go down or stabilize. There is also a possibility that instead of a sustained upward trend, we may experience a new period of long-term oil price stabilization.

An important factor in the way of world energy markets impact on the local energy budgets relates to the practice of state and monopolistic price regulations. Due to this, all countries have very different prices for the same fuels. The premium gasoline price in Venezuela in January 1989 was 10 times less than in Italy (Figure 3.13). This factor has a very large impact on relative energy intensities of GNP. This is about to be further modified as energy prices have to incorporate environmental costs.

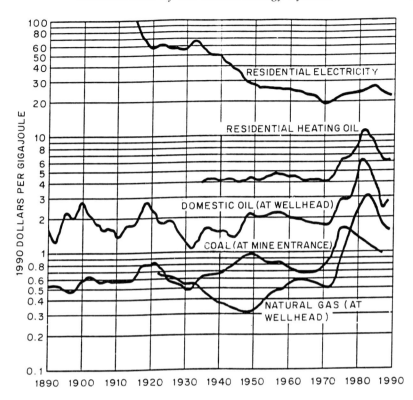

Figure 3.12 Energy costs in the USA: 1890–1990 (Source: Holdren 1990)

ENERGY AND THE ENVIRONMENT

Recently, there has been a growing realization that past concepts of energy pricing are not appropriate for the present and the future, because they do not incorporate externalities related to environmental and sociopolitical expenditures. The environment is a finite resource. Its capacity to absorb anthropogenic pollution and provide the same quality of natural services for humankind is limited. It is now realized that we have to weigh the usefulness and costs of energy services against environmental services. Internalization of ecological impacts has led to increased costs of supplying petroleum products of 25% and of generating electricity from coal and nuclear power by 40% or more during the past 20 years (Davis 1990). Today environmental costs average 2 or 3% of of GNP for most OECD countries (Holdren 1990). Therefore there are not only direct effects of energy utilization on the environment (the air, water, and soil pollution) to consider, but also a

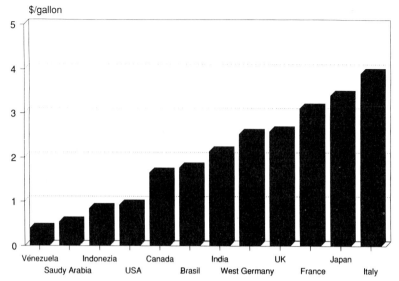

Figure 3.13 Premium gasoline prices (January 1, 1989)

powerful feedback, the costs of the modification of future energy systems for environmental reasons.

Environmental impacts of energy use and production are growing in scale. The Chernobyl accident, the greatest such catastrophe in humankind's history, polluted 2.5 Mha of territory occupied by 3.9 million people and between now and 2000 will cost 180–250 billion rubles (CA 1990). That means that each GJ of energy consumed in the USSR from 1986–2000 will cost an additional 20% because of these external costs.

Now we realize that activities of humans threaten even the global climate. If the present apprehensions of climatologists are confirmed, it is possible that the problem of "tolerable future climate change" will become a decisive force in the revision of current energy development strategy. If biomass is excluded, energy-related CO_2 emissions from 1860–1990 increased by 55 times, and 16.7 times if it is included (Table 3.3). According to Putnam (1953), the cumulative emission was equal to 19 Gt of carbon in the period 1800–1849. Since industrialization, this emission increased sharply and now is close to 300 Gt (including biomass combustion). The trend toward the reduction of carbon emitted per unit of primary energy consumption commenced a long time ago and mirrored the growing improvement of efficient energy production. This ratio has fallen from 27 kg C GJ^{-1} in 1860 to 18 kg C GJ^{-1} in 1990. The inertia of structure of the energy budget is the main reason for this relatively slow decrease.

Let us limit our discussion to only two main greenhouse gases, CO_2 and

Characteristics of the Global Energy Systems

Table 3.3 Growth of global CO_2 emission, 1860–1990

	1860	1900	1950	1970	1990
Annual CO_2 emission* (Mt C)					
excl. biomass	107	597	1622	3999	5928
incl. biomass	403	965	2111	4675	6741
Cumulative CO_2 emission (Gt C)					
excl. biomass	0	12	71	124	225
incl. biomass	0	25	106	171	287
Specific CO_2 emission per GJ of primary energy consumed (t C)					
excl. biomass		2.46	2.46	2.12	1.95	1.71
incl. biomass		2.46	2.46	2.18	1.95	1.81

*Assuming that for the non-energy use, the unoxidized fraction is equal to 0.6 for oil and 0.33 for gas

CH_4, for which more certain information is available. At the present time, anthropogenic emission of these two gases is equal to 8.8 Gt C (Table 3.4) with 92% from CO_2 and 8% from CH_4. The main source of the emissions is fossil-fuel combustion (81%), followed by biomass combustion (11.8%), CH_4 leakage (5%), CH_4 associated with coal production (1.3%), and gas flaring (0.7%). Industrialized countries produce two-thirds of the global emission, led by the U.S. and the USSR. The CANZ region emits 16 times more energy-related CO_2 and CH_4 per capita than southern Asia. This gap is larger than the GNP per capita gap. This region also has the largest per unit of GNP emission, 0.73 kg C \$$^{-1}$, while Japan has the smallest one, 0.18 kg C \$$^{-1}$. It is clear that, due to these inequalities, whatever the global greenhouse-gas emission control targets might be, disaggregating those targets regionally will be very difficult.

Carbon dioxide emissions related to the global energy budget are shown in Table 3.5. Contribution by fuel type can be listed in order: oil, coal, gas, biomass. Electricity generation is the first item in contribution by sector, followed by the residential and commercial sector, the industrial sector, the transport sector, the energy sector and finally the sector of nonenergy use. Such disaggregation of CO_2 emissions by sector is very useful. It shows the challenge of the 20% emission reduction target. Even if humans stop fuel use for transportation, CO_2 emission reduction will be less than 20%. This approach will help to develop least-cost plans for CO_2 reduction.

THE CHALLENGE OF THE FUTURE

To understand the challenge of the future better, it is useful to compare the cumulative numbers for 1860–1990 (in brackets) with the numbers for

Table 3.4 Energy-related CO_2 and CH_4 emissions for 1990 (on CO_2 equivalent basis expressed as Mt carbon). Calculations are based on data from Table 3.1.

	fossil-fuel consumption[1]	CO_2 biomass combustion	gas flaring	coal production[2]	CH_4[2] pipe-line leakage[3]	biomass combustion	Total volume	share %
1. USSR	1019	23	6	15	81	4	1148	15.7
2. U.S.	1460	77	2	21	86	11	1657	22.7
3. Eastern Europe	461	0	1	6	1	0	469	6.4
4. Western Europe	870	31	5	7	35	5	953	13.1
5. Japan	277	0	0	1	0	0	278	3.8
6. Canada, Australia, New Zealand	206	6	1	5	51	1	270	3.7
7. Socialist Asia	655	275	1	25	0	38	994	13.6
8. Africa	189	113	11	5	26	16	360	4.9
9. Latin America	267	83	7	1	37	11	406	5.6
10. Middle East	171	0	9	0	44	0	224	3.1
11. S&E Asia	353	144	10	6	4	20	537	7.4
WORLD	5928	752	53	92	365	106	8793	100.0

[1]Estimates for regional emissions do not include "bunker fuels" used in international transport.
[2]The emissions were converted to a CO_2-equivalent basis using the Global Warming Potential with an integration time 100 years which is equal on molar basis to 21 or weight of methane to carbon in CO_2 basis to 5.73.
[3]Calculations based on data for 1987 from the WRI (1990).

Table 3.5 Global 1990 CO_2 budget (Mt C). Calculations are based on data from Table 3.1.

Energy Consumption Sectors	Energy Sources					sectors' share (%)
	coal	biomass	oil & solid	gas	total	
Primary energy consumption*	2.48	0.75	2.65	0.96	6.84	100.0
Electricity generation	1.25	0.04	0.33	0.26	1.8	27.5
Energy sector	0.12	0.01	0.18	0.13	0.44	6.4
Final energy consumption	1.11	0.70	1.94	0.57	4.52	66.1
Industry	0.73	0.14	0.43	0.29	1.59	23.2
Transport	0.03	0.00	1.03	0.00	1.06	15.5
Residential & commercial	0.35	0.56	0.57	0.24	1.52	25.1
Non-energy use*	0.00	0.00	0.11	0.04	0.15	2.2
Fuels' share (%)	36.3	11.0	38.7	14.0	100.0	

*Assuming that unoxidized fraction is equal to 0.6 for oil and 0.33 for gas.

1990–2030 under a proposal for stable volumes and structure of the world energy budget: coal consumption, 3832 EJ (5100); biomass, 1232 EJ (2528); oil, 5180 EJ (3899); gas, 2816 EJ (1616); hydro, 908 EJ (189); nuclear, 764 EJ (5). The cumulative energy relating CO_2 and CH_4 emissions will amount to 292 Gt C, including 267 Gt C from fossil fuels and biomass combustion (287). Therefore it is obvious that even with these very improbable proposals, energy pressure on the Earth's resource base and environment capacity in the next 40 years will be the same or in some cases larger than in the past 130 years. It is also clear that to manage this situation, there is a need for fundamental changes in energy systems, including unusual international cooperation in improving our knowledge about forces which drive long-term energy development; in creating global energy and environmentally related institutions; in the transfer of energy conservation and environmentally less destructive energy supply technologies from one region to another; and in political stability of international energy trade.

ACKNOWLEDGEMENTS

This paper was written during the period in which I was a visiting scientist at the Battelle, Pacific Northwest Laboratories. I wish to thank the staff of these Laboratories for help and especially William Chandler and James Edmonds for their very useful comments and editorial help.

REFERENCES

Bashmakov, I.A., et al. 1989. The current problems of energy resources conservation. Moscow: VINITI (Russian).

Bashmakov, I.A., et al. 1990. Comparison of the main energy development indexes and energy efficiency in the USSR, the USA and Western Europe in 1971–2000. vol. 1 and 2. Moscow: Energy Research Institute.

CA. 1990. Chernobyl accident–the greatest catastrophe in the history of Earth. *Energy, economy, technique, ecology.* 7:2–7. (Russian).

Davis, G.R. 1990. Energy for planet Earth. *Sci. Amer.* **263**:55–62.

Goldemberg, J., T.B. Johansson, A.K.N. Reddy, and R.H. Williams. 1988. Energy for a Sustainable World. New Delhi, India: Wiley Eastern Limited.

Holdren, J.P. 1990. Energy in transition. *Sci. Amer.* **263** (Sept.):157–163.

Martin, J.M. 1988. L'intensité energetique de l'activité économique dans les pays industrialisés: les évolutions de ties longue période livrentelles des enseignements utilies? *Economies et soc.* **4**:9–27.

Menshikov, S., and L. Klimenko. 1989. Long waves in economy. Moscow:Mezhdunar. otnoshenia. (Russian).

Putnam, P.C. 1953. Energy in the Future. New York: Van Nostrand.

Summers, R., and A. Heston. 1988. A new set of international comparisons of real product and price levels. Estimates for 130 countries, 1950–1985. *Rev. of Income and Wealth* **4**:1–25.

WRI. 1990. World Resources 1990–1991, pp. 345–356. New York: Oxford Univ. Press.

4
Insurance Against the Heat Trap: Estimating the Costs of Reducing the Risks

I.M. MINTZER
Stockholm Environment Institute, Stockholm, Sweden and Center for Global Change, University of Maryland, College Park, MD 20740, USA

ABSTRACT

Human actions are changing the composition and behavior of the atmosphere. In the last five years, scientific uncertainty about the effects of these changes has focused the attention of politicians and citizens in some countries on the risks of rapid climate change. Some have argued that the costs of effective strategies to reduce the risk of rapid climate change will be prohibitively high. This chapter reports on the results of a modeling study that compared a business-as-usual scenario with an aggressive response scenario. The aggressive response scenario substantially reduces the rate of growth in the concentration of key greenhouse gases, relative to the business-as-usual scenario, without a significant reduction in the rate of economic growth in either industrialized or developing countries. These reductions in greenhouse-gas emissions are achieved while reducing the gap in real income per capita between the industrialized and developing countries.

INTRODUCTION

A string of unusual weather events during the last three years has focused the attention of politicians and citizens in some countries on our changing atmosphere. Much of this attention has been directed toward the effects of a continuing buildup of radiatively active gases, popularly called greenhouse gases. The compounds, including carbon dioxide (CO_2), methane (CH_4), nitrous oxide (N_2O), tropospheric ozone (O_3), and various chlorofluorocarbons (CFCs), are transparent to incoming solar radiation but absorb and reemit the thermal radiation released by the Earth's surface. The effect of this reemission is to increase the average surface temperature of the planet. Many scientists now believe that a continuing buildup of these greenhouse gases could raise the average surface temperature by as much as 4–5°C during the next century. A rise of this magnitude could cause rapid changes in regional climate with potentially severe effects on national economies and natural ecosystems.

Concerns about the still uncertain risks of rapid climate change have caused many authors to begin to explore the economic costs of reducing the risks (NRC 1991; OECD/IEA 1989). After accounting for the direct costs (but not the direct and indirect benefits), some have suggested that the expense of reducing the projected future concentrations of greenhouse gases may be prohibitively high (Nordhaus 1990). Others have suggested that the costs of measures which have reduced the historic rate of emissions may have already substantially penalized at least the U.S. economy (Manne and Richels 1990).

The study reported here summarizes a modeling effort implemented at the World Resources Institute (Mintzer and Moomaw 1991). It illustrates the characteristics of a scenario leading to a stabilization of global temperatures at levels below the doubled-CO_2 effect by the middle of the next century. The present study suggests that the preliminary analyses indicating prohibitively high costs may not reflect the complete picture. Indeed, by examining the effects of a comprehensive strategy to promote sustainable development, the present study suggests that increases in greenhouse-gas concentrations can be minimized in a global context where national economic growth continues at historical rates in industrialized countries and at faster than historical rates in developing countries.

TWO VISIONS OF THE FUTURE

This study utilizes a model, called the Model of Warming Commitment (MWC), that was developed to test the effects of policy choices on the rate of future economic growth and greenhouse-gas emission (Mintzer 1987). It incorporates a regionally disaggregated partial-equilibrium economic model developed at the Institute for Energy Analysis (the Edmonds-Reilly model) and various submodels to link future concentrations of the non-CO_2 gases to rates of future economic growth. Emissions of CO_2, N_2O, and CFCs are tracked through to concentrations in the atmosphere. Future rates of growth of CH_4 and tropospheric O_3 are estimated exogenously.

The following sections describe an exploration of two scenarios of future economic growth developed with the MWC. The first scenario, business-as-usual, represents a continuation of current trends in the growth of primary energy use with no special effort made to reduce the rate of greenhouse-gas buildup. The second scenario, aggressive response, describes a world in which national governments make a concerted effort to promote economic growth, especially in the Third World, while minimizing the rate of buildup of greenhouse gases.

The Business-As-Usual Scenario

This scenario illustrates a world in which recent and current trends continue unabated. This continuation of current trends would mean that lifestyles would not change much until forced to do so by an altered climate. New energy technologies might or might not be introduced, depending on the marketplace. The current gaps between rich and poor, North and South, and East and West would continue to divide the population of our planet into competing and increasingly anxious factions.

Economic growth continues in this scenario at close to historical levels. The income gap between rich and poor countries narrows slightly but persists throughout the period. The ratio of the average per capita income in the rich countries to the average per capita income in the poor countries declines from about ten to one in 1975 to about seven to one in 2025.

Urbanization continues as a worldwide trend. In the advanced industrial economies, more of the population moves into suburbs and areas just outside the boundaries of today's largest cities. In the U.S., immigration and internal migration continue well beyond the turn of the century, with the population continuing to gravitate toward the coasts and especially toward the southern and western regions of the country. In western Europe, Japan, and the USSR, the trend is toward evermore extensive industrial centers surrounded by rural areas increasingly dependent on extensive mechanized agriculture.

In many areas of the Third World, urban cores grow radially, surrounded by a dense ring of poorer suburbs. The central cores become oases of modern technology and convenience for those who can afford to live there. The poor communities on the city's edge are pushed further and further out from the "downtown" center, eating into the rich agricultural land that surrounded many cities in 1985. Lack of capital resources at the national and municipal levels inhibits the smooth extension of basic services and public utilities. Pollution, hunger, disease, and poverty strain the social fabric of society.

Growth in personal income continues, and oil prices increase less rapidly than the per capita gross national product (GNP). The global fleet of private cars continues to expand. Automobile production grows in all the advanced economies except the U.S. and increases, although at a slower rate, in developing countries. In the U.S., Germany, France, Great Britain, and several other industrialized countries, the trend is back toward larger, heavier, faster, and more powerful cars.

Developing countries rapidly expand their transportation systems, increasing dependence on cars and light trucks. Most countries evolve vehicle fleets similar to those found today in western Europe although some Third World countries expand on the Los Angeles model, where every adult believes he or she needs a large, fast car. Today's global fleet of about 400 million cars and light trucks grows to over 1 billion by 2025. Virtually all automobiles and light trucks are designed to run on gasoline.

Air pollution also mounts in this world with rapid global urbanization, a growing, mostly unregulated global car fleet, and dramatic increases in coal combustion. As temperatures rise and stratospheric O_3 is depleted by CFCs, the frequency of photochemical smog alerts and the concentration of ground-level ozone increase in urban areas on six continents. In short, the planet becomes hotter, dirtier, and more crowded (although somewhat more affluent). A declining level of natural amenities is shared unequally among a much larger population.

The Aggressive Response Scenario

The aggressive response scenario illustrates a quite different world. In this scenario, a very high priority is given to policies which can ensure sustainable economic growth while protecting the environment. All nations make a commitment to implement policies designed to slow the buildup of greenhouse gases. Free and open trade policies promote international interdependence, exploiting the comparative advantage of different countries to produce various goods and services.

Compared with the world of business-as-usual, in the aggressive response scenario, significant changes in public values accompany newly earned wealth. Government policies emphasize economic cooperation and the expansion of social services at the expense of military hardware and international adventurism.

In aggressive response, governments make a dedicated and systematic effort to promote sustainable economic growth. Public policy instruments are used to incorporate the full economic and environmental costs of energy supply and use into the average price of fuels. Investments in new energy supply options compete for available capital with opportunities to increase the efficiency of energy use on a "level playing field." Public attention and financial resources are redirected from extravagant military adventures and exotic hardware toward cooperative efforts to confront the common enemies of poverty, hunger, disease, and environmental degradation. In this scenario, the gaps in per capita income between North and South, East and West begin to close.

As in the business-as-usual scenario, economic growth continues throughout the simulation period in aggressive response. GNP grows rapidly in this scenario, at slightly less than historical rates in the industrialized countries, but at faster than historical rates in the developing world. In the developing countries of the South, more extensive education, better nutrition, and improved health care in rural areas dramatically increase the literacy rate. This, in turn, leads to a rapid increase in labor productivity. As a result, real GNP grows much faster than occurs under business-as-usual, at a rate of approximately 6% y^{-1}. Thus, in aggressive response, over a 50-year period, the developing country's share of global GNP increases from less than 20% in 1975 to over 50% in 2025. By contrast, in the business-as-usual scenario, the developing country's share of global GNP is only about 40% of the total in 2025.

In aggressive response, incomes are more equitably distributed, among countries and within countries, than they are in the business-as-usual scenario. The gap between rich and poor, North and South closes significantly in this scenario. The ratio of per capita incomes in the industrial countries to those in the developing world falls from 10 to 1 in 1975 to about 4 to 1 in 2025 in aggressive response.

The distribution of population is also somewhat different in this scenario, relative to business-as-usual. Urbanization continues as a global trend, but less intensively. In the advanced industrial countries, cityscapes have more the character of southern England or northern Germany than of southern California. The pattern is one of multiple, urban centers in regional clusters, each surrounded by a ring of suburbs and separated by intensively cultivated agricultural land. This is the alternative to massive Los Angeles-style "slurbs," which sprawl into the countryside, swallowing farmland and wetlands indiscriminately and necessitating the importation of water, fuel, and food over ever-increasing distances.

In the Third World, development follows a new pattern. The introduction of small industrial facilities into rural areas spreads economic growth. Rural development policy focuses on the provision of attractive, economically profitable opportunities in traditional agricultural areas. Mass migration to the cities is discouraged.

First in the industrialized economies and later in the developing world, the new patterns of urbanization and rural development encourage new patterns of energy supply and use. Substantial investments are also made to increase efficiency.

Electric utilities treat investments in electricity conservation and efficiency improvements on an equal footing with investments in new supply. Major programs to improve the efficiency of lighting, refrigeration, and electric motor drives are implemented first in the industrialized countries of the OECD and then duplicated in the East Bloc and the Third World. Improvements in energy-intensive industrial processes (including the introduction of advanced computer-aided process controls, the retrofit of variable-speed motor load controllers on industrial prime movers, and the development of continuous (rather than batch) process lines in basic materials production plants) combine to reduce the energy intensity of industrial production.

The process of technology transfer, with new developments often provided on concessionary terms to Third World countries, is mediated both by national governments and, increasingly, by multinational corporations that operate in both industrial and developing countries. Aggressive efforts to promote the early transfer of the best and most efficient new technologies are viewed by the governments of most OECD and CMEA countries as an essential ingredient in the process of industrialization needed to support sustainable development in the Third World.

The pattern of energy use in transportation is quite different in aggressive response from what is seen in business-as-usual. As cities expand and new regional complexes evolve, provision for mass transit is made from the start. Car use expands less rapidly in this scenario, discouraged by high fuel taxes, heavy parking fees, and high road taxes. Greater emphasis is put on the use of local buses and light rail systems in town, following more closely the Dutch model of urban transport in the 1980s than the American model of the 1960s. Fast intercity trains are combined with high-efficiency jet aircraft and barges for moving passengers and freight over long distances.

Electricity plays an increasingly important role in this scenario. Not only does the share of electricity increase as a fraction of total primary energy in aggressive response, but the mix of electricity sources shifts as well. As concern about the environmental impacts of combustion increases worldwide in these scenarios, nonfossil sources become increasingly important. Electricity grids are expanded less by building behemoth new power plants and long transmission lines and more by the incremental development of decentralized local generators feeding regional mini-grids. Distributed solar power systems, small hydro facilities, and small- to medium-sized cogeneration systems make up an increasing fraction of the new generation in these mini-grids. The mini-grids are redundantly interconnected to ensure system reliability with minimal amounts of spinning reserve. In this politically unrealistic and improbable environment, humankind is able to limit the risks of rapid global climate change and stabilize atmospheric temperatures.

INPUT ASSUMPTIONS FOR THE MWC SCENARIOS

The most important input assumptions driving the MWC are those which represent the human factors in the equations. These include demographic trends, improvements in human capabilities, and behavioral responses to a changing environment. The second most important set of assumptions represents the ways engineering characteristics of energy technologies change over time. The third set of parameters reflects assumptions affecting greenhouse gases other than CO_2. These will each be taken up in turn below.

Human Factors

The basic building block in the MWC is a partial-equilibrium, macroeconomic model developed at the Institute for Energy Analysis (Edmonds and Reilly 1983). This model (referred to below as the Edmonds–Reilly model) estimates future economic growth, energy use, and CO_2 emissions for nine world regions. Economic activity, represented as projected GNP for each region, is determined principally by the rate of growth in population and the rate of improvement in labor productivity of each region.

In both of the two principal scenarios evaluated in this study, global population follows approximately the UN/World Bank mid-range growth projection, stabilizing in about 2075 (Vu 1985). In this projection, population growth is much faster in the developing world than it is in the industrialized countries. Population levels off quickly in the advanced industrial economies but continues to grow in developing countries (see Table 4.1). Most of the growth in population occurs in the period between 2025 and 2075. Worldwide,

Table 4.1 Assumed population growth for the two scenarios (population in billions)

	1975	2025	2075	Annual Growth Rate 1975–2025	2025–2075
Industrialized countries	1.142	1.437	1.502	0.5%	0.1%
Developing countries	2.834	6.782	9.03	1.8%	0.6%
WORLD TOTAL	3.976	8.219	10.53	1.5%	0.5%
U.S.A.	0.214	0.287	0.292	0.6%	0.0%
W. Europe & Canada	0.405	0.522	0.543	0.5%	0.1%
OECD Pacific	0.128	0.155	0.153	0.4%	0.0%
USSR & E. Europe	0.395	0.473	0.514	0.4%	0.2%
P.R. China & C.P. Asia	0.911	1.589	1.706	1.1%	0.1%
Middle East	0.081	0.276	0.409	2.5%	0.8%
Africa	0.399	1.593	2.654	2.8%	1.0%
Latin America	0.313	0.722	0.896	1.7%	0.4%
South & East Asia	1.130	2.602	3.365	1.7%	0.5%
WORLD TOTAL	3.976	8.219	10.532	1.5%	0.5%

Table developed by the author based on projections of the World Bank and U.N. Fund for Population Activities.

population doubles from the current level of about 5 billion to 10.5 billion by 2075.

It is not just the number of people in the world that affects the rate of economic growth in this model. The second human factor that strongly influences GNP and the demand for energy is the rate of increase in labor productivity. For this set of experiments, very different assumptions have been made about the regional rates of growth in labor productivity. In the business-as-usual scenario, relatively modest rates of growth in labor productivity have been assumed, rates close to historical experience. However, in the aggressive response scenario, an assumed increase in investments by the governments of developing countries in health care, nutrition, education, and other social services is expected to result in rapid increases in literacy and in output per worker hour. This belief is represented in the model by an assumption of higher than historical rates of growth in labor productivity for these regions throughout the simulation periods (see Table 4.2).

The presumption of increased investment in education and social services is assumed to have other effects on regional economies as well. In the business-as-usual scenario, the pattern of industrialization follows the conventional western model. The aggressive response scenario assumes, however, that the social investment programs described above will simultaneously change cultural values and consumer behavior as well. In particular, this scenario assumes that as a result of increased public education concerning environmental issues, citizens in developing countries will choose to avoid some of the patterns of conspicuous and wasteful consumption that are so evident in the advanced industrialized economies. This emerging consciousness about the global environment is represented in the model by assuming an income elasticity of demand for energy for developing countries in the aggressive response scenario that is lower than is assumed for these countries in the business-as-usual scenario (see Table 4.3). In light of the higher rate of GNP growth experienced by developing countries in the aggressive response scenario, this assumption has a strong impact on the projection of future energy demand and CO_2 emissions (see Table 4.4).

Table 4.2 Annual rates of increase in labor productivity (annual percentage improvement in output per labor-hour)

	Business-as-Usual Scenario	Aggressive Response Scenario
Industrialized countries	1.5–2.0%	2.0–2.3%
Developing countries	2.5%	2.5–4.0%

Table 4.3 Long-run income elasticity of demand for energy in the WRI scenarios (percentage change in primary energy demand resulting from a 1% change in regional GNP)

	Business-as-Usual Scenario	Aggressive Response Scenario
OECD countries	1.03	0.95
USSR & E. Europe	1.08	1.03
Developing countries	1.60	1.05

Table 4.4 GNP growth in the WRI scenarios (GNP in trillions of 1985 U.S.$)

	GNP (1985 U.S.$)			Annual Growth Rate	
	1975	2025	2075	1975–2025	2025–2075
Business-as-Usual Scenario					
Industrialized countries	9.8	33.8	76.6	2.5%	1.7%
Developing countries	2.3	23.5	128.0	4.7%	3.5%
WORLD TOTAL	12.1	57.3	204.6	7.2%	2.6%
Aggressive Response Scenario					
Industrialized countries	9.8	41.9	116.8	3.0%	2.1%
Developing countries	2.3	45.1	530.4	6.1%	5.1%
WORLD TOTAL	12.1	87.0	647.2	4.0%	4.1%

Energy Technologies

Another set of assumptions which has a strong impact on the protection of future energy demand and CO_2 emissions are the values assigned to the parameters representing technological change in the model. The most important of these is the parameter used in the Edmonds–Reilly model to represent the rate of improvement in the efficiency of energy use in each region. This parameter aggregates the effects of many individual changes made in the key end-using technologies in the residential, commercial, industrial, and transportation sectors. For the two scenarios studies here, this parameter takes on very different values. In the business-as-usual scenario, the parameter takes on a median estimate, close to historical experience. By contrast, in the aggressive response scenario, the presumption of increased investments in the development and use of energy-efficiency improving technologies has led to a much higher assumed value (see Table 4.5). The effect of this assumption is to reduce the projected demand for energy in the aggressive response scenario relative to the energy demand at the same GNP level in the business-as-usual scenario.

The Edmonds–Reilly model also incorporates parameters that affect the relative price of energy from different technologies. One group of these parameters represents the rate of efficiency improvement for each energy supply technology. For the two scenarios, generally similar assumptions have been made about the rate at which such future improvements will occur; however, there are differences worth noting. The rates of improvement in efficiency for nuclear fission and coal electric technologies are slightly more rapid in aggressive response than in the business-as-usual scenario.

Another set of parameters determines the rate at which advanced technologies penetrate the energy markets. For solar technologies and synthetic fuels from coal and shale, the model assumes that prices will fall over time to some minimum level, measured in constant U.S. dollar terms. The minimum price level and the time it takes each technology to reach it are assumptions that are specific to each scenario. In the aggressive response scenario, for example, assumed additional investments in solar energy development reduce the minimum price for solar electricity to a level that is less than half the price assumed in the business-as-usual scenario. By contrast, due to relatively more limited investments in synthetic fuels development, the minimum prices for synfuels are assumed to be about 50% higher in the aggressive response scenario than those in the business-as-usual scenario. The effect of these assumptions is to increase the rate at which solar energy captures market share in the aggressive response scenario compared with the role it achieves in business-as-usual (see Table 4.6).

Another set of factors which affect the relative price of fuels (and thus their market share over time) in these scenarios is the cost assigned to the environmental impacts of energy supply and use. In the aggressive response scenario, policies are assumed to try to capture these environmental costs in the price of energy. In the business-as-usual scenario, much more limited efforts are assumed. These policies are reflected in values assigned to a tax penalty that is applied to the cost of energy from each technology. In the aggressive response scenario, for example, a carbon-based fuel tax is assumed

Table 4.5 End-use efficiency increase in the WRI scenarios (annual rate of improvement)

	Business-as-Usual Scenario	Aggressive Response Scenario
OECD countries	0.80%	1.2–1.6%
USSR & E. Europe	0.80%	1.5%
Developing countries	0.75–0.85%	2.0–3.0%

Table 4.6 Minimum cost of solar energy and synfuels in the WRI scenarios (costs in 1985 U.S.$ per GJ)

	Business-as-Usual Scenario		Aggressive Response Scenario	
	min cost	yr achieved	min cost	yr achieved
Solar electricity	24.00	2010	10.20	2005
Synthetic oil	10.50	2025–30	14.50	2050
Synthetic gas	8.30	2025–30	12.30	2050

to be applied to the costs of energy derived from coal, oil, natural gas, and synthetic fuels. No such tax is assumed in the business-as-usual scenario. In addition, the environmental charge placed on nuclear electricity is increased in the aggressive response scenario by about 35% compared to the charge assumed in the business-as-usual scenario. Finally, starting in 2025, a consumption tax is assumed in the aggressive response scenario on the use of oil, gas, and coal in all industrialized countries (see Table 4.7). The effect of these assumed charges is to increase the cost and decrease the competitiveness of fossil-fuel and nuclear technologies relative to investments in renewable energy systems and energy efficiency improving technologies.

Other Greenhouse Gases

Other model assumptions also affect the rate of greenhouse-gas buildup in the MWC scenarios. Because the model does not explicitly represent activities associated with deforestation and land use changes, an assumption must be made about the rate of biotic CO_2 emissions in each scenario. For the business-as-usual scenario, deforestation is assumed to increase, although at a pace somewhat slower than that seen in the last few years. Thus, biotic

Table 4.7 Environmental taxes on energy supplies imposed in all regions. Taxes are in constant 1985 U.S.$ per GJ

	Business-as-Usual Scenario		Aggressive Response Scenario	
	tax	yr imposed	tax	yr imposed
Gas	0.00	2010	0.80	2010
Oil	0.00	2010	1.50	2010
Coal	0.90	2010	1.80	2010
Unconventional oil	5.00	2010	9.50	2010
Nuclear	15.00	2010	19.00	2010

emissions of CO_2 are expected to more than double between 1985 and 2075. By contrast, in the aggressive response scenario, an active and vigorous program of worldwide reforestation and soil conservation is assumed to reduce the biotic contribution to global CO_2 emissions dramatically. In this scenario, the biota is assumed to become a declining source over time, reaching a neutral state (i.e., no net CO_2 emission to the atmosphere) by 2025. By 2075, in the aggressive response scenario, this process has made the terrestrial biota a net sink for about 0.8 GtC y^{-1}.

Many factors affect the rate of buildup of CH_4 and tropospheric O_3 in the atmosphere. In the business-as-usual scenario, historical rates of growth are assumed for both CH_4 and tropospheric O_3. Methane concentration is assumed to grow by about 1% y^{-1} throughout the simulation period with tropospheric O_3 to increase about 15% by 2030. In the aggressive response scenario, by contrast, efforts to limit CH_4 leakage and to control CO emissions from automobiles and biomass burning could reduce the rate of CH_4 buildup to about half the historical rate, or 0.4% y^{-1}. Tropospheric O_3 is assumed to increase by only about 10% by 2030.

The final set of assumptions concerns the rate of future production and use of CFCs. The processes controlling the use and release of these compounds are well understood, but the policy choices that will limit their ultimate release have not yet been fully implemented. For the business-as-usual scenario, it is assumed that the Montreal Protocol on "Substances that Deplete the Ozone Layer" will be ratified by all countries and implemented as written, with no cheating. It is also assumed that the HCFCs (safer substitutes that still deplete the ozone layer and add to the greenhouse effect, although at a lower rate than the CFCs they replace) will be phased out of commercial use by industrialized countries in 2030. (A sensitivity case was also run in which no further international agreements beyond the Montreal Protocol to control these dangerous gases was assumed during the simulation period.) The aggressive response scenario assumes that substantial additional agreements to control these gases will be implemented and that thus the phaseout of both CFCs and HCFCs will be accelerated. In particular, it assumes that the use of the most dangerous CFCs will be phased out by the turn of the century in the industrialized countries. It also assumes that the compounds that replace those now in use will be carefully controlled. This scenario, in effect, assumes that by 2020, all CFC compounds will be eliminated from use in the industrialized countries (with the exception of some very small-scale medical applications). It assumes that the same compounds will be eliminated from use in the developing countries by 2030. It further assumes that the compounds replacing the current generation of chemicals will, after these dates, be limited to those which have no ozone-depleting potential and no global warming effect.

The assumptions illustrated above are represented explicitly by the mathematical parameters in the MWC. They are generally consistent with the illustrations of two alternative worlds given above. The model results reflecting the impact of these assumptions on the timing and extent of global warming are described below.

MODEL RESULTS

Business-as-Usual Scenario

The business-as-usual scenario roughly corresponds to a continuation of current trends. Economic growth continues in this scenario at slightly faster than historical levels. In the industrialized countries of the North, GNP increases from 1985 to 2025 at a rate of about 2.5% y^{-1}, and at a somewhat slower rate in the following fifty years. In developing countries, economic growth is more rapid during the next forty years, approaching an average annual rate of about 4.7%. The income gap between rich and poor countries narrows slightly but persists throughout the period. In 2025, per capita incomes in the industrialized countries average approximately $23,500 (in 1985 U.S.$) compared with an average of $3,460 in developing countries.

No specific effort is made to limit growth in primary energy demand or to reduce the environmental risks of economic activity. Global primary energy demand continues to grow between 1985 and 2025 at an annual rate of about 2.2%, approximately equal to the historical rate from 1970 to 1985. After 2025, demand grows at a more moderate rate, approximately 1.7% y^{-1} (see Table 4.8).

The business-as-usual-scenario represents a world where more concern is placed on avoiding shortages than on the problems created by energy surpluses. Government planners assume that demand will grow and focus investment capital on increasing supplies of energy. Because energy efficiency increases slowly in this scenario, achieving these levels of energy supply requires vigorous and simultaneous efforts to expand energy production of

Table 4.8 Ad valorem taxes on energy supplies (percentage increase in fuel prices, industrial countries only). Taxes are in constant 1985 U.S.$ per GJ

	Business-as-Usual Scenario		Aggressive Response Scenario	
	tax	yr imposed	tax	yr imposed
Gas	0.00	NA	30%	2025
Oil	0.00	NA	50%	2025
Coal	0.00	NA	65%	2025

every type, in all countries. In the U.S., primary energy demand increases at an annual rate of 1.2% y^{-1} from 1975 to 2025. This compares with a historical growth rate of 1.1% y^{-1} from 1967 to 1987. For the total OECD, the business-as-usual scenario implies an annual rate of growth in commercial energy demand of about 1.2% y^{-1} from 1975 to 2025. For the USSR and the East Bloc, primary energy demand grows in this scenario at an annual rate of 0.5% from 1975 to 2025. Commercial energy demand grows especially rapidly in the developing world, approximately 4% y^{-1}, during the next 40 years. As a result of these regional increases, the global growth rate follows the historical trend in this scenario, with primary energy demand increasing an average annual rate of 2.2% for the next four decades. This scenario implicitly assumes that the future will be smooth and surprise-free during this period, with no big oil price shocks, no supply disruptions, and no global depressions due to bank failures or to the debt crisis.

Meeting the needs of such a high-demand future requires a smoothly managed market for liquid fuels throughout the next forty years. OPEC and non-OPEC oil producers must cooperate wholeheartedly to meet the growing global appetite for petroleum products. World oil production must increase by 50% between 1985 and 2000 in this scenario, then remain largely constant for the next 25 years.

Electrification intensifies in all countries in this scenario, with the most rapid rates of demand growth occurring in developing countries. In the period 2000–2025, electricity growth rates approach 5.6–8.4% y^{-1} for developing countries compared to 1.1–3.2% y^{-1} for the industrialized countries.

In this scenario, the petroleum era gives way to a world fueled by coal, nuclear, and solar power. Increasing reliance is placed on coal to fire the engine of world economic growth. The level of coal consumption grows at an annual rate of about 1% between 1975 and 2025 and a rate of about 4.7% annually in the following fifty years. Coal demand reaches about ten times the 1975 level at the end of the simulation period. The supply of electricity from nuclear fission grows at almost 4% y^{-1} from 1985 to 2025. The output of hydroelectric plants increases by more than a factor of five over this 100-year period.

Energy prices rise steadily in the business-as-usual scenario, with oil prices increasing by about a factor of three between 1985 and 2025. The real price of oil increases by an additional 50% between 2025 and 2075. The price of natural gas increases at an annual rate of nearly 3% for the period from 1985 to 2075. The price of coal more than doubles in constant U.S. dollars by 2025 and increases by a factor of three by 2075, relative to the 1985 levels.

As a result of changing relative prices and declining resource availability, the fuel share of conventional oil declines from 40% of the global total in

1986 to 23% in 2025, and to less than 1% in 2075. The share of conventional gas in the fuel mix declines from about 20% of the total in 1986 to a little over 10% in 2025, and to less than 1% in 2075. The coal share increases in this scenario from 28% in 1986 to 38% in 2025, and to over 65% in 2075. Nuclear energy increases by a factor of more than 20 from 1975 to 2075 but continues to supply only 5% of the world's primary energy, supplying about 6% in 2025 and 15% in 2050 (see Table 4.8).

The impact of this pattern of energy use is a steady increase in the emissions of CO_2 from fossil-fuel combustion. At the global level, total emissions increase at an annual rate of about 2% y^{-1} from 1987 to 2025, doubling from about 5.5 GtC in 1987 to about 11.3 GtC in 2025. The increase accelerates after 2025, growing at a rate of about 1.9% y^{-1}. By 2075, fossil-fuel related CO_2 emissions reach about 29 GtC y^{-1} (see Table 4.9).

Lack of concern about environmental quality leads to increasing biotic emissions of CO_2 as well. In this scenario, a growing rate of deforestation increases biotic emissions. Tropical forests are increasingly converted to pasture lands. Primary growth is burned off, stumps are left to decay, and the land is seeded in grasses. Forest soils are stable and productive in these applications for only 3 to 5 years. After that, each recently cleared area must be abandoned and left to lie fallow. Denuded of forest cover, tropical soils rapidly erode. Their carbon content is washed away or picked up by the wind and quickly oxidized to CO_2.

Total CO_2 emissions from the biota increase from about 1.25 GtC y^{-1} in 1985 to three times that amount by 2075. This estimate does not take into account the releases from CO_2 from soil carbon that result from heat-

Table 4.9 Global energy use by fuel type (exajoules per year)

	Oil	Gas	Solids	Nuclear	Solar	Hydro	Total
Business-as-Usual Scenario							
1975	119	41	76	4	0	19	258
2000	178	66	132	10	1	57	444
2025	277	84	319	29	50	104	764
2050	17	117	766	52	161	118	1230
2075	13	33	1246	89	273	119	1774
Aggressive Response Scenario							
1975	119	41	76	4	0	19	258
2000	137	65	71	10	15	49	346
2025	106	53	77	25	171	91	524
2050	32	58	69	22	329	120	628
2075	27	24	55	30	478	126	740

induced increases in respiration rates by soil bacteria. Even without this potentially strong positive feedback, by 2075, total annual emissions of CO_2 in this scenario are more than 30 GtC, more than five times the current level.

As a result of this combination of biotic and fossil-fuel derived CO_2 emissions, the concentration of CO_2 reaches twice the preindustrial level by about 2050 (see Figure 4.1). This CO_2 buildup alone would commit the planet to a warming of about 1.5–4.5°C. However, the concentrations of other greenhouse gases continue to increase in this scenario as well.

The rapid increase in coal and fertilizer use leads to a nearly 15% increase in N_2O concentration by about 2025 and close to a doubling by 2075. The rate of growth in N_2O concentration in this scenario, about 0.7% y^{-1}, is much more rapid than the historical rate of about 0.2% y^{-1} (see Figure 4.2).

This scenario assumes one further international agreement on CFCs beyond the Montreal Protocol on "Substances that Deplete the Ozone Layer." Business-as-usual assumes that all industrialized countries reduce their emissions by 50% from the 1986 levels and that all developing countries take full advantage of the special provision for low-consuming countries. It further assumes that CFCs are fully phased out by 2000 and that their replacements, the HCFCs, are phased out in the industrialized countries by 2030. As a result, the concentration of CFC-11 increases at an annual rate of about 0.5% y^{-1}, compared with a historical rate of more than 5% y^{-1}.

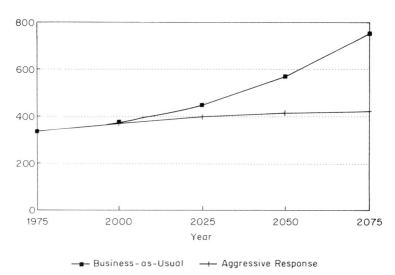

Figure 4.1 CO_2 concentration in the WRI scenarios (concentration in ppmv)

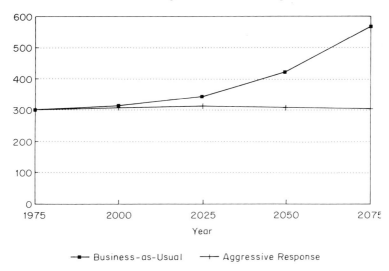

Figure 4.2 N_2O concentration in the WRI scenarios (concentration in ppbv)

Thus, CFC-11 concentration increases from 0.2 ppbv in 1985 to about 0.53 ppbv in 2025 and then declines to 0.3 ppbv in 2075 (see Figure 4.3a).

The concentration of CFC-12 grows slightly less rapidly in this scenario. It increases at a rate of about 0.9% y^{-1} from 1985 to 2075, compared to the historical rate of over 5% y^{-1}. Over this period, the atmospheric concentration grows from about 0.4 ppbv to a peak of 0.9 ppbv in 2025 and then declines to about 0.65 ppbv in 2075 (see Figure 4.3b). The increases in CFC concentration in this scenario could have significant implications for stratospheric O_3 depletion as well as for global warming.

The concentration of CH_4 continues to grow at the historical rate of 1% y^{-1} under business-as-usual. Many factors are assumed to contribute to this rapid CH_4 buildup. The growing human population in developing countries is assumed to increase the demand for paddy rice cultivation. Rising incomes are expected to lead to a growing demand for meat in human diets. Both of these activities result in larger biotic emissions of CH_4.

The growing global demand for natural gas between 1985 and 2050 is assumed to lead to higher losses from pipeline leakage and releases from local distribution systems. At the same time, the growing global car population, the rising demand for fuelwood, and the inefficient burning of biomass contribute to emissions of CO that reduce the atmospheric sink for CH_4. These factors together combine to stimulate a rise in the atmospheric concentration of CH_4 from about 1.65 ppmv in 1985 to over 4 ppmv only 90 years later (see Figure 4.4).

Air pollution is also assumed to increase, leading to higher levels of ozone

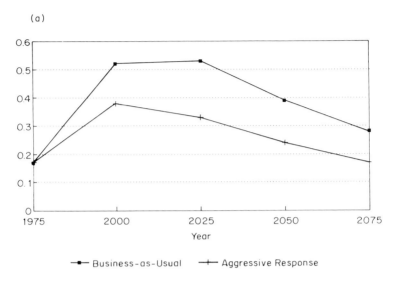

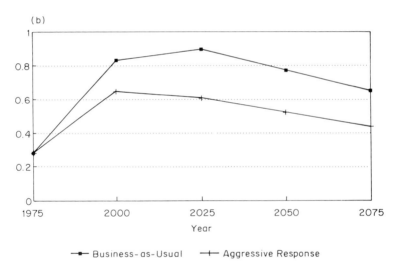

Figure 4.3 (a) CFC-11 concentration and (b) CFC-12 concentration in the WRI scenarios (concentrations in ppbv)

pollution. As temperatures rise and stratospheric O_3 is depleted by CFCs, the frequency of photochemical smog alerts and the quantity of ground-level ozone is expected to increase in urban areas on six continents. The average global concentration of tropospheric O_3 is assumed to increase by about 15% in this scenario from 1985 to 2025.

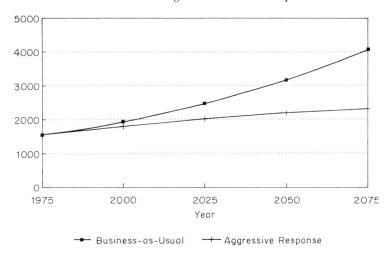

Figure 4.4 Methane concentration in the WRI scenarios (concentration in ppbv)

In business-as-usual, the combined effect of greenhouse-gas emissions commits the planet to a large and rapid global warming. If the trends described in this scenario continue, the planet will be committed to a warming relative to the preindustrial atmosphere that is equal to the effect of doubling the preindustrial concentration of CO_2 alone, a warming of 1.5–4.5°C by about 2010.[1] By 2025, the planet will be committed to a warming of 2.0–5.5°C, relative to the preindustrial atmosphere. By 2075, the commitment to future warming will be approximately 3.5–10°C, relative to the preindustrial atmosphere. The continuing buildup of greenhouse gases throughout this scenario suggests that temperatures would rise quite rapidly. The climate would not be expected to equilibrate at the elevated level of 2075 but rather global surface temperatures would continue to increase into the indefinite future (see Figure 4.5a).

The Aggressive Response Scenario

The aggressive response scenario shares some important similarities with the business-as-usual scenario but represents a very different world. The principal difference is that in this scenario, vigorous, systematic, and widespread efforts are made to ensure sustainable economic growth while minimizing the long-term environmental consequences of energy supply and

---

[1] This takes into account an estimated commitment to warming of 1–2.5°C relative to the preindustrial atmosphere that is already anticipated due to emissions that occurred between 1880 and 1980.

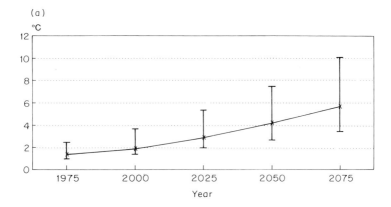

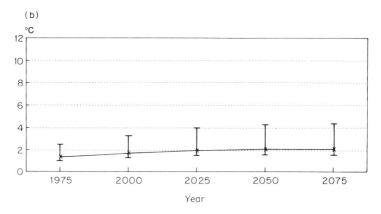

Figure 4.5 Warming commitment in the (a) business-as-usual scenario and (b) aggressive response scenario (both assume 2C sensitivity to $2 \times CO_2 + H_2O$)

use. As a result of this multifaceted policy strategy, world growth in primary energy demand is moderated, the trend toward increasing deforestation is reversed, and global climate stabilizes during the third quarter of the twenty-first century.

The principal similarities between the two futures have to do with population and economic growth. In the aggressive response scenario, as in the business-as-usual scenario, global population grows significantly between 1985 and 2075, ultimately stabilizing at about 10.5 billion people (see Table 4.1). Here again, most of the population growth takes place in the developing world, during the period between 2025 and 2075.

As in the business-as-usual scenario, in aggressive response, economic growth continues throughout the simulation period. GNP grows rapidly in this scenario at slightly higher than historical rates in the industrialized countries and at faster than historical rates in the developing world. In the industrialized North, real GNP grows at a rate of about 2.9% y^{-1} between 1975 and 2025. In the developing countries of the South, assumed investments in more extensive education, better nutrition, and improved health care in rural areas dramatically increase the literacy rate. This, in turn, leads to an assumed rapid increase in labor productivity. As a result, the model projects real GNP to grow much faster during this period, at a rate of approximately 6% y^{-1}. Thus, in aggressive response, over a fifty-year period, the developing countries' share of global GNP increases from less than 20% in 1975 to over 50% in 2025. By contrast, in the business-as-usual scenario, the developing countries' share of global GNP is only about 40% of the total in 2025.

At first this seems to be an unlikely and inconsistent result. How could GNP grow rapidly in developing countries if economic growth slows down in the industrialized world? The answer lies in a fundamental change in the pattern of development. In aggressive response, economic growth in the developing world is not principally directed toward export markets but instead is aimed at meeting basic human needs and building capital infrastructures in the rural areas of each country. This pattern of development is assumed to be facilitated by a program of technology transfer and development assistance from the industrialized world that is designed to promote ecologically sustainable development and to minimize global environmental problems. Motivated by the need to protect the global commons rather than to enrich the metropolitan countries of the First World, this new pattern of industrial development is expected to cause a net flow of capital to the developing world.

Reflecting this shared global commitment to limiting global environmental risks, energy demand is projected to grow much less rapidly in aggressive response than it does in business-as-usual, about 1.4% y^{-1} from 1975–2025 in this scenario compared with 2.2% y^{-1} for the same period in business-as-usual. Primary energy use grows at an annual average of 0.5% in the industrialized countries and about 2.3% y^{-1} in the developing world. At these rates, by 2025, in the aggressive response scenario, global primary energy demand reaches only 525 EJ, about twice the 1975 level compared with more than three times the 1975 level in business-as-usual.

In aggressive response, primary demand grows even more slowly after 2025, reaching only 740 exajoules (EJ)[2] in 2075. This represents a rate of about 0.7% y^{-1} in aggressive response, from 2025 to 2075, compared to almost three times that rate in business-as-usual.

This seemingly counterintuitive result (high rates of GNP growth with slow growth of energy demand in the developing world) results from the lower income elasticity of demand for energy that was assumed in the aggressive response scenario. As a result of comprehensive educational and social welfare programs, the long-run income elasticity of demand is assumed to be approximately 1.05 for developing countries in the aggressive response scenario, compared with 1.6 for the same countries in the business-as-usual scenario. Thus, even though per capita incomes are substantially higher in the aggressive response scenario, aggregate demand for energy is substantially lower than in business-as-usual. Whereas per capita demand for energy grows by about 1.5% y^{-1} in industrialized countries in the business-as-usual scenario, it grows by only 1% y^{-1} between 2000 and 2025 in the same countries in aggressive response.

Because the demand for energy grows more slowly in this scenario, the world oil market is much easier to manage in a nondisruptive fashion. Largely because the size and activity of the world car fleet is limited in aggressive response, oil demand grows only slightly by 2000 (i.e., a little over 10%) over the 1980 level. Growth in oil demand levels off, then declines to about 105 EJ y^{-1} in 2025 (approximately 53 million barrels per day). After 2025, oil demand falls rapidly to less than 23% of the 1980 level in 2075. In this slowly growing market, the geopolitical pressures associated with assuring steady supplies of oil at increasing volumes decline substantially. As in business-as-usual, in the aggressive response scenario the U.S. and Japan are almost completely dependent upon imported oil by 2025. In this scenario, the share of global production coming from Canada, western Europe, and the USSR increases. Nonetheless, by 2025 the Middle East is supplying over 40% of the global supply of conventional oil in aggressive response.

Electricity plays an increasingly important role in this scenario. For the U.S., electricity demand grows from less than 40% of primary energy demand in 1975 to about 78% in 2025. For the USSR, the increase is from about 35% in 1975 to about 70% in 2025. In China, electricity use grows from less than 15% of primary energy demand in 1975 to almost 50% in

---

[2] One exajoule equals 10^{18} Joules or approximately 0.95 quadrillion BTUs.

2025. And for south and east Asia, the increase is nearly as dramatic, from less than 35% in 1975 to about 70% in 2025.

Not only does the share of electricity increase as a fraction of total primary energy in aggressive response, the mix of electricity sources shifts as well. As concern about the environmental impacts of combustion increases worldwide in these scenarios, nonfossil sources become increasingly important. In aggressive response, solar, nuclear, and hydropower represent 55% of the global primary energy supply in 2025 compared with less than 10% in 1975. Solar energy alone contributes almost 33% of the total or about 130 EJ in 2025. By contrast, in the business-as-usual scenario, the contribution from solar, nuclear, and hydropower together represent less than 25% of the total primary energy supply in 2025. The solar fraction alone is about 6% of the total in business-as-usual, or a little more than 50 EJ for 2025 (see Table 4.10).

The demand for gas increases significantly during the next several decades

Table 4.10 Growth in fossil-fuel CO_2 emissions in the WRI scenarios (annual rate of increase, 1975–2025)

	Business-as-Usual Scenario	Aggressive Response Scenario
OECD countries	1.30%	−0.65%
USSR & E. Europe	0.18%	−1.84%
Developing countries	3.77%	1.03%
WORLD TOTAL	2.02%	−0.32%

CO_2 Emissions from Fossil-Fuel Use, G+C as CO_2

	Conventional Oil	Shale Oil	Synoil	Coal	Syngas	Conventional Gas	Total
Business-as-Usual Scenario							
1975	2.2	0.0	0.0	1.8	0.0	0.6	4.6
2000	3.3	0.0	0.0	2.8	0.0	0.9	18.2
2025	3.2	0.0	0.5	6.4	0.0	1.1	30.4
2050	0.2	0.1	6.6	10.4	0.2	1.6	48.2
2075	0.1	0.2	9.8	15.9	2.5	0.5	29.0
Aggressive Response Scenario							
1975	2.2	0.0	0.0	1.8	0.0	0.6	4.6
2000	2.5	0.0	0.0	1.1	0.0	0.9	4.4
2025	1.9	0.0	0.0	0.7	0.0	0.7	3.4
2050	0.6	0.0	0.0	0.0	0.0	0.8	1.4
2075	0.5	0.0	0.0	0.0	0.0	0.3	0.8

in aggressive response. Gas use increases largely because of its environmental advantages relative to coal and oil consumption. In effect, gas in this scenario is used as a "bridge fuel" to an era of dependence on "smokeless technologies." Global gas demand reaches a peak in 2000 at about 55 EJ y^{-1} in this scenario, an increase of about 25% over the 1975 level.

Commercial biomass plays a much larger role in aggressive response than it does in business-as-usual. Intensively cultivated on scientifically managed commercial plantations, biomass supplies an increasing share of global demands for solid fuels. On a regional basis, commercial biomass harvesting and cultivation are controlled so that CO_2 uptake in new growth just matches the CO_2 released when the annual harvest is burned. Major programs to develop energy plantations substantially reforest China, south and east Asia, Latin America, and Africa in this scenario. By 2025, biomass plantations are providing almost four times as much energy annually as are coal mines in aggressive response. By 2050, coal production for energy has been phased out in this scenario.

Energy prices increase in aggressive response, but because overall demand grows less rapidly, most prices remain somewhat lower than in business-as-usual. By 2025, the real price of oil has nearly doubled from the 1988 levels but remains about 25% below the expected price in business-as-usual. The real price of gas in aggressive response is slightly more than twice the 1988 level in 2025. This is about 20% less than the real price in business-as-usual for 2025.

By contrast, the average price of coal (before taxes) and biomass in aggressive response in 2025 is about 15% higher than the price in business-as-usual. The share of commercial biomass in aggressive response is almost 80% of the total supply of solid fuels on a global basis in 2025. The price of commercial biomass for energy is elevated by the competition for land with the enormous global reforestation program that is assumed in this scenario. Beyond this, the price of coal in the aggressive response scenario is pushed up by the environmental taxes which are applied to it.

By contrast, in business-as-usual, coal, which is substantially cheaper to produce, represents about 90% of the total solids supply in 2025. The carbon tax placed on coal supplies is less than one-third the size of the carbon tax applied in the aggressive response scenario.

The demand for coal grows rapidly in the business-as-usual scenario. By 2075 the price comparison for coal between the two scenarios is reversed. Largely because the demand for solid fuels is so high in the business-as-usual scenario compared with aggressive response, the average price of solid fuels is approximately twice as high in constant U.S. dollars in the business-as-usual scenario as it is in aggressive response.

As prices change and new technologies evolve, the mix of energy sources consumed in the aggressive response scenario shifts. The fuel share of conventional oil declines from 40% of the total in 1986 to about 20% of the total in 2025, and to about 3.5% fifty years later. The share of conventional gas declines from about 20% of the global total in 1986 to a little more than 10% in 2025, and to about 3% in 2075. The coal share declines from 28% in 1986 to about 6% in 2025, and to zero in 2075. Solar-derived electric energy, on the other hand, grows from virtually nothing in 1986 to almost 30% of world primary energy supply in 2025, and to over 60 percent of the total in 2075.

The consequence of this change in the fuel mix and the slow growth in primary energy demand in the aggressive response scenario is that CO_2 emissions from fossil-fuel combustion decline dramatically in this scenario. Emissions decline by 26% from 1986 to 2025. In aggressive response, by 2075, CO_2 emissions fall by 85% from the 1976 level, declining to approximately 0.8 GtC y^{-1} (see Table 4.10).

Growing worldwide concerns about environmental quality combined with the strong international program to develop scientifically managed, commercial biomass plantations, lead to a radical reduction in CO_2 emissions from the biota.

Large areas of the U.S., Europe, Asia, and Africa are reforested in this scenario. Multiple-use projects in developing countries provide fuelwood, fodder, and building materials as well as commercial solid fuels from carefully managed woodlots and forests. In the U.S. and other industrial countries, CO_2 offset programs encourage the planting of new forests to offset the emissions of CO_2 from new industrial facilities. Urban tree-planting programs increase CO_2 uptake, decrease the local albedo, and provide shade and comfort in the summer while reducing building cooling loads. Trees planted as wind-breaks and hedgerows stabilize rural soils, reduce erosion losses, and increase carbon storage.

By 2025, biotic emissions of carbon are assumed to fall from 1.25 GtC y^{-1} in 1976 to about 0.5 GtC y^{-1} in aggressive response. By 2075, the biota has become a net sink for atmospheric CO_2, withdrawing 0.8 GtC y^{-1} from the air. At this point the enhanced CO_2 uptake capability of the biota just offsets the fossil-fuel releases each year, resulting in no net injection of CO_2 into the atmosphere.

Because of these controls on CO_2 emissions, the atmospheric concentration of CO_2 grows slowly in the aggressive response scenario. By 2025, the estimated concentration is 399 ppmv in aggressive response, less than 13% higher than in 1976. By 2075, the atmospheric concentration of CO_2 is only slightly more than 420 ppmv, an increase of only about 70 ppmv from current levels. Thus, in aggressive response, the concentration of CO_2 does not exceed twice the preindustrial level during the simulation period. This situation contrasts with that in business-as-usual, in which the atmospheric concentration of CO_2 reaches twice the preindustrial level (approximately

550 ppmv) by about 2050. In business-as-usual, by 2075, the concentration of atmospheric CO_2 is about 750 ppmv, more than twice the 1986 level.

Without any other increases in greenhouse-gas concentration, the buildup of CO_2 in aggressive response would commit the planet to an ultimate warming of only 0.4–1.35°C. However, the concentration of other greenhouse gases also increases in this scenario.

Given the limited use of coal in this scenario and the expectation that fertilizer use will be managed with the goal of limiting emissions of N_2O in many regions, the buildup of N_2O is rather slow in aggressive response. In this scenario, the concentration of N_2O peaks in 2025, and by 2075 it has declined to approximately the 1985 level. As a consequence, the commitment to future warming from N_2O buildup in aggressive response is very small, relative to the 1980 atmosphere.

Aggressive response assumes substantial further reductions in the production and use of the most dangerous CFCs and halons. In fact, this scenario assumes that new agreements to supplement the Montreal Protocol on "Substances that Deplete the Ozone Layer" will result in the full and rapid phaseout of all such compounds. Furthermore, it assumes that the alternative compounds which provide the services now provided by CFCs will be limited to those which have no global warming impacts and no ozone-depleting effects. Aggressive response assumes that the industrialized countries will implement this total phaseout by 2020. By 2030, developing countries will also reduce their use of these infrared-absorbing compounds to nearly zero.

As a consequence of these international control measures, the concentration of CFC-11 grows at about 3.4% y^{-1} until it peaks at 0.39 ppbv in 2005. Following the phaseout, concentrations decline slowly, falling to about 0.17 ppbv in 2075, about equal to the 1980 level. The situation is similar for CFC-12. Concentration increases at about 3.4% y^{-1}, peaking in 2005 and falling thereafter. By 2075, the concentration of CFC-12 in this scenario is approximately 0.44 ppbv, slightly less than three times the 1980 level.

The concentration of CH_4 continues to grow in this scenario, but at an average rate of only 0.4% y^{-1}, significantly slower than historical rates. The slower rates of growth in this scenario are due (a) to reductions in the rate of CH_4 emissions from fossil-fuel extraction, mobilization, and use; (b) to reductions in emissions from inefficient biomass combustion; and (c) to reductions in the rate at which CO emissions destroy the atmospheric sink for CH_4. Methane concentration thus grows at only 0.5% y^{-1} from 1985 to 2025, or about half the historic rate of 1% y^{-1}, reaching an atmospheric level of about 1980 ppbv in 2025. Concentration grows at about 0.2% y^{-1} in the following fifty years, reaching approximately 2175 ppmv by 2075. The combined effect of the greenhouse-gas buildup is to increase the ultimate level of global surface temperatures. In aggressive response, the planet is committed, by 2025, to a warming of 1.5–4.0°C when the effects of a 1–2.5°C warming already "in the pipeline" are taken into account. The interesting thing about this scenario, however, is that the warming effect levels off not much after the middle of the century. By 2045, the planet is committed to a warming of 1.6–4.3°C. However after this date, the combined effects of future trace-gas emissions commit the planet to almost no further temperature increase. That is, sometime after 2050 (and probably before 2075), when the full effects of earlier emissions are observable, in this scenario the climate of the planet will stabilize with respect to the greenhouse effect at a level of about 1.6–4.4°C above the preindustrial average (see Figure 4.5b).

CONCLUSIONS

This exercise illustrates the possibility that the net costs of dramatically reducing the risks of rapid climate change may be small and that there could even be net economic benefits. By measuring the overall effects of such strategies on regional GNP, it is possible to identify combinations of policies which could sustain the prospects for economic growth while minimizing the buildup of greenhouse gases.

Of course, the fact that such a scenario is technically and economically feasible does not suggest that it is politically practicable. Achieving the kinds of results suggested in the aggressive response scenario implies a level of political cooperation that is not presently visible in the world as we know it. Whether the recognition of the potential dangers of rapid climate change could stimulate such a change in political perceptions remains to be seen.

REFERENCES

Edmonds, J.A., and J. Reilly. 1983. A long-term global energy economic model of carbon dioxide release from fossil fuel uses. *Energy Economics* **5(2)**:74–88.
Manne, A.E., and R. Richels. 1990. CO_2 emissions limits—an economic cost analysis for the U.S. *Energy J.* April.
Mintzer, I.M. 1987. A Matter of Degrees: The Potential for Controlling the Greenhouse Effect. Washington, D.C.: World Resources Institute.
Mintzer, I.M., and W. Moomaw. 1991. Escaping the Heat Trap. Washington, D.C.: World Resources Institute, in press.
NRC (National Research Council). 1991. Policy Implications of the Greenhouse Problem. Washington, D.C.: National Academy of Sciences.
Nordhaus, W.O. 1990. Contribution of different greenhouse gases to global warming: A new technique for measuring impact. New Haven, CT: Yale Univ.
OECD/IEA. 1989. Energy technologies for reducing emissions of greenhouse gases. Proceedings of an Experts Seminar. Paris, France.
Vu, My T. 1985. Population Projections: 1985–2100. Washington, D.C.: World Bank.

5

Carbon Dioxide Emissions Control and Global Environmental Change

J. REILLY
Economic Research Service, U.S. Department of Agriculture,
Washington, D.C. 20005-4788, U.S.A.

ABSTRACT

Multiple trace gases contribute to climate change and these trace gases also have other environmental effects. In developing a general expression for an index among trace gases, it is shown that one based only on an instantaneous radiative forcing and gas lifetime will provide an inefficient basis for formulating policy on trace gases when nonclimatic effects of trace gases exist or when the damage from climate change is nonlinear with respect to gas concentrations. Estimates of the cost of reducing carbon dioxide emissions are compared, showing the cost for a 20% reduction in emissions for some countries as estimated to be $1/ton, whereas for other countries the estimates range as high as $500/ton. The factors contributing to this wide range of estimates are discussed.

INTRODUCTION

This chapter begins by identifying carbon dioxide (CO_2) emissions as only part of the problem of climate change which, in turn, is only part of the broader problem of global environmental change. In the next section, the issue is cast in a dynamic cost-benefit framework to focus on the need to consider both non-CO_2 contributors to climate change and non-climate consequences of trace-gas accumulation. Next, the implications of such a formulation of the problem for policy are discussed. Finally, studies estimating the costs of limiting CO_2 emissions are reviewed.

Limiting the Greenhouse Effect: Options for Controlling Atmospheric CO_2 Accumulation
Edited by G.I. Pearman published 1992 John Wiley & Sons Ltd

CARBON DIOXIDE, OTHER TRACE GASES, AND CLIMATE CHANGE

The 1983 U.S. EPA Report, "Can We Delay A Greenhouse Warming?" (EPA 1983), somewhat inadvertently illustrated that trace gases other than CO_2 could, despite virtually whatever was done to limit CO_2 emissions, make it impossible to prevent or even significantly delay climate change. In doing so the report illustrated that if serious efforts were undertaken to limit climate change as induced by human activity, then the efforts could not be limited to CO_2 alone.

One way to consider the relative effect of different gases is to create an index of relative warming potential. Lashof and Ahuja (1990) provided a candidate index for computing the relative contribution of different trace gases to warming, taking into account trace-gas residence times and relative radiative forcing. Scientists working on the Intergovernmental Panel on Climate Change (IPCC) Scientific Assessment of Climate Change (Houghton et al. 1990) offered significant refinements to this index by improving the representation of physical processes. If one is to use an index of trace gases for political bargaining or for identifying a cost-effective tradeoff among gases, the problem is considerably more complex and involves both economic and physical considerations. While full accounting of both economic and physical considerations is controversial, because it requires knowledge of relationships on which there is little information, effective strategies to minimize costs must deal with the tradeoff among gases whether or not an index is explicitly used in the process. In other words, political choices among levels of control of different gases implies a relative valuation.

Trace gases and the processes that lead to trace-gas emissions have nonclimatic consequences as well. Clark (1986) provides one of the broader interpretations of global enviromental change, identifying coal in particular as contributing to several environmental problems ranging from climate change to acid rain to health effects. More narrowly focusing only on radiative gas concentrations and nonclimatic consequences, CO_2 has direct effects on plant growth, chlorofluorocarbons (CFCs) contribute to stratospheric ozone (O_3) depletion, and carbon monoxide (CO) degrades local air quality.

AN ECONOMIC INTERPRETATION OF A TRACE GAS INDEX

Nordhaus (1989, 1990) and Peck and Teisberg (1990) provide recent economic interpretations of the climate change problem. I interpret (Reilly 1990) the index problem as an optimal control problem that has become a

CO_2 Emissions Control and Global Environmental Change

familiar tool in solving dynamic problems of intertemporal resource use and investment. The key insights are:

1. The relevant index should be an index of economic effect of climate/environmental change. This requires adjustments to simple measures of radiative forcing because the gases are removed from the atmosphere at different rates, and marginal damages are likely to vary over time.
2. Gases have economic effects that are nonclimate-related and differ among gases. These should be counted as credits (e.g., direct CO_2 fertilization of crops) or debits (e.g., CFCs as contributors to O_3 depletion).
3. Programs that limit emissions or enhance sinks are not simply described as either a permanent or one-time reduction in emissions (or enhancement of a sink) but rather represent unique dynamic profiles. For example, forestation is an enhanced sink over 30 or 40 years with a neutral effect thereafter (if the forest achieves equilibrium) or a bulge of emissions (if clearing occurs). Similarly, programs of carbon taxes (recognizing that there are rents associated with fossil resources) may be better cast as a delay in fossil-fuel consumption rather than altogether avoidance.

The economic problem can be represented as that of maximizing the present value, V(O), of climate-related resource endowments where net returns from resource sectors at time t, R(t), are discounted by r:

$$V(O) = \int_0^\infty R(t) \exp^{-rt} dt \qquad (1)$$

where R(t) is defined as profits; π, from activities using climate-related resources minus investment expenditures; $P_I(t)I(t)$, where

$$\pi(t) = p(t) [F(C(t), K(t), N(t), G_i(t), E_i; t)] - w(t)N(t) \qquad (2)$$

p and w are prices of output and variable inputs; F is the production function relevant in time, t, relating capital (K), other inputs (N), the atmospheric stock of each trace gas (G_i, i = 1, . . .n, where n is the number of trace gases considered), the stock of climate-related resources (C), and emissions of trace gases (E_i). F_K, F_N, and F_{Ei} > 0 and F_{KK}, F_{NN}, and F_{EiEi} < 0. These conditions are traditionally assumed for conventional inputs K and N. For trace-gas emissions, the interpretation is that control of emissions could be accomplished by increasing capital and other inputs, thus relaxing controls which would lead to increased output.[1] $F_{Gi} < 0$ and $F_{GiGi} > 0$ for gases like CFCs, which have negative consequences for output through effects on stratospheric O_3 depletion, while $F_{Gi} > 0$ and $F_{GiGi} < 0$ for gases like CO_2, where the direct effects of increased concentrations include increased crop yields. Climate-related resources (C) are determined by the aggregate level of trace gases.

By representing the relationships between emissions, concentrations, and climate change over time, the general solution of the expression for the dynamic control problem can be derived. Quantification of the solution requires specific representations of complex physical processes that are poorly known. The simple representations of economic and physical processes used here are typically modeled as complex multi-equation models and include, for example, carbon cycle models, general circulation models (GCMs) of climate, and dynamic general equilibrium models of the economy. Expressions for the value of the constraints on trace gases are:[2]

$$\frac{\alpha_i p F_c + p F_{Gi} + d\lambda_i/dt}{(r + \delta_i)} = \lambda_i \qquad (3)$$

where α_i are the instantaneous gas-specific radiative forcing effects per unit of gas, δ_i are gas-specific dissipation rates, and λ_i are the current value multipliers for trace gases $i = 1, \ldots, n$.

The numerator in Equation 3 includes a term that adjusts the economic impact of trace-gas-induced climate change by the radiative forcing effect (the general presumption is that this term will be negative, i.e., climate change will adversely affect economic output). The second term is the direct effect of trace-gas accumulation on economic activity. This effect can be positive or negative depending on the gas. For example, increased CO_2 concentrations have a positive effect on agricultural yields while increased CFC emissions contribute to stratospheric O_3 depletion and consequent effects on human health and other biological systems resulting from increased UV radiation. The final term is the expected change in the value of the constraint.

The denominator of the expression is a discount factor (or adjusted discount rate) involving the interest rate and the rate of dissipation of the gas. The discount rate will be the same across gases. Significant differences among gases will occur due to differences in the dissipation rate.

To interpret this expression in terms of an index among gases, we can

---

[1] I would like to acknowledge Bruce Larson of the Economic Research Service, USDA for suggesting this representation.
[2] For a complete derivation of the model presented here, see Reilly (1990).

choose gas j as a numeraire and define the denominator of (3) as the gas-specific discount factor, γ_i. As a result, the index, β_i, is:

$$\beta_i = \left[\frac{\alpha_i pF_c + pF_{Gi} + d\lambda_i/dt}{\alpha_j pF_c + pF_{Gj} + d\lambda_j/dt}\right]\frac{\gamma_j}{\gamma_i} \qquad (4)$$

For the sake of example, choosing CO_2 as the numeraire one would note that, in comparison, CFCs are more radiatively active per ton and contribute to ozone depletion whereas CO_2 has a positive effect on agricultural yields. Thus, the index will tend to be larger for CFCs, suggesting that a ton of CFC reduction should be worth more than a ton of CO_2 reduction.

Equation 4 demonstrates the need to consider economic damages in any index. The intuition for this result is fairly clear. To add apples and oranges (climate and nonclimate effects) one needs a common metric. Monetary value is one candidate for a common metric. Note that Equation 4 reduces to a simple expression of relative radiative forcing adjusted by the discount factor if nonclimatic effects of trace gases, the pF_G terms and the $d\lambda/dt$ terms, are ignored.

To calculate relative index values that are explicitly dependent on the effects, Equation 3 can be rewritten as:

$$\lambda_i = \int_0^\infty (\alpha_i pF_c C(t) + pF_{Gi} G_i(t))\exp^{-(r+\delta_i)t} dt. \qquad (5)$$

ESTIMATED TRACE-GAS INDICES

Table 5.1 provides the basic physical data necessary to evaluate Equation 5. There remains considerable uncertainty about physical parameters even regarding measurement of current emissions and atmospheric stocks. Future emissions, radiative forcing, and dissipation each reflect complex physical and economic systems. The values in Table 5.1, drawn from the IPCC (Houghton et al. 1990), attempt to summarize these complex systems, of which there is uncertainty, with a single parameter. There is, for example, a general recognition among carbon cycle scientists that the atmospheric lifetime concept poorly represents the fate of carbon in the atmosphere.

Little evidence of aggregate future damages exists but the economic effects on agriculture may be illustrative. One estimate is that the loss of global welfare associated with agricultural effects of an effective doubling of gases in CO_2 equivalent is on the order of $38 billion (1986 U.S. dollars), including

Table 5.1 Characteristics of important trace gases

Gas	Concentration ppmv	Atmospheric Stock Tg	Emissions in 1990 Tg	Dissipation Rate yr^{-1}	Radiative Forcing mass	Emissions Growth rate
CO_2	354	2710000	26000	0.0083	1	0.006
CO	0.09	430	200	0.1	2.3	0.0085
CH_4	1.717	4800	300	0.1	58	0.005
N_2O	0.3097	1400	6	0.0067	206	0.003
HCFC-22	0.0001	1.51	.1	0.0067	5440	−0.08
CFC-11	0.0028	6.7	.3	0.0167	3970	−0.08
CFC-12	0.000484	10.19	.4	0.0077	5750	−0.08
CFC-113	0.00005	1.63	.15	0.0143	3710	−0.08

Sources: IPCC Scientific Assessment of Climate Change, Tables 2.3, 2.5; EPA (February 1989), pp. II-51-II-58; Wuebbles and Edmonds (1990).
Notes: The dissipation rate is 1/gas lifetime. This simplification ignores the fact that gas lifetime, for CO_2 for example, depends on the gas concentration and other factors. Radiative forcing is the instantaneous radiative forcing effect on a mass basis relative to CO_2. Emissions growth are best estimates of Wuebbles and Edmonds (1990) except CO_2 is the median value from Reilly et al. (1985). CFC emissions growth rates are based on Montreal Protocol target of 50 reduction by 1998 and assumes reductions continue beyond 1998 at the same rate.

the offsetting effects of direct CO_2 fertilization (Kane et al. 1990).[3] Direct CO_2 fertilization offsets about one-half of the yield loss in the U.S. due to climate change when CO_2 contributes about two-thirds of the effective doubling. Assuming this ratio holds for the rest of the world on average and can be directly applied to the welfare change means that CO_2 fertilization may contribute a benefit of $38 billion with the loss associated with climate change alone being at $76 billion (see Appendix 1). There have not been comprehensive studies of economic damages to other resource-dependent activities such as water resources, coasts, forests and fisheries, recreation and other values of natural ecosystems, and on possible population migration and human settlement changes. For illustrative purposes, damages in each of these five resource activities are assumed to be the same as for agriculture. However, the benefits of CO_2 fertilization are assumed to apply only to agriculture. The simplest assumption is that damages will be linear with respect to changes in trace-gas concentrations. Yet, a variety of existing reasons cause one to believe that damages may be increasing. There is a broad temperature range over which crops perform relatively well. However, once this range is exceeded, yields fall off rapidly. Evapotranspiration, contributing to soil moisture loss, increases more than proportionally to temperature increase. Also, the relationship of area inundated per centimeter increase in sea level, as well as the cost of coastal protection, may increase more than proportionally (NCPO 1989).

Based on these assumptions, it is possible to provide a preliminary and still partial evaluation of the relative value of controlling different trace gases.[4]

With the above and the data in Table 5.2, trace-gas indices can be computed. These are compared with the IPCC (Houghton et al. 1990) indices based on purely physical parameters, the direct radiative forcing, and a calculation showing the effect of ignoring the damage calculation but including the discount rate (Table 5.3). The simplest measure of comparison is the instantaneous radiative forcing. This fails, however, to take into account the differences in atmospheric lifetimes.

The IPCC Science report (Houghton et al. 1990) attempted to address this with indices of global warming potential (GWP). In recognizing the infinite horizon of the problem, the report developed indices by truncating

---

[3] Effective doubling of CO_2 refers to increases in radiatively active trace gases such that in combination the increases have the same radiative forcing effect of doubling the current concentration of CO_2.

[4] Function must converge. With damages a function of exponential growth in concentrations, discounting will generate convergence in the linear case (quadratic case) as long as (2 times) the exponential rate of growth is less than the discount rate plus the rate of gas dissipation. In general, these conditions are easily met by most parameterizations. Emissions growth projections are on the order of 0.003 to 0.015 whereas discount rates typically used are on the order of 0.03 to 0.05.

Table 5.2 Damage function parameters

Linear Damage	Quadratic Damage	Linear Benefit
d_1 −543	d_1' −250	b_1' 57.57
d_2 1.538	d_2' −.521	b_2' −.163
—	d_3' .001954	—

Damage parameter estimates were determined by assuming 0 damages (benefits) in 1990 when the "effective" CO_2 concentration (C) was 441 ppm, 76 billion × 6 sectors at 706 ppm, and for the quadratic damage function 152 billion × 6 sectors at 882.5 ppm. Benefit parameters applied only to CO_2 assumed benefit of 38 billion at 0.66 × 706 ppm.

the integration time to a finite, arbitrarily defined number of years.[5] The closest comparison calculations in this paper is with climate effects calculated using a linear damage estimate. With constant marginal costs, damage drops from the index, and the only economic parameter is the discount rate. While comparable to the IPCC calculations, it has the advantage that it more smoothly decreases the importance of future years rather than abruptly delineating years that matter and years that do not. The indices calculated in this paper also include future emissions growth whereas the IPCC calculations examine the relative effects of a unit of gas emitted today.

While the desirability of discounting is sometimes debated, the IPCC calculations, by limiting the time horizon considered, implicitly use a discount rate of 0 for the first 20 or 500 years and a discount rate of infinity for out-years. Decreasing the discount rate is somewhat comparable to increasing the integration period because both changes increase the weight of years in the future. In general, moving from the instantaneous radiative forcing to the integrated effect reduces the value of cutting the emissions of short-lived gases such as methane (CH_4), CO, and HCFC-22.

Under the assumption of a quadratic damage function, the index for shorter-lived gases also increases. Finally, the full index includes a positive offset for CO_2. This has the expected effect of increasing the value of reducing all other gases relative to CO_2.

Because the damage estimates are largely speculative and the physical representations are extremely simplified, these indices must be considered illustrative. However, the exercise provides a relatively simple framework for jointly considering radiative forcing, atmospheric residence, and both

---

[5] In addition to the 20- and 500-year calculations, a 100-year calculation was also made but is not reported here.

Table 5.3 Comparisons of trace gas indices

	Economic-based			Physical Effects Only		
	Climate + CO_2 fert. (quad.)	Climate Effects (quad.)	Climate Effects (linear)	Radiative Forcing	IPCC–GWP	
					20 y	500 y
CO_2	1	1	1	1	1	1
CO	3.7	2.9	.9	2.3	7	2
CH_4	92	74	21	58	63	9
N_2O	260	208	201	206	270	190
HCFC-22	8950	7174	1427	5440	4100	510
CFC-11	6343	5085	1389	3970	4500	1500
CFC-12	9119	7309	2140	5750	7100	4500
CFC-113	5917	4743	1319	3710	4500	2100

Source: Economic-based are calculated. Physical Effects Only are from Houghton et al. (1990).
Note: All values are indices where CO_2 is the numeraire. Radiative Forcing is the instantaneous radiative forcing on a mass basis. IPCC–GWP are "greenhouse warming potential" indices which integrate radiative forcing over time where the integration is 20 and 500 years as indicated by the column heading. Economic-based are calculated as indicated by Equations 4 and 5 (see text) and the data in Tables 1 and 2. As such they are the current value per mass unit, relative to CO_2, of reducing emission by one mass unit.

the climatic and nonclimatic effects of trace gases. It is possible to add somewhat more accurate representations of dissipation and radiative forcing by making these processes functionally dependent on concentrations. In the current formulation, future emissions are characterized by a rate of growth. Other simple formulations, such as linear or log-linear growth (decay), should be explored to determine sensitivity of the index to such assumptions. In particular, effort is required to characterize the economic damage associated with climate and the direct effects of trace-gas accumulation. As with emissions growth, effort must be focused on whether such damages increase linearly, exponentially, or log-linearly with increased climate change, i.e., whether marginal cost is increasing, decreasing, or constant.

Some studies have begun to address these issues (see NCPO 1989). For example, costs of sea level rise have been characterized as mounting at an increasing rate with each centimeter rise. Agricultural effects of climate change may be very small through a doubling but may increase more rapidly thereafter. While such efforts to quantify the index will require speculation and simplification that will unfortunately have limited scientific foundation, writing down the speculations and comparing them within an index framework would be an improvement from the current state of affairs.

Index calculations illustrate the importance of considering the timing of damages in addition to the time profile of gases. In addition, some particular caveats apply. The negative effects of CFCs on stratospheric O_3 were not included. These would significantly increase the index values for CFCs relative to CO_2 and other gases. In addition, CO_2 fertilization was only considered as a positive effect on agriculture; it may benefit forestry, the hydrological cycle, and natural ecosystems as well.

MISCELLANEOUS ISSUES

The appeal to mathematics in the previous section sacrificed the richness of many discussions regarding global change policy. The framework for broadly comparing different gases also provides perspectives on several issues frequently discussed in regard to climate change policy considerations.

No Regrets Policies

This is a popular policy option which encourages taking actions that make sense for reasons other than climate change. The argument is that because we really do not know how Earth systems will respond to increasing emissions nor do we know whether the societal/economic effects of such changes will be severe or even negative, an appropriate response is to do things that are justified on nonclimate grounds which also contribute to reducing emissions. Thus, at a later point, we will not regret having taken such actions if climate change does not occur or is not so bad or even good. In general, one cannot mount an argument against actions that are currently cost-effective, environmentally beneficial, and which coincidentally reduce trace-gas emissions unless climate change actually turns out to be beneficial.

Significant debate exists, however, regarding the extent of trace-gas savings available through no regrets actions. At the extremes are the positions taken by Lovins (chapter 14, this volume), who sees opportunities for limiting trace gases that will also save billions of dollars and traditional econometric modeling approaches (e.g., Jorgenson 1990) that implicitly assume all economically effective technologies are being utilized and only rising fossil-fuel prices or further advances in technology will induce savings.

Scale of Impact, Scale of Action

It seems reasonable not to overburden international bargaining with issues that can be handled within single countries, or with bilateral or regional agreements. This consideration has direct implications for decisions regarding which nonclimatic factors should be included in an international bargaining index. The ozone hole and direct effects of CO_2 on agriculture are global phenomena. The case for their inclusion in an international index calculation is strong. On the other hand, acid rain, nitrate contamination of groundwater, particulates, and direct health effects of gases (such as CO and NO_x) are environmental problems of sub-global dimensions. The scale of action on these issues should be consistent with the scale of impact, excluding them in international negotiations on global change or as part of a gas-index computation. This should not exclude individual countries from including domestic/sub-global environmental or other public policy goals in evaluating tradeoffs among control options at the national or regional level.

What Targets, Which Instruments

Various target emissions levels have been proposed, and there are efforts to estimate the cost of complying with such targets. There are, however, a variety of different points within the global climate-change system at which an international agreement could be negotiated, each point providing a different set of monitoring, enforcement, and efficiency considerations. Emissions targets have appeal because emissions are a point in the system where there are existing, or easily imagined, instruments of public control. However, economically efficient negotiated targets must be broader than emissions. Afforestation to enhance carbon withdrawal is an obvious action excluded from an emissions policy focus. While proposals to fertilize the ocean to enhance carbon uptake or to orbit panels that cut down incoming solar radiation may be farfetched, these actions are among a general class of actions including (1) removal of gases from the atmosphere, (2) climate modification to offset change, and (3) adjustment of production processes that use climate resources that may not be encouraged if international negotiations focus on emissions only.

This perspective suggests that rather than viewing emissions as pollutants, we must adopt the perspective of the mathematical problem in the previous section. This perspective indicates the need for management approaches that maximize the value of the atmosphere/climate resources to society. That is, we think of "selling" the atmosphere as a sink for emissions and "selling" climatic resources to resource using sectors. Such a perspective helps emphasize the interdependencies of these two activities while recognizing that there are a variety of other actions we can take to increase the amount of atmospheric sink and climate resource, ranging from reforestation to climate modification. It also emphasizes the continuing global resource management problem that must be faced over time.

Flexibility to respond to new information will require mechanisms for ongoing management. From this perspective, dealing with the problem in a comprehensive manner is appropriate, however it is difficult to imagine designing a single instrument that can achieve an efficient outcome. Recognizing the need for multiple instruments, even if an ideal gas index could be created, frees us to consider separate instruments for each gas thus eliminating the need for an explicit numerical index. For example, it may well make sense to have a mixed set of internationally negotiated policies with the level and mix of policies adjusted through the international bargaining process as the costs and benefits of action become better identified. Such a mix could include taxes on fossil fuels or tradeable permits in gross carbon emission, identified reforestation goals, further research on ocean fertilization, variety development on rice that may reduce CH_4 emissions, etc. Calculating an index to support the negotiation process would be beneficial but, given the uncertainty of such a calculation, institutionalizing a particular value in a trading scheme may be premature.

Bargaining, Buyouts, and Unilateral Action

Unilateral action is not very effective in reducing trace-gas accumulation, thus the need for international bargaining (Edmonds and Reilly 1983, 1985). Whatever the policy instrument, distributional consequences are unavoidable. Early recognition of this can probably save considerable effort. Rights could be distributed on the basis of population, income, the inverse of income, emissions, or geography. One might suggest the allocation of rights to maintain neutrality with respect to distributional consequences. A component of neutrality should be that the benefits of climate stability and the avoidance of the accumulation of trace gases in the atmosphere, minus the cost of emissions reduction, are "equal" among signees. Given the difference in size among countries (however measured), equal total benefit is unlikely to be considered to be a fair distribution. Thus, equality is likely to be judged in terms of, for example, per capita, per unit of gross national product (GNP), or per value of the natural resource stock.

COSTS OF LIMITING TRACE-GAS CONCENTRATIONS

What are the costs of significant reduction of trace gases? Wuebbles and Edmonds (1990b) have prepared a comprehensive review of existing studies examining the costs of the reduction of emissions of fossil fuels (Table 5.4). Considerably less effort has been invested to consider the cost of emissions reductions of other trace gases or of enhancing the sinks for carbon. In Table 5.4, percent reduction is relative to current levels of emissions.

A number of authors have noted the dependence of cost estimates on the business-as-usual assumption. Some of the studies suggest that emissions can be held constant at no cost in at least some countries (e.g., Chandler 1990). However, most assume that, in the absence of specific control policies, emissions will increase. Thus, they find positive costs in control scenarios where there is no reduction in emissions relative to 1990 (e.g., Edmonds and Barns 1990a).

Table 5.4 reports the period to which the estimates refer. The period can affect costs in several ways. First, trace-gas-emitting activities involve long-lived capital. Rapid reductions in fossil energy use would mean premature retirement of capital stock, adding to the cost of emissions reduction whereas a phased program, for example replacing coal-fired power plants only after they normally are retired, would be less costly. Second, the target date for emissions reduction and the business-as-usual emissions growth path affects the cost estimates. Wide differences exist in what future fossil energy use will be without specific controls to limit trace-gas emissions. Some analysts are very optimistic that cost-effective technology exists and will be adopted, or that energy use will not increase with income. If such assumptions give rise to a 20% reduction in emissions in 40 years without any climate change policy, then the analyst will find that the costs of reaching a target of 20% reduction in 40 years is zero. However, reaching the same target in 20 years will mean additional costs because incentives must then be found to develop and put in place technology faster than it otherwise would occur. Conversely, a business-as-usual scenario featuring exponential growth in emissions would likely find the costs of maintaining a 20% reduction in emissions relative to current emissions escalating over time. And, third, the relative prices of alternatives are likely to change over time.

Table 5.4 The cost of reducing fossil-fuel CO_2 emissions: summary of findings of published studies

Study	Region	Case	Forecast Year	Percent Emissions Reduction[a]	Ref. Year	Avg. Cost (1989 U.S. $/TC)	Mar. Cost (1989 U.S. $/TC)
Boonekamp et al. (1989)	Neth.	*2nd* Stage					
		Gas	2010	(10)	1990	6[b]	n.a.
		Nuc	2010	0	1990	0[b]	n.a.
		Gas	2030	(13)	1990	8[b]	n.a.
		Nuc	2030	6	1990	0[b]	n.a.
		3rd Stage					
		Gas	2030	27	1990	5[b]	n.a.
		Nuc	2030	46	1990	5[b]	n.a.
Gerbers et al. (1990)	Neth.	Trend Case	2020	0	1990	n.a.	1[b]
			2020	20	1990	n.a.	1[b]
			2020	30	1990	n.a.	1[b]
			2020	40	1990	n.a.	2[b]
			2020	50	1990	n.a.	3[b]
			2020	60	1990	n.a.	5[b]
			2020	70	1990	n.a.	12[b]
Yamaji et al. (1990)	Japan		2005	0	1988	n.a.	281[c]
Manne and Richels (1990a)	U.S.A.		2030	0	1990	n.a.	250
			2030	20	1990	210	250
			2030	50	1990	n.a.	250
Manne and Richels (1990b)	U.S.A.	Optimistic Case	2050	0	1990	n.a.	110
			2100	0	1990	n.a.	250
		Pessimistic Case	2050	0	1990	n.a.	250
			2100	0	1990	n.a.	250

Source	Region					
Hogan and Jorgenson (1990)[d]	U.S.A.	2030	0	1990	n.a.	500
		2030	20	1990	n.a.	500
		2030	50	1990	n.a.	500
Jorgenson and Wilcoxen (1990)	U.S.A.	2010	0	1990	n.a.	17[e]
		2010	20	1990	n.a.	46[e]
		2010	0	2000	n.a.	7[e]
Chandler (1990)	U.S.A.	2005	0	1989	0	n.a.
	U.S.A.	2005	20	1989	92	n.a.
	France	2005	0	1989	0	n.a.
	Hungary	2005	0	1989	0	n.a.
	Poland	2005	0	1989	0	n.a.
	U. Kingdom	2005	0	1989	0	n.a.
CBO (1990)	U.S.A.	2000	6–(5)[f]	1988	n.a.	110[e]
		2100	20	1988	n.a.	110–440[e]
Morris et al. (1990)	U.S.A.	2010	0	1990	4	n.a.
		2010	10	1990	15	n.a.
		2010	20	1990	28	39
Chandler and Nicholls (1990)	U.S.A.	2000	0	1989	n.a.	82
Nordhaus (1979)	World	2000	(56)[e,g]	1988[g]	n.a.	25[e]
		2020	(30)[e,g]	1988[g]	n.a.	232[e]
		2040	(58)[e,g]	1988[g]	n.a.	263[e]
		2100	(72)[e,g]	1988[g]	n.a.	422[e]
Nordhaus and Yohe (1983)	World	2025	(52)	2000[h]	n.a.	77[i]
		2050	(43)	2000[h]	n.a.	266[i]
		2075	(93)	2000[h]	n.a.	303[i]
		2100	(151)	2000[h]	n.a.	303[i]
Edmonds and Barns (1990a)	World	2025	0	2000	55	95
		2025	20	2000	65	155
		2025	50	2000	128	405

Table 5.4 Continued

Study	Region	Case	Forecast Year	Percent Emissions Reduction[a]	Ref. Year	Avg. Cost (1989 U.S. $/TC)	Mar. Cost (1989 U.S. $/TC)
			2050	0	2000	83	250
			2050	20	2000	120	450
			2050	50	2000	n.a.	n.a.
Manne and Richels (1990b)	World		2030+	0	1990	n.a.	250
			2030+	20	1990	210	250
			2030+	50	1990	n.a.	250

Notes: [a]Percentage emission reduction is measured relative to a reference year's emission and not the reference case emission for the designated forecast year. Negative values in ( ).
[b]Converted to 1990 constant U.S. dollars from 1990 Dutch Guilders at the rate of 2.25:1.
[c]Tax rate begins at 4,000Y/TC and rises at the rate of 4,000Y/TC/yr until the year 2005 when it reaches the rate of 64,000Y/TC. Tax rate reported is 64,000Y/TC. Yen converted to 1990 U.S. at the rate of 227.68Y.$, the Tokyo exchange rate 22 November 1990. Assumptions incorporated in the trade component of the models differ and are 238.5Y/$, 1985; 127.3Y$/, 1990; 100.0Y/$, 2000; and 90.0Y/$, 2005.
[d]Hogan and Jorgenson (1990) do not employ a model themselves. Rather they examine the work of Manne and Richels (1990a) and conclude that "Simple calculations with the best available estimation results suggests that common assumptions in existing models may be missing as much as half of the total cost of controlling greenhouse gases."
[e]Results reported in short tons of carbon converted to metric tonnes at the rate of 1.102.
[f]Results ranged from 5 above 1988 levels to 6 below 1988 levels.
[g]Nordhaus does not attempt to achieve a carbon emission target. Rather he examines the larger carbon cycle system. The experiment he runs is to constrain cumulative carbon accumulation in the atmosphere. The 1988 reference year is arbitrary and supplied by the authors. The authors use 5.7 GtC y^{-1} 1988 emissions.
[h]As with Nordhaus (1979), Nordhaus and Yohe (1983) do not attempt to stabilize emissions at any particular level. We report the most stringent of the tax cases examined. This tax scenario is initiated in the year 2000. We therefore take the forecast year 2000 emissions rate of 5.54 GtC y^{-1} as a reference emissions rate.
[i]Carbon taxes reported in $1975 constant U.S. dollars per ton of Coal equivalent were converted using 0.7682×10^6 gC/(short ton coal equivalent) and 1 1975$ = 2.13 1989$.

CO_2 Emissions Control and Global Environmental Change

Table 5.4 also reports marginal and average costs. Marginal costs are, conceptually, the cost of the last ton removed and will be higher than average costs.

Technical characterizations of the problems of comparing costs are also based on quite different approaches. Edmonds and Barns (1990b) characterize the two approaches as "bottom-up" and "top-down," suggesting specific biases. Bottom-up studies are those which analyze individual energy-using activities, attempt detailed engineering cost estimates of alternatives, and derive estimates for national or global savings by multiplying by the number of similar activities and adding across all the activities considered. Top-down studies develop stylized relationships among aggregate variables such as population, economic growth, and energy prices usually derived from statistical estimation. Quoting from Edmonds and Barns (1990b):

> "Bottom-up" studies do not in general allow for changes in the cost of energy or nonenergy factors as the scale of activity of a technology grows from marginal to market dominating. Similarly, the "bottom-up" studies frequently do not allow for the heterogeneity of end-use energy services that must be accommodated. "Bottom-up" studies estimate the levelized cost[6] of technologies at market interest rates or social discount rates rather than implied internal rates of return. Because the consumer discount rate on energy investments is much higher than the market rate of interest, technologies penetrate much more rapidly in "bottom-up" studies than they do in "top-down" studies.

They go on to identify that market imperfections may lead to the high implied consumer discount rate and that bottom-up studies implicitly assume such market imperfections can be removed at no cost. I would add that many of these barriers are likely to be one-time costs or costs that may fall per unit of emission reduction as the scale of activity increases. Such would be the case if gaining information, uncertainty about performance, existence of supporting infrastructure (alcohol fuels distribution stations, for example) were part of the reason why technologies have not penetrated. The error in many top-down studies is that they implicitly assume that there are no such economies of scale and thus may overestimate costs.

---

[6] The concepted of levelized cost is to spread capital costs or other one-time costs over the life of the investment. The basic formula is:

$$\frac{\sum_{1}^{t} \frac{C_i}{(1+r)^i}}{\sum_{1}^{t} Q_i}$$

where summation is over i, c_i is the cost in period i, r is the interest rate, Q_i is output (e.g., energy savings), and t is the life of the investment. Detailed calculations would differentially take into account variable versus capital costs, scrappage value, taxes, etc.

Moulton and Richards (1990) estimate that up to 40% of U.S. CO_2 emissions in 1990 could be sequestered through afforestation, with marginal costs ranging from 5 to 50 dollars per ton (1989 U.S.$). Sedjo and Solomon (1988) provide very rough estimates of what they term the "minimal" cost of sequestering global carbon emissions. Annualizing these costs at a 30-year forest life at a 5% discount rate suggests marginal costs of between 6 to 18 dollars per ton in temperate regions and between 1 and 10 dollars per ton in the tropics. Richards (1991), based on an actual profile of carbon sequestration, indicates that the discounted costs could be significantly higher than the Moulton and Richardson estimates. Adams et al. (1990), evaluating the possible impacts on land rents due as larger areas are forested, also provide evidence that the Moulton and Richards estimates (1990) may underestimate costs.

Regarding the costs of controlling other trace gases, one estimate is that the U.S. cost of meeting the Montreal Protocol on CFC reduction would average 5.8 to 12 million dollars per ton of CFC reduction (J. Edmonds, pers. comm.). If the index value for CFCs is 9000, then this is the equivalent of spending about 600 to 1300 dollars per ton of carbon emissions reduction. The motivation for the Montreal Protocol was ozone depletion rather than climate change. Considering the climatic effects of CFCs would lead one to further restrict emissions if climate damage is expected to be significant.

CONCLUSIONS

Limiting climate change by reducing trace-gas concentrations will require reductions in emissions of several different gases. This paper provided an approach for considering tradeoffs among different trace gases and for incorporating nonclimatic effects of trace-gas accumulation into an index. Because the relationship of damages to climate may be nonlinear and gases have different lifetimes in the atmosphere, a simple index based on physical parameters will provide an inefficient basis for trading-off reductions among gases. An explicit index is needed if an emissions trading scheme will be used; however, even if an explicit trading system is not implemented, informed tradeoffs among policies will require an understanding of the relative value of decreasing different gases. In addition, gases have nonclimatic effects which should be considered. In particular, CO_2 has a positive effect on plant growth and thus on agricultural output. Failure to include such effects in decisions regarding control of trace gases will lead to overly severe limitations on CO_2 relative to other gases, such as CFCs and CH_4.

The estimated costs of controlling carbon emissions differ considerably. A key consideration in comparing costs is the likely future growth in emissions in the absence of climate change considerations. If the scenario suggests that emissions will grow very slowly or will fall significantly, then the costs of controlling emissions to a given level will be lower than if the base-case scenario includes rapid emissions growth. While this point seems relatively obvious, many of the differences appear to be due to differences in the base-case assumption. For comparison purposes, it is useful to include so-called no regrets actions as part of the base case because these are actions that would be taken with or without climate change. Adopting this convention would necessarily mean that any further reductions would be costly.

ACKNOWLEDGEMENTS

The views expressed in this paper are solely those of the author and do not reflect the views of the U.S. Dept. of Agriculture or the U.S. government. I would like to acknowledge the helpful comments and suggestions of Bruce Larson, Dick Brazee, and Jae Edmonds while taking full responsibility for any remaining errors.

APPENDIX

There is considerable debate regarding the extent of the CO_2 fertilization effect. Yet, the beneficial effects of CO_2 on plants are well established. Acock and Allen (1985) state that "experiments demonstrating the practical benefits of growing crops in elevated CO_2 concentrations were first reported in France in 1902 (Demoussy). Since then many crops have been tested and none has failed to respond to increased CO_2 concentration." Increased ambient CO_2 levels affect several plant functions. A doubling of CO_2 increases the instantaneous photosynthetic rate by 30–100%, with variability due to other environmental conditions (Pearcy and Bjorkman 1983). While measures of increased photosynthetic rates provide only indirect evidence of the potential effect on yield, direct measure of yield increases under controlled experimental conditions with optimum conditions have indicated crop yield changes ranging from -20% to $+200\%$ over a variety of crops (Kimball 1983). There is reason to believe that those experiments showing yield declines suffered from incomplete control of conditions between the control and experimental groups (Acock, pers. comm.). It is widely recognized that the growth response to elevated CO_2 is greater in C3 crops (e.g., wheat, rice, barley, root crops, and legumes) than in C4 crops (e.g., corn, sorghum, millets, sugar cane) (e.g., Acock and Allen 1985; Kimball 1985).

In addition to the growth response, increases in ambient CO_2 levels reduce stomatal aperture thereby reducing transpiration (increasing water-use efficiency). The water-use efficiency effect is greater in C4 plants than in C3 plants (Kimball 1985). There are uncertainties with regard to what this means for total water use because increased water-use efficiency may be offset by increased leaf area over which water is transpired (Acock and Allen 1985; Kimball 1985).

Plant response to increased CO_2, when conditions are nonoptimal or stressful, differs from the response when other conditions are at optimum. Increased growth under enhanced CO_2 is less if other nutrients are limiting, with the one exception of legumes, where nitrogen fixation increases under enhanced CO_2 (Kimball 1985). A variety of both C3 and C4 species varying in their salt tolerance all have a greater increase in growth response to high CO_2 when grown under high salt concentrations than when grown under low salt concentrations (Acock and Allen 1985). Studies indicate that plant response also may be higher under other environmental stresses (Kimball 1985).

Enhanced CO_2 may also affect plant developmental processes such as flowering (Morrison 1989), increase in starch and sugar content (Bazzaz et al. 1985; Strain 1985), and reduced nitrogen content of leaf, stems, and fruits.

To consider the ultimate effect on agricultural productivity under farm conditions and any differences that might occur across regions requires consideration of interaction among the physiological effects and relevant environmental conditions as well as adaptations that are likely to occur within the agricultural production system. It is sometimes asserted that the predominance of C4 plants in tropical regions will particularly disadvantage developing countries. In terms of metric tons, developing countries produced 75% of the world production of major C4 crops (millet, corn, sorghum, and sugar cane) and 60% of major C3 crops (rice, wheat, barley, oats, roots and tubers, legumes, and vegetables) in 1987 (FAO 1988). Africa, the poorest continent, is closely balanced between C3 and C4 plants, producing 8% of the world's C4 crop in 1987 and 7.4% of C3 crops. Total world tonnage of C3 crop harvests is 50% greater than for C4 crops. Thus, developing countries produced 20% more C3 than C4 crops.

The slight bias in developing countries toward C4 crops need not necessarily lead to a less positive yield effect. The greater water-use efficiency for C4 plants could prove to be beneficial in uncontrolled environments where water supplies may be more restricted than under controlled experiments where conditions are optimal. Salinity is a greater problem in tropical regions because higher temperatures create conditions where salts build up in soils, particularly under continuous irrigation. Both C4 and C3 plants may show greater relative improvement in tropical regions where salinity is currently a yield-limiting factor. Even if C3 plants are favored globally, it need not lead to a relative disadvantage in developing countries because they already produce more than one-half of C3 crops. Production shifts out of C4 crops is likely to include C3 crop production increases in both developing and developed countries.

Sugar cane (a C4 crop of which 92% of 1987 production was in developing countries [FAO 1988]) versus sugar beets (a C3 crop of which 89% of the 1987 crop was produced in developed countries [FAO 1988]) is often cited as a specific case where tropical developing countries would be disadvantaged. Duties, quotas, and other trade policies support sugar beet production in the U.S. and Europe and other temperate areas. For example, raw sugar prices in the U.S. averaged 50.3 cents/kg (22.81 cents/lb) in 1990 compared with world sugar prices of 28.2 cents/kg (12.79 cents/lb) (USDA 1991). Because quotas limit sugar imports, sugar beet production costs would have to fall substantially relative to cane production costs before any production shift between tropical cane producers and temperate beet producers occurred as a result of changes.

Kimball (1985) notes that a variety of adjustments are likely to occur that could increase CO_2 response as CO_2 concentrations increase. In particular, the process of selecting cultivars with superior performance in field trials will naturally lead to selection of cultivars that perform better in higher CO_2 environments, as CO_2 gradually increases even without a conscious effort to select for CO_2 response. Whereas crop breeders have long searched and selected for genetic characteristics that broaden the climate tolerance of crops because of the wide geographic variation in climate, selection for genotypes that respond to higher CO_2 concentrations has not occurred because there has been no practical way to increase ambient CO_2 levels under which agricultural crops are commercially grown. Thus, CO_2 response is an unexploited genetic characteristic.

The limited CO_2 response when other nutrients are limited may be important in assessing whether the standing biomass in unmanaged ecosystems will increase and therefore increase the amount of carbon stored. While the growth-limiting factors of plants in unmanaged systems are not well known, other nutrients may be limiting in many cases. This consideration is largely not relevant to agricultural systems because inorganic and organic fertilizers are used to supply additional nutrients.

REFERENCES

Acock, B., and L.H. Allen, Jr. 1985. Crop responses to elevated carbon dioxide concentrations. In: Direct Effects of Increasing Carbon Dioxide on Vegetation, ed. B.R. Strain and J.D. Currie, DOE/ER-0238, pp. 53–97. Washington, D.C.: U.S. Dept. of Energy.

Adams, R.M., C. Ching-Chang, B.A. McCarl, and J.M. Calloway. 1990. The role of agriculture in climate change: a preliminary evaluation of emission control strategies, 25 pp. Presented at Global Change: Economic Issues in Agriculture, Forestry, and Natural Resources. Washington, D.C., Nov. 19–21. Eugene, OR: Oregon State Univ.

Bazzaz, F.A., K. Garbett, and W.E. Williams. 1985. Effect of increased atmospheric carbon dioxide on plant communities. In: Direct Effects of Increasing Carbon Dioxide on Vegetation, ed. B.R. Strain and J.D. Currie, DOE/ER-0238, pp. 155–170. Washington, D.C.: U.S. Dept. of Energy.

Boonekamp, P.G.M., T. Kram, P.A. Okken, M. Rouw, and D.N. Tiemersma. 1989. Baseline and CO_2 Response Scenarios for The Netherlands. Petten, NE: Energy Study Center.

Chandler, W.U., ed. 1990. Carbon Emissions Control Strategies: Executive Summary. Baltimore, MD: World Wildlife Fund and the Conservation Foundation.

Chandler, W.U., and A.K. Nicholls. 1990. Assessing Carbon Emissions Control Strategies: A Carbon Tax or a Gasoline Tax? American Council for an Energy-Efficient Economy, ACEEE policy paper no. 3.

Clark, W.C. 1986. Sustainable development of the biosphere. In: Sustainable Development of the Biosphere, ed. W.C. Clark and R.E. Munn. Cambridge, MA: Cambridge Univ. Press.

CBO (Congressional Budget Office). 1990. Carbon Charges as a Response to Global Warming: The Effects of Taxing Fossil Fuels. Washington, D.C.: Congressional Budget Office.

Edmonds, J., and D. Barnes. 1990a. Estimating the Marginal Cost of Reducing Global Fossil Fuels CO_2 Emissions, PNL-SA-18361. Washington, D.C.: Pacific Northwest Laboratory.

Edmonds, J., and D. Barnes. 1990b. Factors Affecting the Long-term Cost of Global Fossil Fuel CO_2 Emissions Reductions. Washington, D.C.: Pacific Northwest Laboratory, in press.

Edmonds, J., and J. Reilly. 1983. Global Energy and CO_2 to the Year 2050. *Energy J.* **4(3)**:21–47.

Edmonds, J., and J. Reilly. 1985. Global Energy: Assessing the Future. Oxford: Oxford Univ. Press.
EPA. 1983. Can We Delay a Greenhouse Warming? (Sept. 1983) Washington, D.C.: Office of Strategic Studies.
EPA. 1989. Policy Options for Stabilizing Climate, ed. D. Lasof and D. Tirpak, vol. 1, 372 pp. Washington, D.C.: U.S. Environmental Protection Agency, Office of Policy, Planning, and Evaluation.
FAO. 1988. FAO Yearbook: Production, vol. 41 (1987). Rome: Food and Agriculture Organization of the United Nations.
Gerbers, D., T. Kram, P. Lako, P.A. Okken, and J.R. Ybema. 1990. Opportunities for New Energy Technologies to Reduce CO_2 Emissions in The Netherlands Energy System Up to 2030. Prepared for ETSAP Workshop, Geneva, Oct. 8–12, 1990. Petten, NE: Netherlands Energy Research Foundation.
Hogan, W.W., and D.W. Jorgenson. 1990. Productivity Trends and the Cost of Reducing CO_2 Emissions. Cambridge, MA: Energy and Environmental Policy Center, J.F. Kennedy School of Govt., Harvard Univ.
Houghton, J.T., G.J. Jenkins, and J.J. Ephraums, ed. 1990. Climate Change. The IPCC Scientific Assessment, 365 pp. Cambridge, MA: Cambridge Univ. Press.
Jorgenson, D.W., and P.J. Wilcoxen. 1990. The cost of controlling U.S. carbon dioxide emissions. Presented at a workshop on Economic/Energy/Environmental Modeling for Climate Policy Analysis. Washington, D.C., Oct. 22–23, 1990.
Kane, S., J. Reilly, and J. Tobey. 1990. Climate Change: Implications for World Agriculture. Washington, D.C.: Economic Research Service.
Kimball, B.A. 1983. Carbon dioxide and agricultural yield: an assemblage and analysis of 770 prior observations. WCL Report 14. Phoenix, AZ: U.S. Water conservation Lab.
Kimball, B.A. 1985. Adaptation of vegetation and management practices to a higher carbon dioxide world. In: Direct Effects of Increasing Carbon Dioxide on Vegetation, ed. B.R. Strain and J.D. Currie, DOE/ER-0238, pp. 185–204. Washington, D.C.: U.S. Dept. of Energy.
Lashof, D.A., and D.R. Ahuja. 1990. Relative global warming potentials of greenhouse gas emissions. *Nature* **344**:529–531.
Manne, A.S., and R.G. Richels. 1990a. CO_2 emissions limits: an economic cost analysis for the USA. *Energy J.* **11(2)**:51–75.
Manne, A.S., and R.G. Richels. 1990b. The Costs of Reducing U.S. CO_2 Emissions: Further Sensitivity Analysis. Palo Alto, CA: Electric Power Research Institute.
Morris, S.C., B.D. Solomon, D. Hill, J. Lee, and G. Goldstein. 1991. A least cost energy analysis of U.S. CO_2 reduction options. In: Energy and Environment in the Twenty-first Century. Cambridge, MA: MIT Press.
Morrison, J.I.L. 1989. Plant growth in increased atmospheric CO_2. In: Carbon Dioxide and Other Greenhouse Gases: Climatic and Associated Impacts, ed. R. Fantechi and A. Ghazi, pp. 228–244. Dordrecht: CEC Reidel.
Moulton, R.J., and K.R. Richards. 1990. Costs of Sequestering Carbon through Tree Planting and Forest Management in the United States. U.S. Dept. of Agriculture, Forest Service, General Technical Report WO-5B, Dec., 46 pp. Washington, D.C.: USDA.
NCPO. 1989. Climate impact response function. Report of a workshop held at Coolfront, West Virginia, Sept. 11–14. Washington, D.C.: National Climate Program Office.
Nordhaus, W.D. 1979. The Efficient Use of Energy Resources. New Haven, CT: Cowles Foundation Press.

Nordhaus, W.D. 1989. A Survey of Estimates of the Costs of Reduction of Greenhouse Gases. New Haven, CT: Dept. of Economics, Yale Univ.

Nordhaus, W.D. 1990. A sketch of the economics of global climate change. Paper presented at the Allied Social Science Association Meetings. Washington, D.C.

Nordhaus, W.D., and G.W. Yohe. 1983. Future carbon dioxide emissions from fossil fuels. In: Changing Climate, pp. 87–153. Washington, D.C.: Natl. Academy Press.

Pearcy and Bjorkman. 1983. Physiological effects. In: CO_2 and Plants. The Response of Plants to Rising Levels of Atmospheric Carbon Dioxide, ed. E.R. Lemon, pp. 65–105. Boulder, CO: Westview.

Peck, S.C., and T.J. Teisberg. 1990. A Framework for Exploring Cost Effective CO_2 Control Paths (progress report). Palo Alto, CA: Electric Power Research Institute and Teisberg Associates.

Reilly, J. 1990. Climate Change Damage and the Trace Gas Index Issue. Washington, D.C.: USDA, Economic Research Service.

Reilly, J.M., J.A. Edmonds, R.H. Gardner, and A.L. Brenkert. 1985. Uncertainty analysis of the IEA/ORAU CO_2 emissions model. *Energy J.* **8(3)**:1–29.

Richards, K. 1991. An Assessment of the Potential for Carbon Sequestration in the United States (draft). Pennsylvania: Univ. of Pennsylvania.

Sedjo, R.A., and A.M. Solomon. 1988. Climate and forests. Presented at Workshop on Controlling and Adapting to Greenhouse Warming, Resources for the Future. Washington, D.C.

Strain, B.R. 1985. Background on the response of vegetation to atmospheric carbon dioxide enrichment. In: Direct Effects of Increasing Carbon Dioxide on Vegetation, ed. B.R. Strain and J.D. Currie, DOE/ER-0238, pp. 1–11. Washington, D.C.: U.S. Dept. of Energy.

Tobey, J., and B. Larson. 1990. Public Response to Uncertain Climate Change. Washington, D.C.: USDA Economic Research Service.

USDA. 1991. Sugar and Sweetener Situation and Outlook Report. Economic Research Service SSRV16N3, Sept., 51 pp. Washington, D.C.: USDA.

Victor, D.G. 1990. Calculating greenhouse budgets. *Nature* **347**:431.

Wuebbles, D.J., and J.A. Edmonds. 1990. A Primer on Greenhouse Emissions (draft). Washington, D.C.: Pacific Northwest Lab.

Wuebbles, D.J. and J.A. Edmonds. 1991. A Primer on Greenhouse Gases. Chelsea, MI: Lewis Publishers, in press.

Yamaji, K., R. Matsuhashi, Y. Nagata, and Y. Kaya. 1990. An integrated system of CO_2/energy/GNP analysis: case studies on Economic Measures for CO_2 reduction in Japan. Presented at a Workshop on Economic/Energy/Environmental Modeling for Climate Policy Analysis. Oct. 22–23, 1990, Washington, D.C.

6

To What Extent Can Renewable Energy Systems Replace Carbon-based Fuels in the Next 15, 50, and 100 Years?

C.-J. WINTER
Deutsche Forschungsanstalt für Luft- und Raumfahrt e.V., 7000 Stuttgart 80, F.R. Germany

ABSTRACT

Renewable energies are a very attractive replacement for the fossil energies because they do not require primary energy raw materials and thus cannot give rise to any of the pollutants which have their source in these materials. Renewable energies are closed-loop energy systems, without radioactivity, and they do not supply additive carbon or heat to the geosphere. Ecological consequences can only arise from the conversion technologies.

Energy has both technical and nontechnical aspects. It would be an illusion to think that the contributions to be made by renewable energies are a matter of technology alone. Indeed, the solution of technical problems will not even be in the forefront for the next 15 years. Energy legislation, energy regulations, the complete internalization of external costs, replacement of the now dominant primary energy economy with an end-energy economy, and people's energy-related attitudes and behavior all must be adapted to the special character of renewable energies. Renewable energies are not simply another energy in humankind's energy supply system: There will not be one single "solar energy supplier" alone representing the interests of renewable energies.

In this chapter I will briefly characterize renewable energies, describe their potential in a future world energy economy which reflects new attitudes toward energy consumption and toward the environment, and outline the economic implications and next steps.

Long ago when there were still forests on Attica's slopes, a thick layer of earth absorbed the water and preserved it until it could gradually spread out from the heights and feed the springs. But now this rich, soft earth has been swept away, leaving only the naked skeleton of the land—evoking the bony frame of a body wasted by sickness.

Plato, Kritias, 111b–d (condensed)

INTRODUCTION

In 1989, the world required about 12×10^9 tce ($\approx 10^{13}$ kWh) of primary energy, of which 80% was fossil energy, 7% hydropower, and 5% nuclear energy; 8% was biomass (e.g., wood or sugar cane) and the agricultural, forest, and animal waste that is energetically used, especially in the developing countries on a noncommercial basis.

Modern research and development in renewable energies is 15 years old. Although impressive from an engineering point of view, the contribution of renewable energies to the global energy supply remains marginal (wind energy 2,000 MW_e, solar thermal energy 400 MW_e, while photovoltaic production rates have grown to 40–50 MW_e y^{-1} [1990]).

"New" energies need decades to become established:

1. The energy supply for the next 15 years has already been decided. Energy investment decisions have a decades-long effect!
2. The next 50 years will reflect the classic energetic time constants (e.g., the first controlled nuclear fission took place in 1938, and in 1990 nuclear energy supplies 5% of the primary energy equivalent). In this time period, the contribution renewable energies can really make to the energy supply of humankind will also have to be decided.
3. Sometime within the next 100 years the fluid hydrocarbons, oil and gas, will no longer be available for energy supply. Coal, nuclear energy, and renewable energies will have to take over the entire job; the potential for nuclear fusion is unclear.

In conclusion one can say that renewable energies might supply something like the following shares of the future end-energy requirements of humankind: about 10% in 15 years, discounting the noncommercial use of biomass; 20–30% in 50 years; and, perhaps, most needs in 100 years.

DEVELOPMENTS IN HUMANKIND'S ENERGY SUPPLY

From the beginning of humankind's time on Earth up to the 18th century, it was exclusively the renewable energies, wind, biomass, animal, and human muscle power which assured humankind's very existence and development. Then came coal, which initiated the industrialization of Europe. Mineral oil was the precondition at the end of the 19th century for motorized transport on land, sea, and air. In Europe, natural gas and nuclear energy have been around for barely 50 years (Figure 6.1).

In the future, more than ever before, no step on the conversion chain—from extracting primary energy to making secondary energies and useful end energies available—must be taken without strictly enforced efficient energy conversion and use. This, in effect, turns "efficient energy conversion" into the equivalent of an additional "energy" that can replace other energy sources. There is a high potential for this "energy," especially in the industrial countries, since their large primary energy consumption makes energy conservation more readily possible and since this potential requires only technical know-how and capital in order to be tapped.

The next development is likely to be the more intensive use of renewable energies locally (solar irradiance, wind, hydropower, biomass, ambient heat, ocean thermal enthalpy differences, waves, tides, geothermal sources) as well as the import of renewable secondary energy carriers (electricity, hydrogen) from those parts of the world with a large supply of renewable energies. This may be followed later, perhaps, by fusion toward the end of the 21st century.

There never was only one energy in humankind's energy supply system, and new energies never completely replaced the old ones; there is enough demand for them all. The energy heptagon (Plate I) suggests that there are

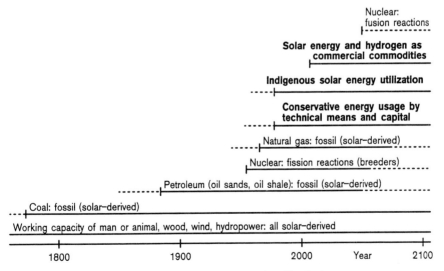

Figure 6.1 History of the world energy economy (qualitative)

not only four classic energies available, (1) coal, (2) oil, (3) natural gas, and (4) nuclear energy, but seven: (5) efficient energy conversion and use, (6) local use of renewable energies, and (7) international trade with secondary energy carriers (electricity, hydrogen) generated from renewable energies.

Plate I shows the energy heptagon for the F.R. Germany in 1973 and 1987, and estimated for the year 2000. One can note:

1. the stagnation in primary energy consumption;
2. the ecologically welcome relative decrease in coal and oil in favor of natural gas (the proportion of carbon per unit of energy in coal:oil:natural gas = 1.8:1.4:1);
3. the development of nuclear energy, now virtually concluded;
4. the immense significance of a more efficient use of energy;
5. a satisfactory although relatively slow increase in local solar energy use;
6. the as-yet marginal role of the import of solar-generated secondary energy carriers.

OPEN- VS. CLOSED-LOOP ENERGY SYSTEMS

The traditional fossil and nuclear systems are open systems (Plate II) with two or more openings: they remove irreplaceable energy raw material from the Earth, change it chemically or isotopically, and release residuals or pollutants elsewhere in the geosphere. These are often toxic, radioactive in the case of nuclear systems, and are associated with an atmospheric oxygen depletion and a CO_2 increase in the case of the fossil fuels. In both cases this occurs with an additive warming equivalent to the energy content of the primary energy put to use. Open systems imperatively require environmental safeguards.

Renewable energies are different: they are closed-loop systems (Plate III) without primary energy raw materials and thus without the associated residuals/pollutants. Solar energy from space is returned to space as thermal energy. Water from the Earth's inventory is split into hydrogen and oxygen and returned when it is later recombined. The kinetic energy of the tides is available as long as the mechanics of Sun, Earth, and Moon continue to function. A constantly renewed geothermal potential is available as long as isotopic decay continues in the Earth's magma, etc. Natural closed-loop systems are in accord with nature and thus do not require environmental safeguards. Natural, closed-loop systems can, however, become environmentally relevant whenever people intervene with technology to make use of them: wherever solar power plants are set up, the albedo of the Earth's surface changes; if hydrogen and oxygen are recombined with air as an oxidizer, nitrogen oxides arise; certain photovoltaic compounds can hardly be recycled today and have to be treated as toxic waste. One thing gives rise to forbearance: being without radioactivity and almost entirely without toxicity, renewable energies are inherently very safe and secure; accompanying hazards and persistent damage are almost impossible.

TECHNOLOGY AND CAPITAL UNLOCK THE RAW-MATERIAL-FREE RENEWABLE ENERGIES

Energy conversion chains (Figure 6.2) have three classic sections: (a) the energy raw material section, (b) the section relating to primary/secondary energy conversion, storage and distribution of secondary energy, and secondary/end-energy conversion, and (c) the residuals/pollutants section. For the raw-material-free renewable energies, sections (a) and (c) do not occur per se. All the relevant technical, financial, sociological, and ecological considerations are limited to section (b). This is where primary/secondary energy conversion occurs in power plants, solar installations, and converters; here is where the secondary energy is stored and transported to the consumer when needed; here is where secondary/end-energy conversion occurs and where there is dissipation of waste heat into space. The cycle is closed (exceptions being geothermal and tidal energies).

Since there are no energy raw materials, any ecological consequences can only arise from the technologies. Only section (b) is relevant when availability of land, material, and energy intensities are evaluated. In the cases of fossil and nuclear systems, sections (a) and (c) have to be evaluated as well.

It is true (Figure 6.3) that the amount of land required for solar power plants is 2–3 orders of magnitude greater than for conventional power plants, although the picture would look quite different if not only power plants (b), but also sections (a) and (c) had to be considered as well. All the same, a thought experiment (Figure 6.4) suggests that it would require less than 1% of the Earth's land surface area to meet the primary energy needs of the whole world with solar energy!

The material requirements of solar power plants (Figure 6.5) are ca. 1–2 orders of magnitude greater than those of conventional power plants, again ignoring the claims of sections (a) and (c). It is to be expected that with continued further development of the renewable energy systems, on the one hand, and the achieving of complete *nuclear containment* and of *complete fossil containment* (CO_2!), on the other, this gap will close further.

Another factor to consider in an energy conversion system is how long it has to operate before it produces an amount of energy equivalent to what was needed for its construction and life-long operation, including dismantling

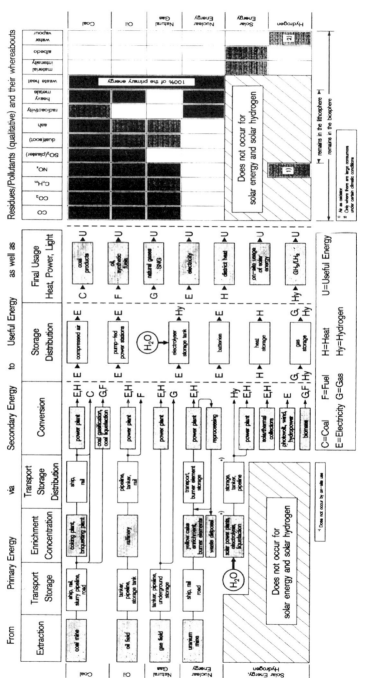

Figure 6.2 Energy conversion chains

Renewable Energy Systems

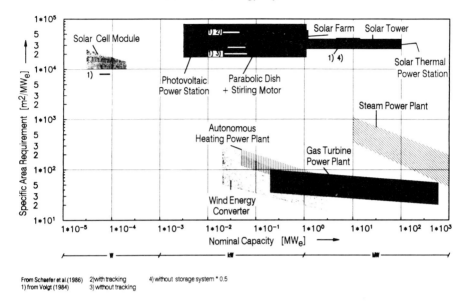

Figure 6.3 Specific area requirement for power plants (source: Schaefer et al. 1986; Voigt 1984)

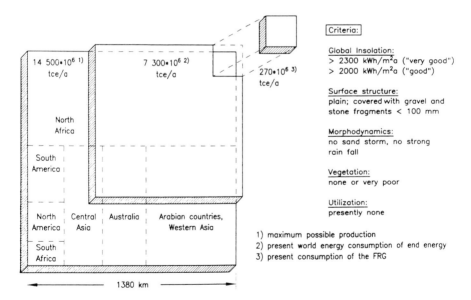

Figure 6.4 Land areas suitable for solar hydrogen production (1.9 mio. km^2 = 5% of global desert area = 1.3% of global land area)

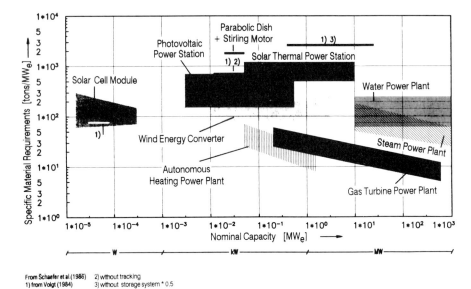

Figure 6.5 Specific material requirements for power plants (from Schaefer et al. 1986; Voigt 1984)

and ultimate disposal. This "energy payback time" has two components, Z_1 and Z_2: Z_1 relates to construction, the original fuel inventory and final dismantling; Z_2 relates to the supply of appropriate fuels and the disposal of residuals and pollutants over the entire service life.

Solar power plants have Z_1 energy payback times that are up to one order of magnitude higher than those of conventional power plants; however, Z_2 payback times are close to zero. If the two are added together for the different energy conversion systems and expressed as a yield factor (How many times more secondary energy is generated by an energy conversion system during its lifetime than was needed for its construction, operation and disposal?), then the following conclusions can be drawn:

1. With yield factors of 15 to 20, the hydropower and wind renewable energy systems beat all the competition (and they will continue to do so!).
2. Coal-fired power plants are out of the running; they may even produce a negative yield if faced with the requirement of CO_2 containment!
3. Today, nuclear and renewable energy systems have similar yield factors, which means that the renewable energy systems will fairly win in the end when their development and mass production potentials have been exploited and when nuclear containment (Pollock-Shea 1986) has finally

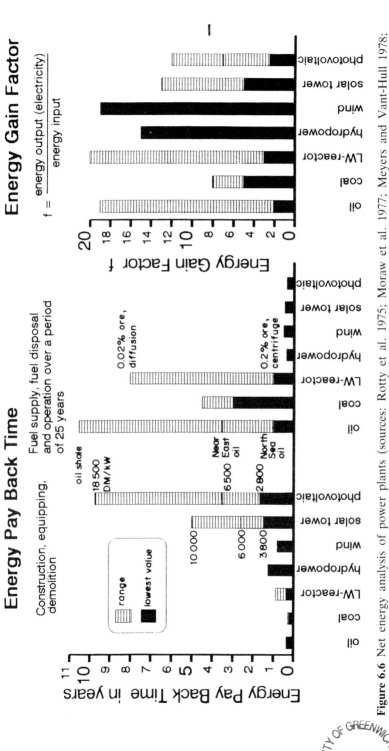

Figure 6.6 Net energy analysis of power plants (sources: Rotty et al. 1975; Moraw et al. 1977; Meyers and Vant-Hull 1978; Wagner 1978, 1986; Enger and Weichel 1979; Sandia Natl. Lab., pers. comm.; Heinloth 1983; Voigt 1984; Aulich 1986; Hagedorn 1989)

been achieved throughout the entire nuclear fuel system. "*Solar containment*" can be ignored with a clear conscience for the renewable energies. However, nowhere on Earth has the imperative need for *nuclear containment* yet been convincingly met!

DOMESTIC AND IMPORTED RENEWABLE ENERGIES

Renewable energies fall into two groups: (a) local (decentral) primary/secondary energy conversion, in which the secondary energy produced can be used where it is generated and does not need to be stored for longer periods of time; (b) central primary/secondary energy conversion at locations of high renewable energy density (highest insolation), with storage and transport of the secondary energy carrier, usually in a chemical form, to the main user areas (of lower insolation).

Local (Decentral) Primary/Secondary Energy Conversion

There is no country on Earth without at least one form, most have several forms, of renewable energy available for local use (irradiance, biomass, wind, hydropower, ambient heat, ocean thermal gradients, geothermal, wave and tidal energies, etc.). For example, under the climatic conditions of central Europe, an area not especially blessed with renewable energies, even a country such as the F.R. Germany has a potential of irradiance, biomass, wind, hydropower and ambient heat which in several decades could meet a quarter of the country's primary energy requirements. At other locations, by contrast, potential renewable energy resources are geographically much more concentrated:

— The really large hydropower reserves are found only in Greenland, Canada, Latin America, Africa and Asia.
— Areas with high wind velocities are mostly located near coasts and on mountain tops.
— Larger enthalpy gradients in the ocean are only found in the Earth's equatorial belt.
— Wave and tidal energies can of course only be used at the coasts, geothermal energy only in the Earth's geologically unsettled zones.

Central Primary/Secondary Energy Conversion

If the amount of locally usable renewable energy is insufficient, there is the opportunity of exporting secondary energies generated from renewable

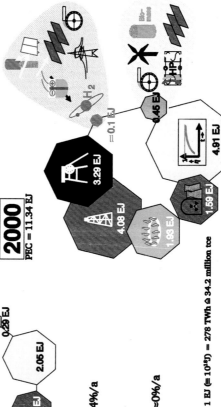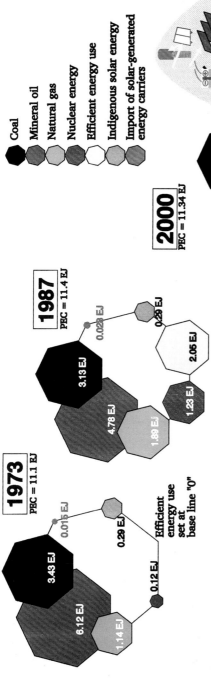

Plate I F.R. Germany: the energy heptagon (sources: AG Energiebilanzen 1973, 1987; ESSO 2000; Statistisches Bundesamt)

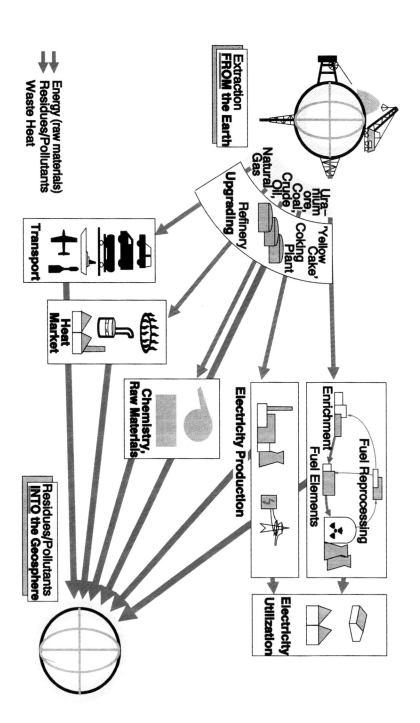

Plate II Coal, mineral oil, natural gas, nuclear energy: open-ended energy systems (exhaustible, environmentally harmful, hazardous)

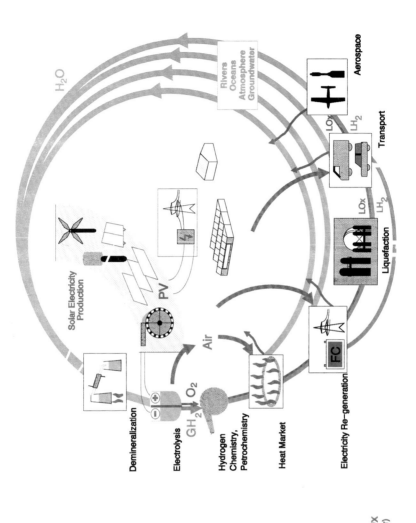

Plate III Solar hydrogen: a closed-loop energy carrier (inexhaustible, environmentally harmless, safe [proper handling presumed])

energies to those areas of the world where consumption is high. This can be electricity if the distances are not too great for high-voltage transmission lines (high-temperature superconductivity is not yet an option). Otherwise a chemical energy carrier is needed, such as hydrogen, wherever large, economically relevant amounts of energy have to be stored and transported, and wherever a chemical end-energy carrier is needed, for example as a transportation fuel. At any rate, only by means of a chemical energy carrier can—for the time being, at least—the transportation sector have easy access to renewable energies. Solar hydrogen makes renewable energies compatible with the traditional world energy trade system, perpetuating it beyond that point in time when fossil and nuclear energies will no longer be available in unlimited amounts (Bockris 1980; Ogden and Williams 1988; Peschka 1984; Winter and Nitsch 1988).

AN END-ENERGY ECONOMY

We should not expect too much from "technologies"!

The paradigm of a demand-oriented end-energy economy (Figure 6.7), one which depends on efficient energy conversion/use and is congenial to renewable energies, must supplement and finally supplant the conventional, now dominant supply-oriented primary energy economy (Figure 6.8) (Winter and Nitsch 1988).

Energetically, nobody really requires coal, oil, natural gas, or uranium. The entire energy economy exists for the sole purpose of meeting the demand for the four useful energies: heat, light, power, and communication services. On their side, they must guarantee the energy services: heating and cooling households, lighting places of work, providing mechanical support to the manufacturing and transportation sectors, enabling worldwide immaterial communication, etc.

Any demand for an end energy should be met by renewable energies first of all locally, then regionally, and only thereafter globally. Regionally and globally, renewable energies should only be supplemented by conventional energies if the amount of renewable energy is insufficient to solve the problem of how to meet a demand for end energy. Whenever end-energy conversion/use takes place, it must be as efficient as possible. What is the sense of laboriously collecting solar energy, only to have it subsequently dissipated to the environment through badly insulated walls? How sensible is it to generate solar hydrogen with great effort, only to use it thereafter in motors which are only 10–15% efficient?

In an end-energy economy, the energetic responsibility of every single energy consumer, community, and region is strengthened without freeing the traditional large interregional energy suppliers of their basic responsibility.

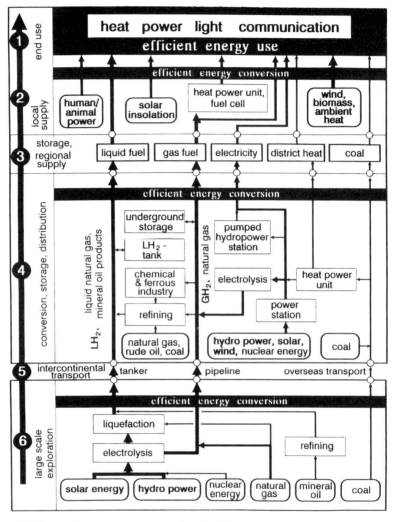

Figure 6.7 The end-use energy system for the 21st century

Solar energy is ubiquitous. Solar energy supply densities are similar to the energy consumption densities of rural areas. In such places, local solar energy utilization is a natural option.

Having an end-energy economy does not mean that the inexpert energy consumer does without the professional services of the energy supplier. However, an end-energy economy properly understood, including strictly efficient energy conversion/use and the utilization of renewable energies, does indeed mean that the classic *energy* suppliers must become energy

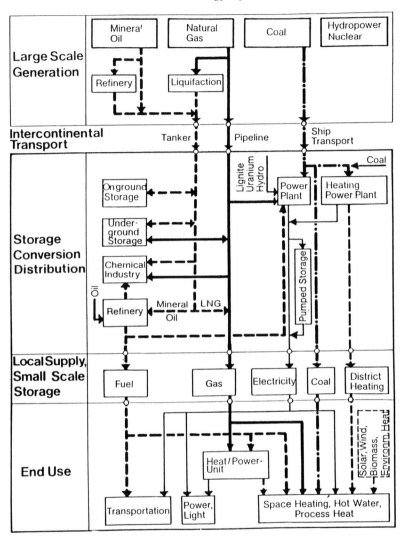

Figure 6.8 The hydrocarbon economy

service suppliers. It is irrelevant for people's comfort whether the temperature of a residence is adjusted by turning up the central heating or by insulating the walls, whether a household is supplied with energy from a photovoltaic system on the roof or by conventional means. What is important is to replace fossil and nuclear energies, which are in finite supply and inherently in conflict with the environment, with efficient energy use and renewable energies. The point is *not* to have to use energy raw materials, but technology and capital instead.

RENEWABLE ENERGIES AND ECONOMICS

Renewable energies will become economic when:

1. their development potential has been realized,
2. economies of scale and mass production of the components needed by the renewable energy technologies are in evidence,
3. the costs to the total economy of each energy is completely calculated and fully debited to that energy, and
4. the economic criterium of the gross national product (GNP), in use since the 1940s and indicative of quantitative welfare, is supplemented and finally supplanted by the criterium of an "ecosocial product" (ESP), which indicates qualitative welfare and is calculated by ecological-economic accounting.

As mentioned in the INTRODUCTION, modern research and development in renewable energies is 15 years old. At this point, 2,000 MW_e from wind energy converters and 400 MW_e from solar thermal power plants are being supplied to grids. Global production of photovoltaic generators is now 40–50 MW_e with annual increases expected $\geq 10\%$. In the F.R. Germany, 50,000 heat pumps have been installed to date (1990). Much has been achieved; however, much remains to be done: Figure 6.9 (Winter et al. 1991; Siemens AG 1988; Strese and Schindler 1988) shows the cost trends for photovoltaic generators, Figure 6.10 (Winter et al. 1991) for solar farm power plants.

To date, production has been in quantities much too small to benefit from the favorable unit costs of mass production. For example (J. Nitsch, pers. comm.), the automobile industry in the F.R. Germany alone produces 4×10^6 automobiles per year, each with a 50 kW "decentralized energy converter" on board. This 200 GW each year is twice as much as the installed power plant capacity (100 GW_e) of the country, and at costs of 400–600 DM/kW, which are up to one order of magnitude below the installation costs of modern power plants (admittedly with service lives of 10,000 h, which are one to two orders of magnitude lower). Such decentralized energy converters are typical for decentral renewable energy systems: most photovoltaic generators, paraboloid/Stirling units, heat pump systems, biomass or wind converters have unit outputs ≤ 100 kW. Figure 6.11 (Winter et al. 1991) provides the data for 100–200 MW_e solar tower power plants. The most important conclusion is that only 30–40% of the

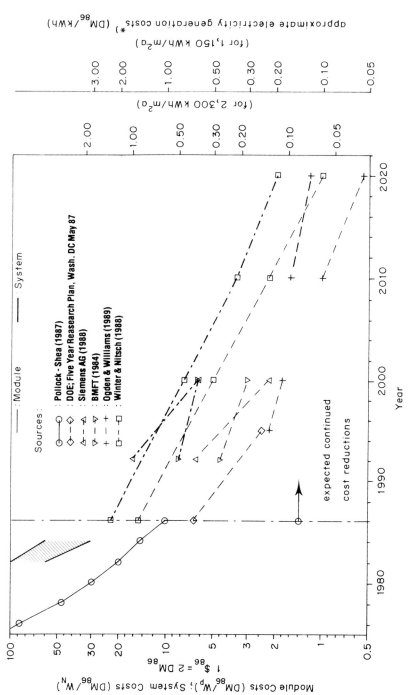

Figure 6.9 Cost trends for photovoltaic modules and systems (7% interest rate, depreciation over 30 years, operating costs 0.05–0.01 DM/kWh)

Characteristics of SEGS I-VIII

Plant	First Full Operating Year	Status	Turbine Capacity (MWe net)	SolarField Temp (°F)	SolarField Size (m²)	Turbine Cycle Efficiency Solar	Turbine Cycle Efficiency Boiler	Annual Output (MWh net)
I	1985	Operational	13.8	585	82,960	31.5*	---	30,100
II	1986	Operational	30	600	188,987	29.4	37.3	80,500
III	1987	Operational	30	660	230,300	30.6	37.4	92,780
IV	1987	Operational	30	660	230,300	30.6	37.4	92,780
V	1988	Operational	30	660	233,120	30.6	37.4	91,820
VI	1989	Construction	30	735	188,000	37.5	39.5	90,850
VII	1989	Construction	30	735	194,280	37.5	39.5	92,646
VIII	1990	Design	80	735	464,340	37.5	37.6	253,380

*Includes Natural Gas Superheating

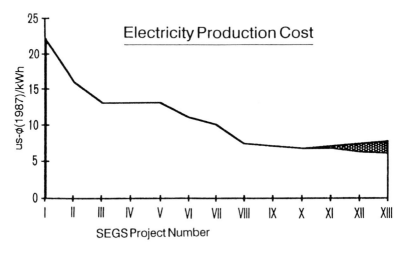

Figure 6.10 LUZ-solar electricity generating systems (SEGS) (source: LUZ 1988)

installation costs are for the "solar specific" components of the heliostat field and solar absorbers. So 60–70% of the power plant costs are related to the normal market activity of the current power plant industry, which means that solar plants are no longer "exotic curiosities."

The history of the energy economy is full of examples for costs which are not internalized in the price of the product and thus have to be paid by the economy at large. These include ecological and social costs. For example, the ecological costs of nonexistent desulfurizing or denoxing in coal-fired power plants were only incorporated into the kilowatt-hour price to the consumer when mandated by the relevant legislation. It is said that the costs of desulfurization/denoxing equipment in a modern 700 MW_e coal-fired plant represent up to one third of the total investment cost for the plant (J. Nitsch, pers. comm.).

If it is true that the noninternalized external costs of renewable energy systems are far below those of fossil and nuclear systems (since they do not require energy raw materials or containment technologies and are inherently safe, being nonradioactive and nontoxic), then fair economic comparisons are only possible on the basis of completely internalized external costs. Table 6.1 (Friedrich et al. 1989; Hohmeyer 1988) gives external costs as calculated in two recent studies in Germany: in the first case, the external costs of *all* energy conversion systems are almost zero; in the second, the electricity costs of conventional power plants would have to be doubled if the external costs were internalized. Both studies ignore CO_2, which otherwise would penalize especially the fossil-fueled installations (Figure 6.12) (San Martin 1989). The first study ignores resource depletion, macroeconomics and subsidies, the second preparatory and follow-up production. Figure 6.13 (Nitsch and Luther 1990) uses the example of the F.R. Germany to show how much CO_2 can be avoided per year for the investment of 1 DM.

The GNP does not make a distinction between intended economic growth and the unintended growth of ecological, social, and health hazards, which is relatively greater. If renewable energies contribute almost in their entirety to the growth which is actually intended, then they will be favored when the results are published of studies by the statistical bureaus in western industrialized countries which make an ecological-economic calculation (ESP), in addition to calculating the GNP (e.g., Hölder 1989). The difference between the two indicates the high cost of nonintended damages, which would drop to zero or at least be far lower if renewable energies, with their higher environmental compatibility and social acceptability, were used more widely (Binswanger et al. 1988; Fisher 1930; Flavin 1989; Kapp 1958; Leipert 1989; Bundesamt für Konjunkturfragen 1985; Reich and Stahmer 1983; Ruckleshaus 1989; Wicke 1986, 1989).

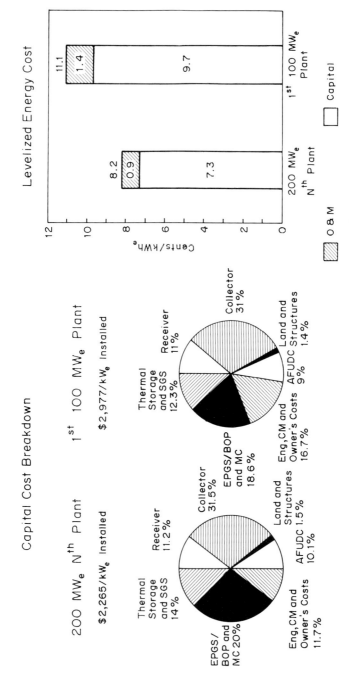

Figure 6.11 Solar central receiver technology, 1987 state-of-the-art (source: PG&E 1987)

Table 6.1 External costs of selected electricity generating methods (DM_{1988}/kWh)

Category	Coal	Nuclear Energy	Wind Converters	Photovoltaic
According to Friedrich et al. 1989:				
Health	0.0018	0.0002–0.0006	0.0002–0.0005	0.0006–0.0009
Possible accidents	—	0.0001–0.0007	—	—
Forest	0.0019	—	—	—
Noise	—	—	0.0001	—
Fauna	0.00002	—	—	—
Flora	0.0002	—	—	—
Material	0.0005–0.0008	—	—	—
R&D	0.0006	0.0045	0.0017–0.0034	0.0050–0.0125
Subsidy	0.0058	0.0003	—	—
Total	0.0108–0.0112	0.0050–0.0061	0.0020–0.0040	0.0056–0.0134
According to Hohmeyer 1988:				
Without macroeconomic side effects	0.0294–0.0854	0.0971–0.2448	0.0027–0.0053	0.0096–0.0148
Net benefit including external effects			0.0026–0.0041	0.0144–0.0521

THE FUTURE ROLE OF RENEWABLE ENERGIES

When adapting any new energy to humankind's energy supply system, we must recognize that:

1. 15 years from now is already "tomorrow" and thus the relevant decisions have already been made,
2. 50 years is the time span needed for the adaption to become self-sustaining,
3. for the time 100 years from now, only general trends can be indicated and "if . . . then" statements made.

15 Years

What will happen in 2005 has already been decided as far as energy supply is concerned. The time constants for energy investments are decades, even half-centuries; the turnover time for buildings is 50–100 years. Even such impulses as those coming from war in the Middle East, the oil price crises

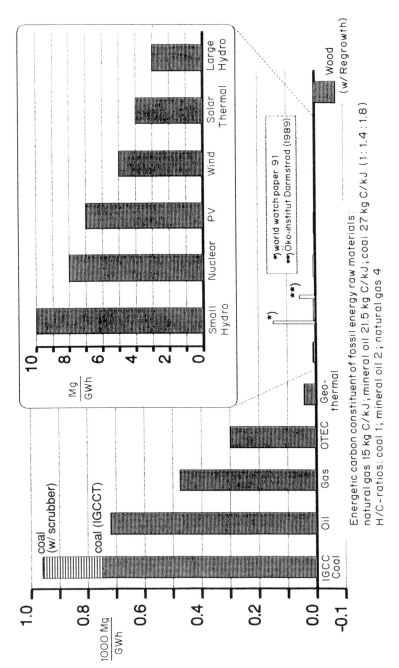

Figure 6.12 CO_2-emissions in electricity generation (source: San Martin 1989)

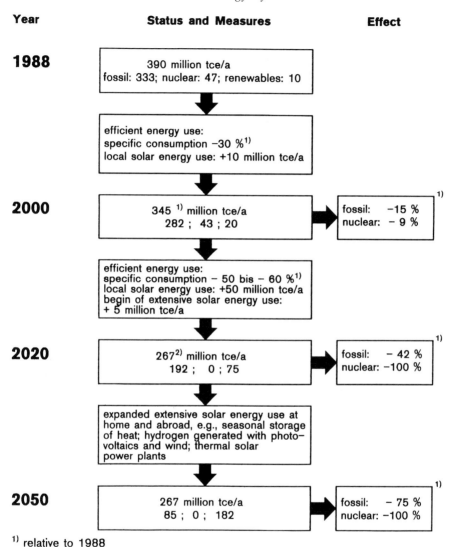

Figure 6.13 F.R. Germany: reducing demands for fossil and nuclear energy (from Nitsch and Luther 1990)

of the 1970s, or reactor disasters like Chernobyl have dissipated in one to two decades. The task of the next 15 years is therefore to increase and share knowledge about renewable energies, to provide the legal and regulatory preconditions for their utilization, and steadfastly to continue research, development, and demonstration in order to improve efficiency and service life, reduce costs, and initiate the needed market dynamics.

Above all, the commercial-scale demonstration of mature technologies has yet to take place. Since, at present, renewable energies have to be introduced to a market which is unfavorable, the State is challenged to provide the necessary framework and to subsidize demonstration facilities which are not yet economically viable.

The goals are:

1. to adapt building codes to solar city planning and solar architecture,
2. to adopt operating and licensing regulations for renewable energy installations,
3. to conclude international treaties and agreements which favor equal treatment of renewable energies and conventional energies under international law.

Specifically, among others, they must include measures:

—to reduce the cost of a collector-thermal kilowatt-hour by a factor of two with the help of marketing incentives,
—to reduce the cost of a photovoltaic kilowatt-hour by one order of magnitude (F.R. Germany "1,000 Rooftops Program" [see Bundesforschungsbericht 1989]),
—to reduce the cost of a wind-generated kilowatt-hour by a factor of two to three (F.R. Germany "100–200 MW_e Program" [see Bundesforschungsbericht 1989]),
—to agree on and subsequently to construct solar thermal storage facilities in the low-temperature area $\leq 100°C$ for solar district heating,
—to construct second generation solar tower demonstration power plants (100–200 kW_e),
—to design and construct daily and seasonal thermal storage systems for base load solar farm and solar tower power plants in the 400–800°C temperature range,
—to develop parabolic troughs, heliostats, and parabolic dishes which can be mass produced for prices $\leq$ 30 US$/m^2, $\leq$ 50 US$/m^2, and $\leq$ 100 US$/m^2, respectively,
—to design and construct second generation absorbers for solar tower power plants (volumetric air receiver, direct absorption particle receiver),
—to design and construct second generation parabolic solar power plants,

— to design and construct solar-dedicated Stirling motors and gas turbines (≤ 100 kW$_e$) which can be mass-produced for parabolic power plants,
— to design and construct sorption and compressor heat pump installations with annual mean coefficients of performance ≥ 1.5 for output levels up to several tens of MW,
— to construct the next generation of geothermal power plants and tidal power plants,
— to intensify research, development and demonstration in ocean thermal energy conversion (OTEC), ocean wave power plants, solar fuels and chemicals, photolysis and photocatalysis, and in photobiological processes,
— to design and construct biomass converters which can be mass-produced for generating low-temperature/medium-temperature heat, biogas, bioalcohol,
— to develop further and construct second generation solar-dedicated electrolyzers in the 100 kW$_e$ to MW$_e$ range,
— to intensify research, development and demonstration of fuel cells, including light, medium, and heavy weight versions; for low-temperature/medium-temperature/high-temperature applications; based on alkaline, phosphoric acid, molten carbonate, and oxide ceramic principles,
— to develop further and construct catalytic hydrogen/air diffusion burners,
— to continue research, development and demonstration of H_2/O_2 peak load power plants,
— to design and construct magnetocaloric hydrogen liquefaction equipment.

50 Years

The next 50 years will reveal that the fluid hydrocarbons oil and gas are no longer available in unlimited quantities for *energetic* applications. Coal, nuclear energy, and renewable energies will have to assume more of the energy supply burden; nuclear fusion will not yet be available.

Clear evidence of global, anthropogenic, energetically caused ecological damage will intensify the demand for increased use of renewable energies. Table 6.2 (Nitsch and Luther 1990) uses the example of the F.R. Germany to show a share of renewable energies just under 30% for 2020, and just under 70% for 2050. In both cases the share of nuclear energy has dropped to zero.

The industrial preconditions for the manufacture of renewable energy technologies will exist; the typical financial requirements for energetic investments of 1–2% GNP/yr will not be exceeded. Because they do not have ecological costs, their financing will even be facilitated ("dual mode" technologies).

Table 6.2 How much CO_2 reduction for an investment of 1 DM? (From Enquête Kommission [1988; 1990], Baden-Württemberg [1987], BMFT [1984], and Nitsch and Luther [1990]. F.R. Germany: Existing power plants 0.623 kg CO_2/kWh_e; existing structures for heat supply 0.298 kg CO_2/kWh_{th}

	Specific fuel/electricity savings $\dfrac{kWh/a}{DM}$		Specific CO_2 reduction $\dfrac{kg\,CO_2/a}{DM}$	
	today	potential	today	potential
Fuel savings				
Insulation	1.77		0.53	
Solar collector	0.35	0.58	0.104	0.172
Oil	0.31		0.09	
Gas	0.98		0.29	
Biogas	1.37	2.28	1.41/0.68	
Electricity savings				
WEC (250–500 kW)	0.92	1.28	0.57	0.80
PV Central Europe	0.05	0.38	0.03	0.24
Southern Europe	0.09	0.67	0.06	0.42
Solar farm/tower plants	0.25	0.61	0.16	0.38
Hydropower plant (10 MW) new	1.00		0.62	
modernized old	2.00		1.24	
LWR	1.65		1.03	

100 Years

Who is able to see 100 years into the future? (Who would have imagined in 1890 what the world would be like in 1990? Bismarck had just resigned as Chancellor; two World Wars and almost half a century of Cold War were ahead; mankind developed the ability to destroy itself, and perhaps also insights into how to prevent that destruction; the United Nations was established and the idea of a "common future" and the need for "sustainable development" [WCED 1987; Enquête-Kommission 1988, 1989; Pestel 1988] arose; now the awareness is growing that people are capable of destroying their ecological basis for life all over the world.)

Nevertheless, a *subjective*, admittedly optimistic, attempt is made below to describe the role of renewable energies and their future context:

1. The atomic or genetic self-destruction of humanity will not have taken place.
2. The world's population will level off at some point between 10 and 15 $\times\,10^9$.

3. The North/South disparity will be less, although still present.
4. A closed-loop (energy) economy will have basically replaced the now dominant open (energy) economy.
5. Criteria of qualitative growth and of ecological-economic accounting (which will lead to an ESP) will have replaced those of quantitative growth and a GNP.
6. The manufacturing sector will continue to decline with respect to the service sector, and within the manufacturing sector the trend will continue toward products with lower energy intensity and lower specific weight (ceramics instead of steel, biotechnology instead of bulk chemistry, immaterial communication instead of transportation, etc.).
7. Life paradigms will have become post-materialistic. What will count will be health and sports, charitable and religious activities, artistic creativity, science and education. The discrepancy in the value attributed to paid and unpaid labor will have further shifted to the benefit of unpaid labor.
8. The fluid hydrocarbons will not be available any longer for energy supply, and coal only in oligopolistically limited amounts. Nuclear energy will no longer be used; fusion energy will not have been achieved.
9. The renewable energies will meet most of mankind's specific end-energy requirements, which will be reduced because of rigorously enforced energy conservation and efficiency (3 kW/cap ESP). The local and regional potentials for renewable energy will have been exploited; the equatorial belt $\pm$ 40° N/S will have become the world's energy supplier. For ecological and safety reasons and because of a lack of public acceptance, solar power satellites will not be realized.

Summary

Of the future end-energy requirements of humankind, renewable energies will supply 10% over the next 15 years, 20–30% in 50 years, and possibly most needs in 100 years.

WORLD ENERGY POLICY

There is so far no firm international political commitment to the use of renewable energies. The following decisions could be steps on the way to that goal, and ultimately to a responsible world energy policy:

1. Solar energy utilization and solar power systems require a global commitment and identification with the United Nations. It is recommended that a United Nations subunit—UNRE (United Nations

for Renewable Energies) and/or ISEA (International Solar Energy Authority)—be established following the pattern of IAEA (International Atomic Energy Agency).
2. Forty years after the Treaties of Rome were signed, the European Community will create a common economic unit with open borders in 1992; a common energy economy should be one of the components. The *Montanunion* was agreed for coal and steel as part of the original Treaties of Rome. In 1957, a treaty was concluded on the peaceful utilization of nuclear energy, *Euratom*. Next, an agreement should be made to found a European solar community, *SolarEurop*, for the purpose of joint research, development, demonstration and utilization of solar energy in the European Community.
3. The World Energy Council should set up a program committee for renewable energies, parallel to its committees on traditional energies in order to demonstrate its commitment to identify with this new form of future energy free of energy raw materials.

UNSOLVED PROBLEMS, OPEN QUESTIONS

After 15 years of research and development, renewable energies are still far from being in a position to play an established commercial role in an energy supply system. There is a large number of unsolved problems and open questions reflecting both technological and nontechnological realities which act as a hindrance and are in urgent need of attention. The following list is only a selection.

1. No mass production of solar technologies has so far taken place, which keeps their price high and public awareness low.
2. No firm national/international political commitment has been made to solar energy utilization (with some exceptions), which hinders the establishment of broad-based pilot and demonstration projects.
3. No solar-appropriate framework of laws, regulations, taxes, etc. has been drawn up. For example, in the F.R. Germany, "clean" energy derived from wind converters is fed into grids whose electricity is sold at prices which includes a subsidy for the local coal industry, which is otherwise not competitive, i.e., "wind subsidizes coal."
4. Also, no answer has yet been offered to the question, "How do you tax a non-fuel energy?"
5. No significant commitment to solar energy has been made by the UN or the European Community.

6. No adequate experience could be gathered to date with demonstration facilities of the appropriate size (with some exceptions).
7. No material accounting, surface area accounting, or energy accounting is being carried out for energy conversion chains, thus precluding real comparisons.
8. No policy is in evidence concerning the need to replace cut trees to an extent which assures sustainable forest growth.
9. No significant integration of equipment that utilizes renewable energies is evident in urban areas, e.g., in rail and road transportation systems, factories, private homes, etc.
10. Little cooperation between environmentalists and (solar) energy experts is taking place to interpret successfully to each other their respective goals and methods.
11. No agenda of global politics really contains issues related to the role of renewable energies.
12. There is inadequate information and awareness of the problems associated with energy supply which can be addressed by the renewable energies. There is a deficit of educational and training options.

REFERENCES

Aulich, H. 1986. Energierücklaufzeit—ein Kriterium für die Wirtschaftlichkeit der Photovoltaik. *Zeitschrift Sonnenenergie* **6**:14–17.

Baden-Württemberg Wirtschaftsministerium. 1987. Energiegutachten: Perspektiven der Energieversorgung. Gutachten im Auftrag der Landesregierung von Baden-Württemberg. Stuttgart.

Binswanger, H.C., et al. 1988. Arbeit ohne Umweltzerstörung. Strategien für eine neue Wirtschaftspolitik. Frankfurt: S. Fischer.

BMFT (Bundesministerium für Forschung und Technologie). 1984. Abschätzung des Potentials erneuerbarer Energiequellen in der BRD. Bonn: BMFT.

BMFT. 1989. Bundesforschungsbericht 1989. Bonn: BMFT.

Bockris, J.O'M. 1980. Energy Options—Real Economics and the Solar-Hydrogen System. ISBN 0 85552 0884. Sydney: Australia & New Zealand Book Co.

Bundesamt für Konjunkturfragen. 1985. Qualitatives Wachstum, Studie Nr. 9. Bericht der Expertenkommission des Eidg. Volkswirtschaftsdept.

Enger, R.C., and H. Weichel. 1979. Solar energy generating system resources requirements. *Solar Energy* **23**:255–261.

Enquête-Kommission. 1988. Vorsorge zum Schutz der Erdatmosphäre. Study Commission of the 11th German Bundestag. Bonn: Economica Verlag, Verlag C.F. Müller.

Enquête-Kommission. 1990. Technikfolgen-Abschätzung und -Bewertung. Study Commission of the 11th German Bundestag, Bundestagsdrucksache 11/7993. Bonn: Economica Verlag, Verlag C.F. Müller.

Fisher, I. 1930. The Nature of Capital and Income. New York: Macmillan.

Flavin, C. 1989. Slowing global warming: a worldwide strategy. Worldwatch Paper 91. Washington, D.C.: Worldwatch Inst.

Friedrich, R., U. Kallenbach, E. Thöne, A. Voβ, H. Rogner, and H. Karl. 1989. Externe Kosten der Stromerzeugung. Frankfurt: Verlags- und Wirtschaftsgesellschaft der Elektrizitätswerke mbH (VWEW).
Hagedorn, G. 1989. Kumulierter Energieverbrauch und Erntefaktoren von Photovoltaik-System. *Energiewirtschaftliche Tagesfragen Heft* **11**:712–718.
Heinloth, K. 1983. Energie. Reihe Studienbücher Physik, ISBN 3-519-03-057-8, 447 pp. Stuttgart: Teubner.
Hohmeyer, O. 1988. Social Costs of Energy Consumption (also available in German: Soziale Kosten des Energieverbrauchs). Berlin: Springer.
Hölder, E. 1989. Ziel und Aufbau einer Umweltökonomischen Gesamtrechnung. Bonn: German Federal Statistics Office.
Kapp, K.W. 1958. Volkswirtschaftliche Kosten der Privatwirtschaft. Tübingen: J.C.B. Mohr.
Leipert, C. 1989. Die heimlichen Kosten des Fortschritts—Wie Umweltzerstörung das Wirtschaftswachstum fördert. Frankfurt: Fischer.
Meyers, A.C., and J.L. Vant-Hull. 1978. The new energy analysis of the 100 MW_e commercial solar tower, vol. 2.1, pp. 786–792. Proc. of 1978 Denver Meeting of the Am. Sec. of the Intl. Solar Soc.
Moraw, G., et al. 1977. Energy investments in nuclear and solar power plants. *Nuclear Tech.* **33**:174–183.
Nitsch, J., and J. Luther. 1990. Energieversorgung der Zukunft—Rationelle Verwendung und Erneuerbare Quellen, 93 pp. Berlin: Springer.
Ogden, J.M., and R.H. Williams. 1988. Solar Hydrogen: Moving Beyond Fossil Fuels. World Resources Institute, Washington, D.C., New York: Pergamon.
Peschka, W. 1984. Flüssiger Wasserstoff als Energieträger. Berlin: Springer.
Pestel, E. 1988. Jenseits der Grenzen des Wachstums, ISBN 3 421 06393 1. Bericht an den Club of Rome. Stuttgart: DVA.
PG&E (Pacific Gas and Electric Co.). 1987. Solar Central Receiver Technology Advancement for Electric Utility Applications, Phase 1, Final Review. San Ramon, CA: PG&E.
Pollock-Shea, C. 1986. Decommissioning: nuclear power's missing link. Worldwatch Paper 69. Washington, D.C.: Worldwatch Inst.
Pollock-Shea, C. 1987. Renewable energy: today's contribution, tomorrow's promise. Worldwatch Paper 81. Washington, D.C.: Worldwatch Inst.
Reich, U.-P., and C. Stahmer. 1983. Gesamtwirtschaftliche Wohlfahrtsmessung und Umweltqualität, Campus Forschung Bd. 333, ISBN 3 593 33145 4. Frankfurt/New York: Campus Verlag.
Rotty, R.M., et al. 1975. Net Energy from Nuclear Power. Report No. IEA 75–3, Inst. for Energy Analysis. Oak Ridge, TN: Oak Ridge Assoc.
Ruckleshaus, W.D. 1989. Towards a sustainable world. *Sci. Am.* **260**: 114–120.
San Martin, R.L. 1989. Environmental emissions from energy technology systems: the total fuel cycle. Washington, D.C.: U.S. Dept. of Energy.
Schaefer, H., et al. 1986. Dezentrale Energieversorgung: Aspekte und Chancen. *Energiewirtschaftliche Tagesfragen* **8**: 610–619.
Siemens AG. 1988. Photovoltaik—eine Stellungnahme zu den Kosten für die großtechnische Fertigung kristalliner PV-Module und die Stromerzeugung aus PV-Kraftwerken. Bericht einer internen Arbeitsgruppe der Siemens-Kraftwerkunion (March 1988). Erlangen, F.R. Germany.
Strese, D., and J. Schindler. 1988. Kostendegression Photovoltaik, Forschungsbericht der Ludwig-Bölkow Systemtechnik GmbH, erstellt im Auftrag des Bundesministers für Forschung und Technologie (May 1988). Ottobrunn, F.R. Germany.

Voigt, C. 1984. Material and energy requirements of solar hydrogen plants. *Int. J. Hydrogen Energy* **9**:491–500.
Wagner, H.J. 1978. Der Energieaufwand zum Bau und Betrieb ausgewählter Energieversorgungstechnologien. Bericht JÜL-1516. Jülich: KFA.
Wagner, H.J. 1986. Der Energiebedarf zum Bau und Betrieb von Energieversorgungsanlagen. Manuskript 7, Hochschultage Energie. Universität Essen.
WCED (World Commission on Environment and Development). 1987. Our Common Future (The Brundtland Report). Oxford: Oxford Univ. Press.
Wicke, L. 1986. Die ökologischen Milliarden—Das Kostet die zerstörte Umwelt—So können wir sie retten. Munich: Kösel.
Wicke, L. 1989. Umweltökonomie: Eine Praxisorientierte Einführung. Munich: Franz Vahlen.
Winter, C.-J., and J. Nitsch. 1988. Hydrogen as an Energy Carrier, Technologies, Systems, Economy. Berlin: Springer.
Winter, C.-J., L. Vant Hull, and R. Sizmann. 1991. Solar Power Plants. Berlin: Springer.

7
The Potential Role of Nuclear Power in Controlling CO_2 Emissions

W. FULKERSON, J.E. JONES, J.G. DELENE,
A.M. PERRY, and R.A. CANTOR[1]
Oak Ridge National Laboratory, Oak Ridge, TN 37831, U.S.A.

ABSTRACT

Nuclear power currently reduces CO_2 emissions from fossil-fuel burning worldwide by about 8% (0.4 GtC y^{-1}). It can continue to play an important role only if it can grow substantially in the next 50 years. For such growth to occur, public confidence will need to improve throughout the world. That might happen if (a) other nonfossil alternatives are inadequate to meet electricity demand growth, (b) the risks to society from global warming are perceived to be very high, (c) nuclear technology improves substantially, and (d) an international institutional setting is devised to manage the nuclear enterprise so that the technology is available to all nations while catastrophic accidents and proliferation of nuclear weapon capabilities are avoided. It seems feasible that the necessary technological and institutional advances can be devised and tested over the next 20 years. It is also plausible that the direct costs of electricity produced by the system would be in the range of 50–100 mills/kWhr (1990 U.S. dollars) delivered to the grid. In other words, the direct costs of nuclear power should not be greater than they are today. Achieving such an outcome will require aggressive technical and institutional research, development, and demonstration performed in a cooperative international setting. If rapid growth of nuclear power can begin again in 15–20 years it could supply 30–50% of world electricity in 50 years and cut CO_2 emission rates by up to 2.5 GtC y^{-1}. This would be a substantial contribution to controlling greenhouse gases, but it is not sufficient. Improved efficiency and various renewable energy sources must also grow rapidly

---

[1] The submitted manuscript has been authored by a contractor of the U.S. Government under contract No. DE-ACO5-840R21400. Accordingly, the U.S. Government retains a nonexclusive, royalty-free license to publish or reproduce the published form of this contribution, or allow others to do so, for U.S. Government purposes.

Limiting the Greenhouse Effect: Options for Controlling Atmospheric CO_2 Accumulation
Edited by G.I. Pearman published 1992 John Wiley & Sons Ltd

if CO_2 emission rates from electricity generation are to be reduced from the current value of about 2 GtC y^{-1}.

INTRODUCTION

Controlling CO_2 emissions,[2] when it becomes necessary to do so, will be a difficult, protracted, and expensive process. Changing world energy systems from predominant dependence on fossil fuels[3] to nonfossil sources will take decades to accomplish, even if it is pursued vigorously. This would be true because of the slowness and cost of capital stock turnover, even if nonfossil substitutes were competitive. They are not. Each alternative is limited by at least one of five factors:

1. resource base such as for hydroelectricity and biomass,
2. geography such as for geothermal and wind,
3. cost such as for direct solar thermal, photovoltaics, ocean thermal energy conversion, and wave power,
4. technical feasibility such as for fusion, and
5. public acceptance of larger-scale deployment such as for nuclear fission.

It is fair to conclude, therefore, that we are currently not very well prepared with technology to cope with curtailing the use of fossil fuels (Fulkerson et al. 1989a, 1989b).

There is no single fix for controlling CO_2 emissions. Four major strategies can be important: efficiency improvement; changes in the fossil fuel; recovering and sequestering or recycling CO_2; and substituting nonfossil energy sources. In our opinion, all these strategies should be pursued, particularly to the assessment of the potential and relative costs and effects.

Nevertheless, in the near-(10–15 y) to mid-term (10–30 y), the most effective strategy is to improve the efficiency of fossil-fuel conversion and end use. Often this can be done at less cost than the investment in new supply. Furthermore, efficiency is always an attractive strategy. When it can

---

[2] Our concern here is control of CO_2, which is the principal greenhouse gas impacted by the world energy system. Already all Organization of Economic Cooperation and Development (OECD) nations (except the U.S. and Turkey) have pledged some degree of greenhouse-gas control.

[3] 89% of world commercial energy is supplied by fossil fuels. Fossil fuels supply probably about 70–75% if noncommercial biomass sources are included. Although these noncommercial, nonfossil fuels produce CO_2 when they are burned, it is possible to manage biomass sources so that their net emissions are zero (or even negative) if as much (or more) carbon is removed by growing plants as by burning them. Most estimates indicate that the world biomass reservoirs are being burned faster than they are growing, so that biomass is a net source not a sink. However, the worldwide imbalance is not known.

be practiced economically, it brings extra benefits of reducing environmental impacts from increasing supply which are generally greater than those of the technologies for more efficient use. Efficiency improvement also lessens the stress on world oil markets and supply, and it can improve the competitiveness of nations which practice it. Not only is efficiency improvement already important, but the potential for continuing improvement through research and development also appears large (Fulkerson et al. 1989a, 1989b; Carlsmith et al. 1990).

Another strategy is the substitution of natural gas for coal, which may be a useful approach in some areas of the world, particularly Europe. In the era of Peristroika, we can imagine Europe being able to turn off coal plants by using Soviet gas. The USSR is, after all, the Saudi Arabia of natural gas, with resources estimated to be equal to about 60 years supply for the USSR and Europe (east and west) at present use rate plus substituting for all current coal use. Greater use of gas could improve the Soviet economy and the local and regional environments across Europe as well as reducing CO_2 emissions, particularly if gas is used more efficiently than coal, and if methane leakage from the natural gas system is kept very low ($< 2\%$) (Fulkerson et al. 1990).

A third possibility is recovering CO_2 from fossil-fuel combustion and sequestering it in biomass reservoirs, the deep ocean, or depleted gas reservoirs. The growing of new forests to offset CO_2 emissions is attractive in this day of deforestation; it is currently relatively inexpensive (Fulkerson et al. 1990), but it is likely to be limited by land-use priorities. However, growing biomass for energy directly rather than as an offset is likely to be a more productive CO_2 control measure. Recovering CO_2 from large point sources such as power plants appears to be expensive, but some recent calculations (Hendriks et al. 1990) based on an oxygen-blown integrated gasification combined cycle (IGCC) process indicate a much lower cost than previous estimates (Baes et al. 1980; Steinberg and Albanese 1980; Steinberg and Horn 1982; Steinberg 1983; Steinberg et al. 1984; Steinberg 1985; Steinberg 1986; Steinberg 1987; Wolsky and Brooks 1989).

The fourth and ultimate strategy is, of course, substitution of nonfossil for fossil sources. This includes research and development to improve the nonfossil sources. Of the nonfossil sources which have the potential to be expanded significantly, nuclear fission is the nearest to realization. It is already used in 26 countries to produce 16% of the world electricity generation (EIA 1989). However, opposition to its greatly expanded use is growing and can be found even in countries which are its strongest devotees.

We do not pretend to understand all the reasons for public opposition. We suspect opposition is driven partly by concern over the risk of a disastrous accident such as Chernobyl, partly by the fear of radioactivity leading to great concern about the proper handling of wastes not only for this but future generations, partly by the connection between nuclear weapons and nuclear power, and partly by a loss of trust in the institutions constructed to develop and manage the nuclear enterprise (Rayner and Cantor 1987). We know the technology can be improved significantly. Correcting or mitigating weaknesses in the technology is surely a necessary, if not sufficient, condition in gaining a system which enjoys greater public support and confidence. Without that improvement, the possibility for large-scale expansion of nuclear power seems remote. A number of papers on this topic have recently been published, and we have benefited greatly from them (Weinberg 1990; Häfele 1989; Häfele 1990; Murray 1990; Keepin and Kats 1988).

In the second section we will review the situation relative to improving nuclear power. Then, in the third section we will examine how much impact nuclear power might have over the next 50 years or so if the weaknesses could be reduced to acceptable levels.

TOTAL SOCIAL COSTS OF NUCLEAR POWER

We will discuss improving nuclear power by reducing the total social cost of the fuel cycle. Total social costs include both the direct ones for building, operating, and decommissioning facilities and also those indirect costs such as risks of accidents or environmental impacts. These indirect costs are primarily in three areas: power plant accidents, environmental impacts due to waste management (particularly the inter-generational issue), and proliferation of weapons, including possible diversion of fissionable materials.

Indirect Cost

In order to discuss total social costs of developing the nuclear power option, we have looked at the contributions of each major element of the fuel cycle. In Table 7.1, we list the fuel cycle elements and give our judgement on a qualitative estimate of the social costs of each as measured in terms of impacts on the environment, human health, safety, proliferation, and direct costs. For each element in the matrix, we indicate whether we think current plans for improved technology increase or decrease impacts, risks, or direct costs. Comparisons should be made by looking down a particular column of cost, not across a row. These estimates represent our judgement and are used to suggest a method of analysis rather than the results of any scholarly study. Nevertheless, some interesting features seem to emerge.

Environmental problems of normal operations are centered primarily in mining, milling and processing, and waste management. Human health impacts (from normal operations) tend to be dominated by mining, milling,

Table 7.1 Our judgement on contributions to indirect or external and direct costs of nuclear power and the potential for improved technology to reduce costs by various stages of the fuel cycle

	Indirect or External Costs				
Fuel Cycle Elements	Environmental Impacts (normal operations)	Human Health Impacts (normal operations)	Safety (accidents)	Proliferation	Direct Costs
Mining, milling, and ore processing	xxx(+)	xxx(+)	x(+)		x(o)
Enrichment	x	x	x	xxx(−)	x(+)
Fuel fabrication	x	x	x	xx(?)	x
Nuclear power plant operation					
—Burners	x	x	xxx(+)	x	xxx(+)
—Breeders	x	x	xxx(+)	xxx	xxx(+)
Spent fuel reprocessing	xx	xx	xx	xxx(+)	xx(+)
Spent fuel or high level radioactive waste storage/disposal	x(+)				
Other waste disposal	xx(+)	x	x	x	x
		x	x		x
Decommissioning and decontamination	x	x	x		x
Transportation	x	x	xx(+)	xx(+)	x

Improved Technology
(+) reduces impact, risk, or direct cost
(o) neutral
(−) increases impact, risk, or direct cost
(?) uncertain

Fuel cycle elements should be compared down any column not across a row. Then, for each column, the fuel cycle element with the highest cost under that column is ranked to have a triple x (xxx) cost and other elements are compared on a relative basis. If a fuel cycle element has no score, it is judged to have no cost in that cost category.

and processing, whereas safety is dominated by reactor accident risk. Proliferation risks are influenced by enrichment, breeder reactors, reprocessing, and transportation. Direct costs are dominated by the construction and operation of nuclear power plants.

All improved technologies development tends to reduce (or effect neutrally) direct costs and generally reduces risks or adverse impacts with one notable exception. Improved enrichment technologies are designed to reduce the cost of nuclear fuel[4] and also have lower investment costs and smaller scale than gaseous diffusion (Goldsworthy 1990). Therefore they are likely to be more affordable to a larger number of countries than current enrichment options and tend to increase the risk of proliferation.

However, Table 7.1 does not represent the whole story. Many of the indirect costs grow as the nuclear enterprise grows. These include the risks of power plant accidents and proliferation by diversion of fissile materials as these become items of international commerce. Hence, to maintain the same overall level of risk would require that the technology or institutional control improve in proportion to system growth (Komanoff 1981). That is, each new facility in the system must be better. It is plausible to argue that the level of external costs that exist today is about at the tolerable level. This certainly would appear to be the case relative to reactor safety. For some nations, such as Sweden,[5] the limit has already been exceeded, while for others it has not; e.g., France, Japan, and Canada. Furthermore, safety is a system-wide problem. An accident anywhere is the same as an accident everywhere.

Following the Three Mile Island accident, the Institute of Nuclear Power Operations (INPO) was formed in the U.S. by the electric power industry to improve the safety systems and operating procedures across the U.S. nuclear community. This was a voluntary action bolstering the oversight effort of the Nuclear Regulatory Commission. Now INPO is an international organization that aims to avoid reactor accidents everywhere in the world. It is joined now by the World Association of Nuclear Operators (WANO), which fosters exchanges between the USSR and eastern European countries and the OECD countries.

Over the past 12 years there have been two serious reactor accidents. Only one, Chernobyl, caused deaths and significant contamination of the environment; the other, Three Mile Island, ruined a reactor and cleaning it up has cost more than a billion 1990 U.S. dollars (Peach 1982). That is, the world has experienced two serious accidents for some 2000 GW years of integrated power reactor operating experience. It is tempting to divide these two numbers to get a rough measure of the probability of such

---

[4] Target reductions in fuel production cost range from 5 to 75% depending on the technology.
[5] Sweden may be rethinking her decision to phase out nuclear power (Hibbs 1990).

accidents, but that calculation is statistically unjustifiable. Probabilistic risk assessments (PRA) have been used, however, to obtain a forecast of reactor safety. For light water reactors, the probability of an accident like Three Mile Island (i.e., core melt) is in the range 10^{-3} to 10^{-5} per reactor year for the 95th and 5th percentile, respectively (Garrick 1989). That is, the probability of a core melt is $< 10^{-3}$ y^{-1} with a confidence of 95% and it is $< 10^{-5}$ y^{-1} with a confidence of 5%. Since worldwide nuclear capacity is about 312 GW_e, this implies an accident every 3 to 300 years if the system does not grow. For a world where this capacity has increased by a factor of 10, it would mean a possibility of a serious accident once every 0.3 to 30 years. That appears unacceptably high since the recent experience of two in a decade appears unacceptable. Thus, reactor safety must be improved by at least one order of magnitude and probably two orders of magnitude (a factor of 10 to 100).

Public perceptions about safety are also related to the problem of reliability and the effects of continual operating problems with nuclear power. Although only two serious accidents have occurred, numerous less serious ones occur much more frequently. Even trivial problems at nuclear plants tend to make the national news and this level of visibility serves to reinforce suspicions that the technology is unsafe or poorly managed.

Risk of proliferation or clandestine diversion of weapons-grade materials is another external cost which depends on how the nuclear enterprise is operated as a worldwide system. As technology for enrichment and spent fuel reprocessing improves and as their use increases, the risks will grow unless safeguards are also improved. The once-through "throw away" fuel cycle is one approach in reducing the risk because it eliminates spent fuel reprocessing and hence trade in plutonium (or ^{233}U in the case of thorium breeding). On the other hand, fuel processing will be necessary if the breeder is needed to expand the uranium resource base. This circumstance is, of course, more likely if nuclear power is to be a major nonfossil energy source in a greenhouse-constrained society.

For the large-scale nuclear system, it may be necessary to have breeder reactors co-located with reprocessing plants and to have these centers under international control. In fact, internationalization of enrichment and spent fuel storage is also attractive and may be necessary. The problem of safeguards was recognized since the beginning of the nuclear era (Acheson-Lilienthal Report of 1946). Recently, Williams and Feiveson (1990) have proposed five criteria they believe must be met to reduce the risk of proliferation adequately. These are:

1. Restrictions on sensitive nuclear technologies and materials shall be nondiscriminatory among nations (i.e., the same restrictions will apply to all nations).

2. Fissionable weapons-usable material (outside spent fuel) and facilities to enrich uranium or to separate plutonium shall not exist outside international centers.
3. To the extent possible, fissionable weapon-usable material (outside spent fuel) shall not be produced even at the international centers.
4. Spent fuel shall be stored and disposed of at international centers.
5. Reactors under national authority shall be designed so as to reduce to very low levels the production of weapons-usable material in spent fuel (of the order of a critical mass or less per year per gigawatt of capacity).

Meeting criteria such as these would require an extraordinary restructuring of the nuclear enterprise. Before the unprecedented changes in the world triggered by Peristroika in the USSR, such a restructuring was unlikely. Now, however, some of these criteria may be achievable and worth exploring as part of an overall worldwide effort to reduce and control nuclear weapons—particularly criteria 2, 3, and 4.

Although significant strengthening of safeguards may prove necessary, the once-through fuel cycle may well remain the most cost-effective way to organize nuclear power for decades even if a rapid growth in reactor deployment occurs. We believe uranium resources are much more extensive than once thought, and economic methods of recovering uranium from very extensive but low grade ores seem probable. Eventually uranium may even be extracted from seawater (Best and Driscoll 1986). We return to the issue of uranium requirements in the section on REDUCTION OF CO_2 EMISSIONS.

Total Direct Cost of Nuclear Power: Current and Improved Technology

Although we cannot calculate the total social cost of nuclear power quantitatively, we can say something about the total direct costs and how these may change with advancing technology.

First, we should review what has been the cost experience to date. The direct cost experience of nuclear power in the U.S. today varies widely from plant to plant. As shown in Table 7.2, the average cost (based on 1988 data adjusted to 1990 U.S. dollars) is 65 mills/kWh but the range of experience around that average is very wide (20–200+ mills[6]/kWh for plants with capacity factors greater than 40%). World projections of costs for power from nuclear plants expected to start operation in 1995 to 2000 range from 24–82 mills/kWh, also shown in Table 7.2. The upper range of costs here may be low since current estimates for the cost of power from the Sizewell pressurized water reactor (PWR) plant in England are now 60–100 mills/kWh

---

[6] One mill is one tenth of a cent or 0.001 of a U.S. dollar.

Table 7.2 Direct costs for current generation nuclear plants (1990 U.S. dollars mills/kWh$_e$)

	Current U.S.[a]		Projected World Range[d] for 1995–2000
	Average[b]	Range[c]	
Capital	43	8[e]–146	16–53
Operation & maintenance[f]	14	7–37	4–13
Fuel[g]	8	5–20	4–16
Decommissioning[h]	0.5	0.3–1	
Total	65	2–200	25–75

[a] Based on internal Oak Ridge National Laboratory analysis of actual 1988 industry data for all operating nuclear power plants with capacities greater than 400 MW$_e$. Data were adjusted to 1990 U.S. dollars.
[b] Average defined as the sum of the total cost divided by the total energy generated.
[c] Based on actual cost range adjusted to 1990 U.S. dollars. Plants with capacity factors less than 40% removed from data set.
[d] OECD/NEA (1989). Range of national responses adjusted to 1990 U.S. dollars.
[e] Some early plants with subsidized construction cost removed from data set.
[f] Includes the cost of low level waste disposal.
[g] Includes the cost of mining and milling of uranium ore, conversion to UF$_6$, enrichment, fuel fabrication and high level waste disposal.
[h] Decommissioning cost range of $100–300 million based on *projected* costs for U.S. plants.

or more (Marshall 1990). The largest projected cost variation is in the capital cost component; however, operation and maintenance and fuel costs also have wide variations. Experience in the U.S. indicates that operation and maintenance costs increased rapidly in the early to mid 1980s (Hewlett 1988).

Advanced reactor technology with passive[7] safety features holds the promise of not only increasing safety (perhaps by the one or two orders of magnitude needed for a much larger-scale nuclear enterprise) but also of reducing costs. The cost reductions are achieved because passively safe systems are inherently simpler than redundant active systems, and some passive reactors may not even require secondary containment although they will be located below grade. Such reactors may be forgiving of equipment failures as well as operator errors (Forsberg and Weinberg 1990).

Such concepts now under development in the U.S., Japan, and Sweden

---

[7] Passive safety means that safety features operate without depending on external input (i.e., external mechanical or electrical active controls). For example, a reactor might suffer a sudden loss of circulation in the core cooling system. In a passively safe reactor, natural conduction of heat to the surrounding ground or to the atmosphere would cool the core sufficiently to avoid reactor damage or release of radioactive materials and the reactor would be shut down without any external mechanical or electrical active intervention.

include an improved version of the current light water reactor (LWR), a smaller LWR (~600 MW$_e$) with passively safe features, the Swedish Process Inherent Ultimate Safety LWR, the Modular High-Temperature Gas-Cooled Reactor (MHTGR), and the Advanced Liquid Metal Reactor (ALMR). Projections of the power generation cost from these advanced concepts are shown in Table 7.3. The cost estimates shown apply to the technologies once they are fully established. First-of-a-kind plant costs may be significantly greater. In our judgement, concept development cost, including first-of-a-kind design and licensing costs, are expected to fall in the $250–500 million range for the advanced LWR reactors and $500 million to $1 billion each for the ALMR and MHTGR.[8] These costs are important deterrents to the introduction of advanced reactors.

The uncertainties in the costs shown in Table 7.3 at the current level of development are such that no real difference between the various concepts can yet be assigned with any confidence. The estimates are based on detailed vendor estimates but may be very optimistic at this early stage of development. Furthermore, vendors are likely to use the LWR cost experience as a target ceiling to offer competitive alternatives; therefore, we are not surprised by the clustering in cost estimates.

Developers claim that these passively safe reactors are much less likely to experience a serious accident than the current LWRs, which depend on

Table 7.3 Power generation costs for advanced nuclear plants (1990 U.S. dollar Mills/kWh$_e$)[a]

Plant Units × MW$_e$/Unit	Current[b] LWR 1 × 1144	Improved LWR 1 × 1280	Passive LWR 2 × 600	MHTGR[c] 2 × 538 (module block)	ALMR[d] 3 × 465 (modular block)
Capital	37	27	27	32	24
Operation & maintenance	10	10	11	8	8
Fuel[e]	6	6	6	9	12[f]
Decommissioning	0.5	0.5	1	1	1
Total	54 (50–110)	44 (35–57)	45 (39–67)	50 (42–75)	45 (36–72)

[a]Range shown in parentheses indicates our best guess adjustment for uncertainties.
[b]Current light water reactor based on U.S. construction and operating experience with best plants.
[c]Modular High-Temperature Gas-Cooled Reactor
[d]Advanced Liquid Metal Reactor: General Electric PRISM concept.
[e]Includes cost of mining, milling, conversion, enrichment, fabrication, and disposal.
[f]Initial actinide fuel from LWR fuel reprocessing plant at cost of $37/gram fissile Pu.

---

[8] Estimates based on unpublished ORNL review of vendors' projections.

redundant active systems. This remains to be verified by testing and experience, but some critics are skeptical (MHB Technical Associates 1990). Nevertheless, these are the kinds of reactor systems which may achieve one or two orders of magnitude safety improvement over the technology currently in use.

The front end of the fuel cycle should not pose any great technical or economic problem for the advanced LWRs or even the MHTGR. These concepts use enriched uranium fuel just as does the present generation of power plants. Also, the fabrication of fuel assemblies is a well-established technology even for the MHTGR with its graphite blocks. Enrichment technology is well developed and advanced technology may reduce production costs below $50/Separative Work Unit (SWU)[9] or less than half of today's costs charged by the U.S. Dept. of Energy (Goldsworthy 1990). Some of the advanced technologies, such as the atomic vapor laser isotope separation process, cause proliferation concerns because of the potential to produce highly enriched uranium more easily.

The current U.S. Dept. of Energy reference design for the ALMR is the General Electric PRISM (General Electric 1990) concept which is, or can be, a breeder reactor. This is a metal-fueled reactor using plutonium for both the start-up and recycle fuel. The concept envisions using an on-site integral recycle facility being developed by Argonne National Laboratory. The costs associated with this fuel cycle are very uncertain, hence the wide uncertainty indicated in Table 7.3. The source of start-up fuel for the ALMR is of concern. Highly enriched uranium can be used, but plutonium is a much better fuel neutronically. On a kilogram basis, less fissile plutonium is required by the ALMR than if ^{235}U were used. The source of this start-up plutonium remains a question, with reprocessing of spent LWR fuel the prime candidate, but also the burning of excess weapon material is a possibility.

The back-end of the fuel cycle (i.e., reprocessing and high-level waste disposal) continues to be a problem which must be solved if nuclear power is to grow substantially. Of course, the waste problem must be solved whether or not the nuclear enterprise expands. In the U.S., spent fuel assemblies are currently stored in pools of water at the reactor sites. Current plans are to transport these assemblies to a repository in Nevada (Yucca Mountain), where they will be permanently sequestered in a volcanic tuff formation deep underground. Entombment is also being pursued by other nations (e.g., by Sweden), where spent fuel may be deposited in a mined rock formation under the Baltic Sea (Carlsson and Hedman 1986). The

---

[9] The SWU is the unit by which uranium enrichment services are bought and sold. Approximately 5 "SWUs" are required to produce 1 kg of low-enriched (3.2% ^{235}U) material from natural (0.7% ^{235}U) feedstock.

entombment approach may suffer severe political opposition, particularly in the areas where repositories are sited. Concerns range from fear of accidents to uncertainties about the long-term integrity of the geologic formations which must survive and remain stable for millennia until the radioactive materials in the spent fuel have decayed sufficiently so they are no longer dangerous.

An alternative to current final storage plans is to keep spent fuel in so-called monitored retrievable storage. Pool storage at reactor sites represents such a strategy, but the mature development would involve a number of carefully engineered, monitored, and protected facilities in each country to which spent fuel would be shipped. Retrievable storage provides more time to develop more permanent solutions without irreversible commitment, and it could become a sort of bank for plutonium to be withdrawn as uranium becomes expensive.

Currently, electric utilities in the U.S. are charged 1 mill/kWh to be paid to the high-level waste disposal fund. This charge may be increased in the future as the cost of ultimate disposal continues to increase. The current cost estimate for the U.S. waste management program is about $30 billion. To this should be added the total cleanup cost attributable to commercial reactors (e.g., the proportionate share of costs of cleaning up at enrichment and fuel reprocessing plants, national laboratories, etc. amounts to a possible addition of $50 billion). If this cost were to be spent over a 20-year period, it could add about 3 mills/KWhr to the current waste disposal fee, and the total costs of waste disposal and cleanup would be about 5 mills/kWhr assuming 100 GW_e capacity at 75% capacity factor). The range of projections by various nations for the back-end of the fuel cycle cost from the NEA/IEA study (1989) was about $400–$1700 per kg of heavy metal (HM), which translates to about 1–5 mills/kWhr. Some of these back-end costs may include the cost of the reprocessing of the spent fuel.

The cost of reprocessing large quantities of spent fuel is uncertain. The French company, Cogema, reportedly has signed contracts at a price of FF 5000/kgHM (about $600–1000/kgHM depending on exchange rates) for the reprocessing of spent fuel in the post year 2000 period (MacLachlan 1990). There is currently no commercial fuel reprocessing in the U.S. nor is any planned.

Recently a tie-in with high-level waste disposal and spent fuel reprocessing has been proposed. The ALMR fuel cycle could be modified so as to use mixed actinides (plutonium and trans-plutonium elements) from LWR spent fuel as its initial fuel. By reprocessing LWR spent fuel, with very efficient separation of actinides from the resultant high-level wastes, the job of high-level waste disposal may be made easier and less costly. The ALMR would have its initial fuel, and the stored waste would only be a radioactive threat for hundreds, rather than thousands, of years. Also, plutonium would not be separated in a pure stream but would remain in a highly radioactive mix with the other actinides and some fission products. The proliferation concerns about reprocessing may therefore be reduced. A large reprocessing plant would be needed for the recovery of the actinides from spent LWR fuel (spent fuel stored at monitored retrievable storage sites). The capital cost of such a plant could be $4–10 billion with unit costs of $250–1000/kgHM depending on ownership and financial assumptions. The cost of the actinide recovery could be borne jointly by the ALMR and by the waste disposal fund; the latter could be charged to the extent that the reprocessing reduced waste disposal costs. The acceptability, feasibility, and economics of such a system remain to be seen, but it is an interesting possibility.

In summary, the direct costs of nuclear power vary widely, but they are probably in the range of 50–100 mills/kWh around the world. Advanced technologies hold the promise of providing greater safety and closing the fuel cycle at costs in the same range. In fact, if costs cannot be kept in this range, nuclear power will not be economically viable. Given adequate uranium resources, advanced LWR and MHTGR technologies may permit economic expansion of nuclear power in the immediate future. In the longer term, the introduction of ALMR breeders which can also burn actinides of the spent fuel from thermal reactors is an attractive option to pursue. Nevertheless, these advanced technologies are not fully developed or tested, and accelerated research, development, and demonstration are needed, as is refinement of the cost estimates.

However, even with favorable evidence on the projected direct costs, a major role for nuclear power in a global strategy to address the greenhouse effect depends on a willing marketplace. There remain a number of considerations regarding international cooperation and coordination that warrant further examination to project this role. We have mentioned the issues of proliferation and safety, but other considerations to be addressed by international cooperation include capital financing, the transfer of technical expertise and equipment, and the transfer of sound operating and siting practices. Unfortunately, at this time, there is insufficient research in the area of institutional designs and acceptable large-scale nuclear power deployment to suggest the best options for global strategies.

REDUCTION OF CO_2 EMISSIONS

In considering the potential role of nuclear power in limiting future worldwide CO_2 emissions, we focus primarily on the generation of electricity. We do not assume that this is the only useful or potentially large-scale application of nuclear power. However, it is likely to remain the principal one for several decades and over that period can serve in our analysis as a surrogate for other potential applications such as the use of nuclear heat for producing liquid fuels from coal or methane.

Demand for Electricity

We expect worldwide demand for electricity to continue to increase at nearly the rate of economic growth, even with substantial progress in improving the efficiency of energy utilization throughout the economy. From 1973 to 1987, demand for electricity in OECD countries increased at an average rate of 2.9% y^{-1}, in non-OECD countries at 5.9% y^{-1}, and in the world as a whole at 4.0% y^{-1} (IEA 1988). In a recent report for the U.S. Department of Energy on the potential of energy efficiency (Carlsmith et al. 1990), it was estimated that U.S. electricity demand will increase at 1.8 to 2.2% y^{-1} from 1990 to 2010, the lower figure corresponding to full realization of cost-effective conservation based on life-cycle costs. This study assumes no real increase in the price of electricity over the 20-year period; therefore, these demand projections do not reflect the price changes that may follow supply increases. In calculations using the Edmonds–Reilly energy-economy model (Fulkerson et al. 1989a; Edmonds and Reilly 1986), a base case approximating the IIASA Low Scenario (IIASA 1981) showed worldwide electricity demand increasing from 1975 to 2025 at an average rate of 2.9% y^{-1}, even while overall energy use, relative to gross world economic product, was decreasing by 1% y^{-1}. A high-efficiency case designed to approximate the scenario of Goldemberg and colleagues (Goldemberg et al. 1988) showed worldwide demand for electricity increasing over the same period at 1.8% y^{-1}.

In the present analysis, we consider a high electricity demand case and a low demand case with electricity consumption increasing respectively at 3% y^{-1} and 2% y^{-1} from a nominal value of 12×10^{12} kWhr y^{-1} in 1990. Both cases are intended to represent substantial progress in overall energy conservation. By 2040, the total worldwide electricity use would total 53×10^{12} kWhr y^{-1} and 32×10^{12} kWhr y^{-1} for the high and low electricity cases, respectively.

Possible Contributions of Nuclear Power to Electricity Supply

For two decades nuclear power has been, in relative terms, by far the fastest growing component of world electricity supply, averaging about 17% y^{-1} (United Nations 1989). However, that period of growth, which (following Alvin Weinberg [Weinberg et al. 1985]) we may call the first nuclear era, is drawing to a close. Planned construction programs in the U.S., France, and elsewhere are nearing completion and, in general, few new orders are foreseen in the next few years. Other major programs (e.g., in the USSR and eastern Europe) may soon follow suit, and some earlier reactors (e.g., in Sweden) may be forced into early retirement (although Sweden may be reconsidering such action). Thus, the prospect for the next 10 to 15 years is for generating capacity, now about 312 GW_e (mid-1990), to peak at about 390 GW_e around the end of the century and then begin a slow decline as increasing retirements overtake declining construction of new reactors. Even with anticipated increases in average capacity utilization factors (currently around 65%) to 70% to 75%, generation of electricity, now at about 1.8×10^{12} kWhr y^{-1}, will probably level off at about 2.4×10^{12} kWhr y^{-1} in the period 2000–2005.

Of course we do not know what the longer-term future of nuclear power may be. For the purposes of this analysis, we consider two widely divergent scenarios, defining a low nuclear (LN) and a high nuclear (HN) scenario. The former represents the continuing, slow demise of the first nuclear era as a result of scheduled retirements of reactors already in place. We assume service lifetimes of 30 years for reactors built in the early 1970s, rising to 40 years for those built in the early 1980s, and to 50 years for those built in 1990 or later years. We do not assume forced early retirements on a major scale, although that is a possibility.

However, the main point at issue in this paper is how much nuclear power could contribute to reducing future emissions of CO_2. Therefore, our HN scenario is designed to represent a very robust revival of nuclear power in what Weinberg has called the second nuclear era. One condition for such a revival is likely to be the development and demonstration of improved, passively safe reactors such as advanced LWRs, MHTGRs, and, eventually, ALMRs. We judge that construction of such reactors in numbers comparable to those reached in the 1980s in the first nuclear era (e.g., 20–30 GW_e y^{-1}) could hardly be expected before 2010. As a simplified representation of this renewed construction activity, we assume a linear increase in annual capacity additions, starting from 0 in 2005 and reaching 100 GW_e y^{-1} in 2030 (i.e., new additions increasing by 4 GW_e each year).

Although such numbers are startling at first glance, it must be remembered that *net* annual additions of all types of generating capacity averaged 84 GW_e y^{-1} from 1970 to 1987, that the assumed additions will be to a generating system perhaps 2–3 times larger than at present (5–8 times as large as in 1970), and that the world economy is assumed to be several times larger than at present. In fact, generating capacity of all types must increase at much faster rates to supply total electricity needed for either the low (2%) or high (3%) scenarios we consider.

Characteristics of the assumed nuclear scenarios are shown in Figure 7.1. Our HN scenario reaches an installed capacity of 1458 GW_e in 2030 (1300 GW_e of new reactors and 158 GW_e remaining from the first nuclear era) and reaches 2600 GW_e in 2040, which is approximately equal to total world

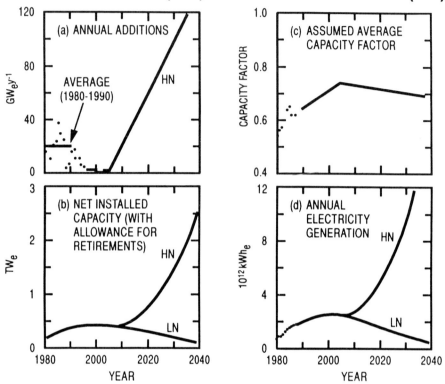

Figure 7.1 Nuclear power scenarios: high nuclear (HN) and low nuclear (LN): (a) assumed annual additions of second generation passively safe nuclear plants; (b) net installed capacity with allowances for retirements; (c) assumed average capacity factors worldwide; (d) annual electricity generation for HN and LN

generating capacity of all types in 1987. For comparison, Häfele (1990) assumes an increase of nuclear capacity to 2000 GW_e in 2030. Our HN scenario represents a very aggressive program (if not a maximum) and is suitable for estimating the potential displacement of fossil-fired plants and consequent reduction of CO_2 emissions.

Potential CO_2 Reductions

At this point, we would like to estimate the reductions in CO_2 emissions made possible by following the HN scenario rather than the LN scenario.

To do this, however, we need an estimate of the CO_2 emissions from fossil-fired stations generating the same amount of electricity. This is not a fixed characteristic of these stations, but varies with the mix of fuels and with the efficiency of the stations. These factors will not be the same for all possible future scenarios, of course. Nevertheless, we adopt a single trend line of CO_2 emissions from fossil-fired stations ($GtC/10^{12}kWhr$), based on a 2% annual growth in output. To satisfy this increased demand, we assume a more rapid growth in gas-fired units than coal or oil, i.e., 2.8% y^{-1} for gas vs. 1.7% y^{-1} for coal and oil. We also assume significant improvement in the efficiency of new units, reflected in an assumed linear increase in the average efficiencies of all operating units, reaching 40% in 2040 for coal-fired units, 47% for oil, and 47% for gas. (The best new units may have higher efficiencies than these, e.g., greater than 50% for gas-fired turbines; the assumed efficiencies given above represent nominal averages for all units in operation, new and old.) These assumptions yield a twofold increase in coal and oil consumption for electricity and nearly a threefold increase for gas. They also yield a trend in CO_2 emissions decreasing approximately linearly from 0.23 $GtC/10^{12}kWhr$ in 1990 to 0.16 $GtC/10^{12}kWhr$ in 2040.

Reductions in CO_2 emissions, calculated under these assumptions, are

THE POTENTIAL OF NUCLEAR POWER TO CONTROL CO₂ EMISSIONS

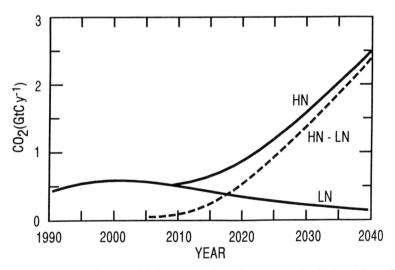

Figure 7.2 CO_2 emissions avoided by use of nuclear power by high nuclear (HN) and low nuclear (LN) cases

shown in Figure 7.2 for the HN and LN scenarios and for the difference between them. The difference (HN–LN) is shown to indicate the potential contribution of the advanced nuclear technologies expected to supersede current technologies. Nuclear power already reduces CO_2 emissions by about 0.4 GtC y^{-1}. (For comparison, worldwide CO_2 emissions associated with electricity generation are now approximately 1.8 GtC y^{-1} and, without the nuclear reduction, would be about 2.2 GtC y^{-1}.) However, it may also be seen that a renewed commitment to nuclear power is likely to make very little difference in CO_2 emissions before about 2010. Thereafter, it could become an increasingly important tool for limiting CO_2.[10]

Figure 7.3 indicates how much of the total electricity requirement can be met by the HN case. For the low electric scenario, nuclear power could supply up to 50% of electrical requirement worldwide by 2040 and about

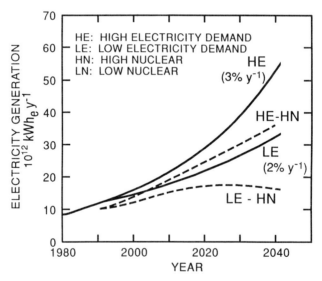

Figure 7.3 Nuclear power under the high nuclear scenario could supply up to 50% of world electricity needs by 2040 (low electric case [LE] or 30% high electric case [HE]). The remainder indicated by the dashed curves must be supplied by renewables and fossil fuels

---

[10] Weinberg (1990) points out that while nuclear plants do use fossil fuel during construction and preparation of fuel, this added CO_2 burden is far less for a nuclear plant than a coal plant of the same capacity. A 1000 MW$_e$ nuclear plant (PWR, with no recycle) releases at most only 13% as much carbon as the same size coal plant for lifetime operation.

30% for the high electric scenario. The remainder must be supplied by renewables and fossil fuels.

Estimating the penetration of renewable sources is beyond the scope of this chapter; however, a recent analysis in the U.S. (SERI 1990) indicates they could produce as much electricity as nuclear power for our HN case if research, development, and demonstration paid off and/or fossil sources were severely penalized. The SERI report details a very aggressive scenario, at least as aggressive as our HN scenario. On this basis, if both HN and high renewables were developed, fossil-fuel sources needed to supply electricity would remain at about present levels for the high electric demand scenario, but fossil sources could be phased out altogether for the low electric case.

Thus, an important conclusion is that neither nuclear power, nor efficiency improvement, nor renewables separately can significantly reduce CO_2 emissions from electricity generation. All three will be required to do that. They are complementary, not in competition.

Uranium Requirements

The HN scenario is an ambitious one and would require a large amount of uranium if it were restricted to the once-through fuel cycle for LWRs (or fuel cycles of comparable material efficiency in other reactor types). In Figure 7.4 we display the cumulative uranium requirements for the HN scenario, based on the once-through fuel cycle, in three different ways: (1) the actual cumulative consumption of uranium up to a given year; (2) cumulative uranium consumption up to a given year plus forward commitments for 20 full-power years of operation for each reactor at the time of start-up; (3) cumulative consumption plus forward commitments for 20 full-power years at the time the reactor is ordered, here taken to be ten years prior to start-up.

Prospective reactor owners may be hesitant to order new reactors unless they are reasonably assured of a lifetime supply of fuel. It would appear from Figure 7.4 that even a resource base of 20–25 Tg of natural uranium[11] would not be sufficient to sustain our HN scenario much past 2040 and that steps would be required, even prior to that date, to assure a larger supply of uranium or to make much more efficient use of the 20–25 Tg, e.g., via breeding. Breeding, however, should multiply uranium productivity by at least a factor of 50.

---

[11] OECD/IAEA projections at $130/kg ($60/lb) are: 6 million metric tons reasonably assured and estimated additional resources, and 24 million metric tons of speculative resources (OECD/NEA 1983).

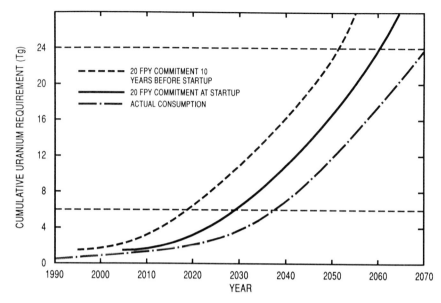

Figure 7.4 Cumulative uranium requirements for the high nuclear scenario and a once-through system. The lower curve is actual cumulative uranium consumption; the middle curve adds 20 full power years (FPY) of fuel forward commitment; the upper curve allows for 20 FPY commitment made 10 years before start-up. The horizontal dashed lines indicate current OECD/IAWA uranium resource projections of 6 and 24 million metric tons. The former is reasonably assured resources at $130/kg whereas the latter represents speculative resources

Other Considerations

In this crude analysis we have not examined many of the geographical problems with the expansion of nuclear power. For example, much of the growth of electricity demand will be in the developing world. Penetration of nuclear power in these nations is constrained by availability of capital and trained people. Equity and political considerations may require very drastic reduction of CO_2 emissions by the industrialized world (90% or more)[12] so that developing nations can use increasing amounts of fossil fuels. We assume, in any event, that nuclear power must be an internationally

---

[12] This derives from simple arithmetic. For CO_2 concentrations in the atmosphere to be stabilized, carbon cycle models suggest that emission rates must be reduced to less than half of what they are today. Suppose we want to achieve this stability in fifty years. Today OECD nations emit about half of the 5.5GtC from fossil-fuel burning but they comprise only 17% of the world population. By the middle of the next century they may comprise only 10%. If the annual allowable fossil emission of ~2.5 Gt is distributed on a per capita basis that means the OECD share would be 0.25 Gt y^{-1} or only 10% of today's emission rate. The OECD nations would need to reduce emission rates by 90% by the middle of the next century.

controlled enterprise and that it will be accessible to all nations under the conditions that all accept international regulations. It is, of course, not obvious that such a world system can be established at reasonable costs; however, if it cannot, the role of nuclear power for controlling CO_2 is likely to be very constrained.

CONCLUSIONS

From the preceding discussion we derive the following conclusions regarding the nuclear power contribution to limiting CO_2 emissions over time:

1. Nuclear power certainly can help reduce CO_2 emissions; in fact, it has already. However, even if nuclear power grows substantially in the next 50 years it can be only part of the means to reduce CO_2 emissions. Both much improved efficiency of energy use and aggressive deployment of renewable energy sources are also essential.
2. For nuclear power to continue to be an important factor, a much larger-scale enterprise worldwide will have to evolve (larger by a factor of 5–10 in 50 years). Such an enterprise will require:
    a) safe operation of the present system (no serious accidents);
    b) for the next generation, much safer reactors which are forgiving of equipment failure and operator error and which can be used by developing nations as well as industrialized countries;
    c) strengthened international institutions to control nuclear materials perhaps by putting enrichment, fuel reprocessing, and breeders (when they are needed) in centers under international control as well as the transportation and monitoring of reactor fuel assemblies new and spent; as yet, we have limited information of the kinds of international arrangements and institutions that would be the most effective to deploy a large-scale nuclear enterprise;
    d) developing better methods to dispose of waste, perhaps by burning actinides as part of breeder and reprocessing technologies.
3. These requirements are not likely to come easily. They can be met only by the most concerted efforts of international cooperation. This cooperation should include joint research, development, and demonstration on both hardware and institutional arrangements.
4. The will to make such efforts will depend in part on the extent of the perceived need. That is, it will depend on how seriously nations perceive the need to reduce CO_2 emissions and the cost, limitations, and other aspects of social acceptance of alternatives to the nuclear option. These other energy options were also the subject of this conference. As we said in the beginning, no single option is likely to emerge as a panacea.

At this stage, it is prudent to pursue all options vigorously, including nuclear. The eventual outcome will be determined by the relative success of research, development, and demonstration.

REFERENCES

Baes, C.F., S.E. Beall, D.W. Lee and G. Marland. 1980. Options for the collection and disposal of carbon dioxide, Report ORNL-5657, 35 pp. Oak Ridge, TN: Oak Ridge Nat. Lab.

Best, F.R., and M.J. Driscoll. 1986. Prospects for the recovery of uranium from seawater. *Nuclear Tech.* **73**:55–68.

Carlsmith, R., W. Chandler, J.E. McMahon, and D.J. Santini. 1990. Energy efficiency: how far can we go? Report ORNL TM 11441, 75 pp. Oak Ridge, TN: Oak Ridge Nat. Lab.

Carlsson, A., and T. Hedman. 1986. Tunnelling of the Swedish undersea repository for low and intermediate reactor waste. *Tunnelling and Underground Space Tech.* **1**:243–250. The facility described is now fully operational.

Edmonds, J.A., and J. Reilly. 1986. The IEA/ORAU long-term global energy CO_2 model: personal computer version A84/PC. Report ORNL/CDIC-16, CMP002/PC, 299 pp. Oak Ridge, TN: Oak Ridge Nat. Lab.

EIA (Energy Information Administration). 1989. International Energy Annual, 1988. Report DOE/EIA-0219 (88), 185 pp. Washington, D.C.: U.S. Dept. of Energy.

Forsberg, C.W., and A.M. Weinberg. 1990. Advanced Reactors, Passive Safety, and Acceptance of Nuclear Energy. *Ann. Rev. Energy* **15**:133–152.

Fulkerson, W., S.I. Auerbach, A.T. Crane, D.E. Kash, A.M. Perry, and D.B. Reister. 1989a. Energy Technology R&D: What Could Make a Difference? A Study by the Staff of the Oak Ridge National Laboratory. Part I: Synthesis Report. Report ORNL-6541/V1, 123 pp. Oak Ridge, TN: Oak Ridge Nat. Lab.

Fulkerson, W., S.I. Auerbach, A.T. Crane, D.E. Kash, A.M. Perry, and D.B. Reister. 1989b. Global Warming: An Energy Technology R&D Challenge. *Science* **246**:868–869.

Fulkerson, W., R.R. Judkins, and M.K. Sanghvi. 1990. Energy from Fossil Fuels. *Sci. Am.* **263**:128–135.

Garrick, J. 1989. Lessons learned from 21 Nuclear Plant Probabilistic Risk Assessments. *Nuclear Tech.* **84**: 319–330.

General Electric. 1990. ALMR Summary Plant Design Description, ed. G.L. Stimmell. Report GEFR-00878. San Jose: General Electric Nuclear Energy.

Goldemberg, J., T.B. Johansson, A.K.N. Reddy, and R.H. Williams. 1988. Energy for a Sustainable World, 517 pp. New Delhi: Wiley Eastern, LTD.

Goldsworthy, M.P. 1990. The Possibility of a Commercial Enrichment Venture in Australia. Paper presented at USCEA Fuel Cycle 90 Conference, Nashville, TN March 27, 1990.

Häfele, W. 1989. Energy systems under stress. In: World Energy Horizons 2000–2020: 14th Congress of the World Energy Conference, Conservation and Studies Committee, Montreal, September 1989. J.-R. Frisch, project director. Paris: Editions Technip for the World Energy Conf.

Häfele, W. 1990. Energy from nuclear power. *Sci. Am.* **263**:136–145.

Hendriks, C.A., K. Blok, and W.C. Turkenburg. 1990. Technology and cost of recovery and storage of carbon dioxide from an integrated gasifier combined cycle plant, submitted to *Applied Energy*.

Hewlett, J.G. 1988. An Analysis of Nuclear Power Plant Operating Costs. U.S.

Energy Information Administration Report DOE/EIA 0511, 84 pp. Washington, D.C.: U.S. Dept. of Energy.
Hibbs, M. 1990. Swedish government backing off early nuclear plant phase-out. *Nucleonics Week* **31**:1, 8.
IEA (International Energy Agency). 1988. IEA Statistics, World Energy Statistics and Balances. Paris: International Energy Agency, Organization for Economic Cooperation and Development.
IIASA (International Institute for Applied Systems Analysis). 1981. Energy in a Finite World: A Global Systems Analysis Report by the Energy Systems Program Group of the International Institute of Applied Systems Analysis, 837 pp. W. Häfele, program leader. Cambridge, MA: Ballinger Publishing Co.
Keepin, B., and G. Kats. 1988. Greenhouse warming: comparative analysis of nuclear and efficiency abatement strategies. *Energy Policy* **16**:538–61; also *Science* **241**:1027.
Komanoff, C. 1981. Power Plant Cost Escalation Nuclear and Coal Capital Costs, Regulation and Economics, 323 pp. New York: Komanoff Energy Associates.
MacLachlan, A. 1990. Cogema gets reprocessing contracts covering 15 West German utilities. *Nuclear Fuel* **15**:5–6.
Marshall, P. 1990. U.K. to finish Sizewell despite cost hike to over 2-billion pounds. *Nucleonics Week* **31**:1, 12.
MHB Technical Associates. 1990. Advanced Reactor Study, 189 pp. San Jose: prepared for the Union of Concerned Scientists.
Murray, Jan. 1990. Can nuclear energy contribute to slowing global warming? *Energy Policy* **18**:494–499.
NEA/IEA (Organization for Economic Cooperation and Development Nuclear Energy Agency/International Energy Agency). 1989. Projected Costs of Generating Electricity from Power Stations for Commissioning in the Period 1995–2000, 206 pp. Paris, Washington: OECD Nuclear Energy Agency/International Energy Agency.
OECD/NEA. (Organization for Economic Cooperation and Development Nuclear Energy Agency/International Energy Agency). 1983. Uranium: Resources, Production and Demand. Paris, Washington: OECD Nuclear Energy Agency/International Energy Agency.
Peach, J.D. 1982. Impact of Federal R&D Funding on Three Mile Island Cleanup Costs, Report EMD-82-28. Washington, D.C.: U.S. General Accounting Office. 19 p.
Rayner, S., and R. A. Cantor. 1987. How fair is safe enough? The cultural approach to societal technology choice. *Risk Analysis* **7**:3–9.
SERI (Solar Energy Research Institute). 1990. The potential of renewable energy. An interlaboratory white paper, SERI/TP-260-3674, 202 pp. Golden, CO: Solar Energy Research Institute.
Steinberg, M. 1983. An Analysis of Concepts for Controlling Atmospheric Carbon Dioxide. DOE/CH/00016-1, 66 pp. Upton, NY: Brookhaven Nat. Lab.
Steinberg, M. 1985. Recovery, disposal, and reuse of CO_2 for atmospheric control. *Env. Progress* **4**:69–77.
Steinberg, M. 1986. Production of a Clean Carbon Fuel Derived from Coal for Use in Stationary and Mobile Heat Engines. Brookhaven National Laboratory Report BNL-38086, 25 pp. Upton, NY: Brookhaven National Laboratory.
Steinberg, M. 1987. Clean Carbon and Hydrogen Fuels from Coal and Other Carbonaceous Raw Materials. Report BNL-39630, 38 pp. Upton, NY: Brookhaven Nat. Lab.
Steinberg, M., and A. Albanese. 1980. Environmental control technology for

atmospheric carbon dioxide. In: Interactions of Energy and Climate, eds. W. Bach, J. Pankrath, and J. Williams, pp. 521–551. Dordrecht: Reidel.

Steinberg, M., H. Cheng, and F. Horn. 1984. A Systems Study for the Removal, Recovery, and Disposal of Carbon Dioxide From Fossil Fuel Power Plants in the U.S. Report DOE/CH/0016-2, 76 pp. Upton, NY: Brookhaven Nat. Lab.

Steinberg, M., and F. Horn. 1982. Possible Storage Sites for Disposal and Environmental Control of Atmospheric Carbon Dioxide. Report BNL 51597, 23 pp. Upton, NY: Brookhaven Nat. Lab.

United Nations. 1989. 1987 Energy Statistics Yearbook (and earlier years), 485 pp. New York: United Nations.

Weinberg, A.M. 1990. Nuclear energy and the greenhouse effect. Nuclear Safety: Power Production and Waste Disposal, pp. 5–16. Conf. Proc. Midwest Universities Energy Consortium, Inc.

Weinberg, A.M., I. Spiewak, J.N. Barkenbus, R.J. Livingston, and D. Phung. 1985. The Second Nuclear Era, A New Start for Nuclear Power, 439 pp. New York: Praeger.

Williams, R.H., and H.A. Feiveson. 1990. Diversion-resistance criteria for future nuclear power. *Energy Policy* **18**:543–549.

Wolsky, A., and C. Brooks. 1989. Recovering CO_2 from large stationary combustors. In: Energy Technologies for Reducing Emissions of Greenhouse Gases. Proc. of an Experts' Seminar, Paris, April 12–14, vol. 2, pp. 179–185. Paris: Intl. Energy Agency OECD.

8
The Likely Roles of Fossil Fuels in the Next 15, 50, and 100 Years, With or Without Active Controls on Greenhouse-Gas Emissions

R.L. KANE[1] and D.W. SOUTH[2]
[1]Office of Fossil Energy, U.S. Department of Energy, Washington, D.C. 20585, U.S.A.
[2]Policy and Economic Analysis Group, Argonne National Laboratory, Argonne, IL 60439, U.S.A.

INTRODUCTION

Since the industrial revolution, the production and utilization of fossil fuels have driven economic and industrial development in many countries worldwide. This source of economic growth is expected to continue; most forecasts of future global economic activity in developed and developing countries are based on a continued reliance on fossil fuels. An extensive institutional infrastructure and fossil-fuel resource base exists around the world to support the continued (and expanded) use of this fuel source. However, future reliance on fossil fuels has been questioned due to emerging concerns about greenhouse-gas emissions, particularly carbon dioxide (CO_2), and its potential contribution to global climate change.

While substantial uncertainties exist regarding the ability to predict climate change and the role of various greenhouse gases accurately, some scientists and policymakers have called for immediate action. As a result, there have been many proposals and worldwide initiatives, such as the Intergovernmental Panel on Climate Change (IPCC), to address the perceived problem. In many of these proposals, the premise is that CO_2 emissions constitute the

Limiting the Greenhouse Effect: Options for Controlling Atmospheric CO_2 Accumulation
Edited by G.I. Pearman published 1992 John Wiley & Sons Ltd

principal problem, and, correspondingly, that fossil-fuel combustion must be curtailed to resolve this problem.

This paper demonstrates that the worldwide fossil-fuel resource base and infrastructure are extensive and thus will continue to be relied on in developed and developing countries. Furthermore, in the electric-generating sector (the focus of this paper), numerous clean coal technologies (CCTs) are currently being demonstrated (or are under development) that have higher conversion efficiencies, and thus lower CO_2 emission rates, than conventional coal-based technologies. As these technologies are deployed in new power plant or repowering applications to meet electrical load growth, CO_2 (and other greenhouse gas) emission levels per unit of electricity generated will be lower than that produced by conventional fossil-fuel technologies.

Thus, deployment of advanced fossil-fuel technologies will automatically reduce greenhouse-gas emissions relative to conventional technologies while continuing to use the extensive worldwide fossil-fuel resource base. In addition, if it is determined that CO_2 emission levels should be reduced, fossil fuels and fossil-fuel-based technologies can also play a role in the near and long term, through various fuel substitution and CO_2 scrubbing options.

THE GREENHOUSE EFFECT: ROLE OF FOSSIL-FUEL EMISSIONS

The greenhouse effect is a popular term used to describe the roles of water vapor, CO_2, and other trace gases in keeping the Earth's surface warmer than it would be without their presence. These "radiatively active" gases are relatively transparent to incoming short-wave radiation but relatively opaque to outgoing long-wave radiation. The long-wave radiation, which would otherwise escape to space, is partially absorbed by these gases in the lower atmosphere. The subsequent "re-radiation" of some energy back to the Earth's surface produces surface temperatures higher than those that would occur if the gases were absent. It is the continued buildup of these higher surface temperatures over time that is postulated to produce climatic changes. While relatively easy to explain conceptually, there are numerous uncertainties in our ability to document the role of each greenhouse gas, model the current greenhouse and global climate change effect, and accurately forecast potential climate change based on various energy scenarios.

There are four principal greenhouse gases: CO_2, nitrous oxide (NO_x), methane (CH_4), and a group of chlorofluorocarbons (CFCs). The sources of each gas and their current contribution to global climate change vary substantially as illustrated in Figure 8.1. While fossil-fuel combustion

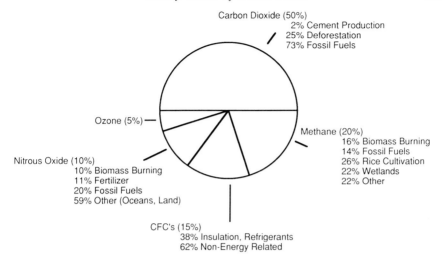

Figure 8.1 Greenhouse-gas contribution during 1980s to increment in global climate change (EPA 1988)

contributes to a substantial portion of the global climate change increment due to CO_2, it contributes much smaller shares of CH_4 and NO_x; fossil-fuel combustion does not contribute directly to CFC emissions.

The contribution of each gas to the greenhouse effect is based on its atmospheric concentration and lifetime, together with its radiative properties. Although CO_2 has the highest current atmospheric concentration among the greenhouse gases (358 parts per million), it is the least potent in terms of global warming potential on a molecule-for-molecule basis. Figure 8.2 illustrates the global warming potential of each greenhouse gas on a molecule-for-molecule basis, relative to CO_2; it demonstrates, for example, that a CFC molecule is approximately 20,000 times more potent than a molecule of CO_2 in terms of its potential contribution to global climate change.

GLOBAL CLIMATE CHANGE: PRINCIPAL UNCERTAINTIES

While the U.S. Congress has introduced a number of bills aimed at addressing climate change (see Hootman and South 1989), and the IPCC and other groups have been discussing potential options to reduce greenhouse-gas emissions (Houghton et al. 1990; EPA 1989; Jäger 1988), substantial

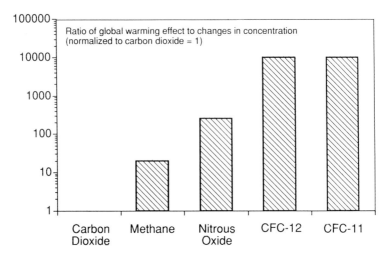

Figure 8.2 Global warming effect of greenhouse gases relative to CO_2 (derived from Ramanathan 1988)

uncertainties remain about the potential magnitude and temporal dimensions of climate change due to greenhouse-gas accumulation. Most of the uncertainty falls into three broad categories: determining future emissions and concentrations of greenhouse gases; predicting the climate based on general circulation models (GCMs) and their associated scientific base; and detecting changes in the historical climate record (see South et al. [1990a] for a more detailed discussion). Given these uncertainties, actions to curtail fossil-fuel usage would incur substantial economic and social costs that might later be determined to have been unnecessary.

One area of uncertainty requiring resolution is the future rate of fossil-fuel-based CO_2 emissions and other trace-gas emissions, together with their associated atmospheric concentrations. It is difficult to predict the technical developments, economic factors, and policy decisions on which future emission rates depend. In the projected data uncertainties are present on both energy use and its determinants. There is also a spectrum of potentially available technologies and fuels, as well as a host of difficult-to-predict geopolitical, economic, social, and demographic factors that must be addressed. Included in these techno-economic uncertainties is the nature of any policy response to climate change, which is only dimly perceived at present.

The rate of global economic growth, which is only one determinant, will depend heavily on the rate of economic development in both developed and less-developed countries, and on the energy demands and choices of energy technologies associated with that development. Potential changes in both supply and demand technologies and the possibilities for interfuel substitution must also be considered, because such factors add to forecasting difficulties. End-use energy efficiency improvements also rank high on the list of important determinants. The remainder of this paper will focus on existing fossil-fuel supply options for the electric-generating sector. Demand technologies and end-use efficiency improvements to reduce CO_2 emissions are addressed elsewhere in this volume (Ashton et al., chapter 12; Jenney, chapter 11; Jochem, chapter 13; Yamaji, chapter 10).

FOSSIL-FUEL COMBUSTION AND CO_2 EMISSIONS

Between 1950 and 1980, worldwide CO_2 emissions increased by almost 219%, an average growth rate of 7.3% y^{-1} (from 5.82 Gt y^{-1} of CO_2 in 1950 to 18.55 Gt y^{-1} in 1980). A further increase of 12.8% took place between 1980 and 1988, although the annual rate of change was substantially less, 1.6% y^{-1}. Current worldwide CO_2 emissions from fossil fuels and cement facilities total 20.9 Gt y^{-1}.[1] Figure 8.3 shows that the rate of growth in fossil-fuel-related CO_2 emissions has varied considerably by geographic region. (The top figure illustrates the CO_2 emissions pattern for developed or industrialized regions; the lower figure shows the pattern for developing regions.)

After exhibiting steady growth between 1963 and 1974 (4.5% y^{-1}), CO_2 emissions in North America leveled off until 1986; however, the annual growth rate since then has returned to the 1963–1974 rate. Centrally planned European countries and centrally planned Asian countries have maintained rapid growth rates throughout the period 1950–1988. Since 1950, CO_2 emissions in centrally planned European countries have grown 405.7%, or 10.7% y^{-1}. Despite a slight pause between 1978 and 1981, the growth rate in CO_2 emissions has returned to the long-term trend. While CO_2 emissions for centrally planned Asian countries have expanded rapidly since the 1960s, the increase has been the greatest since 1974; between 1974 and 1988 emissions increased by more than 116%, or 7.7% y^{-1}. After having grown by 75% between 1959 and 1973 (5.4% y^{-1}), western Europe's CO_2 emissions have fallen by 330 Mt y^{-1}, or 10.3%, the only region to do so.

Carbon dioxide emissions in developing countries are expected to increase relatively rapidly because the energy consumption rate in these countries, both in aggregate and for fossil fuels, is increasing faster than in industrialized nations. Warrick and Jones (1988) found that energy consumption has grown by 6.2% y^{-1} in developing countries but by only 0.5% y^{-1} in industrialized

---

[1] Carbon emissions increased from 1.59 Gt y^{-1} in 1950 to 5.06 Gt y^{-1} in 1980. As of 1988, carbon emissions were 5.7 Gt y^{-1}.

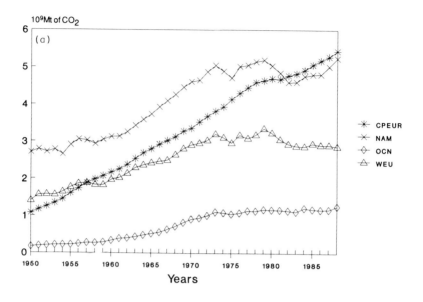

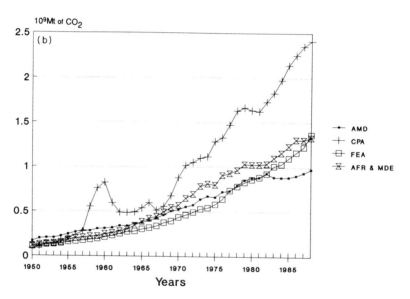

Figure 8.3 Trends in fossil-fuel-related CO_2 emissions by major geographic area: developed (top) and developing (bottom) regions (derived from Marland et al. 1989). AMD = South/Central America and islands; AFR = Africa; CPA = centrally planned Asia; CPEUR = centrally planned Europe; FEA = Far East; MDE = Middle East; NAM = North America; OCN = Oceania and Japan; WEU = Western Europe

Likely Roles of Fossil Fuels

nations. The higher growth rate is attributable to the fact that many developing countries, including China, have initiated plans for accelerated development and industrialization. In most cases, these policies entail a rapid expansion in the use of energy, and the most readily available, cost-effective, and practical source is coal. Based on energy consumption trends from 1973 to 1980, the largest increase in fossil-fuel use in the next few decades is expected to be in the Third World (Warrick and Jones 1988).

While developed countries have historically relied on fossil-fuel-generated electricity, centrally planned economies and less-developed countries are rapidly increasing their dependence on coal-fired units to develop their economic and industrial infrastructures. This dependency is likely to continue both for economic and political reasons; that is, fossil fuels, particularly coal, are indigenous or easily accessible.

Globally, electricity generation is the source of approximately one-third of all the CO_2 emitted. As of 1988, more than 60% of worldwide electricity generation is produced by fossil fuels, with approximately 45% being supplied by coal (WEC 1989; DOE 1989a). Table 8.1 illustrates that although the share of fossil-fuel-generated electricity varies by country, there is a predominant reliance in most countries on fossil fuels for the production of electricity, heat, and power. Fossil-fuel-generated electricity is greater than 60% in all regions except Central and South America and western Europe, which have 30% and 45%, respectively.

It has been estimated that to replace all fossil-fueled power plants in the world with nuclear power would entail commissioning a new 1000 MW nuclear plant every one to three days until 2025; furthermore, even with this action, CO_2 emissions would still continue to grow (Keepin and Kats 1988). Since this is not a likely or feasible scenario, reliance on fossil fuels is likely to continue. Fossil fuels currently have the advantage of extensive worldwide recoverable reserves and an established international infrastructure for production, distribution, and consumption. In addition, alternative energy sources are not presently economically competitive or capable of directly replacing fossil-fuel-generated electricity.

GLOBAL FOSSIL-ENERGY RESOURCE BASE

Fossil fuels presently constitute over 88% of the world's total primary energy sources and produce over 60% of the world's electricity (DOE 1989a). This reliance on fossil fuels is based on extensive recoverable reserves; globally there is approximately 40 to 60 years' worth of oil and gas remaining at present production rates, and more than 325 years worth of coal (see Figure 8.4).

Table 8.1 Electricity generated (10^9 kWh) by region and fuel type for 1988 (from DOE 1989a)

Fuel Type	Region							
	North America	Central and South America	Western Europe	Eastern Europe and USSR	Middle East	Africa	Far East	Total
Hydro	546.4 (0.16)[a]	314.2 (0.59)	512.3 (0.24)	251 (0.12)	8.6 (0.05)	43.5 (0.18)	377.1 (0.18)	2053.1 (0.20)
Nuclear	605.1 (0.18)	5.4 (0.01)	654.7 (0.31)	260 (0.12)	0 (0.00)	10.5 (0.04)	237.2 (0.12)	1772.9 (0.17)
Coal, Gas and Oil	2188.3 (0.66)	134.2 (0.30)	954.1 (0.45)	1585.1 (0.76)	169.2 (0.95)	194 (0.78)	1437.5 (0.70)	6662.4 (0.64)
Total	3339.8	453.8	2121.1	2096.1	177.8	248	2051.8	10488.4

[a] Fuel share of total electricity generated in region.

Oil

According to recent estimates of recoverable reserves, the world's supply of oil can last 40 to 50 years at current production/consumption rates. However, this lifetime is predicated on continued oil production from the Middle East, which has approximately 63% of the world's known recoverable oil reserves (see Figure 8.5).

Natural Gas

World proven reserves of natural gas currently total 109 Tm^3 and constitute approximately 60 years of supply at 1987 consumption rates. This reserve estimate is not static: proven recoverable reserves of gas have been doubling every 10 years. Some estimates indicate that the ultimate natural gas resources will be between 250 and 350 Tm^3 (WEC 1989).

Figure 8.6a indicates the regional concentrations of raw natural gas (excluding natural gas liquids); most of the recoverable reserves are located in centrally planned economies or the Middle East. Figure 8.6b further delineates the distribution of these reserves by country. The USSR has the largest single reserve, almost four times the second-ranked country, Iran. Unlike oil reserves that are more geographically concentrated, gas reserves are somewhat more globally dispersed; each continent has several countries with a significant level of proven gas reserves.

Increased availability of natural gas to the developed world, which also possesses the existing infrastructure to handle gas and the capital to expand

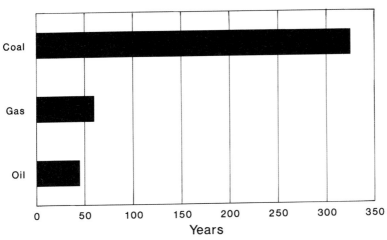

Figure 8.4 Duration of recoverable fossil-fuel reserves at current production rates (source: derived from DOE 1989a; WEC 1989)

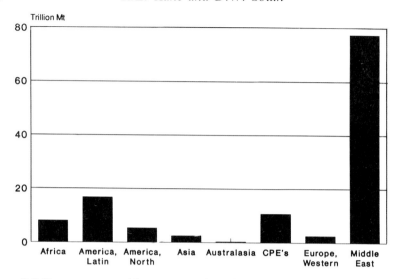

Figure 8.5 Proven recoverable reserves of crude oil and natural gas liquids, 1987 (WEC 1989)

that infrastructure, has led a number of Organization for Economic Cooperation and Development (OECD) countries to increase their use of gas instead of coal and nuclear energy. This has been primarily due to environmental concerns (IEA 1989).

Coal

As shown in Figure 8.7, most of the world's coal reserves are concentrated in eight countries. China, the U.S., and the USSR contain more than 75% of worldwide recoverable reserves. However, coal reserves exist or coal is generally available to all continents and nations.

Most of the world's coal reserves consist of anthracite, bituminous, and subbituminous, which are cleaner burning and more fuel efficient than lignite (see Figure 8.8). While lignite reserves are found in most countries, Australia, China, Germany, and the USSR have relatively large quantities. In Australia, Germany, and the USSR, lignite comprises 40% or more of the total coal reserve.

At current production rates, economically recoverable coal reserves would last roughly 325 years. Figure 8.9 illustrates the lifetimes of economically recoverable coal reserves by region, assuming 1987 production rates. Somewhat lower reserve lifetimes (defined with respect to current production rates) arise when projected demand growth is included or if coal trade with developing countries without reserves increases dramatically; even with these

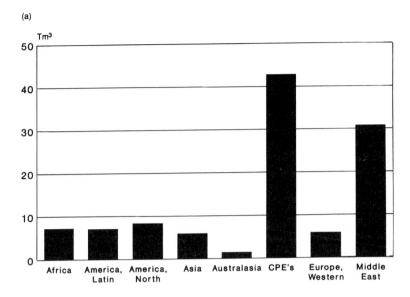

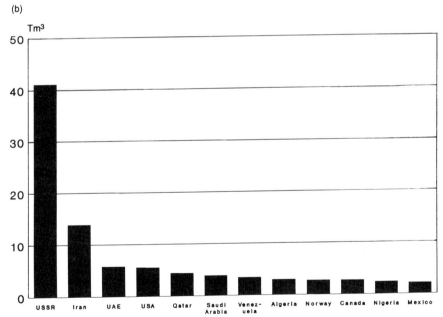

Figure 8.6 (a) Proven worldwide recoverable reserves of natural gas by major region, 1987 (WEC 1989). (b) Proven worldwide recoverable reserves of natural gas by country, 1987 (WEC 1989)

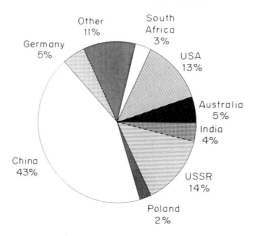

Figure 8.7 Proven worldwide recoverable coal reserves by country for 1987 (from WEC 1989)

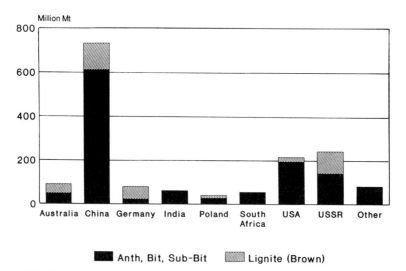

Figure 8.8 Estimated recoverable coal reserves by type and country for 1987 (from WEC 1989)

considerations, however, more than 100 years of recoverable reserves should be available in the year 2100.

Because of extensive availability, assured supply and lower prices, coal will likely continue to play a major role in the world's energy picture. In the U.S., 24% of all energy and approximately 57% of all electricity was

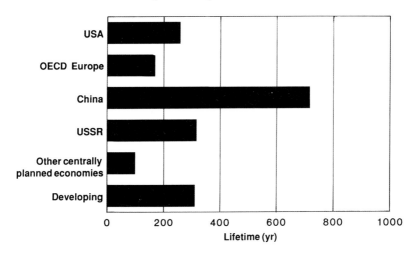

Figure 8.9 Remaining lifetime (recoverable reserve/current consumption) of recoverable coal reserves by region for 1987 (derived from WEC 1989)

produced from coal in 1988 (IEA 1990a). Worldwide, 25% of all primary energy production, and 44% of all electricity, is produced from coal (WEC 1989). For most countries, because of compelling economic and energy security reasons, it would be extremely difficult, if not impossible, to replace coal with another fossil fuel or a non-CO_2-emitting fuel source.

Looking into the future, reliance on fossil fuels is expected to expand worldwide, but the greatest growth is expected in developing countries. The recent history and projected usage of fossil fuels can be seen in projections produced by the Edmonds–Reilly global CO_2 model.[2] In a 1990 reference case forecast, the expected growth rate for fossil-fuel consumption in the developed world between 1975 and 2025 is 65%; in comparison, the growth rate in developing countries is projected to be over 428% (see Table 8.2). In 1975 the developing world consumed only 28% of that by the developed world. By 2025 that percentage is expected to increase to 90%.

While the largest percentage increase in demand is for natural gas, the largest heat content expansion will be in coal: 123 EJ for coal versus 86 EJ for oil and 92 EJ for gas. Table 8.3 shows the projected growth in fossil-fuel utilization by developing nation/region. Coal and oil are projected to increase more than three-fold between 1975 and 2025, with natural gas increasing by more than ten-fold. Table 8.4 shows that if this pattern of

---

[2] The Edmonds–Reilly (ER) model has been used to forecast worldwide energy patterns and CO_2 emissions, and to examine alternative control policies in a multitude of global climate change studies. For more information on the ER model and some of its applications, see Edmonds and Reilly (1983a, 1983b, 1986 and 1987), and Edmonds et al. (1986).

Table 8.2 Growth of fossil-fuel use by developed and developing countries for 1975–2025 as projected by Edmonds and Reilly (1990)

	Fossil Fuel Consumption (EJ)			1975–2025 Growth (%)	Average Annual Rate (%)
	1975	2000	2025		
Developed Countries					
Coal	37.55	76.98	77.27	105.8	2.12
Oil	60.33	86.37	80.73	33.8	0.68
Gas	63.76	72.87	109.13	71.2	1.42
	161.64	236.22	267.13	65.3	1.31
Developing Countries					
Coal	21.55	47.17	104.85	386.5	7.73
Oil	19.36	44.38	84.35	335.7	6.71
Gas	4.58	13.98	51.16	1017.0	20.34
	45.49	105.53	240.36	428.4	8.57

fossil-fuel demand in developed and developing countries is extended to the year 2100, current proven reserves of oil and natural gas will be essentially depleted and coal will remain the only fossil fuel of choice, capturing 95% of the market.

Table 8.5 illustrates the projected rapid growth in the utilization of coal in developing countries. By 2025 developing countries will constitute 57.8% of the world's coal demand. By 2100, that share is projected to increase to 67.2%.

Fossil-fuel consumption for electricity generation follows a similar pattern to that exhibited for total energy, particularly for coal (see Table 8.6). However, when expressed in fossil-fuel equivalent terms, there is also substantially increased use of "other" sources of electricity, primarily nuclear and hydroelectric. Nonetheless, as stated above, the level of fossil-fuel usage for production of electricity continues to increase in real terms, with oil increasing 11.54 EJ, natural gas 29.22 EJ, and coal 46.16 EJ.

As indicated, fossil-fuel consumption and, particularly, coal are projected to increase dramatically in both developed and developing countries. In developing countries, population growth and pressures for a higher standard of living are driving the need for more energy and electricity. With the extensive fossil-fuel resource base and infrastructure, it would be extremely difficult, if not impossible, to displace fossil-fuel utilization in the near and midterm. As discussed in the next section, CCTs offer a means to continue use of fossil fuels worldwide while minimizing CO_2 emissions.

Table 8.3 Projected growth in fossil-fuel use by developing countries for 1975–2025 (from Edmonds and Reilly 1990)

	EJ			Growth (%) (1975–2025)	Average Annual Rate (%)
	1975	2000	2025		
Coal					
China	15.37	30.08	54.06	251.72	5.03
Middle East	0.08	0.31	3.03	368.75	7.38
S. and E. Asia	3.30	9.32	22.30	575.76	11.52
Africa	2.32	4.95	13.49	481.47	9.63
Latin America	0.48	2.51	11.97	2393.75	47.88
	21.55	47.17	104.85	386.54	7.73
Oil					
China	3.16	9.27	18.93	499.05	9.98
Middle East	2.63	6.86	13.44	411.03	8.22
S. and E. Asia	4.74	9.73	12.20	157.38	3.15
Africa	2.31	4.60	14.21	515.15	10.30
Latin America	6.52	13.92	25.57	292.18	5.84
	19.36	44.38	84.35	335.69	6.71
Natural Gas					
China	0.24	1.87	10.19	4145.83	82.92
Middle East	1.06	2.43	8.25	678.30	13.57
S. and E. Asia	0.75	1.60	8.53	1037.33	20.75
Africa	0.22	1.89	9.07	4022.73	80.45
Latin America	2.31	6.15	15.12	554.55	11.08
	4.58	13.98	51.16	1017.03	20.34

Table 8.4 Project energy demand (EJ) for fossil fuels and coal in developed and developing regions (from Edmonds and Reilly 1990)

	1975	2000	2025	2050	2075	2100
Total Fossil Fuel	207.13	341.35	507.49	777.93	376.69	1229.23
Coal	59.10	124.15	182.12	334.38	697.68	1168.54
% Coal	28.53	36.32	35.89	42.98	79.58	95.06

FOSSIL-ENERGY OPTIONS TO REDUCE CO_2 EMISSIONS

Several options exist that could reduce the amount of CO_2 emitted from fossil-fuel combustion. Conservation, conversion to lower-CO_2-emitting fuels, or substitution to non-CO_2-emitting fuels are three possible strategies.

Table 8.5 Projected demand for energy (EJ) from coal in developing countries for 1975–2100 (from Edmonds and Reilly 1990)

	1975	2000	2025	2050	2075	2100
China	15.37	30.08	54.06	71.67	131.41	228.96
Middle East	0.08	0.31	3.03	13.5	24.74	46.77
Africa	2.32	4.95	13.49	46.01	63.05	120.50
Latin America	0.48	2.51	11.97	37.69	97.48	185.56
S. & E. Asia	3.30	9.32	22.30	50.98	114.78	203.84
Total	21.50	47.17	104.85	219.85	431.46	785.63
LDC %	36.38	37.99	57.57	65.75	61.84	67.23

Table 8.6 Projected fuel consumption for electricity generation (EJ) in developing and developed countries (from Edmonds and Reilly (1990)

	1975	% of Total	2025	% of Total
Developed Countries				
Coal	17.62	30.24	43.09	25.14
Oil	10.55	18.11	13.86	8.09
Gas	12.99	22.30	31.13	18.16
Other	17.10	29.35	83.33	48.61
Total	58.26	100.00	171.41	100.00
Developing Countries				
Coal	4.03	35.07	24.72	17.74
Oil	2.36	20.54	10.59	7.60
Gas	1.15	10.01	12.23	8.78
Other	3.95	34.38	91.81	65.88
Total	11.49	100.00	139.35	100.00

These potential solutions, however, would take time to have an effect due to (a) the maturity of the alternative technologies, (b) the limited availability of replacement fuels, and (c) the rate of replacement fuels and turnover for existing combustion systems in developing and industrializing economies that are heavily dependent on fossil fuels.

There are, however, three feasible options that involve the continued use of fossil fuels. The first is to combust more oil and natural gas in cases where this option is available and economical since both of these fuels generate less CO_2 than coal.[3] Stated differently, more electricity could be generated per ton of CO_2 produced by burning natural gas or fuel oil than by burning coal (see Table 8.8).

The second option involves the utilization of advanced fossil-fuel technologies, such as CCTs being developed by the U.S. Department of Energy (DOE). These CCTs have lower CO_2 emission rates because of their higher thermodynamic conversion efficiencies: 40–45% for near-term CCTs (50–60% for midterm CCTs) versus 30–35% for conventional technologies with SO_2 controls.[4]

Figure 8.10 contrasts, for a single plant, the conversion efficiencies and CO_2 emissions of these advanced fossil-energy technologies with those of a conventional pulverized coal power plant having a flue-gas desulfurization unit. The data for each technology are based on a 500 MW power plant that burns coal with a sulfur content of 2.8% and a heating value of 27 Mt/kg, and that operates at a 65% capacity factor.

The advanced technologies include atmospheric fluidized-bed combustion, pressurized fluidized-bed combustion, integrated gasification combined cycle

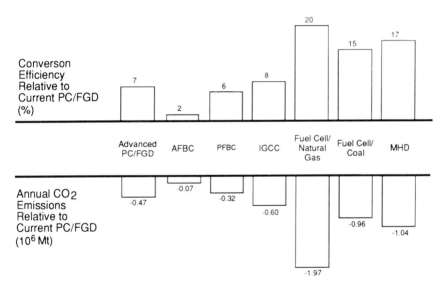

Figure 8.10 Comparison of conversion efficiencies and CO_2 emissions from advanced and conventional fossil-fuel technologies (from WEC 1989).
PC/FGD—pulverized coal/flue-gas desulfurization; AFBC—atmospheric fluidized-bed combustion; PFBC—pressurized fluidized-bed combustion; MHD—magneto-hydrodynamics

---

[3] Co-firing 10 to 20% natural gas with coal could reduce the amount of CO_2 emitted from coal-fired plants by the difference in CO_2 emission rates and share of gas co-fired.
[4] Near-term is defined to mean 1995–2005; midterm is 2005–2015.

technology, fuel cell with natural gas technology, fuel cell with coal gasification technology, and magnetohydrodynamic technology. The data in Figure 8.10 clearly demonstrate that increased efficiency produces lower CO_2 emissions. Thus, deployment of these advanced fossil technologies in either new or repowering/retrofit applications would reduce current and future CO_2 emission levels below those achievable with conventional technologies.

Although commercial-scale CCTs are not currently operational in the U.S., an aggressive demonstration program sponsored by the U.S. DOE is currently underway to make them available at the earliest possible date.[5] Table 8.9 indicates the CCT demonstration projects that have been selected as part of the DOE CCT Program; approximately 40 alternative CCTs are currently being demonstrated. In addition, a number of U.S. electric utilities are considering construction of CCTs to meet future electric load growth requirements; the Electric Power Research Institute has conducted 10 integrated coal gasification/combined cycle studies for utilities needing new capacity (Gluckman et al. 1989).

Through the CCT demonstration program and subsequent initial deployment, these technologies should be available some time after the year 2000. By that time, fossil-fuel-based fuel cell technologies and magnetohydrodynamic technologies with conversion efficiencies of 45–60% will probably become available. The adoption of these technologies will provide an even greater reduction in CO_2 emissions, allowing fossil-fuel use to be maintained.

The U.S. DOE is also actively involved in pursuing the international transfer of CCTs. Cooperative agreements have recently been signed with Costa Rica and Chile. The objective is to enhance energy security through the use of coal and CCTs; other benefits include fuel diversity, increased economic cooperation, and expanded trade opportunities. It is also recognized that deployment of these advanced technologies will have environmental benefits, such as reduced CO_2 emissions.

The third option to reduce CO_2 emissions is CO_2 scrubbing via tail-gas cleanup technologies. To apply this technique in the electric power industry would involve the adaption of acid-gas removal technologies that are currently used by the petroleum and petrochemical industries to remove sulfur dioxide (SO_2). While several techniques are possible, they are presently not cost effective and have disposal problems (South et al. 1990a; Smelser and Booras 1990).

---

[5] The Clean-Coal Technology Demonstration Program is a $5 billion industry and government initiative to demonstrate and deploy commercially innovative, low-emission, coal-based technologies. The program is a result of former U.S. President Reagan's commitment to Canadian Prime Minister Mulroney to address the transboundary acid rain problem and comply with the recommendations of the joint report of the special envoys on acid rain (DOE 1986).

Substitution between Fossil Fuels

One strategy to reduce CO_2 emissions from power plants without displacing fossil fuels is to alter the fuel mix. This can be accomplished by switching to less-CO_2-intensive fossil fuels, such as natural gas.

Utilities typically decide on a fuel mix on the basis of long- and short-term considerations. A long-term choice is made when a utility decides whether to fill its need for new capacity by building a coal, oil, gas, nuclear, or hydroelectric power plant. In many cases, once a plant has been built for a particular fuel type, there is little opportunity to switch to an alternate fuel without incurring significant costs. Generally speaking, it is easier (i.e., less costly) to switch from coal to oil or gas than to switch in the opposite direction (EIA 1985; Power 1982; Fay et al. 1986; Flour 1983).

In some cases, fuel flexibility is inherent in the technology at the plant. For example, some boilers are designed to burn both oil and gas (the selection at any given time depends on local fuel availability and price), and some are designed to burn both coal and oil. Table 8.7 indicates the number and share of multi-fired boilers that exist in North America, OECD Europe, and OECD Pacific. Multi-fired boilers comprise between 18 and 24% of total 1988 worldwide capacity, with the majority of the boiler capacity designed to burn oil and natural gas or coal and oil. Only a small proportion (1.2–3.4%) of the existing boilers can switch from coal to natural gas. In the short term, a utility decides which units and fuels in an existing mix will be used to meet a given level of demand. These short-term decisions are primarily based on fuel costs, operating costs, and maintenance requirements.

Every fuel has advantages and disadvantages, depending on the issue of concern. Oil is the most versatile fuel and natural gas is the cleanest. Coal has continued to be the cheapest fossil-fuel option for meeting base-load demand, but it emits significantly more CO_2 (and other pollutants) than oil and gas. It does have the largest resource base of the fossil fuels, however, as pointed out earlier (see section on GLOBAL FOSSIL ENERGY RESOURCE BASE).

As illustrated in Table 8.8, coal has the highest carbon content per unit of energy produced of any of the widely used fossil fuels (with the exception of synfuels, which are really synthesized coal). Natural gas has the lowest carbon content, and thus can generate the most electricity per ton of CO_2. Thus, if it were feasible to switch from coal to natural gas, the rate of CO_2 emissions from energy generation could be decreased significantly.

Implementation of any fuel-switching strategy would not be easy, especially in developing countries. Whereas coal is relatively easy to transport and requires only a minimal amount of specialized equipment, natural gas and oil require more sophisticated transportation systems and equipment. In the

Table 8.7 Share of multi-fired capable boilers in selected regions for 1988 (IEA 1990b)

Classification by Fuel	North America			OECD Europe			OECD Pacific		
	Capacity (MW)	% of Total Capacity	% of Total Conventional Thermal	Capacity (MW)	% of Total Capacity	% of Total Conventional Capacity	Capacity (MW)	% of Total Capacity	% of Total Conventional Thermal
Total Capacity	772510			498045			207023		
Total Conventional Thermal Capacity	519929	67.30		245215	49.24		129949	62.77	
Multi-Fired									
Solids/Liquids	26608	3.44	5.12	39493	7.93	16.11	11296	5.46	8.69
Solids/Natural Gas	26412	3.42	5.08	6007	1.21	2.45	1885	0.91	1.45
Liquids/Natural Gas	124434	16.11	23.93	32697	6.57	13.33	16053	7.75	12.35
Solids/Liquids/Nat. Gas	3542	0.46	0.68	11965	2.40	4.88	880	0.43	0.68
Total	180996	23.43	34.81	90162	18.10	36.77	30114	14.55	23.17

Likely Roles of Fossil Fuels 209

Table 8.8 CO_2 emitted by fossil-fuel combustion (after Steinberg and Cheng 1988)

Fuel	Heat Content (MJ/kg)	CO_2 Generated (kg CO_2/GJ)	Electricity Generated (kWh/kg CO_2)
Bituminous coal	29.5	87	1.24
Fuel oil/gasoline	55.8	76	1.54
Natural gas	55.8	49	2.23
Synthetic natural gas from coal	55.8	140	0.75

U.S., the locations of many power plants were chosen to be near a coal transportation network, and an oil or gas network may not be accessible. Oil and gas networks and infrastructure may be costly to develop for the low-density populations in the Third World. Furthermore, many parts of the Third World and China have abundant supplies of coal, whereas natural gas and oil have to be imported.

Switching from coal to natural gas and oil is not a feasible long-term means to reduce greenhouse-gas emissions. The reservoirs of gas and oil are relatively small when compared with those of coal; approximately 90% of the recoverable reserves of fossil fuel are in the form of coal. Moreover, at current production rates, these oil and gas reserves may last only a few decades; any substitution that would increase their rate of utilization could reduce their reserve life unless supplemented with new discoveries.

Another problem associated with natural gas and oil is that these fuels are concentrated in a few areas and are subject to political and economic manipulation. Finally, any reduction in the use of coal (unless carried out on a global basis) would increase the supply of this fuel, thereby decreasing its price and making it even more attractive to countries not participating in the switch to oil or gas. The increased use of coal would then add to CO_2 emissions, offsetting any potential decrease in CO_2 emissions resulting from the countries that might have switched to less carbon-rich fuels.

Given the economic importance and vast deposits of coal in many nations, coupled with the limited reserves of natural gas and oil, coal will undoubtedly play a major role in the energy future of the world. Other nonfossil-fuel options could play a role in the energy sector while producing little or no CO_2 at the power plant. The material, transportation, and land requirements and other needs of these technologies must be examined to ascertain their overall CO_2 (and other greenhouse-gas) impact. Combinations of technologies, such as the gasification of coal via nuclear power, also offer

Table 8.9 Classification of demonstration projects into characteristic technology categories (from DOE 1990)

Technology Category	Demonstration Project	Selected CCT-I	Selected CCT-II	Selected CCT-III	Total Projects
Advanced Combustion	Advanced cyclone combustor	x			4
	Advanced slagging coal combustor	x			
	LNS burner for cyclone-fired boilers		x		
	Healy power project			x	
Coal Preparation	Advanced coal conversion process	x			3
	Coal quality expert project	x			
	OTISCA fuel		x		
Flue-Gas Cleanup: Combined SO_2/NO_x Control	Gas reburning/sorbent injection	x			6
	Limestone extension and coolside Demonstration project	x			
	SOX-NO_x-ROX box flue-gas cleanup		x		
	WSA-SNO_x flue-gas cleaning		x		
	NO_xSO SO_2/NO_x removal flue-gas cleanup			x	
	Confined zone dispersion flue-gas desulfurization			x	
Flue-Gas Cleanup: NO_x Control	Advanced tangentially fired combustion		x		7
	Advanced wall-fired combustion		x		
	Coal reburning in cyclone-fired boilers		x		
	Selective catalytic reduction		x		
	Low NO_x cell burner retrofit			x	
	Wall-fired boiler gas reburning and low NO_x burner			x	

Category	Project					Count
	Integrated dry NOx/SO₂ emission control system				x	
Flue-Gas Cleanup: Sulfur Control	Advanced flue-gas desulfurization			x	x	4
	CT-121 FGD					
	Gas suspension absorption flue-gas desulfurization				x	
	LIFAC flue-gas desulfurization				x	
Fluidized-Bed Combustion: Circulating Fluidized-Bed	Nichols CFB repowering				x	4
	Nucla CFB		x			
	Alma PCFB repowering				x	
	CFB (Tallahassee)		x			
Fluidized-Bed Combustion: Pressurized Bubbling-Bed	PFBC utility demonstration				x	2
	Tidd PFBC		x			
Integrated Gasification Combined-Cycle	Clean energy IGCC		x			3
	IGCC repowering				x	
	IGCC demonstration			x		
Industrial Processes	Cement kiln gas cleaning				x	4
	Coke oven gas cleaning				x	
	Prototype commercial coal/oil coprocessing		x			
	Blast furnace granulated coal injection				x	
New Fuel Forms	Liquid phase methanol				x	2
	Mild gasification				x	

some potential for reduced CO_2 emissions; many technical, economic, and societal considerations, however, must be addressed and resolved before such combinations can be expected to make a significant impact on CO_2 emissions (Green 1988).

Advanced Fossil-energy Technologies

Many advanced fossil-fuel technologies that are currently under development in the U.S.—by the government, principally the U.S. DOE, and in some cases in the private sector through the Electric Power Research Institute—could reduce CO_2 emissions. Several of these technologies are based on the continued use of fossil fuels, principally coal. The reduction in CO_2 emissions is accomplished by increasing the efficiency of the coal-to-electricity (or other marketable output) conversion over that typically attained in current, commercial fossil-fuel technologies. Although other coal-based technologies are under development, six representative options are summarized in the following paragraphs.[6] Figure 8.10 provides a comparison of conversion efficiencies and total CO_2 emissions of advanced fossil-fuel technologies relative to a conventional fossil-fuel technology.

In the following sections, greenhouse-gas emissions from each of these options are compared with those from a reference 500 MW power plant operating at an annual capacity factor of 65%. This reference plant uses a wet limestone flue-gas desulfurization system to remove 90% of the SO_2 generated, by burning a coal with 3.5% sulfur and a higher heating value of 27 Mt/kg.

Reference Case, Conventional Coal Combustion with Wet Limestone Flue-Gas Desulfurization

Coal-fired plants using the conventional, commercial technology burn pulverized coal and use a wet limestone flue-gas desulfurization system to remove the SO_2 from the flue gas before releasing it into the atmosphere. The net efficiency for this type of power plant is approximately 30–35% (a value of 33% will be used for comparative purposes). Based on the above assumptions and on the assumption of 100% combustion efficiency, 0.4 kg of coal would be burned for every kilowatt-hour (kWh) of saleable electricity. The corresponding CO_2 emission rate from the coal is approximately 87 kg/GJ of energy released, or approximately 0.94 kg/kWh of saleable electricity.

---

[6] A more detailed characterization of each technology with a discussion of its cost, performance, and availability is found in South et al. (1990b).

As noted above, conventional pulverized-coal-fired units typically use a wet limestone flue-gas desulfurization system to control SO_2 emissions. Because limestone ($CaCO_3$) is a carbonate, some additional CO_2 is released when the limestone is calcined into lime (CaO). Based on typical values of calcium-to-sulfur ratios used in this type of system, the amount of CO_2 released in this process is approximately 0.02 kg/kWh. The total CO_2 released from the power plant boiler is thus about 0.96 kg/kWh. The annual amount of CO_2 released from the reference plant would be approximately 3.02 Mt. This estimate is only for releases that result from coal combustion and does not include releases that occur during the mining of coal or limestone, transportation, waste disposal, or any other phase of the total fuel cycle. Small quantities of other greenhouse gases are also released as a consequence of coal combustion. Table 8.10 delineates the emission rates for these other gases.

Advanced Coal Combustion and Flue-Gas Desulfurization

A considerable amount of research and development is being done to improve the operating characteristics of conventional power plants such as the one described above. With respect to CO_2 emissions, improvements could be realized through an increase in the efficiency of the plant and through greater utilization of sorbent materials in the flue-gas desulfurization system. Efficiency improvements could be realized by using supercritical steam, higher initial steam temperatures, or multiple reheat. It has been estimated that the efficiency of an advanced pulverized coal-fired power plant could be increased to about 40% in the near term (EPRI 1989). This degree of efficiency represents a substantial improvement over the reference-case efficiency of 33% for conventional plants.

Table 8.10 Approximate greenhouse-gas emissions for selected advanced fossil-fuel technologies (kg/kWh)

Technology	CO_2	NO_x	N_2O	CH_4
Pulverized coal/flue-gas desulfurization	0.96	0.003	0.00006	0.0009
Advanced pulverized coal/flue-gas desulfurization	0.81	0.002[a]	0.00005[a]	0.0007[a]
Advanced fluidized-bed combustion	0.94	0.0011	0.0007	0.0010
Pressurized fluidized-bed combustion	0.86	0.0010[a]	0.0006[a]	0.0009[a]
Integrated gasification combined cycle	0.77	0.0004	0.00006	0.0009

[a]Estimate based strictly on efficiency improvements.

Improvements in the flue-gas desulfurization system could be achieved through regenerating the sorbent, using noncarbonate sorbents, or changing the temperature, humidity, and other parameters that influence the use of sorbent in the flue-gas desulfurization system. Figure 8.11 shows a schematic of an advanced pulverized coal plant with an advanced flue-gas desulfurization system. Although the use of lime in place of limestone would reduce CO_2 emissions at the power plant, an equal amount of CO_2 would be released in the production of commercial lime, since lime is produced by calcining limestone.

The rate of CO_2 emitted from coal combustion in the advanced plant would be approximately 0.79 kg/kWh from the carbon content of the coal and approximately 0.02 kg/kWh from the limestone. The latter number was estimated based on the assumption that near-term applications of advanced flue-gas desulfurization systems would be based on improved sorbent utilization rather than wide-scale sorbent regeneration. The total rate of CO_2 emissions from this advanced coal-fired plant would thus be approximately 0.81 kg/kWh, and the annual emissions would be about 2.55 Mt, based on the same assumptions for plant size and annual capacity factor. Emissions of other greenhouse gases are shown in Table 8.10.

Atmospheric Fluidized-Bed Combustion

One of the technologies being developed as part of the U.S. DOE CCT program uses atmospheric fluidized-bed combustion. In this technology, pulverized coal (or some other fuel), an inert bed material such as sand or ash, and a sorbent such as limestone are suspended (fluidized) by an upward flow of combustion gases and air. This process provides a high level of combustion efficiency while keeping temperatures low enough to minimize NO_x formation and to provide for near-optimal SO_2 capture by the sorbent. Heat is removed from the combustion zone by producing steam in water-filled tubes passing through the fluidized bed and/or the hot gas stream. The steam is sent to a conventional steam turbine to generate electricity. Particulates (consisting of ash, spent sorbent, and unreacted sorbent) are removed from the gas stream with conventional cleanup devices such as baghouses, cyclones, or electrostatic precipitators (South et al. 1990).

This technology can be used both for new generating facilities or as a repowering option in refurbishing old power plants (DOE 1987). When used in new facilities, the efficiency of such a power plant is expected to be approximately 35%, which represents an improvement of about two percentage points over the conventional pulverized coal plant with flue-gas desulfurization (EPRI 1989).

Based on the earlier assumptions about coal characteristics, approximately 0.90 kg of CO_2 would be created per kWh of electricity generated. Because

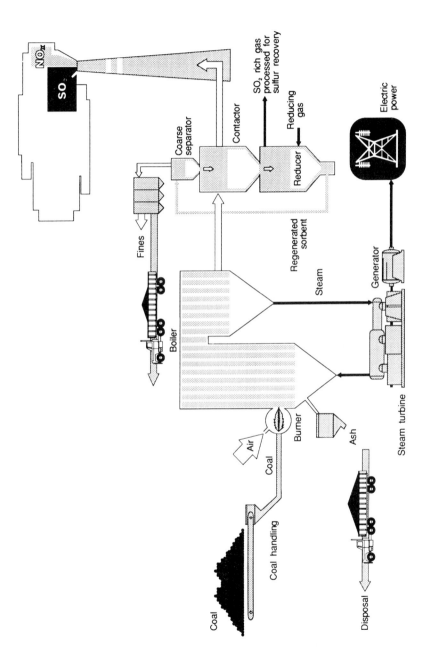

Figure 8.11 Principal component of a power plant with advanced flue-gas cleanup (from DOE 1987)

limestone is used as the SO_2 sorbent in atmospheric fluidized-bed combustion units, additional CO_2 is released during this process at a rate of 0.04 kg/kWh of electricity generated. The total CO_2 release from a new facility of this kind is thus estimated at 0.94 kg/kWh or about 2.95 Mt/yr. It must be emphasized that this value, which is 0.07 Mt/yr less than that of the reference case, is dependent on the calcium-to-sulfur value needed for the required level of SO_2 control. The other greenhouse gases emitted are comparable to those of the reference plant (see Table 8.10).

Pressurized Fluidized-Bed Combustion

Another version of the fluidized-bed concept uses pressurized fluidized-bed combustion. In this technology, the combustor is maintained at a pressure of 8–16 atmospheres as opposed to the near-atmospheric pressure maintained in atmospheric fluidized-bed combustion units. Particulates are removed from the pressurized gases exiting the combustor in a hot-gas cleanup system. The cleaned gases are then expanded in a gas turbine to produce electricity. Steam that is generated by cooling the combustor and in wasteheat steam generators at the exit of the gas turbines is expanded in a conventional steam turbine to produce additional electricity. This combined-cycle aspect of pressurized units yields an overall efficiency of approximately 39% (EPRI 1989) and the amount of CO_2 created is about 0.81 kg/kWh.

As in atmospheric fluidized-bed combustion units, SO_2 control in pressurized units is accomplished in the fluidized bed through the addition of a calcium-based sorbent. For many years, it was believed that dolomitic limestone (a mixture of approximately equal molar quantities of $CaCO_3$ and $MgCO_3$) rather than regular limestone ($CaCO_3$) must be used in pressurized fluid-bed combustion units. Therefore, even though some recent evidence suggests that the necessary SO_2 removal rates can be achieved with regular limestone, the following estimates are based on the use of dolomitic limestone.

A calcium-to-sulfur ratio of about 1.5 is believed necessary to achieve 90% SO_2 removal from a high-sulfur coal. Because both the calcium and the magnesium are in the form of a carbonate, twice the amount of CO_2 will be released from dolomitic limestone as from an equal number of moles of regular limestone. Based on a calcium-to-sulfur ratio of 1.5, CO_2 is released at approximately 0.05 kg/kWh of electricity generated. The total rate of CO_2 emissions from a pressurized unit is thus 0.86 kg/kWh. An annual CO_2 release of about 2.70 Mt, or 0.32 Mt less than that of the reference case, is estimated. This estimate is dependent on the required calcium-to-sulfur ratio and whether dolomitic or regular limestone is used. Other greenhouse gases released from pressurized fluidized-bed combustion units are similar to those from atmospheric fluidized-bed combustion units but are reduced by the plant efficiency ratio (see Table 8.10).

Integrated Gasification Combined-Cycle Technology

Integrated coal-gasification combined-cycle technology is under development for use in new power plants and for repowering old ones. In this technology, coal is partially combusted under substoichiometric conditions so that it is gasified. The gas is then cleaned and burned in a combustion turbine, where it produces electricity. Although there are several concepts based on different gasification processes and gas cleanup systems, steam can be produced in waste-heat steam generators, the gasification chamber, and the gas cooling system. This steam is then sent to a conventional steam turbine, where it is expanded to produce additional electricity. A schematic of such a power plant is shown in Figure 8.12.

The combined-cycle aspect, when combined with advances in combustion turbines, yields an expected efficiency of approximately 41% for some systems (EPRI 1989; South et al. 1990a). Based on the assumption that all the carbon in the coal is eventually oxidized to CO_2 and released at a rate of 0.77 kg/kWh of saleable electricity, the annual release of CO_2 would be approximately 2.42 Mt, or about 0.60 Mt less than that of the reference case. The other greenhouse-gas emissions are shown in Table 8.10.

Unlike the technologies described thus far, most integrated gasification combined-cycle units do not use a carbonate sorbent to control SO_2. Instead, they use special solvents that selectively absorb the hydrogen sulfide (H_2S) and carbonyl sulfide (COS) along with some CO_2 and H_2O. The sulfur is then recovered and can be sold if an appropriate market exists. As a result, no significant additional CO_2 releases are associated with SO_2 control as they are with the technologies described previously. Such hot-gas cleanup systems are undergoing extensive research and development with the objective of being commercially available within the next few years. An exception to this type of gas treatment exists with a fluidized-bed gasifier, in which in-bed SO_2 control is accomplished with limestone in a way similar to that used in atmospheric fluidized-bed combustion units.

Fuel Cells with Natural Gas

One of the advanced technologies currently undergoing research and development is the fuel cell. An advantage of this technology is that it produces electricity directly from the combustion of fuel, thus eliminating the need for the intermediate steps of conventional technologies (fuel combustion with the subsequent conversion of the heat to steam, to

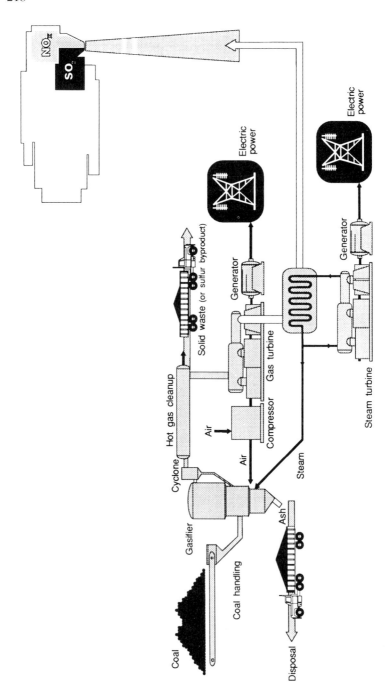

Figure 8.12 Principal component of a power plant with integrated gasification combined-cycle technology (from DOE 1987)

mechanical energy, and then to electricity). As a consequence, the efficiency of fuel-cell power plants can be quite high. Several types of fuel cells are under development: phosphoric acid and molten carbonate fuel cells have been receiving a great deal of attention.

Early versions of fuel cells to be used in the production of electricity will probably be based on the use of a light distillate oil or natural gas as the basic fuel. Efficiencies for this type of fuel cell are estimated to be as high as 53% (DOE 1988). When natural gas is the base fuel, the CO_2 released from this power plant would be about 0.34 kg/kWh, based on the assumption that the natural gas releases 49 kg/GJ of CO_2. The annual release from a power plant using this concept, based on the reference case assumptions, would be about 1.05 Mt. No estimates of non-CO_2 greenhouse-gas emissions are available; however, the use of natural gas would result in some increase in CH_4 losses with respect to the reference technologies.

Fuel Cells with Coal Gasification

As an alternative to the use of relatively scarce (when compared with coal) natural gas and light distillate fuels, a technology that combines coal gasification with fuel cells is undergoing research and development. In this concept, coal is gasified in a manner similar to that used in integrated gasification combined-cycle units to produce a hydrogen-rich synthesis fuel. Because fuel cells have an extremely low tolerance to sulfur, the synthesis gas is subjected to a gas cleanup system, where almost all the sulfur compounds are removed. Depending on the type of fuel cell, other impurities such as CO are also removed before the synthesis gas enters the fuel cell.

Due to its coal gasification and gas cleanup steps, the fuel-cell/coal gasification technology would probably be slightly less efficient than fuel-cell technologies based on natural gas or light distillate. An efficiency of about 48% has been estimated for a fuel-cell/coal gasification concept using the Texaco gasification process (EPRI 1989). Based on this efficiency, it is estimated that the CO_2 released from this power plant would be about 0.66 kg/kWh, or about 2.06 Mt/yr. Estimates for the release of other greenhouse gases have not been made.

Magnetohydrodynamics

In this process, coal is burned at temperatures high enough (about 2760°C) to dissociate the resulting combustion gases into highly charged particles. This flow of charged particles (plasma) is then passed through a magnetic field, which results in the production of electricity. The hot gases exiting the magnetic field are sent to a heat recovery steam generator to produce steam, which is expanded in a conventional steam turbine to produce additional electricity. This combined-cycle type configuration is expected to yield overall efficiencies of 50% or more (DOE 1989b).

Based on the assumption of 50% efficiency, the rate of CO_2 released from a magnetohydrodynamic power plant would be about 0.63 kg/kWh or about 1.98 Mt/yr. This value is 1.04 Mt/yr less than that released by a conventional pulverized coal plant with wet limestone flue-gas desulfurization. Estimates for emissions of other greenhouse gases are not available.

Summary

The technologies discussed above are representative of the advanced fossil-fuel options being demonstrated or under development for use in electric utilities. These options are the subject of significant research and development efforts in both the public and private sectors. The anticipated conversion efficiencies and the corresponding CO_2 emissions reductions are summarized in Figure 8.10, where the values for the advanced technologies are compared with those of a conventional pulverized coal-fired power plant equipped with a wet limestone flue-gas desulfurization system for SO_2 control. For a rough approximation, the estimates are presented in order of anticipated commercialization (see Figure 8.13). Those technologies on the left-hand side of Figure 8.10 are those believed to be currently available or those that could reasonably penetrate the electric utility sector by the year 2000. Those technologies on the right-hand side of the figure are not anticipated to play a large role until after the turn of the century, at the earliest.

Removal, Recovery, and Disposal of CO_2

The concentration of CO_2 in the atmosphere can, in theory, be controlled by removing it either before or after it has been released from a power plant to the atmosphere. Removal mechanisms designed to extract CO_2 directly from the atmosphere have been proposed but are usually considered to be too energy consuming, expensive, and generally impractical. Furthermore, the additional energy consumed by these processes would, if generated from fossil fuels, release additional CO_2 and other greenhouse gases into the atmosphere, thereby offsetting the emission reductions achieved by the CO_2 process. The control of CO_2 releases before they are released to the atmosphere appears to be a somewhat more feasible option, and several techniques have been proposed. Some of the basic features of some of these techniques are discussed below.

At the present time, the only option available for CO_2 control would involve a tail-gas cleanup (i.e., CO_2 scrubbing) system that could be adapted from the acid-gas removal technologies used in the petroleum and petrochemical industries. In general, this type of control technology consists of four steps:

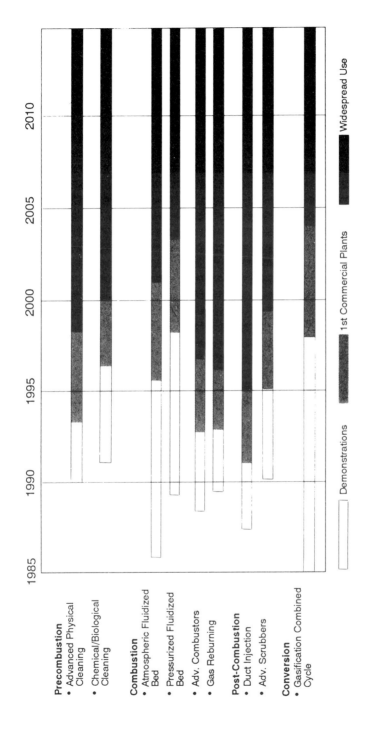

Figure 8.13 Commercial readiness of clean coal technologies (from DOE 1989b)

recovery, concentration, liquefaction, and disposal or reuse of the CO_2. The various CO_2 scrubbing concepts (discussed below) emphasize one step over another; however, most involve some version of each of the four steps.

The most advanced CO_2 scrubbing techniques are based on CO_2 absorption. This process is widely used by the petroleum industry in acid-gas cleanup systems. It is based on the concept that certain liquid solvents (e.g., amines or seawater) can be used to absorb gases selectively in a stream consisting of several different gases. It has been projected that 90% of the CO_2 in a gas stream can be absorbed in this way but also that the net electrical output from a typical power plant would be reduced by 30% (Smelser and Booras 1990). Work is underway to develop more concentrated solutions that would improve the efficiency of the removal process so that the net power reduction would be about 20%.

A second general technique makes use of the adsorption properties of materials such as clay. In this process, CO_2 would be adsorbed in the clay until it became saturated, at which time the material would be removed for storage in clay pits. Removal efficiencies of up to 90% have been estimated; however, a severe power penalty of up to 50% has also been estimated for this level of removal (Steinberg and Chang 1988).

A third technique is based on chemical and biochemical processes. Plankton or algae are exposed to CO_2 in a controlled environment where photosynthesis techniques are used to capture CO_2 and convert it to a useable form such as cellulose. This technique is still being developed and is not expected to be available until after the turn of the century.

The fourth general technique is expected to be capable of removing 90% of the CO_2 by condensing the gases. Power requirements are currently estimated to be 20–30% of the plant output. This is the least mature of all the generic types of scrubbing options discussed and, as such, has very high levels of uncertainty associated with it.

There are at least three fundamental difficulties with each of these scrubbing options. First, the costs are anticipated to be quite high. Flour Daniel, Inc., under contract to EPRI, recently completed a detailed engineering and economic evaluation of CO_2 removal, recovery, and disposal from a 500 MW pulverized coal-fired power plant with flue-gas desulfurization and from a 400 MW integrated coal gasification combined-cycle power plant (Smelser and Booras 1990). It was assumed that ocean disposal would be used; the disposal costs reflect a 640 km pipeline (480 km overland and 160 km offshore for disposal at 450 m), the typical distance from most power plants in the eastern U.S. The evaluation assessed the incremental impact of reducing CO_2 emissions (nominally 90%) on the design, thermal efficiency, capital, and operation and maintenance cost for both plants.

As depicted in Table 8.11, the study found that net power plant output of the pulverized coal/flue-gas desulfurization plant will be reduced by

Table 8.11 Capacity, performance, and cost impact of CO_2 controls on pulverized coal/flue-gas desulfurization (PC/FGD) and integrated gasification combined cycle (IGCC) power plants (Smelser and Booras 1990)

	Net Electrical Output (MW)	Heat Rate (kJ/kWh)	Coal-to-Power Effic. (%)	In-Plant Power Consump. (MW)	Total Plant Cost (1990/M$)	$/kW	Incremental Cost ($/kW)[a]
PC/FGD							
Base Case	513	10,300	34.8	41	580	1,129	—
Retrofit w/CO_2 control	336	16,000	22.8	111	1,282	3,830	1,876
Greenfield w/CO_2 control	338	15,700	22.9	109	1,173	3,469	1,660
IGCC							
Base Case	432	10,200	35.4	70	691	1,600	—
Greenfield w/CO_2 control	389 (Net)[a]	12,700	28.5	117	1,142	3,015	1,226

[a] Total incremental capital requirements are the difference between the cost of the CO_2 removal plant and the cost of the base case plant, plus the replacement power costs, divided by the net power produced by the original base case plant.

approximately 35%, with a net heat rate of 14,965 Btu/kWh. In comparison, the net power output from the integrated coal gasification plant would be reduced by only 12%, with a net heat rate of 11,975 Btu/kWh. Incremental plant capital costs, including the cost of replacement power,[7] are $1660/kW and $1226/kW for CO_2 controls on new pulverized coal and integrated coal gasification plants, respectively. Corresponding CO_2 control costs for a retrofit pulverized coal/flue-gas desulfurization plant would be $1876/kW. Incremental plant costs for integrated coal gasification CO_2 control are much lower than costs for pulverized coal plants because CO_2 removal from high pressure syngas requires less energy and uses lower cost processes than CO_2 recovery from low pressure flue gas (Smelser and Booras 1990).

The second general problem is that of efficiency. Again, depending on the specific process and level of removal, up to 50% of a plant's electrical output could be consumed by the CO_2 control system.

The final problem is that of the disposal or use of the concentrated CO_2. The vast quantities of CO_2 to be removed would require very large markets for reuse and/or have special storage techniques. Several disposal options have been considered; however, they have limited applications and, in some cases, very high costs. One example of a disposal technique is the storage of the CO_2 in depleted gas wells. This option is constrained by how close the depleted wells are to the coal-fired power plants and by how much capacity is available in the wells. It has been estimated that the current U.S. well capacity is 48 Gt or approximately 25 years of CO_2 production at current CO_2 emission rates (Steinberg and Cheng 1988). Yet even this estimate is based on the assumption that there are no constraints that would result from the relative locations of the CO_2 sources and the well storage sites.

Another storage option that has been suggested is to pipe the CO_2 to ocean depths of 500 to 3000 m, where the natural absorptive characteristics of the sea could be used for long-term storage of CO_2. The application of this option appears to be limited to power plants near the sea, and there is a great deal of technical uncertainty about the process itself.

A process that can use the CO_2 removed from power plants is enhanced oil recovery. In this process, pressurized CO_2 is pumped into those oil wells in which little or no oil can be recovered through conventional techniques. High pressure CO_2 mixes with the oil, thereby decreasing its viscosity so that it can flow and be pumped out of the well. Information compiled by Wolsky and Brooks (1988) suggests there is a potential market for 30–90 Gm^3/y^{-1} of CO_2 in the enhanced oil recovery area. However, this quantity of CO_2 is equivalent to the amount that could be recovered from

---

[7] Replacement power was prorated by multiplying the lost power by $1475/kW, the capital cost of a new nuclear power plant (EPRI 1990).

about 50 GW of fossil-fuel capacity, which is only about 10% of the equivalent U.S. capacity. The respective locations of the power plants and the oil wells could also limit the practicality of using recovered CO_2 for this purpose.

SUMMARY AND CONCLUSIONS

Energy forecasts indicate a large increase in worldwide energy consumption as population growth and industrialization expands, particularly in developing countries. One model, the Edmonds–Reilly global CO_2 model, projects an expanded role for fossil fuels, and particularly coal, in the next century. Substantial increases in fossil-fuel consumption are projected for developing countries. Any projected growth in fossil-fuel consumption could be readily accommodated, especially for coal. Up to 325 years of proven coal reserves remain worldwide, and 40–50 years of oil and gas, at current production rates.

While concerns about greenhouse-gas emissions, particularly CO_2, are emerging, it may be premature to take any immediate actions to curtail fossil-fuel consumption given the large scientific and modeling uncertainties that remain unresolved. Regardless of the outcome, fossil fuels can continue to fuel economic growth and development worldwide by reliance on CCTs. Clean coal technologies reduce CO_2 emissions per kWh of electricity generated due to their improved efficiencies. If it is determined that greenhouse-gas emissions should be reduced, fossil fuels and fossil-fuel-based technologies can continue to play a role. Fuel substitution to lower carbon-content fossil fuels is a feasible option, CO_2 scrubbing of power plant flue gases is also theoretically possible; however, at 90% control, the incremental cost of this option is currently estimated to be very expensive ($1200–$1800/kW); however, lower incremental capital costs are expected if only a 20% reduction (for example) is required. Any alternative fuel source or mitigation strategy to reduce CO_2 emissions should be carefully compared with the fossil fuel options to ensure that all aspects and costs of supplying energy and power are appropriately considered.

ACKNOWLEDGEMENTS

Work supported by the U.S. DOE, Assistant Secretary for Fossil Energy, under Contract W-31-109-ENG-38. The authors wish to recognize the research and production support provided by B. Kinzey (Battelle/Pacific Northwest Laboratory), and T. Rogers, J. Gillette and L. Kickels (Argonne National Laboratory).

REFERENCES

DOE (Dept. of Energy). 1986. Report to Congress on the relationships between projects selected for the clean coal technology program and the recommendations of the joint report of the special envoys and acid rain. Report DOE/FE-0072. Washington, D.C.: Office of Fossil Energy.

DOE. 1987a. America's clean coal commitment. Report DOE/FE-0083. Washington, D.C.: Office of Fossil Energy.

DOE. 1987b. The Role of Repowering in America's Power Generation Future. Report DOE/FE-0096. Washington DC: U.S. Dept. of Energy.

DOE. 1988. Energy technologies and the environment. Report DOE/EH-0077. Washington, D.C.: Argonne Natl. Lab.

DOE. 1989a. International Energy Annual, 1988. Report DOE/EIA-0219 (88). Washington, D.C.: Energy Info. Administration.

DOE. 1989b. Clean coal technology, the new coal era. Report DOE/EIA-0149. Washington, D.C.: Energy Info. Administration.

DOE. 1990. Clean coal demonstration program, annual report to Congress DOE/FE-195P. Washington, D.C.: Asst. Sec. for Fossil Energy.

Edmonds, J.A., and J.M. Reilly. 1983a. A long-term global energy economic model of carbon dioxide release from fossil fuel use. *Energy Econ.* **5 (2)**: 74–88.

Edmonds, J.A., and J.M. Reilly. 1983b. Global energy production and use to 2050. *Energy J.* **4 (3)**: 21–47.

Edmonds, J.A., and J.M. Reilly. 1986. The IEA/ORAU long-term global energy-CO_2 model: personal computer version A84PC. Report ORNL/CDIC-16 (CMP-002/PC). Washington D.C.: Inst. for Energy Analysis.

Edmonds, J.A., and J.M. Reilly. 1987. Uncertainty analysis of the IEA/ORAU CO_2 emissions model. *Energy J.* **8 (3)**: 1–29.

Edmonds, J.A., et al. 1984. An Analysis of Possible Future Atmospheric Retention of Fossil Fuel CO_2 Report DOE/IR/21400-1 (TR013). Washington DC: U.S. Dept. of Energy.

Edmonds, J.A. et al. 1986. Uncertainty in future global energy use and fossil fuel CO_2 emissions 1975 to 2075. Report DOE/NBB-0081. Washington, D.C.: U.S. Dept. of Energy.

EIA. 1985. Fuel Choice in Steam Electric Generation: Historical Overview. Washington, D.C.: Energy Information Administration, U.S. Dept. of Energy.

EPA (Environmental Protection Agency). 1988. The potential effects of global climate change on the U.S. Report to Congress. Washington, D.C.: EPA.

EPA. 1989. Policy options for stabilizing global climate. Report to Congress. Washington, D.C.: EPA.

EPRI (Electric Power Research Institute). 1989. Technical Assessment Guide—Electrical supply. Report P-6587, vol. 1. Palo Alto: EPRI.

Fay, J.A., D.S. Golomb, and S.C. Zachariades. 1986. Feasibility and Cost of Converting Oil- and Coal-Fired Utility Boilers to Intermittent Use of Natural Gas. Cambridge, MA: Massachusett Inst. of Tech. Energy Laboratory.

Flour Engineers. 1983. Technical and Economic Evaluation of Retrofitting and Repowering Oil Fired Boilers with Gas From Coal. Report EPRI AP-2854. Palo Alto, CA: Electric Power Research Institute.

Gluckman, M., R. Wolk, and G. Touchton. 1989. Power generation to meet the competitive challenge. AEIC Meeting, April 6, Palo Alto: AEIC.

Green, L. 1988. Alternatives to coal combustion. Presented at the American Assoc. for Advancement of Science, Boston, Feb. 11–15.

Hootman, H.A., and D.W. South. 1989. A comparison of global climate change initiatives in the 101st Congress. Report ANL/EAIS/TM-1. Argonne, IL: Argonne Natl. Lab.

Houghton, J.T., G.J. Jenkins, and J.J. Ephraums, ed. 1990. Climate Change. The IPCC Scientific Assessment, 365 pp. Cambridge: Cambridge Univ. Press.

IEA (Intl. Energy Agency). 1989. Energy and the environment: policy overview. Paris: Organization for Economic Cooperation and Development.

IEA. 1990a. Coal Information. Paris: Organization for Economic Cooperation and Development.

IEA. 1990b. Energy statistics of OECD countries, 1987. 1988. Paris: Organization for Economic Cooperation and Development.

Jäger, J. 1988. Developing policies for responding to climatic change, a summary of the discussions and recommendations of the workshops held in Villach (Sept. 28–Oct. 2, 1987) and Bellagio (Nov. 9–13, 1987). Beijer Institute, Stockholm. WMO/UN Environ. Prog. Report WMO/TD-No. 225. Geneva: WMO.

Keepin, B., and G. Kats. 1988. Greenhouse warming, comparative analysis of nuclear and efficiency abatement strategies. *Energy Policy* **16 (6)**: 538–561.

Marland, G. et al. 1989. Estimates of CO_2 emissions from fossil fuel burning and cement manufacturing, based on the U.N. energy statistics and the U.S. Bureau of Mines cement manufacturing data. Report ORNL/CDIAC-25 (NDP-030). Oak Ridge, TN: Oak Ridge Natl. Lab.

Power Magazine. 1982. Oil-Designed Boilers Can Burn Coal, pp. 88–91. New York.

Ramanathan, V. 1988. The radiative and climate consequences of the changing atmospheric composition of trace gases in the changing atmosphere. In: Changing Atmosphere, ed. T.S. Rowland and I.S.A. Isaksen. Dahlem Workshop Report PC 7. Chichester: Wiley.

Smelser, S.C., and G.S. Booras. 1990. An engineering and economic evaluation of CO_2 removal from fossil fuel-fired power plants. Presented at the 9th annual conference on Gasification Power Plants. Palo Alto: EPRI.

South, D.W. et al. 1990a. A fossil energy perspective on global climate change. Report DOE/FE-0614. Washington, D.C.: Dept. of Energy, Office of Fossil Energy.

South, D.W. et al. 1990b. Technologies and other measures for controlling emissions: Performance, costs, and applicability. In: Acidic Deposition: State of Science and Technology, Report 25. Washington, D.C.: Natl. Acid Precipitation Assessment Program.

Steinberg, M., and M.C. Cheng. 1988. Advanced technologies for reduced CO_2 emissions. Proc. of the 1988 meeting (88–87.3). Dallas: Air Pollution Control Association.

Warrick, R.A., and P.D. Jones. 1988. The greenhouse effect: impacts and policies. *Forum for App. Res. & Public Policy* **3**:48–62.

Wolsky, A.M., and C. Brooks. 1988. Recovery and use of waste CO_2 in enhanced oil recovery. Proc. of a workshop in Denver, March 19–20, 1987. Report ANL/CMSV-TM-186. Argonne, IL: Argonne Natl. Lab.

WEC (World Energy Conference). 1989. 1989 survey of energy resources. London.

Standing, left to right:
R. Klingholz, C.-J. Winter, J. Reilly, R.J. Swart, F. Staiß, G. Pinchera
Seated, left to right:
W. Fulkerson, G.I. Pearman, I. Mintzer, S.T. Boyle

9
Group Report: What Are the Economic Costs, Benefits, and Technical Feasibility of Various Options Available to Reduce Greenhouse Potential per Unit of Energy Service?

S.T. BOYLE, Rapporteur
W. Fulkerson, R. Klingholz, I.M. Mintzer, G.I. Pearman,
G. Pinchera, J. Reilly, F. Staiß, R.J. Swart, C.-J. Winter

INTRODUCTION

Prior to the industrial revolution, humankind utilized essentially "closed-loop" energy systems by capitalizing on natural energy flows, predominantly biomass, running water, and the sun itself. Since the beginning of industrialization some 200 years ago, humankind has increasingly utilized "open-loop" energy systems based on finite fossil fuels and, more recently, uranium. These systems entail the removal of finite resources from the Earth's crust, changing them chemically and releasing residuals (pollutants) into the geosphere. Some of these pollutants, the so-called greenhouse gases such as carbon dioxide (CO_2), are now thought to be leading to unprecedented climatic change (Houghton et al. 1990).

Limiting the Greenhouse Effect: Options for Controlling Atmospheric CO_2 Accumulation
Edited by G.I. Pearman © 1992 John Wiley & Sons Ltd

Energy related greenhouse-gas emissions are currently estimated to have contributed more than 60% of the global warming in the 1980s (Houghton et al. 1990; EPA 1989). Carbon dioxide emissions, of which fossil fuels constitute approximately 75%, and biotic emissions 25% contributed half of the estimated global warming in the 1980s (Houghton et al. 1990). When it is noted that some of the biotic-related emissions are due to the unsustainable burning of biomass for energy purposes, it is clear that energy-related CO_2 emissions must be a key focus for policy initiatives in reducing the risks of climate-induced impacts.

Our group interpreted its task as adding to the papers produced for this volume by Fulkerson et al., Kane and South, Winter, Reilly, Yamaji, Jenney, Ashton et al., Jochem, Lovins and Lovins, and Schipper (see chapters 5–8, 10–15). These provide a current assessment of the range of technologies and policies which can reduce CO_2 emissions, as well as reflecting a perception of the energy-related aspects of the climate change problem. Given that reductions in the greenhouse potential per unit of energy service can come about both through changes in energy supply and reductions in overall energy demand through energy conservation, supply and demand aspects of the energy equation were assessed. The emphasis of group discussions was on identifying important interdisciplinary and methodological issues raised by the papers, key informational gaps, and recommending new research. The latter is essential if remaining scientific uncertainties are to be resolved and if policymakers are to be given clear guidance on a range of actions which can actually reduce emissions.

Some of the themes brought up on a regular basis during discussions included the importance of regular interaction between climatologists, energy specialists, and social scientists, and the fact that some of the technologies and policies being discussed have multiple benefits. For example, advances in fossil-fuel combustion technologies not only reduce CO_2 emissions per unit of electricity generated, they significantly reduce acid emissions, they lessen energy security problems for countries heavily dependent on oil and coal imports, and they reduce energy resource tensions.

The report is divided into two parts: the first section deals mainly with technical issues, while the second section covers implementation mechanisms for the range of CO_2 abatement options. By clarifying the consensus on a technical basis, we were thus able to identify areas with and without common agreement, and the gaps in research and information which need to be filled. We decided that forestry conservation needed to be included in the technical section, as this was closely related to energy consumption in developing countries and has the potential to reduce future CO_2 emissions dramatically in these countries. It was commonly held by all participants, however, that reducing CO_2 emissions is not simply a technical issue. Methods of comparing the costs and benefits of various technologies and policies are needed, as well as an ability to facilitate, assess, and amend international instruments such as technology transfer and multilateral and bilateral funding. The need to develop relevant international organizations for a range of current and future tasks was a theme in the latter stage of our discussions.

TECHNICAL ISSUES

How Might One Best Assess the Maximum Technical Potential of Various Options Including Efficiency, Renewables, Nuclear, Fossil Fuels, and CO_2 Removal?

A clear consensus emerged that there is, in principle, no technical limitation to significant reductions in CO_2 emissions on the scale suggested by both the Intergovernmental Panel on Climate Change (IPCC working group 1; Houghton et al. 1990) and the Scientific Declaration of the Second World Climate Conference (SWCC 1990). Arhennius (pers. comm.) has suggested that "appropriate scientific understanding of the ecosystem and climate changes, and absence of appropriate technical solutions if applied in an integrated fashion, are not the main constraints for achieving an efficient reduction of the disruptive consequence due to greenhouse-gas buildup in the atmosphere in the medium term. The constraints lie in the absence of appropriate ways to implement this knowledge in the socio-economic sector."

Definitions of "technical feasibility" and "technical limits" vary, but we concluded that technical feasibility has four dimensions: technological, cost, quantity, and time. Technical feasibility means that the technology can currently produce some amount of carbon savings at a finite (reasonable) cost. Within this definition there is clearly a wide latitude for interpretation with regard to the definition of "currently," "some amount," and "reasonable." There is little disagreement that in the developed nations it is technically feasible to reduce net CO_2 emissions by 20% or more over the next few decades. If one includes forestry, alternative fuels, energy conservation, and changes in practices and lifestyle, it is likely that it is technically feasible to have a net uptake of CO_2 emissions in the atmosphere (Krause et al. 1990).

Describing these levels as "technically feasible", however, says little about the costs of achieving the reductions. As reviewed by Reilly (see chapter 5, this volume), estimates of the marginal cost of a 50% cut in U.S. emissions from current levels by the year 2020 range from a zero cost per ton of carbon (TC) to $450 per TC. There are widely varying estimates for countries where more than one estimate has been made using different types of methodology. Lovins and Lovins (chapter 14) present estimates that emissions could be cut by 20% or more at a net saving to the U.S. Other studies for Europe show a range of cost estimates, from a net negative cost for the stabilization of CO_2 emissions and cuts of up to 20% from current levels, to a cost of up to $120 per TC saved (Chandler 1990; Leach and Nowak 1990; Danish Ministry of Energy 1990; McKinsey and Consultants 1989).

To stabilize climate will require much higher levels of CO_2 cuts globally than 20%. In the longer term, this may require cuts of up to 90% from current levels in industrialized countries. The cost estimates of essentially transforming the current energy economy, which is dominated by finite fossil fuels, to one which is effectively based on non-fossil fuels, over 50–75 years, are necessarily speculative and heavily dependent on assumptions about the future path of technology, discount rates, economic growth rates, and the type of economic growth experienced. A key message to emerge from this perspective is that a wide range of technologies and policy measures, linked to a greatly enhanced research and development program, is likely to be needed to achieve such reductions.

Assumptions on technological impact are a major impact in the variations in costs and outcomes of the various studies referred to above (Manne and Richels 1990; Zimmerman 1990; Williams 1991). Analysis of these assumptions, plus methodological and other reasons for the above variations are important areas for additional research.

In looking at the "outer limits" of various technological options and the comparative contributions from each, the importance of scale on the speed of implementation was noted. Energy supply options tended to be slower to implement than demand-side options because of the longer lead and construction times; for example, building a power station. This notion, however, may not be applicable to some of the smaller-scale renewable and fossil-fuel options now emerging, such as combined-cycle gas co-generation plants, landfill gas plants, or wind parks. Some of these technologies take only a few years to construct. There was also a wide consensus that none of the supply-side options could really achieve the CO_2 abatement options necessary in the absence of an aggressive energy efficiency program (Manne and Richels 1991; Lovins and Lovins, chapter 14, this volume; Keepin and Kats 1988).

Though it appeared that there were no real technical limits in reaching high CO_2 abatement levels, detailed analysis of the degree of integration that is required to achieve this is limited. A number of CO_2 abatement scenarios assume a high level of technology penetration to achieve reductions in carbon emissions. The extent to which the effective penetration levels of the various technologies are potentially much lower, under current market and social conditions, is one of the key differences between researchers favoring "bottom-up" end-use analysis and those favoring "top-down" macro-economic analysis. This remains an important research question.

There was broad consensus that "technology-fix" or simple "hardware" approaches to the problem alone were unlikely to achieve very large CO_2 reductions that the IPCC assessment indicate may be needed to stabilize greenhouse-gas concentrations in the atmosphere. In some cases, however, a combination of new hardware and intelligent operator behavior could have the potential of achieving significant CO_2 reductions. It is clear, for example, that the differences in energy intensities between nations (e.g., between Japan and the U.S.) are a factor of varying infrastructures, social aspects, and energy systems and not just the use of less or more efficient technology (Schipper 1991). A series of measures, including "technology-fixes", new fuels and associated infrastructures, modifications to planning systems, and changes in lifestyle, are just some of the interlocking options needed. Figure 9.1 shows such a range of options and some illustrative costs for reducing CO_2 emissions from automobiles.

Research has suggested that a broad appreciation of the scale of improvements in energy efficiency needed to bring specific CO_2 reductions can be ascertained by utilizing the formula in Figure 9.2 (Holdren 1990). This links population, affluence, and energy intensity and shows how much reduction in energy intensity would be required to compensate for the expected growth in affluence and population. Further analysis using this type of approach may be useful for assessing the scale of reductions required

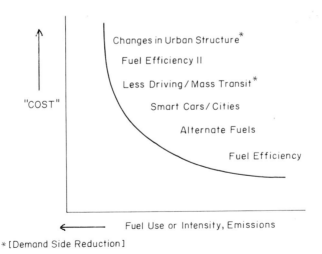

Figure 9.1 Conceptual cost curve for reducing fuel use and emissions indicating the level and effectiveness of selected options

in specific countries and regions, and in comparing the potentials of differing technical options.

Of the CO_2 abatement options, least is known about both the technical feasibility and costs of CO_2 removal from power stations. More research is required to clarify the technical feasibility of CO_2 removal and disposal, the associated efficiency reductions of power stations, and the overall system costs. Though some analysis has suggested that efficiency reductions and the costs would be high, this was by no means certain (Hendriks et al. 1989; Wolsky and Brooks 1989; Williams 1990).

How Would Potential Changes in the Technology and Structure of the Energy System, e.g. Increasing Electrification in Energy End-use Patterns and Changes in Transport Energy Systems, Affect Greenhouse-Potentials per Unit of Energy Service?

Two energy sectors, electricity and transport, were the focus of much discussion. This reflects the relative importance of the sectors in their contribution to CO_2 emissions (some 40 to 60% of total Organization for Economic Cooperation and Development [OECD] emissions), the higher historical and current rate of growth of emissions when compared with other energy sectors, and the complex mix of questions and research needs generated by the problems they generate.

The Electricity Sector

The electricity sector is clearly important in relation to CO_2 abatement. It consumes between 25 and 40% of primary energy in OECD and eastern European countries, uses 3 to 6 units of primary fuel to produce a single

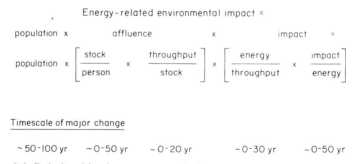

Figure 9.2 Relationship between population, consumption levels, durability of products, efficiency of resource use, and specific impacts (emission factors) (based on Holden 1990; Lovins and Lovins, chapter 14)

unit of electricity actually delivered to the customer, often constitutes the largest single source of CO_2 emissions in a country, and consumes a large proportion of the investment capital available in both developed and developing countries (one-fourth in the latter case) (British Petroleum 1990; pers. comm. with C. Cheatham, Pacific Energy Development Program, UN Development Program 1990).

It was acknowledged that electricity is a powerful symbol of progress, civilization, growth, and development for the public. There is a question over the extent to which there might be a mindset from utility managers and policymakers in over-emphasizing the role of electricity in the overall fuel mix. A current energy paradigm, which has quite strong support within the energy community, is that increasing electrification in all countries is both inevitable and desirable (WEC 1990). Assessing whether this is a suitable paradigm or not in relation to future emissions of CO_2 is problematical. It is not electricity as such that is required by consumers but the services it provides, i.e. light, power and communications, etc. There has been a lack of evaluation of the record of utilities worldwide and the related CO_2 emissions, and only limited understanding as to why there should be such a wide variation in the relative role of electricity in different countries. U. Colombo (pers. comm.), for example, has shown that among OECD nations, electricity contributes between 22.7 and 45.5% of the primary energy consumption. The results of such analysis might provide an understanding of the relative importance of political imperatives, historical precedents, and technological and environmental issues in the current and future roles of electricity in society.

A number of technical and data issues remain to be solved before some of the above questions can be answered and the future role of electricity assessed. These include increased understanding of the future role of advanced electrical technologies for industries (e.g., electro-arc induction), in contributing to increasing electricity demand and related CO_2 emissions, when compared with current nonelectrical technologies. They also include the potential role and impact of biotechnology and catalysts in industrial processes on electricity demand. Though conventional power stations remain relatively inefficient, there is the potential for technologies to improve such efficiencies substantially. For example, advanced steam injection turbines (Williams 1990), integrated combined cycle power plants (for coal and gas), co-generation, and commercial-scale electrical heat pumps can improve the overall efficiency of electricity production. They need, however, to be assessed against direct fuel use, in gas condensing and conventional fossil fuel boilers, plus solar space and water heating (possibly in combination with advanced heat pumps). As well as comparisons between such differing supply options, the significant potential of improved electricity efficiency, which could allow a major increase in electricity services while leading to a static or falling electricity demand, needs to be recognized and assessed in combination with supply-side options (Lovins and Lovins, chapter 14, this volume).

The potential future role of electricity in the transport sector may have a significant impact on future electricity demand and consumption levels. This needs to be assessed in conjunction with long-term transport scenarios and plans. Given the importance of energy use in areas such as minerals mining and waste disposal, there was a clear consensus that any analysis should always include the total fuel cycle for CO_2 emissions and other greenhouse gases (Eyre 1990; San Martin 1989).

Planning the future role of the electricity sector should be integrated into an analysis of the total energy system, taking into account economic, technical, social, and environmental aspects. A final area of discussion thus focused on the future role of electrical utilities. Their role now appears to be in need of redefinition, given that some of the original premises and objectives of the industry no longer hold good in a future where CO_2 emissions restraint is important. Simply responding to increasing electricity demand by building more power stations, or even actively marketing electricity in order to increase demand, is no longer a key objective among many U.S. utilities (Fickett et al. 1990). Such a transformation in thinking may spread to utilities throughout the rest of the world. Key questions thus include:

1. Should we and can we integrate a social role for the utility into the current structure and activities?
2. Should we attempt to turn the utilities into energy service companies, where demand-side management and pollution reduction are an integral part of their business?
3. Should gas and electricity utilities be integrated, either at a local or national level, to improve both economic and environmental efficiency and flexibility between fuels and technologies, or would this risk increasing their often monopolistic structure?

The Transport Sector

Transport is the fastest-growing sector for CO_2 emissions in OECD countries, currently increasing at a rate of between 2 and 5% (Jenney, chapter 11, this volume). Though there is a large potential for improving fuel efficiency, the volume of vehicles and distance travelled will, on the basis of current trends, swallow up much of the improvements (Mackenzie and Walsh 1990; Jenney, chapter 11, this volume). A special characteristic of this sector is its almost total dependence on limited sources of liquid fossil fuels. In a scenario of rising vehicle ownership and miles traveled, significant reductions in CO_2 emission can only occur where large efficiency improvements are assumed (Lovins and Lovins, chapter 14, this volume). In a global sense this will be more difficult to achieve since both developing countries and eastern European countries have the potential to follow the OECD model of transport development, both at local urban and national level. If action to reduce overall CO_2 emissions is required, the transport sector needs to be a priority area for new initiatives and a wide range of multidisciplinary research and analysis.

There is a wide range of technological options available which can reduce CO_2 emissions in the transport sector. These range from improving overall fuel efficiency in private vehicles, mass transit, and airplanes, switching from private transport to more efficient mass transit, to switching to a number of alternative fuels such as hydrogen and electricity. In order to make optimum policy choices, there is a need to improve basic data in the transport sector, including accurate fuel consumption of the whole vehicle fleet as against new vehicles measured in standard fuel-efficiency tests, vehicle occupancy and usage, and identification of the reasons for vehicle trips.

There is wide acceptance of the medium-term potential for significant fuel-efficiency improvements in automobiles (Bleviss 1988). There remains some disagreement, however, on the potential in the shorter term, taking into account the inertia of car manufacturers, transport systems, and vehicle life expectancy (Difiglio et al. 1990; Plotkin 1989). At a recent OECD/IEA meeting on transport, experts indicated that it is possible to bring about significant efficiency improvements over the next 5 to 7 years. Through a combination of technological improvements and the sacrifice of some vehicle performance, they concluded that vehicle consumption could be reduced by 20 to 50% from today's new vehicle levels. The higher figures were assumed to be reached by a shift to lower engine capacity (OECD/IEA 1990). Lower figures have been presented by industry representatives.

A range of policies are available to accelerate the normal vehicle stock turnover. "Feebates," for example (fees or rebates on new vehicles depending on their efficiency), could be designed to speed up the scrappage of older, inefficient models, hence improving the prospects of moving more quickly to higher fuel-efficiency models. Since it requires fossil fuels to mine and produce the metals, plastics, and rubber needed in each vehicle, premature scrapping of the vehicle implies additional CO_2 pollution due to this inherent energy. Recycling materials may mitigate this effect, but the implications of a major scrapping program need to be considered, since pollution levels may worsen over the short term before improving once again.

A wide range of fuels other than the current oil-based ones (gasoline/diesel) exist, though with a few exceptions (e.g., the ethanol program in Brazil) these are only available, in most countries, in relatively small quantities. These fuels have varying impacts on relative CO_2 emissions (Figure 9.3; Sperling 1989); indeed, some fuels, such as methanol derived from coal, produce higher emissions per vehicle mile traveled (Wang et al. 1990). On the basis of current analysis, it appears that only hydrogen (if produced from nonfossil-fuel sources of electricity), nonfossil-fuel sources of electricity (such as solar), several of the biomass-based fuels (e.g., ethanol), and possibly direct solar energy by photolysis may offer the potential for significant CO_2 reductions.

When assessing the relative contributions of alternative vehicle fuels to climate change, it is important that greenhouse gases other than CO_2 be assessed. This should include, for example, the role of carbon monoxide, hydrocarbons, and ozone. A number of initiatives which have the aim of reducing local air pollution may not have assessed the impacts of proposed policy changes on greenhouse-gas emissions. For example, the Southern California Air Quality Management Plan (SCAQMP 1989), which aims to replace substantially gasoline-driven vehicles with methanol derived from natural gas over the next decade, is at best likely to have a minimal effect on reducing CO_2 emissions. Unless the methanol fuel is based on sustainable biofuels, the initiative may, under some assumptions, actually increase emissions (SCAQMP 1989; Sperling 1989).

Transport Systems and Structures

Reducing CO_2 emissions from the transport sector is not simply a technological issue. It is linked to planning, urban design, consumer behavior,

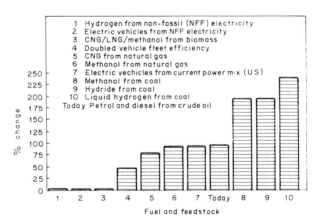

Figure 9.3 CO_2 emissions from vehicle fuels (% change from current gasoline/diesel-based vehicles). From Sperling (1989)

income and choice, energy price and availability, and political philosophy. One important challenge for research is to deliver attractive options for the personal mobility of citizens from developing and transitional economies, without encouraging a rapid increase in greenhouse-gas emissions and extensive urban sprawl which is dependent on large and inefficient automobiles, such as in the Los Angeles model. Another important area of research is the interaction between communication and transportation. It is not at all clear whether all new developments in the communications sector lead to an increase or decrease in transportation needs. For example, the rapid spread of computer and electronically linked workplaces could, in theory at least, lead to a reduction in commuter travel. We may require multidisciplinary research projects to assess the sociological as well as technological aspects of work and communications to determine whether this will occur.

The factors behind urban density and transport choice and usage are varied. Figure 9.4 shows the results of recent research into urban density and gasoline usage. Of the many variables tested in this study, there appears to be a strong correlation between vehicle densities and usage, consumer income, and fuel prices (Newman et al. 1987). It was suggested in our discussions that the correlation may also be due to the relative gasoline costs and incomes of the city inhabitants concerned (Schipper, chapter 15, this volume). Such a correlation would show that low gasoline prices, linked to high disposable income, tends to lead to very low urban density and high vehicle usage. Other factors such as the existence of planning/land zoning controls and the availability of cheap building land may also be important in many instances. Enhanced research that looks at historical and current trends will greatly help decision-makers in avoiding future decisions, in both developed and developing countries, which will lead to large greenhouse-gas emissions.

An examination of the financial aspects of transport is important in order to understand the range of capital flows and sources (e.g., in transport infrastructure), and the subsidies provided for private and mass transit. Low fuel prices, linked to a lack of readily available mass transit systems, and generally low-density urban development are features of the U.S. automobile-dominated transport system. The extent to which modifications to pricing structures (e.g., through a high carbon or energy tax or removal of workplace parking spaces, company cars, tax breaks, and mileage allowances) might alter current transport intensities and modal splits is an interesting area for research. This would lend itself to a range of case studies on different communities, looking at both success stories and failures.

Though 75% of the world's vehicles (and associated vehicle kilometers traveled) are associated with the U.S., western Europe and Japan, the growth potential in eastern European and developing countries is significant

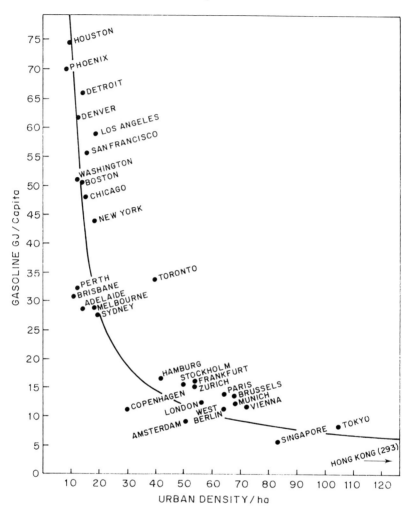

Figure 9.4 Gasoline consumption versus urban density in principal world cities in 1980. From Newman et al. (1987)

(Flavin and Lenssen 1991; Jenney, chapter 11, this volume). The extent to which land zoning and the planning of more dense urban settlements, in order to allow successful and cost-effective mass transit systems and pricing structures, might influence future demand projections is important as a focal point for multidisciplinary research. Such research has significant implications for developing and transitional economies, in identifying appropriate transport models for these countries to follow.

The issue of "market" instruments versus "command and control" policies is the focus of recent debate (Chandler 1990). The appropriate mix of policy measures, in order to achieve optimum econonomic efficiency commensurate with rapid reductions in carbon emissions, is likely to vary between countries. Energy taxes and the removal of subsidies unwarranted on environmental grounds are key elements in assessing an appropriate mix.

The extent to which city and national authorities might intervene to reverse current trends of congestion and air pollution is not a purely scientific question. For example, though it is clear that reducing traffic congestion in the short term should improve the fuel efficiency of vehicles, it may be that such efforts encourage additional trips or vehicle owners. This is just one example of the multifaceted nature of transport and CO_2 emissions.

Given the wide range of research needed to resolve the problem of how to reduce CO_2 emissions in the transport sector, a question remains over the potential funding for this. A significant amount of transport-related research is already carried out by vehicle manufacturers, airlines, and oil companies. Little, however, is related to the above issues. To what extent can they be encouraged to fund this work in the future?

What Information and Methods Are Needed to Assess the Realistic Potential of Forestry Conservation Efforts to Reduce CO_2 Accumulation?

Though forestry carbon flows and sinks clearly have wider implications than just energy issues, they are major components of current CO_2 emissions in many developing countries (Houghton et al. 1990). Nonfossil-fuel sources and quantities of anthropogenic carbon emissions are, however, much less certain than fossil-fuel-related emissions. The focus of our discussion in this area was on forestry and soil conservation and, in particular, on biomass systems for energy.

Forestry

Biomass energy systems have the potential to replace fossil fuels as a fuel source with no net carbon emissions. In many parts of the world, however, such biomass systems are not being operated in a sustainable manner (Houghton et al. 1990). In addition, there is a growing tendency for many developing nations to switch from biomass to commercial fossil fuels. There have been a number of attempts to improve the combustion efficiency of biomass systems, for example in the area of woodstoves, but the success rate has been mixed (Leach and Mearns 1990). More advanced biomass combustion technology such as gasifiers (Williams 1989; Larson 1990) and the "whole tree burner" concept (Ostlie 1989), particularly if linked to the utilization of organic wastes and sustainably farmed plantations of varying size (Hall et al. 1990; Marland 1991), may help redefine the future role of biomass.

The present scientific understanding is that around 1.6 ± 1.0 GtC y^{-1} are emitted to the atmosphere due to deforestation, and that this could be reduced or stopped by halting deforestation and initiating major reforestation programs (Houghton et al. 1990). Reference was made in our discussion to the ambitious target of the 1989 Noordwijk conference in the Netherlands for planting 12 Mha y^{-1} of trees by the year 2000. This target was subsequently criticized as somewhat unrealistic at the Second World Climate Conference in Geneva (SWCC 1990). The German Enquête Kommission has suggested a target of halting net deforestation by the year 2010, before recovering to 1990 forestry levels by 2030 (Enquête-Kommission 1990).

There are multiple benefits from both halting deforestation and initiating sustainable reforestation programs. These include increasing the economic potential of marginal lands, reducing soil erosion, providing income for local populations, providing a sustainable and local fuel supply, improving biodiversity, preventing flooding and droughts, improving water management, and sequestering carbon. Thus forestry conservation cannot be considered in isolation as a CO_2 abatement option. Further, as even mature forests may actually act as significant sinks of carbon as CO_2 levels increase in the atmosphere, greater protection of forested areas might be important. Quantifying this possible feedback effect is extremely difficult. All that can be said at present is that there exists some evidence that forest growth may have been much more important in controlling the growth of CO_2 levels than previously thought (Enting and Mansbridge 1989; Tans et al. 1990).

Estimates of deforestation levels are still poor. The Enquête-Kommission has suggested that the rate of closed forestry deforestation was 205,000 to 495,000 km^2 per annum in the early 1980s, and that for open forests was some 100,000 to 294,000 km^2 per annum (Enquête-Kommission 1990). Brazilian scientists have suggested that the figures assumed for Brazil are seriously flawed, however, and argue that the true deforestation rates may be a factor 3 to 4 lower than those suggested by the UN Food and Agricultural Organization (FAO), Myers, and World Resources Institute (WRI) (Pereira da Cunha 1988). Further research is needed to give more accurate figures. A variety of methods are available to achieve this, including soil surveys, remote sensing, ground studies, and forestry models. A longer-term perspective of 50 to 100 years suggests that the potential of carbon sequestration due to reforestation is at least an order of magnitude lower than the potential carbon releases from fossil fuels during this period. In the next 30 to 50 years, there may be a potential to sequester perhaps, as a maximum, some 5 to 10% of total anthropogenic carbon emissions. This broad brush assessment requires a great deal of further work to validate (EPA 1989).

Research in the U.S. with plantation-scale woody and herbaceous biomass for energy crops indicates a significant potential at costs comparable to coal (SERI 1990; Williams 1990; Wright and Ehrien 1990). The sustainability of trees grown on short rotation cycles (ca. 6 years), however, has yet to be demonstrated. If such production could be coupled with efficient conversion processes to produce liquids for transportation or electricity, biomass could be important on a global scale for both developed and developing nations. The implications of such biomass energy systems, for example in the significant land areas required (Marland 1988) and the necessary infrastructure, need greater assessment.

Recommended Research

There are presently four high priority information needs that are essential to either a global climate change convention or a freestanding international forest agreement. These are as follows:

1. Policies to limit deforestation and enhance reforestation can only be effective if these are well embedded in plans and goals, with respect to national social and economic development. Additional interdisciplinary research (a) to clarify the role of forests in national and international social and economic systems and (b) to assess the implication of forest protection and reforestation versus other goals is urgently needed.

2. Operational Biomass Burning Systems. There is an urgent requirement for accurate annual estimates of CO_2 and other trace-gas emissions from biomass burning. The estimates are needed at the national level as part of the improvement of knowledge of the biogeochemical cycles of CO_2 and CH_4, in particular. Such estimates will improve the predictability of climate change resulting from the accumulation of these gases and will identify options for intervention in emissions. During the next year a cooperative international effort involving the IGBP and national/regional centers should be established, with actual data analysis beginning in the 1992 calendar year. The initial system could be quite crude, relying on existing satellite platforms and the best-available estimates of biomass and emission factors. It is anticipated that the accuracy of such a system could be slowly improved over the next five to ten years as better information becomes available from laboratory emission tests, small test burns under field conditions, and improved space-based platforms. The evolution of such a system fits in closely with the need for a compliance system once any target agreements for forests are agreed.

3. Cost-Effective Management Practices. The implementation of forest management plans requires improved information on the cost-effectiveness of agroforestry practices in different parts of the world. An initial first step should include the compilation of existing information in a manual for

planners, policy analysts, research managers, and development assistance funding agencies. This effort needs to be complemented by expanded efforts to demonstrate and document agroforestry practices in the field (Leach and Mearns 1989).

4. National Forest Plans. Any future international forest agreement is likely to rely on the development of national forest plans to integrate diverse objectives, including sustainable forest use, biomass energy systems, biodiversity, increased carbon storage, and economic development. The potential competition between biomass energy production and other land-uses, such as agricultural production, forestry and nature conservation, needs to be taken into account in the development of such plans.

An improved process to develop national plans for the boreal, temperate, and tropical forests is needed. The development of such a process should build on an improved Tropical Forest Action Planning process through the cooperation of the World Bank, FAO, national governments, and nongovernmental organizations. It should include extensive inventories.

Other informational needs for both forestry and soil conservation include the impact of photochemical and acid pollution on boreal forests and, in particular, their ability to sequester carbon and the wide range of nontechnical issues affecting deforestation (e.g., land ownership, the marketing of forest products, cultural and social variations between regions and countries, national subsidies and incentives in the forestry sector, debt, etc.). Though the former area of research is likely to be relatively long term, the latter could be completed within the next few years.

IMPLEMENTATION MECHANISMS FOR CO_2 ABATEMENT OPTIONS

The introduction of appropriate CO_2 abatement options does not just entail technological assessment. Methods of comparing the costs and benefits of technologies and policies are necessary to evaluate optimum strategies involving solutions which are appropriate to local, national, and international circumstances. In addition, there is a need to facilitate technology transfer, to apply international economic/political instruments, and to develop relevant international organizations for such activities as administering funds, collating data, and carrying out research.

It was implicit in our discussions that implementation problems for developed nations in many instances were significantly different from those confronted by developing nations. OECD nations have largely stabilized CO_2 emissions since the oil price hike of 1973–1974, and all except the U.S. and Turkey have committed themselves to the stabilization or reduction of current emissions (SWCC 1990). Much is already known about the technical and social feasibility of the strategies for achieving reductions, although additional research and development (R & D) is needed in the areas discussed below.

The situation for the developing and transitional countries (eastern Europe and the USSR) is somewhat different. Emissions in many developing nations may increase rapidly from current levels in the absence of concerted policy action, though high levels of inefficiency currently exist in the energy systems (Munasinghe, chapter 18, this volume). Emissions in transitional countries could move in a number of directions. Though a low level of energy service is currently provided to consumers, there are significant inefficiencies in the energy systems (Bashmakov, chapter 3, this volume). Large increases in the energy services provided could thus be achieved without increasing energy demand or CO_2 emissions, though there is a large latent demand for energy services such as personal transport, which could increase overall energy demand rapidly. The research agendas for these countries are thus much less defined. Technology transfer and adaptation of current energy and social systems are likely to be major features of such an agenda, and these issues formed the focus of our discussions on implementation mechanisms for CO_2 abatement options.

What Information and Methods are Needed to Compare the Costs and Benefits of Various Strategies to Reduce CO_2 Accumulation?

The range of estimated costs for CO_2 abatement is large. Reilly indicates a range between zero and $450 t^{-1} of carbon abated (Reilly, chapter 5, this volume). Other analyses suggest the range may be even greater than this, including some analyses which suggest negative overall costs for stabilizing and reducing emissions in some countries (Lovins and Lovins, chapter 14, this volume; Leach and Nowak 1990; Chandler 1990). One of the main reasons for such a range of views is the differing methodologies and assumptions used by researchers.

There are currently two main methods of comparing costs and benefits of the various CO_2 abatement strategies. These are:

1. Engineering cost accounting or "bottom-up" analysis, which looks at potential CO_2 reductions in a sector-by-sector analysis, producing relative and net costs for the various options.

2. A "top down" econometric analysis, heavily based on mathematical relationships between energy prices, economic growth, income, and the relative elasticities between each; these relationships must be constructed from data sets based on past trends, particularly those where variations in energy prices occurred (e.g., 1973–1985).

These two disciplines have a different philosophical perspective that contributes to the differences. The engineering approach is a prescriptive one that attempts to find the best solution for a specific problem or individual application. The economic approach is a descriptive approach, holding that individual agents/consumers know best what the situation is and which solutions to choose. Within the economic approach is the recognition that individuals will seek out or react to the prescriptive solutions offered by engineering approaches. Within the engineering approach is an appreciation of the complexity of the market imperfections that must be addressed and, particularly in the more sophisticated analyses, of the power of market implementation mechanisms.

In applying these approaches for estimating the likely future energy use or the costs or benefits of changing fuel use and technologies, each approach has both strengths and weaknesses. Focusing on the weaknesses first, the engineering approach, which looks at any technology altering energy use and supply, may fail to consider some aspects of the new technology that are relevant to the technology user. The value that technology users (both final consumers and industry) place on convenience, labor requirements, and fitting the new technology into existing structures and patterns of use is difficult to evaluate without market tests. Failure to incorporate fully all such trade-offs between the performance of the old and new technology may also lead to an overestimate of the new technology's market potential. One additional concern is whether this approach can evaluate other characteristics completely, e.g., the long-term equilibrium effect of reducing energy use and relative costs, which could lead to a rebound effect as consumers use free capital for other energy utilizing expenditures. The strengths are (a) that the detailed sectoral analysis allows a policymaker to discern important trends (such as technological change and market saturation), which may not be picked up by macroeconomic analysis, (b) that the analysis addresses specific solutions to a specific problem, and (c) that the analysis is not dominated by past data sets that are potentially inappropriate for the current situation.

The econometric approach attempts to evaluate the actual willingness of consumers to purchase new technologies by extrapolating past data trends and relationships. This approach cannot, however, easily consider untested, new technologies, and it implicitly treats market barriers as a higher unit cost. This tends to lead to an overestimation of the costs of carbon abatement. Under the econometric approach, a question often arises as to why many zero or near-zero cost opportunities do not get adopted. In this approach there is less acceptance of the many market failures and barriers which are behind the low uptake of the technologies. The econometric approach will also tend to overestimate the costs because it treats barriers as a permanently higher cost, whereas the cost of the technology may be lower once the barrier is overcome. For example, a high start-up cost for a new technology may fall rapidly once large-scale production starts or full information dissemination is completed. The major strength of the macroeconomic approach is that CO_2 abatement and related costs are integrated into the wider economy, hence long-term equilibrium costs to the economy can be measured (Chandler and Nichols 1990; Dower and Repetto 1990; Pearce 1991). Potential benefits for industrial investment, unit costs, and hence competitiveness with other nations can thus be measured.

Research Gaps: Confusions and Controversies

How can a comparison between the two approaches be carried out? One important point is in recognizing the different way costs are reported, including the total investment cost, the fuel cost, the net additional cost/benefit relative to existing technology, the annualized levelized cost, the inclusion of environmental costs, and, finally, the role of the expected prices of the analyst versus the price potential of actual adoption, which may be higher or lower. Another is to address the remaining informational gaps in both types of analyses (see Table 9.1).

The question of reconciling both approaches is an important one. Though it is currently difficult to achieve a full consensus on this question, many of those present felt that such a reconciliation was a crucial step forward in

Table 9.1 Information gaps in current energy/CO_2 modeling exercises

- Costs of reducing CO_2 in the industrial sector
- Successful policies in the industrial sector
- Remaining disagreements on the cost-effectiveness of fuel efficiency improvements
- The ancillary benefits of reducing fossil-fuel use, e.g., reductions in methane emissions from coal mines.
- The costs of new infrastructure, e.g., in eastern Europe with the dissemination and production of energy efficiency equipment (controls, insulation, etc.)
- The synergies of energy efficiency investments at a macroeconomic level, e.g., reducing the need for steel in coal mines hence freeing up scarce capital
- The relative costs of efficiency and other CO_2 abatement strategies in different countries and regions, e.g., OECD, eastern Europe, and developing countries
- Full fuel-cycle analysis for a range of options
- The lack of quantification of market failures and barriers
- Understanding the sociological and marketing aspects of consumer choices in technology, reflecting the nonrational (in a pure economic sense) aspects of such choices
- The validity or otherwise of baseline scenarios

the current research and policy debate. It was noted that a number of groups are already working together to integrate some aspects of their work, e.g., between Stanford University and Lawrence Berkeley Laboratory; meetings convened in late 1990 and early 1991 by Dr. Michael Grubb on behalf of UNEP and involving a range of experts from both modeling communities; and meetings between the U.S. and the European Community. So-called "hybrid" models, incorporating both econometric and engineering considerations, to some extent have been developed for both energy supply and demand sides of the equation. Examples include the Edmonds–Reilly model, the EPA energy model, and the Oak Ridge Industrial Model. Utilization of the results of end-use analysis by macroeconomic models is already taking place. Detailed criticism of analysis in 1990 by Manne and Richels, which suggested very high costs for reducing global and U.S. CO_2 emissions (Manne and Richels 1990; Williams 1990), has led to modifications in some of the original assumptions. The greater range of sensitivity testing subsequently used by these analysts has suggested that there is considerable potential under a range of circumstances for low-cost CO_2 emission reductions in the U.S. and that enhanced research and development programs can change the outcomes significantly (Manne and Richels 1991).

Other research gaps include:

1. The Role of Subsidies. In particular, questions to be addressed include: Which fuels and efficiency technologies are benefiting from subsidies and which are penalized? What is the basis of the subsidy? Does it address a perceived externality or market failure, or are there other irrational reasons for it?

2. Externalities. The calculation of social costs, particularly environmental externalities, is well advanced (Ottinger et al. 1990). However, there is as yet no consensus as to how these costs might be expressed in monetary terms and applied in the models. Further research is needed in this area.

3. Modifications to the Current Energy/CO_2 Models. The econometric models need to be developed to analyze the impacts of policies other than energy/carbon taxes. "Bottom-up" engineering models so far do not incorporate equilibrium effects. Is it possible to amend them to analyze this?

Two final observations were made by the group. Part of the difference between the costs in the two approaches may not be a fundamental difference but only a difference of where and how the costs are accounted. The econometric approach, by its very nature, accounts everything as a dollar or monetary value, whereas the engineering approach only counts certain costs and ascribes the lack of penetration of technologies to market barriers, hence needing further information, innovative financing, regulations, standards, etc. Putting a value on some of these barriers may help narrow the difference between the two estimates.

How Can the Technical, Institutional, and Social Feasibility of Various Options and Strategies Be Best Evaluated?

The methodology for assessing the technical feasibility of a wide range of options has been well established over the past 15 years, and a broad consensus has been reached. Institutional and social feasibility is, however, less easy to measure, and there remains a major problem in applying the results of technical analysis.

Total fuel-cycle costing is a key approach for evaluating the non-economic issues of CO_2 abatement. A great deal of work has been and is being produced on total fuel-cycle costs, including the externalities (Ottinger et al. 1990; Hohmeyer 1988). There appears to be a growing consensus on the externalities of the various CO_2 abatement options within an order of magnitude, at least for nonnuclear options. For nuclear power there is still major disagreement, however, particularly when dealing with low-risk/high-consequence catastrophic accidents (Hohmeyer 1988). There also remains a problem as to how actually to apply the externalities for systematic policy analysis.

Continuing controversies of where to draw the energy system boundaries remain. For example, does one include the embodied energy in related products, and can one evaluate the overall impacts of a major program of technology introduction on a country's economy? The development of extensive input–output tables covering the wide range of possible variables is one possible option for clarifying the issue.

The assessment of the social acceptability of new technologies and policies requires considerable additional analysis. There is still resistance by the public to new technologies, particularly energy supply-side options in their local area. Though some evidence is available from the social sciences on how people react to new technologies and systems, these will not always have appropriate variables built in as far as CO_2 abatement is concerned. New research and demonstration programs are needed to evealute public/consumer reaction to CO_2 abatement options, particularly those which will have substantial impacts on lifestyle and social organizations. Work on the importance of more democratic and/or participatory decision-making processes in choosing the options and the applications of them, is a necessary addition to these demonstration programs.

Two large-scale demonstration ideas were strongly supported. The first would entail turning an electrical/gas utility into a full energy service company, and assessing both the process, as the utility changes its modus operandi, and the eventual outcome in economic, social, and environmental terms. The second relates to modifying transport systems in several cities in order to influence modal shifts, mobility, travel distance patterns, and new transport technologies. In addition to such demonstrations, international comparisons of energy-conscious/energy-autonomous communities (e.g., Davis [U.S.], Rottweil [Germany], Milton Keynes [UK] could provide some important lessons for policymakers and other communities.

What Can Be Done in the Way of Technology Transfer to Help Developing Countries and Transition Economies to Reach a Low Energy, Low Emission Path to Sustainable Growth?

Though many developing countries are relatively low CO_2 emitters in the energy sphere at present, they are like to be higher CO_2 emitters in the future as a result of both a shift to commercial fossil fuels and population growth. Analysis of the energy intensity of a range of countries and country groups over time shows that developing countries have not yet peaked. Figure 9.5 shows how, with the onset of a country's industrial revolution and the transition to "commercial fuels," energy intensity increases rapidly before it stabilizes and then falls off in a sustained manner (Colombo, pers. comm.; Boyle 1990). It is noticeable that the later in time a country commences its industrialization phase, the lower the energy intensity peak. Such trends reflect the ongoing development of more efficient machines and energy transforming equipment, and the availability of a wider range of fossil fuels that can be utilized more efficiently.

A key issue affecting future trends of global CO_2 emissions will be the level of energy intensity which developing countries reach in future decades as they industrialize. This, in turn, will depend on their access to efficient and low CO_2-polluting technology from the industrialized North, the development of indigenous low CO_2-polluting technology, and the patterns of economic development followed. The role of technology transfer will thus be important. It should be accepted that such technology and the transfers of technology systems can proceed in both directions. In developing and centrally planned economies are technologies and approaches that can enhance the low CO_2-emitting nature of an economy (e.g., solar architecture, recycling, and mass transit systems).

Discrete country-by-country and sectoral analysis within each country is important in understanding the unique characteristics of the potential recipients of technology transfer. Historical, cultural, and social dimensions to this research are as important as current fuel mixes, trends and market conditions in determining the potential success of the technology in the country. CO_2 abatement may not be the key criterion in relation to technology transfer for many countries, especially where current and even future emissions will be relatively low. Where the low CO_2 technology can

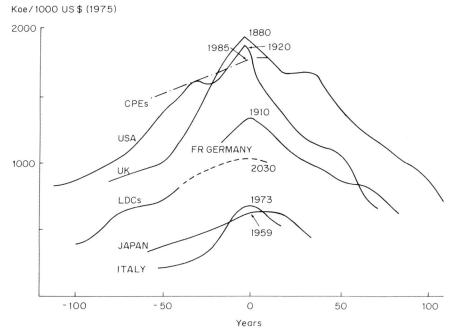

Figure 9.5 Patterns of commercial energy intensity during the development process. From Colombo (pers. comm.)

have multiple benefits, however, including reductions in relative CO_2 emissions, it should have an added attraction.

In the past, many well-intentioned efforts at technology transfer have failed (pers. comm. with C. Cheatham, Pacific Energy Development Program, UN Development Program, 1990). Unfortunately many of these programs and efforts have not been analyzed adequately, hence the reasons why good technology (in an engineering sense) failed are not always clear. Some of the ingredients are known, however, which suggests that the failures were a function of many things, including social and cultural issues, lack of adequate training and maintenance, and inappropriate scales of technology and associated institutions.

Setting up local energy efficiency, renewable, and environmental protection centers in developing countries is an important new initiative. There are a growing number of initiatives of cooperation between industrialized and developing nations which could provide a test-bed for research, e.g., the U.S. EPA/Battelle funded energy efficiency centers in eastern Europe and an Italian funded environmental center in Egypt. The establishment of local and possibly regional "centers of excellence," for the research and deployment of renewable energy and energy efficient technologies, would greatly enhance the penetration of these technologies in the developing countries. Tailored to local needs and possibilities, these centers could be organized and fashioned to some extent along the lines of the Consultative Group on International Agricultural Research centers that were instrumental in guiding the green revolution.

The need to evaluate the success or failure of technology transfer is problematic and requires additional research. One technique that will be helpful to both aid agencies and recipient countries for energy technology, however, is "least cost planning" with associated "demand side management" programs and subsequent evaluation (Goldman et al. 1989; NARUC 1988).

The question of developing countries "leapfrogging" technologies (defined here as technology which has been tried and tested in northern industrial countries but which may not yet have widespread application) or simply following technologies of industrialized countries, raises theoretical and practical issues. Though "leapfrogging" has attractions in allowing developing countries to move ahead and reduce relative CO_2 emissions quickly, it may lead to setbacks when unplanned technical hitches and problems occur. The lack of a support infrastructure may seriously impair the developing country's ability to keep the advanced technology working at optimum levels. Research is needed into this issue.

Successful technology transfer involving the private sector is essential. There is a clear need to break the current cycle of inefficient and polluting technologies being exported to developing countries. A promising initiative might be to establish a clear Code of Conduct for industrialized countries and companies to follow, for example, building on recent efforts in relation to hazardous waste and the Basel Convention, which prohibits the exports of such wastes to developing countries. Previous efforts by the UN Center for Trade and Development (UNCTD) on the transfer of technology in the mid-1980s did not succeed due to differences in objectives and approach by differing groups of countries. Such a process can only evolve by involving all parties in the technology transfer. Research as to whether this can now proceed and whether it will need to be supplemented with other measures, such as compensatory funds to pay for more expensive technology and trading arrangements through General Agreement on Trade and Tariffs (GATT), for example, is needed.

A key emphasis for future work is on demonstration and commercialization rather than simply research programs. The areas where additional work is most needed include energy efficiency technologies and techniques and evaluations of past and current technology transfers (Geller 1990). Such evaluations need to be transparent and where possible, carried out independently. They will need to highlight both the success stories and

What Criteria Are Appropriate for Choosing National and International Levels of Research, Development, and Demonstration on Strategies to Reduce Accumulation of CO_2?

Much of our discussion centered on the role and focus of research, development, and demonstration (R,D & D). Notwithstanding the identification of a wide range of new research requirements, a consistent theme to emerge was the need for the active implementation of technologies and related systems already well-understood in a technical sense, but having had limited application and use so far. It was agreed that different kinds of investments are required to facilitate the rapid introduction of a new technology; investments directly concerning the technology (conventional R & D), and indirect investments to build up social acceptance and the necessary infrastructure to implement the new technology. Both types of investment have to be made if successful implementation is to occur. These are important messages for policymakers, in that CO_2 abatement options can be embarked upon almost immediately, even while additional research is being carried out, as long as associated investments are adequately planned for by both government and the private sector.

There are three main types of R, D & D: that related to technology, that to the social sciences, and basic/cross-cutting research (Fulkerson et al. 1989). Where capital is limited, a systematic set of criteria for deciding on R, D & D priorities is needed. The following criteria for making key choices on where to allocate resources, and in deciding the absolute level of resources, emerged from discussions:

- The size of the potential CO_2 reductions—"the biggest bang (CO_2 reduction) for the buck"—relative to the timescale on which this could be achieved. It is recognized that results would change quite significantly on the timescales of 3–4 years, 15, 50 and 100 years.
- The economic feasibility of the option, including the social costs (e.g., externalities), once again recognizing that this may change over varying planning and time horizons.
- The environmental acceptability of the option, e.g., low or zero carbon intensity.
- The extent to which options are robust, in that they make sense to adopt under a range of conditions (the "no regrets" policy) or have multiple benefits, e.g., reduce acid emissions, CO_2 emissions, and energy security problems.

- The potential size of the consumer market for the new products and services, both domestically and overseas.
- The extent to which the option lends itself to international cooperation with developing and other nations and helps to ease international tensions/problems, e.g., a technology which increases the chances of peace or reduces indebtedness.
- The degree of applicability to developing countries and the extent to which developing countries can be integrated more into national and international R, D & D activities.
- The contingency value of the option as an insurance policy, just in case some of the preferred or more attractive current options fail to achieve the desired objective.
- The extent to which options not yet known about might prove to be important. This calls for a certain proportion of the research budget to be for basic research.

To achieve optimum strategies, assessments of success and failure stories in areas such as the interaction of technology with receiving communities, and evaluations of these are important. Hence a further important criterion should be:

- The extent to which technologies require associated social science research to improve the potential of successful uptake, application, and organization of the technology system. Such research could include, for example, the role of disseminating policies that work, the role of institutions in the success or failure of certain technologies, and understanding the social consequences and acceptability of technologies and programs.

On the crucial question of the absolute amounts or percentages of gross national product (GNP) or energy system costs, which should be spent on R, D & D, no specific numbers were presented. However, two approaches were outlined. The former is based on budgetary criteria, i.e., it relates the proportion of GNP or turnover spent on R & D to levels spent in the private sector for example, or in other nations with apparently successful R & D programs. The other approach looks at the opportunity costs of the R & D expenditure by assessing the potential benefits of the expenditure and contrasting this cost-benefit analysis with one where no expenditure takes place. The potential benefits of R & D can be extremely large if such factors as energy security and the development of new industrial technologies for export are included (SERI 1990).

Determining which of these is appropriate is a further area of research and is dependent on national circumstances. One analysis suggests that in OECD countries, a government expenditure of 1% of energy systems costs (as in the U.S.) is too low and that a figure of at least 3% could be justified (Fulkerson et al. 1989). It was noted that in OECD countries, R, D & D levels had generally fallen substantially in real terms for all energy technologies since 1984, particularly for renewables and energy efficiency.

International Policy Instruments and Institutions[1]

Discussions centered around three questions: (a) What international instruments are available to facilitate CO_2 reduction? (b) What tools are required to monitor the progress and maintain compliance of CO_2 reductions? (c) What roles should international institutions be required to take in order to achieve reductions in the rate of accumulation of CO_2?

Though the implementation of international agreements is carried out by national organizations, the role of current and potential new future international organizations is a crucial area for the successful implementation of CO_2 abatement policies. Some of the main points to emerge were:

1. There are a wide range of organizations currently involved in energy and potential CO_2 abatement issues, including UN agencies, OECD, European Commission, the World Bank and other multi-lateral development banks (MDBs), GATT, International Monetary Fund (IMF), World Health Organization (WHO), the Group of 7 major industrial nations (G-7), and nongovernmental organizations (NGOs).

2. Most of these organizations have either limited remit, limited resources, or other limitations which prevent them from fulfilling a comprehensive role in relation to CO_2 abatement. For example, there is no reference to energy efficiency or renewable energy in the Treaty of Rome and its amendments; the OECD/IEA have almost no resources to work on renewable energy, and the IAEA only works on nuclear power.

It appears anomalous that no major organization in the energy field has a remit for either energy efficiency or renewable energy. The desirability and possible effectiveness of a new organization, for example an International Energy Efficiency/Solar Energy Agency attached to the UN, should be assessed.

3. Though past precedents (e.g., ozone depletion, acid rain) are useful in helping us decide on appropriate international instruments and organizations, the uniqueness of the CO_2 abatement issue should not be underestimated.

4. Research into the key elements of successful organizations needs to be carried out, e.g., the perception by some of the success of UNEP in areas such as the negotiations of the Montreal Protocol and the Regional Seas

---

[1] Additional consideration of these issues is given in Chapters 20 and 24.

Program, may in part be a reflection of its limited size and resources, as well as its role as a catalyst organization.

5. There is a need for an international organization to verify data and CO_2 statistics to monitor and set environmental change targets. There may also be a need for additional mechanisms to administer funding and technology transfer.

6. Research into the effects of national, regional, and international instruments (such as carbon or energy taxes, or emissions trading regimes) has been of a varying quality. Improved research on this and on new initiatives and instruments such as a World Resource Tax is needed. This research must of practicality recognize the national sovereignty issue, as so far no country appears prepared to talk seriously about international environmental taxes.

7. Greater transparency on data (e.g., national CO_2 emission trends) as well as systematic and consistent data collection is needed. A data gathering organization (or organizations) is required, providing, for example, a database on best available technologies, environmentally sound energy options, forestry and agricultural technologies, etc.

8. Nongovernmental organizations have an important role in monitoring, enforcement, setting standards, and consensus-building through media exposure.

9. New organizations may evolve from current ones, e.g., a climate convention and/or climate change funding secretariat which uses the precedent of the Montreal Protocol Fund. This may be more likely to gain acceptance than setting up completely new organizations.

The interactions and networking of a range of organizations is an important dimension to the success of international actions. Research into this (e.g., looking at the respective roles of the World Bank, UNEP, GATT, OECD, etc.) or the potential of a new network of research organizations, particularly in the developing world, would be valuable as the world enters a phase of negotiations on a Climate Change Convention and possible protocols, on CO_2 emissions for energy efficiency and forestry, for example.

CONCLUSIONS

Our discussions revealed the wealth of data and analysis that has been carried out on CO_2 abatement options, particularly over the past few years. Despite the variable quality and the range of views and methodologies revealed, it is clear that technology per se is not the limiting factor for significant CO_2 reductions per unit of energy service.

Problems remain over definitions of costs, and hence related cost-benefit analyses, and appropriate methodologies for integrating engineering studies into wider economic analyses. However, there exists promising recent progress on this and on clarifying definitions and modeling assumptions. In contrast to a perception of widely differing views and conclusions on the costs, benefits, and technical feasibility of reducing the greenhouse potential per unit of energy service, a strong conclusion to emerge from our discussions was that there is a very wide level of agreement on many aspects of this problem. The multiple benefits of many of the potential options were a constant theme.

Data gaps remain, particularly in the forestry and related biomass energy areas. The most urgent areas for new research work are, however, on the nontechnical aspects of the issue. The multi-faceted nature of transport, for example, requires input from planners, sociologists, marketing specialists, and urban designers as well as energy planners and vehicle designers.

Implementation mechanisms for greenhouse-gas reduction are more complicated than technical issues and are particularly so in the developing and transitional countries, which have a high growth potential for greenhouse gases. Once again, nonenergy-related expertise and scientific disciplines will be important in contributing to an optimum mix of policy options. International aspects requiring special attention are the appropriate institutions and mechanisms needed to ensure data accuracy, CO_2 reduction compliance, technology transfer, and funding.

REFERENCES

Bleviss, D.L. 1988. The New Oil Crisis and Fuel Economy Technologies. Westport, CT: Quorum Books. 268 pp.

Boyle, S.T. 1990. Business unusual: Developing countries and energy futures. *Oxford Energy Forum* (**2**):4–6.

British Petroleum. 1990. Statistical Review of World Energy. London: British Petroleum.

Chandler, W., ed. 1990. Carbon Emissions Control Strategies: Case Studies in International Cooperation. Washington, D.C.: WWF/The Conservation Foundation.

Chandler, W., and A.K. Nichols. 1990. Assessing Carbon Control Strategies: A Carbon Tax or a Gasoline Tax? Washington, D.C.: Am. Council for an Energy-Efficient Economy.

Danish Ministry of Energy. 1990. Energy 2000—A Plan of Action for Sustainable Development. Copenhagen: Danish Ministry of Energy. 127 pp.

Difiglio, C., K.G. Duleep, and D.L. Greene. 1990. Cost effectiveness of future fuel economy developments. *Energy J.* **11(1)**:65–86.

Dower, R., and R. Repetto. 1990. Use of the Federal Tax System to Improve the Environment. Evidence to U.S. House of Representatives Committee on Ways and Means. Washington, D.C.: World Resources Institute. 14 pp.

Enquête-Kommission. 1990. Protecting the Tropical Forests—A High Priority International Task. Second Report of the Study Commission of the 11th German Bundestag. Bonn: Economica Verlag, Verlag C.F. Müller. 968 pp.

Enting, I.G., and J.V. Mansbridge. 1989. Seasonal sources and sinks of atmospheric CO_2 from direct inversion of filtered data. *Tellus* **416**:111–126.
EPA (Environmental Protection Agency). 1989. Policy options for stabilizing global climate: report to Congress, vol. 1. Washington, D.C.: GPO.
Eyre, N. 1990. Gaseous emissions due to electricity fuel cycles in the United Kingdom. Energy Technology Support Unit. Harwell, UK: UK Dept. of Energy. 31 pp.
Fickett, A.P., C.W. Gellings, and A.B. Lovins. 1990. Efficient use of electricity. *Sci. Am.* **263**:64–74.
Flavin, C., and N. Lenssen. 1991. Beyond The Petroleum Age: Designing A Solar Economy. Worldwatch Paper 100. Washington, D.C.: Worldwatch Institute. 65 pp.
Fulkerson, W., S.I. Auerbach, A.T. Crane, D.E. Kash, A.M. Perry, D.B. Reister, and C.W. Hagan, Jr. 1989. Energy Technology R & D: What could make a difference? Vol. 1. Oak Ridge, TN: Oak Ridge Nat. Lab. 123 pp.
Geller, H.S. 1990. Electricity Conservation in Brazil: Status Report and Analysis. Washington, D.C.: Am. Council for an Energy-Efficient Economy. 181 pp.
Goldman, C., E. Hirst, and F. Krause. 1989. Least Cost Planning in the Utility Sector: Progress and Challenges (Report No. ORNL/CON-284). Berkeley, CA: Applied Science Division, Lawrence Berkeley Laboratory and Oak Ridge Nat. Laboratory, Oak Ridge, TN. 38 pp.
Hall, D.O., H.E. Mynick, and R.H. Williams. 1990. Carbon sequestration vs. fossil fuel substitution: alternative roles for biomass in coping with greenhouse warming, (PU/CEES Report No. 255). Princeton, NJ: Princeton Univ.
Hendriks, C.A., K. Block, and W.C. Turkenburg. 1989. The recovery of carbon dioxide from power plants. In: Climate and Energy: The Feasibility of Controlling CO_2 Emissions. Dordrecht: Kluwer. 267 pp.
Hohmeyer, O. 1988. Social Costs of Energy Consumption. Berlin: Springer. 126 pp.
Holdren, J.P. 1990. Energy in transition. *Sci. Am.* **263(3)**:156–163.
Houghton, J.T., G.J. Jenkins, and J.J. Ephraums. 1990. Climate Change: The Scientific Assessment. Cambridge: Cambridge Univ. Press. 365 pp.
Houghton, R.A. 1990. The future role of tropical forests in affecting the carbon dioxide concentration of the atmosphere. *Ambio* **19(4)**:204–209.
Keepin, B., and G. Kats. 1988. Greenhouse warming: comparative analysis of nuclear and efficiency abatement strategies. *Energy Policy* **16**:538–561.
Krause, F., W. Bach, and J. Koomey. 1990. Energy policy in the greenhouse. Vol. 1. Report to Dutch Ministry of Housing, Physical Planning and Environment. El Cerrito, CA: International Project for Sustainable Energy Paths.
Larson, E.D. 1990. Biomass gasifier/gas turbine applications in the pulp and paper industry: An initial strategy for reducing electric utility CO_2 emissions. Paper for Ninth EPRI Conference of Coal Gasification Power Plants, Palo Alto, CA, USA.
Leach, G., and R. Mearns, 1988. Beyond the Woodfuel Crisis: People, Land and Trees in Africa, 309 pp. London: Earthscan.
Leach, G., and Z. Nowak. 1990. Cutting Carbon Dioxide Emissions from Poland and the United Kingdom. Stockholm: Stockholm Environment Institute.
MacKenzie, J.J., and M.P. Walsh. 1990. Driving Forces: Motor Vehicle Trends. Washington, D.C.: World Resources Institute. 49 pp.
Manne, A.S., and R.G. Richels. 1990. Global CO_2 emission reductions—the impacts of rising energy costs (draft paper circulated widely in the energy research community for comments), Palo Alto, CA: EPRI and Stanford Univ.
Manne, A.S., and R.G. Richels. 1991. Global CO_2 emission reductions—the impacts of rising energy costs. *Energy J.* **12(1)**:87–107.
Marland, G. 1988. The Prospect of Solving the CO_2 Problem through Global

Reforestation (TR039, DOE/NBB-0082), Oak Ridge Associated Universities, for the U.S. Dept. of Energy. Oak Ridge, TN: DOE.

Marland, G. 1991. Why should developing tropical countries plant trees? A look to biomass fuels to approach zero net CO_2 emissions. Climatic Change, *in press*.

McKinsey and Consultants. 1989. Protecting the Global Atmosphere: Funding Mechanisms. Second Interim Report to Steering Committee for Noordwijk Ministerial Conference on Atmospheric Pollution and Climate Change. Amsterdam: McKinsey. 68 pp.

NARUC (National Association of Regulatory Utility Commissioners. 1988. Least-cost utility planning handbook for public utility commissioners. Vol. 2. The Demand Side: Conceptual and Methodological Issues. Washington, D.C.: NARUC. 88 pp.

Newell, R.E., H.G. Reichle, Jr, and W. Seiler. 1989. Carbon monoxide and the burning earth. *Sci. Am.* **260**:82–88.

Newman, P.W.G., and J.A. Kenworthy. 1989. Cities and Automobile Dependence: A Sourcebook. Aldershot, Hants, UK: Gower Technical.

OECD/IEA. 1991. Low Consumption/Low Emission Automobile: Proceedings of an Experts Panel Meeting, 14–15 February 1990. Paris: OECD.

Ostlie, L.D. 1989. The whole tree burner concept: a new technology in power generation, *Biologue*, 7–9.

Ottinger, R., D.R. Wooley, N.A. Robinson, D.R. Hodas, and S.E. Babb. 1990. Environmental Costs of Electricity. Pace U. Center for Environmental Legal Studies, Harvard Univ. New York: Oceana. 759 pp.

Pearce, D. 1991. The role of carbon taxes in adjusting to global warming. Economic J. **101**:935–948.

Pereira da Cunha, R. 1988. National Space Research Institute (INPE), Survey of 10 years deforestation data using Landsat Thematic Mapper. Sao Paolo: INPE.

Plotkin, S.E. 1989. Increased potential for the U.S. fleet of highway passenger vehicles. Statement before the SubCommittee on Oversight and Investigations Committee on Energy and Commerce. Washington, D.C.: U.S. House of Representatives. 36 pp.

San Martin, R.L. 1989. Environmental emissions from energy technology systems: the total fuel cycle. Paper from OECD Seminar on Energy Technologies for Reducing Emissions of Greenhouse Gases. vol. 1, pp. 255–272. Paris: OECD/IEA. 633 pp.

SCAQMP (Southern California Air Quality Management District). 1989. Air Quality Management Plan. 184 pp. Los Angelos: SCAQMP.

Schipper, L. 1991. Improved energy efficiency in the industrialized countries: past achievements—future prospects. *Energy Pol.* **19**:127–137.

SERI (Solar Energy Research Institute). 1990. The Potential of Renewable Energy, Interlaboratory White Paper, SERI/TP-260-3674. Golden, CO: SERI. 181 pp.

Sperling, D. 1989. New Transportation Fuels: A Strategic Approach to Technological Change. Davis, CA: Univ. of California.

SWCC (Second World Climate Conference). 1990. Scientific Conference Statement and Ministerial Declaration. Geneva: World Meteorological Organization.

Tans, P.P., I.Y. Fung, and T. Takahashi. 1990. Observational constraints on the global atmospheric CO_2 budget. *Science* **247**:1431–1438.

Wang, Q., M. Deluchi, and D. Sperling. 1990. Emission Impacts of Electric Vehicles. Davis, CA: Transportation Research Group. Univ. of California, 42 pp.

WEC (World Energy Conference). 1989. World Energy Horizons 2000–2020. Paris: Editions Techniq. 362 pp.

Williams, R.H. 1989. Biomass gasifier/gas turbine power and greenhouse warming.

In: OECD Seminar on Energy Technologies for Reducing Emissions of Greenhouse Gases, vol. 2, pp. 197–248. Paris: OECD/IEA.

Williams, R.H. 1990. Hydrogen from Coal with Sequestering of the Recovered CO_2 Final draft paper. Princeton, NJ: Princeton Univ.

Williams, R.H. 1991. Low-cost strategies for coping with CO_2 emission limits. *Energy J.* **11(4)**:35–59.

Wolsky, A.M., and C. Brooks. 1989. Recovering CO_2 from large stationary combustors. In: OECD Seminar on Energy Technologies for Reducing Emissions of Greenhouse Gases, vol. 1, pp. 179–186. Paris: OECD/IEA, 633 pp.

Wright, L.L., and A.R. Ehrenshaft. 1990. Short rotation woody crops program: annual progress report for 1989, ORNL-6625. Oak Ridge, TN: Oak Ridge Natl. Lab.

Zimmerman, M.B. 1990. Assessing the costs of climate change policies: the uses and limits of models. Workshop on Limiting The Greenhouse Effect. Washington, D.C.: The Alliance to Save Energy, 22 pp.

10
Prospects for Efficiency Improvements in the Electricity Sector

K. YAMAJI
Central Research Institute of Electric Power Industry
1-6-1 Otemachi, Chiyoda-ku, Tokyo, Japan

ABSTRACT

The electricity sector is expected to play a central role in the efforts of reducing CO_2 emission because it consumes far more fossil fuels than any other single industry, and electricity generation is the only large-scale means to convert nonfossil energy sources (such as nuclear power and renewable energies) to a useful form for humankind. Since electricity is not only a clean form of energy but is also of the highest quality, in terms of energy-service efficiency, environmental impacts of energy use could be significantly reduced by substituting electricity for other energy forms. These general benefits associated with electricity have been partly realized; however, there may remain further potentials to improve efficiency in the electricity sector itself.

This chapter identifies technical potentials for efficiency improvements in both supply- and demand-sides of electricity; prospects for achieving the potentials are then discussed. As a result, substantial technical potentials are identified in many directions. We could anticipate steady improvements in supply-side efficiency particularly through combination of technologies and/or integration of energy systems, but it would generally take a long time to realize the supply-side potentials since most cost-effective improvements have been already done. On the other hand, many technical potentials for efficiency improvements on the demand-side apparently seem to be realizable in a rather short time period; however, it would be a difficult challenge for us to mobilize and/or invent effective incentive schemes to affect numerous consumers in selecting efficient end-use technologies.

ROLE OF THE ELECTRICITY SECTOR

The electricity sector is expected to play a central role in future efforts to reduce carbon dioxide (CO_2) emissions. There are several reasons for this:

- electricity generation consumes far more primary fuel than other single industry;
- electricity generation is the only large-scale means of making nonfossil energy sources, such as nuclear power and renewable energies, useful to consumers; and
- the electric power industry, which is normally organized as a public utility corporation, is susceptible to governments' policies and regulations.

Since the birth of the electric power industry, the growth of electricity use in most industrialized countries has been faster than that of total energy requirements. As a result, electricity accounted for 15–20% of total final energy consumption by the late 1980s in most Organization for Economic Cooperation and Development (OECD) countries (Figure 10.1). Electricity, as a percentage of total energy consumption, continues to increase in all sectors other than the transportation sector. The increase in the relative

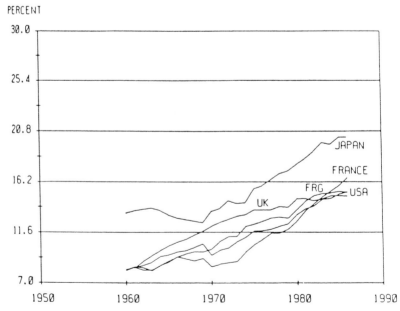

Figure 10.1 Share of electricity in final energy in selected OECD countries (Nakicenovic et al. 1990)

role of electricity during the last few decades has been influenced by the following factors:

- rapid market penetration of electric home appliances to increase the quality of life;
- shifts of industrial structure towards less energy-intensive but more electricity-intensive industries, such as the change from basic material industry to machinery and service industries;
- increasing requirements for precision and ease of control in industrial processes.

While electricity generation involves unavoidable energy losses according to the second law of thermodynamics, electrification, particularly that in the industrial sector, usually contributes to overall energy efficiency improvements, as suggested in Figure 10.2.

For OECD countries, the percentage of each energy form being converted to electricity is plotted in Figure 10.3. Hydro and nuclear energy are exclusively converted to electricity. Seventy-five percent of coal is converted to electricity, with the remaining 25% used in direct form, primarily in steel and cement industries. Reflecting the policies in OECD countries to reduce oil dependence since the first oil crisis in 1973, the share of oil products converted to electricity has decreased to a level of around 10%. As to natural gas, around 20% has been constantly used for electricity generation over the last decades. These statistics clearly indicate the importance of efficiency improvement in the electricity sector to reduce CO_2 emissions

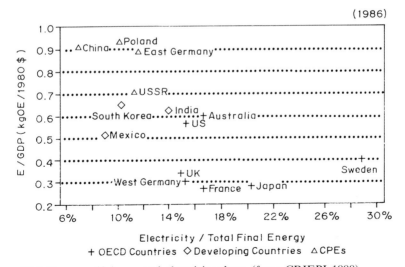

Figure 10.2 Energy efficiency and electricity share (from CRIEPI 1990)

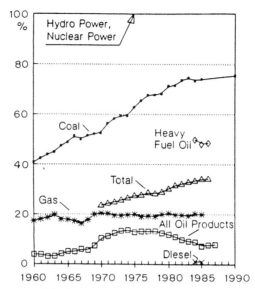

Figure 10.3 Percentage of indirect use of some fuels via electricity to total use in OECD countries (WEC 1988). From Nakicenovic (1990)

from fossil-fuel burning. In the following sections, technical potentials for efficiency improvements in both supply- and demand-side of the electricity sector are identified; prospects for achieving the potentials are thereafter discussed.

TECHNICAL POTENTIAL FOR EFFICIENCY IMPROVEMENTS IN ELECTRICITY SUPPLY

Efficiency of Electricity Generation

Technologies for fossil-fired power plants are well developed. The efficiency of electricity generation in conventional thermal power plants, which employ boilers and steam-turbine generation systems, has been improved to a value around 40% at the transmission end. This is close to the limit of the steam-cycle technology. As shown in Figure 10.4, the average thermal efficiency of operating steam-power plants in Japan improved rapidly from less than 20% to more than 35% during the 1950s and 1960s. The rate of improvements in average efficiency, however, has been reduced to stagnation for the last 20 years, reflecting the technical difficulty for further efficiency improvements, energy losses involved in the attached pollution control equipment, and lowered replacement rate of old plants with more advanced and efficient

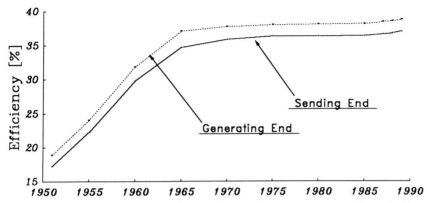

Figure 10.4 Average thermal efficiency of operating steam power plants in Japan (from FEPC 1990)

ones. Similar trends of efficiency improvements can also be found in other parts of the world, as seen in Figure 10.5. However, when one examines efficiencies of electricity generation actually achieved, there are still significant performance gaps even among developed countries (see Table 10.1). Technology transfer and development could be used to decrease these differences and hence contribute to the global average efficiency of electricity generation.

In technically advanced countries like Japan, there seems to be little room for efficiency improvements in the conventional steam-power generation. Technical potentials for breakthroughs in electricity-generation efficiency, however, can be found in plants which combine steam-cycle technologies with other generation technologies. A typical example, which is already commercialized, is the combined cycle generation that utilizes the waste heat from a gas turbine for boiling water to drive a steam turbine in combined cycle power plants using liquefied natural gas (LNG), with 43% of efficiency at the sending end is achieved in Japan. As shown in Figure 10.6, efficiency is expected to be raised to more than 50% by developing a ceramic blades gas turbine, and then close to 60% by combining with advanced fuel cells such as molten carbonate fuel cell (MCFC) and solid oxide fuel cell (SOFC). Gasified coal as well as natural gas will be used for such advanced combined cycle generation technologies. In the long run, other direct power generation technologies, such as magnetohydrodynamic (MHD) generation and thermoelectric conversion, might also be used. Besides, where LNG is stored, cold heat and expansion energy associated with LNG can also be used for electricity generation. More than ten such LNG cold heat generation plants, with about 60 MW of total capacity, are now operating in Japan.

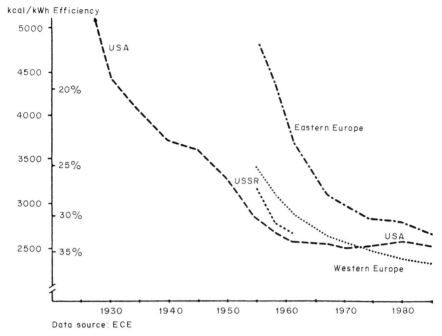

Figure 10.5 Improvements in power plant efficiencies (from Nakicenovic et al. 1990)

Table 10.1 Efficiency of electricity generation and transmission/distribution (%). From FEPC (1990)

	Japan	U.S.	Canada	U.K.	France	FRG	Italy
Thermal Efficiency of Generation (1988*: 1987)	38.6	33.0	34.7*	33.7	34.7	39.5	37.7
Loss in Transmission/ Distribution System (1988*: 1987)	5.7	6.3*	9.1*	8.2	7.3	3.9	8.2

Under the prospects for technology development described above, the average efficiency of thermal power plants operating in 2050 is expected to increase to a level of 50–55% in Japan. Together with the increase of the share of nonfossil electricity generations, about two-thirds of total electricity generation in 2050, CO_2 emissions from the electricity sector in Japan could

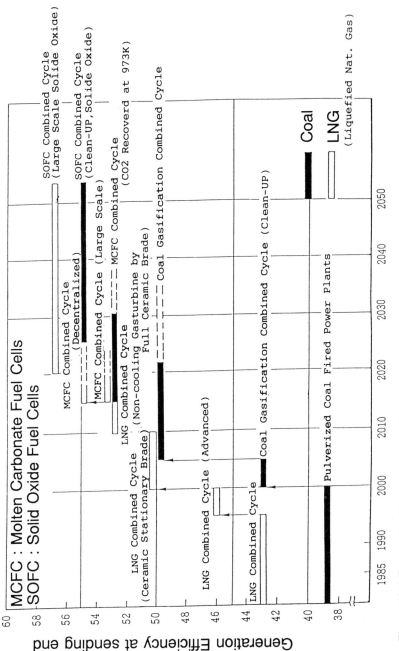

Figure 10.6 Expected efficiency growth of advanced fossil-fuel generation system (from CRIEPI 1990)

be kept constant or even less while supplying more than twice as much electricity as today (CRIEPI 1990).

Losses in Transmission and Distribution Systems

Technologies for transmission and distribution of electricity have also matured. As shown in Figure 10.7, the transmission and distribution loss factor in Japan has been reduced from about 25% in 1950 to less than 6% in 1980. The significant reductions in losses were achieved mainly through raising voltages of transmission and distribution lines. The loss factor, however, has been constant for the last ten years. While many 500 kV transmission lines are now in operation and even a 1000 kV line is about to start servicing, the effects of reducing losses are partly cancelled because they are used for connecting load centers with nuclear power stations and/or pumped-hydro stations located in remote sites.

As in the case of generation efficiency, similar trends of efficiency improvements in transmission and distribution systems can be found in other parts of the world. However, the innovations of the network are rather slow, particularly for a system with a low growth rate, and transmission efficiency is significantly affected by geographical conditions. Thus there exist great differences in the efficiencies among countries. As shown in Table 10.1, even among developed countries loss factors of transmission and distribution systems range from about 4% to 9% (FEPC 1990). In the long run, differences will be lessened, and then global average efficiency will increase by a few percent.

We can identify several technical potentials for breakthroughs in efficiency improvements of transmission and distribution systems, for example, a

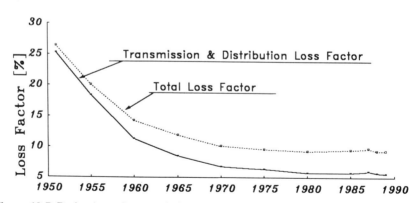

Figure 10.7 Reduction of transmission and distribution losses in Japan: total loss factor includes electricity consumed in power plants (from FEPC 1990)

superconductive network, amorphous core transformers, higher distribution voltage, and further penetration of power electronics. Contributions of the latter three options, however, will be marginal, probably less than 1%, while the penetration of power electronics would significantly enhance the flexibility of electricity networks. On the other hand, the effect of a superconductive network that includes superconductive cables, transformers, and generators could be revolutionary. If superconductivity makes it possible to transport electricity to anywhere in the world, without losses and with a reasonable cost, then the effect will certainly be beyond that of efficiency improvements of electricity transport. Many energy resources currently unused, such as hydropower resources in remote areas and gas and oil reserves in the arctic regions, could be tapped. Unstable natural energies, such as solar and wind power, might be made reliable power sources by linking and controlling them globally. Also, through globally linked electric grids, aggregate demand profile would be levelled so as to make electricity supply more efficient.

Electricity Storage Systems

It might sound strange that electricity storage systems could contribute to an efficient electricity supply. Indeed, storage systems involve energy losses during charge/discharge and storage itself; the efficiency of pumped-storage hydropower plants, currently the most available popular large-scale method of electricity storage, is around 70%. Introduction of storage systems is usually justified by the economic argument that they can keep high capacity factors for capital-intensive, baseload power plants such as nuclear power plants. Storage systems, however, could also contribute to overall efficiency of electricity supply by making it possible for thermal power plants to operate constantly at rated capacities. Thermal power plants are designed to be most efficient when operating at their rated capacities.

In addition, since most storage systems have excellent dynamic operating performances, they are used to follow load fluctuations. Therefore, the spinning reserves of thermal power plants which consume extra fuels can be reduced.

The efficiency of storage systems can be improved in two ways: (a) to improve storage efficiency itself and (b) to locate storage systems close to load centers, thus reducing transmission losses. Among technical potentials for improving the efficiencies of storage systems are compressed air energy storage (CAES), advanced storage batteries, hydrogen production, and superconductive magnetic field energy storage (SMES).

Co-generation Systems

Co-generation systems are energy-efficient systems which supply both electricity and heat. In industrial processes such as those in paper and pulp, iron and steel, and chemical industries, co-generation systems have been commonly used. A significant portion of the capacity of nonutility electricity generation, which is around 17 GW in 1989 and represents about one-tenth of total power plant capacity in Japan, is considered to be supplied through co-generation systems. Prospects of these industrial co-generation systems depend on the future of individual industries. Co-generation systems recently focused on are those with relatively small capacities which provide customers in commercial sectors with electricity, hot water, heating, and cooling. The small scale co-generation systems began penetrating into the Japanese energy market in the early 1980s, and the number of installations has increased rapidly to about 350, with a total capacity of around 150 MW by 1989. These co-generation systems employ gas engines, diesel engines, or gas turbines for generating electricity. Fuel cells will also be used in the future. The efficiency of current co-generation systems is 20–30% for electricity generation and 40–50% for heat recovery; thus the integrated efficiency is expected to be as much as 70–80%. Actual efficiency, however, depends on the profiles of electricity and heat loads. A relatively large and constant heat load is an important condition for achieving high efficiency and good economy.

The effects of co-generation systems on the efficiency of electricity generation should be carefully examined. Co-generation systems pursue the efficiency of combined supplies of heat and power within a relatively small network, the size of which is limited by heat-supply technologies. On the other hand, scale economies of electricity networks still exist at least in combining many electricity demands while scale economies of generation technologies are not so certain. Thus, penetration of co-generation systems into electricity supply might erode the inherent efficiency associated with a large-scale electricity network. Institutional arrangements, such as a prudent rate schedule for the trades between co-generation systems and electricity network, would play an important role. It is important, in the long run, to construct an efficient and flexible integrated energy system to realize ideal cascaded exergetic utilization.

TECHNICAL POTENTIAL FOR EFFICIENCY IMPROVEMENTS IN ELECTRICITY USE

Electric End-Use Technologies

In a general sense, electricity is the most efficient form to meet any energy services because theoretically it can be converted to "work," the highest quality of an energy form, with 100% efficiency. However, the inherent efficiency of this refined energy form does not necessarily guarantee the highest efficiency in actual energy use. Depending on the end-use technologies for using electricity, the efficiency varies significantly.

To evaluate the maximum technical potential for energy savings, many attempts have been done recently to identify the most efficient electric end-use technologies. According to the results of Faruqui et al. (1990), the maximum savings of electricity in the U.S. in 2000 could range 24 to 44%, even only when currently available technologies are taken into consideration. According to the study, in the residential sector, efficient technologies for lighting, space heating, and water heating offer the greatest opportunities; in the commercial sector, technologies for lighting and space cooling have the largest potential for efficiency improvements; and, efficient technologies for motor drive are most promising in the industrial sector. But we cannot apply the results obtained under the U.S. conditions to the rest of the world because the structure of end-use electricity demands differs among countries, and the evaluation of technical potentials for electricity savings depends strongly on the performances of end-use technologies currently used.

Reflecting the difficulty to define energy services, categorization of the electric end-use technologies is not well established. Though the electric utility industry began its business by providing lighting services, fields of its energy services were soon diversified to motive power, electrochemical reaction, and heating. Now it also covers air-conditioning, telecommunications, computing, and so on. It is not an easy task to analyze the efficiencies of so many diversified end-use technologies. Besides, efficiency improvements of electricity end-use have been done not only through efficiency improvements of individual components, such as motors and compressors, but also through the advances in control systems, such as those using microcomputers and power electronics. In the following sections, several important technology developments in electric end-use efficiencies are reviewed.

Lighting

As is well-known, in 1879 Thomas Alva Edison invented the incandescent light bulb using a carbon filament made of Japanese bamboo. He founded the basic structure of the electric utility industry in 1882 by starting to supply electricity to some 100 buildings for lighting the bulbs. Since then, end-use technologies for lighting have been developed. Lighting still represents a significant portion of electricity end-use services, about 18% of total residential electricity use now in Japan.

Development of the efficiency of light sources is shown in Figure 10.8. As shown in the figure, efficiency improvements are achieved through the

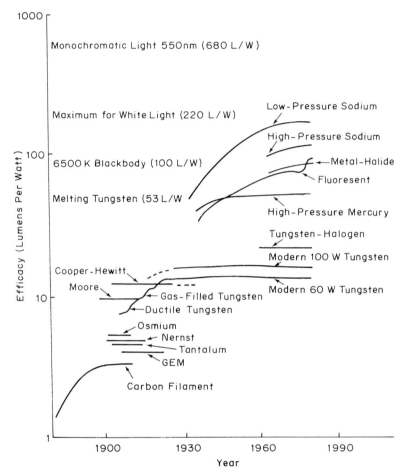

Figure 10.8 Changes in the efficacies of various light sources over time (from Anderson and Saby 1979)

evolution of lighting performance within each technology, as well as the progress from one technology to another. Although the evolution of the efficiency of the incandescent bulb had reached saturation more than 50 years ago by the introduction of the tungsten filament, the incandescent bulb remains a viable technology for lighting, competing with the fluorescent lamp. That the incandescent bulb still keeps a large market share, in spite of the fact that its energy efficiency is less than one-third of fluorescent lamp and that its life is less than one-fifth, suggests a complicated structure of consumers' preference in selecting lighting technology.

Efficiency of the fluorescent lamp has been improved over the past decades, and several recent innovations are now being introduced. Ball-shaped fluorescent lamps with plugs compatible to an incandescent bulb have been developed, and efficiency has been increased by introducing a compact tube with rare earth fluorescent materials and low temperature design. Both efficiency and quality of light have also been improved by the development of three wave-enhancing lamp and high-frequency electronic ballasts with inverters. Thus, the current state-of-the-art fluorescent lamp is more than 50% efficient than the conventional fluorescent lamp, and now rather close to a half of the theoretical maximum for white light, 220 lumens per watt.

The sodium lamp, which is more efficient than the fluorescent lamp, is increasingly used for lighting highways, streets, and parking lots. There are many other technical factors which play important roles for enhancing the efficiency of lighting. Among them are reflecting covers for lamps, dust cleaning of lamp surfaces, light reflections of ceiling/wall/floor, and automatic lighting control using a sunlight sensor.

Home Electric Appliances

Electricity consumption by the residential sector in Japan has increased rapidly and now represents some one-fourth of the total electricity demand. The rapid growth was brought about by the penetration of electric appliances such as refrigerators, televisions, and room air-conditioners. As shown in Figure 10.9, the energy efficiencies of these electric appliances have been improved substantially after the first oil crisis in 1973.

For freezers/refrigerators, the electricity consumption was reduced to a level of about one-third of that in 1973. This efficiency improvement is the combined result of many design changes, such as more efficient compressors and motors, thicker insulation, elimination of various heaters, placing the fan motor outside, etc. In addition, electricity consumption of the refrigerators of larger capacities, which have been more popular due to modernization of living standards, has been reduced more rapidly than that of small capacities (see Table 10.2).

For color televisions, efficiency of electric power consumption was improved from 140 watts in 1973 to 83 watts in 1985 for the 19–20 inch class, as shown in Table 10.3. Substantial efficiency improvements have been done even before the first oil crisis by the use of transistors instead of vacuum tubes; electric power consumption was reduced to about half by transistors. Further efficiency improvements were realized by the developments of circuit elements, from transistors to integrated circuit, then to the large-scale integrated circuit and cathode-ray tube or Brawn tube with less energy consumption. However, electricity consumption by televisions is increasing again in Japan as larger televisions (around the 30

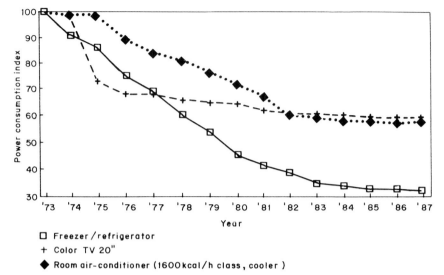

Figure 10.9 Efficiency improvements of household electric appliances in Japan (indices are in watts except for the refrigerator, which is in MWh/month). From Aisaka (1989)

inch class) are becoming popular. As for the technical potentials for further efficiency improvements, using liquid crystal display instead of cathode-ray tubes, which has been realized in small portable televisions, is promising; however, the effect would be marginal.

Room air-conditioners have gradually made their way into Japanese households and are now diffused in around 90% of households, except in Hokkaido where cooling demand is extremely low. Almost half of the room air-conditioners used now are the heat pump variety, which are used as coolers in summer and as heaters in winter. Passing through two oil crises, the electric power consumption of a room air-conditioner was reduced to a level of about 60% of that in 1973 by the adoption of rotary compressors, improvement of heat exchangers, etc. Still further energy savings are being promoted through the efforts such as use of microcomputers for fine controls and non-step capacity controls using inverters. Thus, current state-of-the-art heat pumps are about 20–30% more energy efficient than conventional ones. For still further efficiency improvements, house insulation is considered to be most effective in Japan because only around 20% of the housing stock is insulated, although more than 70% of newly built houses are insulated.

For other home electric appliances, energy efficiencies have been improved after the first oil crisis, for example, electric power consumption of the vacuum cleaner was reduced by 17% during 1973–1988. However, most

Table 10.2 Monthly electricity consumption of refrigerators (kWh/month). From Tsuchiya (1988)

Volume	1973	1974	1975	1976	1977	1978	1979	1980	1981	1982	1983	1984	1985
170 L	80	73	69	60	55	48	43	36	33	31	30	27	26
230 L	—	91	79	66	58	52	46	37	31	26	24	24	24
260 L	—	—	—	—	—	53	49	42	36	29	27	25	24

Table 10.3 Efficiency improvement of television (electricity consumption: watts). From Tsuchiya (1988)

Class	1967	68	69	70	71	72	73	74	75	76	77	78	79	80	81	82	83	84	85
13–14 inch	—	123	107	95	88	83	83	77	71	64	66	65	60	60	57	55	55	55	55
19–20 inch	325	300	233	183	138	150	140	138	102	95	95	92	90	90	87	85	85	84	83

technical efficiency improvements leveled off in the 1980s, while on the other hand, consumers preferred the higher performances and the larger capacities of electric appliances. Thus electricity consumption in the Japanese residential sector is still increasing substantially. Penetration of modern, efficient electric appliances into new end-use fields, such as heat pumps for space heating and the microwave oven and magnetic cooker using eddy current for cooking, has also contributed to the increase of electricity consumption. Such development of electric home appliances would improve the overall energy efficiencies and enhance the quality of living.

Advanced Heat Pumps

In addition to the application to home air-conditioning, equipment using heat pump systems has been applied over a wide range through technological developments and is rapidly penetrating into various fields. For commercial buildings and apartment housing, heat pumps with energy storage systems using inexpensive nighttime electricity are becoming popular. Large-scale heat pumps could supply hot water as well as space cooling and heating. Heat pump systems using urban waste heat, such as warmed river water and the exhaust heat of underground substations, are attracting more attention as a viable means to enhance overall energy efficiency. The introduction of heat pump systems is also making progress in industrial fields.

Development of a super-heat-pump energy accumulation system, which is promoted as a part of the Moonlight Project (the Japanese government's research and development program for energy conservation), would make more advanced heat pump technologies available to various fields. Two types of target performances are set for the super-heat-pump developments: (a) a high efficiency heat pump with a coefficient of performance (the ratio of heat extracted to the input work; AIP 1975) of around 8 and an output temperature of around 85°C, and (b) a high temperature heat pump with an output temperature of 150–300°C and a coefficient of performance of more than 3. The former would be used for air-conditioning and hot water supply, while the latter would be for industrial heating.

Motors

Motive power is by far the largest end-use service supplied by electricity. The efficiency of motors themselves is believed to be more than 80% and there seems to be no more room for improvements. According to Tsuchiya (1988), however, various ways can be identified to reduce losses of energy in motor use. He indicates the following points:

1. Keep the rated voltage. If the voltage is high, torque and power are proportional to the voltage. Excess power is unnecessary.
2. Decrease idling use. Usually a motor is connected to some rotary component of machineries; thus, the energy loss of idling is two or three times more than the idling loss in the motor itself. The idling loss could be reduced through ingenious switching, automatic control of motors, reform of production processes, and so on.
3. Optimum load. Motors can work most effectively at the 80–100% load of rated output. The motor load checker is commercially available and is useful in finding inefficient motor use.
4. Improve motor drive. The efficiency of transmission of mechanical work is 100% for direct coupling, 96–97% for belt, 93–96% for gear, and 85–90% for worm gear. The best device should be selected and good maintenance is required for keeping efficiency high.
5. Power factor control in motor use. The power factor in motor use increases as the load ratio increases. An advance phase condenser can be used to improve power factor in motor use. Since Japanese electricity rate schedules include penalty and/or discount for power factors with the reference of 85%, the introduction of such advance condensers would be cost effective. The improvements in power factors would also improve the efficiency of distribution systems.
6. Speed control in motor use. Although placed last, speed control is the most important point for efficiency improvements in motor use. Speed or rotary control plays the most essential part to improve efficiency in driving blowers, fans, and/or pumps used for variable loads.

Conventional methods to control fluid flow would be to install a damper to adjust the area of flow to load while running a motor at a rated output. Such control of flows involves a lot of energy losses. The losses can be avoided by adjusting motor speeds or capacities. The more efficient methods for adjusting to variable loads are achieved by fluid coupling and a frequency control of motors. During the period immediately after the second oil crisis, when energy conservation was an urgent matter in Japan, use of inverters was as widespread variable speed devices for both large and medium capacities of motors in order to achieve energy conservation of machines handling liquids and gases in industrial fields. The performances and functions of transistor inverters and gate turn-off thyristor inverters have been raised remarkably, and rotation control systems of induction motors using inverters are rapidly spreading the range of applications from manufacturing industries to the transport sector, and then to commercial and residential sectors. For example, motive power for elevators has been reduced significantly by ingenious control; and, as shown in Figure 10.10, by introducing the most efficient variable, voltage and variable frequency electric motors, the efficiency of street cars was improved from 2.1 kWh per car-km to 1.6 kWh per car-km.

Industrial Electric Furnaces

Industrial electric furnaces are widely used and have great potential to save electricity and related costs. One of the most simple methods to save energy in industrial furnaces is to reduce the weight of the product. Not only the product, but also the weight of containers carried into the furnaces can be reduced, and this would contribute to energy savings. In addition, a more ingenious approach to thermal insulation, fine temperature control, and continuous production would certainly contribute to further energy savings. These measures for improving heating efficiencies are relatively easily realized in electric furnaces because electricity is a convenient form for fine control.

For metal smelting in Japan, the specific electricity consumption of arc furnaces was reduced from about 550 kWh per ton of steel ingot in 1970 to about 400 kWh per ton of ingot in 1985. This energy savings was achieved by oxygen injection and other measures described above. In the future, DC arc furnaces with thyrister control could reduce electricity consumption by five to ten percent compared with AC arc furnaces. The laser beam is still expensive but has a great potential to save energy for spot heating demand.

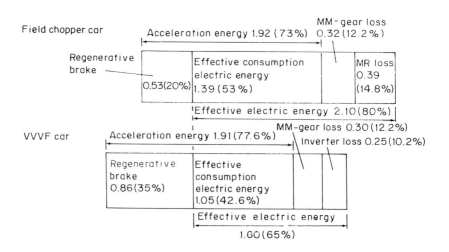

Figure 10.10 Efficiency improvements in streetcars in Japan (from Asaumi 1988). VVVF = variable voltage variable frequency; MM = machine/machine

Microwave and far-infrared rays would improve the efficiency by heating and drying directly inside the materials. The effects of these emerging technologies remain to be realized in the future.

PROSPECTS FOR REALIZING THE TECHNICAL POTENTIALS

Efficiency improvements have an absolute value since the same service is supplied with less consumption of resources, thus less impacts on the environment. As stated above, the technical potentials for efficiency improvements in the electricity sector exist in many directions, and the maximum amount of potential electricity savings is huge. There are, however, many barriers to realize the technical potentials.

Energy efficiency is among the many factors which are taken into account when entrepreneurs and/or housewives adopt a new technology. There seems to be a considerable difference in criteria for adopting technical options in electricity efficiency improvements between supply-side and demand-side. For supply-side options, cost effectiveness is the satisfactory condition for adoption. However, for demand-side options, cost effectiveness is only the minimum requirement. We could anticipate steady improvements in supply-side efficiencies through market forces, but the gains of efficiency improvements would be marginal since most cost-effective improvements have been already realized. On the other hand, to realize, in a timely manner, the large potential for efficiency improvements in demand-side technologies, we must introduce some intentional incentive schemes such as demand-side management. There are different opinions on the "natural" rate of efficiency improvements in energy use.

There are, indeed, several encouraging estimates on the cost effectiveness of demand-side options for electricity efficiency improvements. As shown in Figure 10.11, while a considerable difference exists between the two estimates, huge potentials are identified for cost-effective electricity savings in the U.S. In the estimate by Rocky Mountain Institute, nearly one-fifth of the total electricity demand in 2000 can be saved with benefit while some three-fourths can be saved at a cost less than 4 cents per kWh. The Electric Power Research Institute is not so optimistic; however, still around 30% of total electricity demand is estimated to be saved at a cost less than 4 cents per kWh (Fickett et al. 1990). We should, however, be careful to accept these cost estimates on electricity savings because many assumptions, explicit or implicit, are involved in the cost estimates. For example, the discount rate assumed for cost evaluation may be too low for residential customers, and subtle but quite sensitive conditions for selecting end-use technologies such as the space for installation may be ignored in the estimates. Besides,

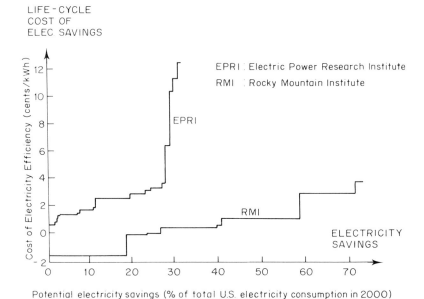

Figure 10.11 Example of cost estimates for electricity efficiency improvements in the U.S. (from Fickett et al. 1990)

there are many barriers other than cost, such as safety, reliability, lack of information, insufficient supply capacity, etc.

According to Japanese experiences in successful promotion of energy conservation during the oil crises, price mechanism or market force are important but not sufficient instruments; meticulous administrative measures such as efficiency standards, inspection, and education have a sure effect; and a long-term approach for research and development with clear goals in the public sector stimulates technology innovations in private sectors.

REFERENCES

AIP (American Institute of Physics). 1975. Efficient Use of Energy. AIP Conference Proceedings, No. 25. New York.

Aisaka, K. 1989. Development of Energy Conservation Technologies in Japan and International Cooperation, 4-4, pp. 471–493. Proceedings of the Third Symposium on Pacific Energy Cooperation, Tokyo.

Anderson, J.H., and J.S. Saby. 1979. The electric lamp: 100 years of applied physics. *Physics Today* **10**:32–40.

Asaumi, S. 1988. Control and Maintenance of VVVF Cars in Tokyo Express Railways. *Science for Electric Cars* **7**:18–21 (in Japanese).

CRIEPI (Central Research Institute of Electric Power Industry). 1990. Energy in the 21st Century (in Japanese). Tokyo: CREIPI.
Faruqui, A. et al. 1990. Efficient Electricity Use: Estimates of Maximum Energy Savings, CU-67T46, EPRI.
FEPC (Federation of Electric Power Companies). 1990. Concise Statistics of Electric Power Industry (in Japanese). Tokyo: FEPC.
Fickett, A.P., C.W. Gellings, and A.B. Lovins. 1990. Efficient use of electricity. *Sci. Am.* **263**:3.
Nakicenovic, N., A. Gruebler, L. Bodda, and P.V. Gilli. 1990. Technological Progress, Structural Change and Efficient Energy Use: Trends worldwide and in Austria. IIASA. Luxemburg.
Tsuchiya, H. 1988. Electricity conservation in Japan. In: Demand-Side Management and Electricity End-Use Efficiency, pp. 595–606. Kluwer Academic Publishers.

11

Reducing Greenhouse-Gas Emissions from the Transportation Sector

L.L. JENNEY[1]
Transportation Research Board, National Research Council, Washington D.C. 20418, U.S.A.

ABSTRACT

Combustion of transportation fuels is a major source of carbon dioxide (CO_2) and other greenhouse gases discharged into the atmosphere. In the near term, over the next 15 years, exploitation of presently available technology and application of conservation measures to reduce transportation fuel consumption could have a beneficial but small effect on global production of CO_2, perhaps enough to slow the present rate of growth. Over the next 50 years substantial, further reductions in CO_2 emissions could be made in the transportation sector, and current trends could be reversed. This could be accomplished primarily through improvements in automobile technology: more efficient engines, switching to fuel with a lower carbon content, and increased use of electric vehicles. To achieve the full reduction of CO_2 now thought necessary over the long term, the transformation of the transportation system begun during the period 2005–2040 would have to be continued even more vigorously, such that by the end of the 21st century the transportation sector would have to rely for propulsion power on a combination of nonpetroleum fuels (hydrogen or biomass-derived liquids) and electricity, which could be used either directly for motive power or indirectly to produce hydrogen motor fuel.

EVOLUTION OF THE AUTOMOBILE TRANSPORTATION SYSTEM

To gain a perspective on how transportation technology has developed and the degree of change that might be expected over the coming century, it

---

[1] The views expressed in this paper are the author's and not necessarily those of the Transportation Research Board or the National Research Council.

Limiting the Greenhouse Effect: Options for Controlling Atmospheric CO_2 Accumulation
Edited by G.I. Pearman © 1992 John Wiley & Sons Ltd

may be helpful to look back for an equivalent period of time and sketch the evolution of what has become the dominant transportation mode of today: the motor vehicle. This is not to neglect rail, air, and marine transport technology. Remarkable strides have been made in all forms of passenger and goods transportation, and all will have to be improved in the interest of lessening the generation of CO_2. Nevertheless, the automobile system must be the center of attention because of its size, complexity, and all-pervasive nature. The automobile accounts for about three-quarters of all CO_2 emissions generated by the transportation sector, and it presents both the greatest challenge and the greatest opportunity to curtail the consumption of fossil fuels for transportation.

Infancy

In 1890, one hundred years ago, the principal means of passenger and freight transport on land were horse-drawn vehicles and steam-powered locomotives fueled by wood or coal. Waterborne travel and commerce also relied largely on steam power. The automobile was a technological curiosity. The airplane was regarded as little more than a dream. Space travel was the province of Jules Verne and H.G. Wells. It was a widely held scientific opinion that speed in excess of 150 km/hr, if attainable by any form of transportation, would likely be fatal to human beings. The prospect that humankind could cause global atmospheric damage, through technology in general and transportation in particular, was unthinkable. Only a few had foresight, such as the Swedish chemist and director of the Nobel Institute for Physical Chemistry, S.A. Arrhenius, who at the end of the 19th century first called attention to the possible climatic effects of increasing the CO_2 concentration in the atmosphere.

Within 25 years, land transportation began to undergo startling change. Automotive transportation emerged from the laboratory and inventor's workshop and began to revolutionize the cities of Europe and North America. In 1914, on the eve of World War I, the world fleet of automobiles numbered about 2.5 million, of which 80% were in the U.S. In 1915, automobile production in North America alone reached almost 1 million vehicles (MVMA 1990a).

Expansion

In the period between World War I and World War II, little more than two decades, the number of automobiles in the world grew to over 46 million (37.2 million passenger cars and 9.0 million trucks and buses). This growth was largely concentrated in North America, which accounted for 70% of the world motor vehicle fleet. The remainder was distributed about equally between Europe and all other countries in the world (MVMA 1990a).

Thus, for the first 50 years of its history, the automobile was largely a U.S. phenomenon stimulated by the geographic extent and relatively low population density of the country, the diffuse pattern of urban development, the abundance of domestically produced petroleum, and, above all, the enormous production capacity of the U.S. automobile industry.

It was also during this period that automotive technology became standardized throughout the world. After an early period of experimentation with different propulsion systems and fuels, the design of the automobile converged into a single dominant type: a four-wheeled, rubber-tired vehicle equipped with an internal combustion engine (spark-ignition or diesel) fueled by some type of petroleum distillate. The course of development since then has been simply to refine and optimize this engine and fuel combination in terms of performance, utility, efficiency, reliability, durability, and manufacturability.

Dominance

In the half century since World War II, automotive technology has completely transformed passenger and freight transportation in all parts of the world. There are now over 550 million motor vehicles in operation worldwide. At the present rate of growth (2%/yr), the fleet could exceed 1 billion in 2025.

The regional distribution of automobiles has shifted substantially (see Figure 11.1). Today the U.S. accounts for only about one-third of the world

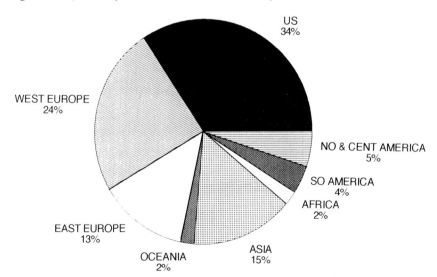

Figure 11.1 Regional distribution of motor vehicles, 1988 data (from MVMA 1990a)

fleet. Western Europe has the second highest concentration, 24%. The automobile population of Asia is growing rapidly; Japan alone has now over 52 million vehicles or just under 10% of the world total (MVMA 1990a). Together, these three centers of automobile concentration, which constitute just under 10% of the world's land mass, have 68% of the world's motor vehicles.

Some appreciation of the degree to which automobile transportation now dominates the movement of people and goods can be obtained by comparing the motor vehicle fleet to population on a regional basis (Table 11.1). Worldwide, there is one passenger car for each 12.2 persons and one motor vehicle (passenger car, truck, or bus) for each 9.4 persons. The distribution by region varies widely, from 1.4 persons per vehicle in the U.S. and Canada to 51 persons per vehicle in Africa. The intensity of automobilization in western Europe and Japan (2.4 and 2.3 persons per vehicle, respectively) approaches that of the U.S. and Canada. If one motor vehicle per four or fewer persons is taken as the index, the fleets of the highly automobilized nations of the world constitute 77% of all the motor vehicles in operation.[2]

Table 11.1 Global motor vehicle registration and population by region for 1988 (from MVMA 1990a)

REGION	VEHICLES (000)[1] Cars	Trucks & Buses	Total Vehicles	POPULATION Number (000)	Per Car	Per Vehicle
Africa	7,996	4,354	12,351	630,261	78.8	51.0
Asia	46,991	35,468	82,460	2,839,193	60.4	34.4
Central America[2]	8,047	3,268	11,315	139,039	17.3	12.3
Europe[3]	168,477	29,698	198,175	836,195	5.0	4.2
Oceania	8,969	2,368	11,336	20,911	2.3	1.8
South America	19,275	4,622	23,896	312,210	16.2	13.1
U.S. & Canada	153,152	47,105	200,257	272,832	1.8	1.4
World Total	412,907	126,882	539,790	5,050,641	12.2	9.4

[1]Totals may not sum due to rounding
[2]Includes Mexico
[3]Includes USSR

---

[2] There are 41 such countries, principally in western Europe, North America, and the Arabian peninsula, plus Japan.

THE TRANSPORTATION SYSTEM TODAY

Energy Use

Transportation accounts for roughly 22% of global energy consumption and 25% of the energy derived from fossil fuels (White 1990; Davis 1990). The percentage varies widely from region to region, largely as a function of the level of economic development. In developed countries such as the U.S., EEC nations, and Japan transportation represents up to 25–30% of all fossil-fuel use (90% or more petroleum); in less-developed areas the figure is as low as 10 to 15%.

The preponderance of transportation fuel use is by motor vehicles. In the U.S., for example, motor vehicles (passenger cars, trucks, buses, motorcycles, and off-highway vehicles) consume about 77% of transportation fuels. Aviation uses almost 9%, and all other modes about 15% (see Table 11.2). The modal shares in other economically advanced nations are roughly the same. Rail and bus account for slightly more in western Europe and Japan, due to the highly developed and extensively used public transportation systems in these countries. Air, water, and pipeline transportation fuel use in Europe and Japan, however, is somewhat less. Consequently, 75% is a reasonably accurate estimate for the motor vehicle share of transportation fuel use on a worldwide basis.

Emissions

Transportation, which is powered almost exclusively by fossil fuels either directly or indirectly, is a major source of CO_2 discharged into the

Table 11.2 U.S. Transportation fuel use by mode for 1987 (from DOT 1990; NRC[*] 1990)

Mode	Percent
Motor Vehicle	76.7
Auto and motorcycle	(40.4)
Truck	(32.6)
Bus	(0.7)
Off-highway	(3.0)
Air	8.6
Water	6.0
Pipeline	3.5
Rail	2.2
Military	2.9

atmosphere. Currently worldwide, all forms of transportation produce on the order of 1,340 Mt of carbon annually, almost one quarter of the yearly output of 5,700 Mt from all sources.[3] Motor vehicles are estimated to account for between 60 and 70% of the transportation share, i.e., about 14–17% of the world total (Paaswell 1990; MVMA 1990b).

The use of fossil fuels in the transportation sector also produces several other air pollutants that contribute either directly or indirectly to global warming. Chief among these are carbon monoxide (CO), nitrogen oxides (NO_x), tropospheric ozone (O_3), methane (CH_4),[4] and volatile organic compounds (unburned hydrocarbons). Table 11.3 provides data on air pollutants (including CO_2) generated by motor vehicles and other transportation in the U.S.

A detailed breakdown of air pollutants emitted by transportation (and specifically by motor vehicles) on a global basis is not available. A simple approximation based on the transportation share of total world petroleum consumption would suggest that transportation emissions worldwide are about four to five times the U.S. figures, but this could be inaccurate for several reasons. Emission control technology and standards vary widely throughout the world. Many highly automobilized countries have standards comparable to those of the U.S. In less-developed countries, standards are generally more lax or, in many cases, nonexistent. Further, the proportion of petroleum consumption devoted to transportation varies widely on a

Table 11.3 Atmospheric contaminants from transportation in the U.S. for 1989 (from MVMA 1990b)

Sources	ATMOSPHERIC CONTAMINANTS (million metric tons)				
	Carbon Dioxide	Carbon Monoxide	Nitrogen Oxides	Sulfer Oxides	Volatile Organics
Motor vehicles[1]	252.8	34.3	6.1	0.6	4.8
Other transportation	76.8	10.4	1.9	0.2	1.5
All U.S. transportation[2]	329.6	44.7	8.0	0.8	6.3
Percent of total emissions from all sources	23	73	40	4	34

[1]Passenger cars, trucks, buses, motorcycles and off-highway vehicles (MVMA 1990b)
[2]Totals may not sum due to rounding

---

[3] These calculations are mine based on data from White (1990), Mintzer (chapter 4), Davis (1990), and NRC (1990).
[4] The Enquête-Kommission (1990) estimated that direct emissions from diesel and spark-ignition engines fueled by CH_4—combined with indirect emissions due to losses during fuel production, storage, and distribution—amount to about 5 Mt/yr worldwide, or 2 to 5% of total CH_4 emissions from all sources.

regional and country-by-country basis depending on economic development, local geography, and population distribution. Climate and meteorological conditions also differ in ways that affect the quantity of various petroleum combustion products.

IMPLICATIONS FOR FUTURE ACTION

Historically, the growth of motor vehicle production and use correlates closely with economic development. As economies expand and standards of living improve, the need for transportation of people and goods increases. This is particularly evident in the patterns of automobile ownership and use. The increase in personal income as a result of economic growth tends to give rise to increased private ownership of automobiles and growing preference for (and reliance on) the automobile for personal transportation.

This trend has reached an extreme in the U.S., where automobile registrations are virtually equal to the number of persons of driving age (16 or older) and the private passenger car accounts for 82% of all person trips and a like percentage of all passenger kilometers of intercity travel (FHA 1985; MVMA 1990b). Automobile dependence in western Europe, Japan, Australia, and New Zealand is only slightly less and seems to be approaching that in the U.S. A corollary of this growing reliance on motor vehicles is a decline in the share of public modes of passenger transportation and stagnation of goods movement by rail and inland waterways. For both people and goods, increasing reliance is being placed on road and air transportation.

It seems likely that, unless there are strong forces to the contrary, the emerging nations of the world will choose to recapitulate the experience of the most economically advanced nations. If so, the coming decades will see a rapid rise in the motor vehicle fleets of less-developed countries. Accentuating this trend will be rapid population growth in these countries. The annual population growth rates in the Middle East, Africa, Latin America, and Asia are forecast to be three to five times greater than those of North America and Europe. If the rise of population is accompanied by economic growth and increase in personal income, the motor vehicle fleet could be in excess of 1 billion by 2025, with a large part of this growth occurring in developing countries (Mintzer, chapter 4).

Because the primary fuel for motor vehicles is gasoline or other petroleum distillates, petroleum consumption for transportation would rise sharply unless there were a major improvement in fuel efficiency. In the absence of an international convention to protect the atmosphere from greenhouse-gas emissions, increased gasoline consumption could lead to both a rise in the price of crude oil and an increasing burden on the economies of developing countries, few of which have domestic sources of petroleum. It may well be that the lure of the automobile, which has become synonymous with economic well-being, personal mobility, and freedom to travel, will become the impediment to economic development in the regions most in need of raising their standard of living.

Another clear implication of increasing automobilization is the harmful effect this would have on global warming. By 2025 a fleet of 1 billion motor vehicles would discharge as much as 1,800 Mt of carbon into the atmosphere annually.[5] This would be the equivalent of about one-third of the 5,600 Mt of carbon emissions from all sources today. Other motor vehicle emissions (notably NO_x, CO, and volatile hydrocarbons) would also increase correspondingly. The vision of the future set forth in the "business-as-usual" scenario (see Mintzer, chapter 4) for this conference is grim when set against the climatologists' estimates that atmospheric protection requires reduction of carbon emissions from today's level of 5,600 Mt to 3,800 Mt in 2020 and to 2,700 Mt by 2040–2050.

NEAR-TERM ACTIONS: THE NEXT 15 YEARS

There are several measures that could be undertaken in the near term to reduce the production of greenhouse gases by the transportation sector. Virtually all involve changes in motor vehicle technology or its use. The opportunities in other modes of transportation are few and of slight impact, with the possible exception of aviation, where reductions in fuel consumption could be attained through improved engines and aerodynamics and more efficient aircraft use. Rail, water, and pipeline transport already have a relatively high level of energy efficiency, and the overall impact of gains that might be made are minute because of the negligibly small percentage of fossil fuel consumed by these modes. As a whole, the effects of these actions are not likely to make a dramatic change in present trends by 2005. The most that can be expected is slowing the rate of growth of atmospheric contamination. Reducing the growth rate to zero, much less bringing the discharge of CO_2 and other greenhouse gases from mobile sources down below present levels, lies beyond our reach in the next 15 years.

The near-term mitigation measures for automobiles can be grouped in two categories: technological and behavioral, each involving actions to reduce engine emissions directly, to conserve fuel and thus reduce emissions

---

[5] These estimates are based on data from White (1990), DOT (1990), and Mintzer (see chapter 4).

indirectly, or to make use of alternate fuels. Table 11.4 lists these measures by category and type of action.

Technological Improvements

Emission Controls

Many of the most highly automobilized nations have set standards for control of motor vehicle emissions through some combination of exhaust gas treatment and redesign of engines. Typically, these standards apply to NO_x, CO, and volatile hydrocarbons (HC). Most of these standards have been in effect for 10 to 20 years, and they have led to substantial reduction of tailpipe emissions. As a result of emission control technology, in combination with increased fuel economy, new model U.S. passenger cars emit 96% less CO and HC and 76% less NO_x (measured in grams per mile) than the pre-control cars of 20 years ago. Similar, although slightly smaller, reductions have been made in light- and heavy-duty trucks. Further reductions, especially in NO_x, might be achieved; however, these measures are increasingly expensive, technologically difficult, and show less and less favorable ratios of cost to environmental and social benefit.

Establishing comparable control measures elsewhere in the world, notably in eastern Europe, Asia, and the more highly automobilized countries of Latin America, would be beneficial, but the overall effect would be small since their fleets constitute only a minor fraction of the world total.

Note also that emission controls would not affect the production of CO_2, which is the principal greenhouse gas. However, the reduction of NO_x and HC (which are precursors to the formation of O_3) and CO (which is a toxic gas and a form of gaseous carbon in the atmosphere) would contribute substantially to improvement of air quality in urban areas.

Table 11.4 Near-term measures to reduce transportation emissions

Type of Action	Technological	Behavioral
Emission control	Cleaner engines	Improved maintenance
		Anti-tampering devices
Fuel conservation	Improved fuel economy	Reduced travel
	Traffic management	Higher vehicle occupancy
		Greater use of mass transit
Alternate	Non-petroleum fuels	Fuel or energy taxes
	Electric vehicles	

Fuel Conservation

Improving the fuel efficiency of motor vehicles would have a proportionate effect on CO_2 production. If the motor vehicle fleet could achieve a 20% overall reduction in fuel consumption (and if there were no offsetting increase in the average amount of travel per vehicle), there would be an equivalent reduction of CO_2 emissions. This is perhaps the most promising avenue in the near term. The technology is available. Vehicles capable of 6 l/100 km or better are on the market. Experimental models with fuel economy in the range 3 to 4 l/100 km have been demonstrated (OTA 1982; Seiffert and Walzer 1984; Greene 1990). The chief barrier is not technology but public acceptance of automobiles that are smaller, less powerful, and lacking in some of the amenities of the typical passenger car of today.

Reduction of fuel consumption can also be achieved through improved traffic management technology. Idling vehicles stalled in traffic jams continue to burn fuel without moving. If average vehicle speed in urban areas could be increased by better signalization, traffic metering, and other such management of traffic flow, fuel savings on the order of 10% could be obtained (Bleviss and Walzer 1990). This would also translate directly into CO_2 reduction.

Reducing highway congestion, however, can have a perverse effect. Experience has shown that measures to alleviate congestion, either through building new roads or improving traffic movement on existing routes, have typically led to increased highway usage that quickly negates the immediate beneficial effects of smoother and more efficient traffic flow. To prevent this, traffic management improvements would have to be accompanied by transportation policies that encourage public transportation and nonmotorized means of conveyance (Pisarksi 1987; Institute of Transportation Engineers 1989).

Alternate Energy Sources

Several substitutes for gasoline or diesel fuel could be used by motor vehicles. These fuels are either gaseous (such as CH_4, propane, and hydrogen) or liquid (methanol, ethanol, and coal-derived hydrocarbons). All have different combustion properties and hence different emission characteristics, some more favorable than conventional motor fuels. With the exception of hydrogen, all produce CO_2 when burned, although CO_2 stemming from combustion of biomass fuel is not considered a greenhouse-gas emission since it would not add to the net atmospheric burden provided it is subsequently reincorporated into biomass through photosynthesis. The amount of CO_2 generated by vehicles using these petroleum substitutes would be somewhere between 5 and 15% less per kilometer than that from an equivalent amount of gasoline (Paaswell 1990).

Advanced vehicles fueled by gaseous or liquid petroleum substitutes could begin to enter the market in significant numbers during the coming 10 to 15 years. Gray and Alson (1989) estimate that a methanol-powered car equipped with a continuously variable transmission, a flywheel to minimize engine operation, and a recirculating system to convert exhaust heat into useful work could achieve very high fuel economy, perhaps 3 l/100 km. This, combined with a fuel that yields less CO_2 than gasoline when burned, could result in as much as 80% less CO_2 emitted for a given amount of travel. The benefits of such a redesigned vehicle, however, would not be fully realized for at least two decades, depending on the rate and extent of fleet penetration.

All substitute fuels necessitate some modification of conventional automobile engines, and they create performance problems, notably cold start capability, long-term engine durability, and possibly harmful emissions (OTA 1982; Seiffert and Walzer 1984; Bleviss and Walzer 1990). The most promising choices for the near term are liquefied petroleum gas, compressed natural gas (CH_4), ethanol, and methanol.

Hydrogen would be the most attractive alternative, but it cannot be produced in quantity at a competitive price now or in the near future. The prospects for hydrogen as a longer-term alternative to petroleum are discussed later in this paper.

Electric power has been proposed as a substitute for the internal combustion engine. In the early days of the automobile, electric vehicles outnumbered gasoline-powered cars, but they faded from competition as cars with spark-ignition and diesel engines improved in power, range, reliability, and durability (Flink 1975; Lewis and Musciano 1977). Small fleets of battery-powered electric vehicles are in use today in urban settings, but they have not proven generally acceptable due to their limited range, small payload, low speed, and cumbersome recharging requirements. Although electric vehicles themselves emit no CO_2 or other atmospheric pollutants, the principal source of electricity to charge their batteries is power plants that burn fossil fuels. The net effect of increased use of electric vehicles would be only a small reduction of CO_2 emissions, due almost entirely to the greater thermal efficiency of power plants compared to automobile engines (Hamilton 1980, 1982; Bosch 1978). Long-term applicability of electric vehicles is discussed later.

Behavioral Changes

Behavioral changes are those measures that do not involve vehicular technology per se. Instead, they relate to how technology is used within the transportation system. They consist of managerial, regulatory, economic, or policy actions taken to alter the way people use transportation system components.

Emission Control

The performance of emission control devices is heavily dependent on the quality of maintenance that the vehicle receives over its service life. A poorly tuned or neglected engine will produce more NO_x, CO, and HC than a properly maintained vehicle by a factor of as much as 4 or 5. Improper tuning can also reduce fuel economy by as much as 20 to 30% and therefore produce more CO_2 emissions for a given amount of travel (OTA 1982). Periodic inspection and mandatory maintenance or repair can help prevent emission control and fuel economy from deteriorating unnecessarily during vehicle lifetime.

Some motorists bypass or otherwise defeat pollution controls in the interest of improving performance (chiefly acceleration) or in the often mistaken belief that it will increase fuel economy. Some forms of tampering can actually decrease fuel economy (Passwell 1990). Periodic inspection and installation of special anti-tampering devices can prevent this form of abuse.

Fuel Conservation

The most direct and obvious way to conserve fuel is to travel less. Better trip planning, elimination of unessential travel, and other ways to economize on vehicle use would have a direct payoff in fuel savings and hence CO_2 production. These measures can be undertaken voluntarily or they can be induced by increasing the price of fuel, restricting vehicle use at certain times and places, or by fuel rationing. The success of these measures depends on value judgements about short-term benefits and convenience versus long-term damage due to global climate change. They might be considered coercive, but they would be highly effective and could be adopted in some circumstances.

Other conservation measures that may be more palatable involve incentives to increase vehicle occupancy (reserved lanes for passenger cars with three or more occupants, preferential parking for car-pool vehicles, or encouraging the use of multipassenger vans). These measures work best in urban areas for people whose place of employment or daily activity allows some convenient form of ride-sharing.

Modal Shift

In cities with well-developed public transportation systems, a substantial share of personal travel by automobile might be shifted to buses or rail transit. Incentives to use public transport could be applied (e.g., reduced fares, preferential treatment of public vehicles, or bans on automobiles in areas of high population density). The entire transportation infrastructure could be reshaped to lessen automobile dependence. Several cities in economically emerging countries (e.g., Singapore and São Paulo and Curitiba in Brazil) have done so by adopting policies that expressly favor public transit as an alternative to automobilization.

In highly automobilized countries, notably the U.S. and to a somewhat lesser degree Canada, Mexico, Argentina, Australia, and New Zealand, substantial shifts to public transit could not be accomplished without extensive and very costly investments in new infrastructure. Public transit in the U.S., for example, has atrophied to the point where it could not absorb a large influx of new passengers in the short term. A 10% shift of passenger travel away from automobiles would translate into a 200% increase in transit ridership. Existing public transport systems in most major U.S. cities would be swamped, and to provide the necessary new capacity would involve massive investments that could take 10 to 20 years to put in place even at a forced pace.

Thus, it may be that the places most amenable and appropriate for future development of less automobilized alternative transportation systems will not be the present centers of high automobile concentration but emerging nations where transportation infrastructure is generally lacking and land-use patterns are not yet irrevocably shaped by dependence on private motor vehicles. Certainly, enormous investments would be required to create highly organized and extensive public transportation systems from scratch in less-developed countries. However, if long-term social costs (including environmental effects and dependence on imported energy) are factored in, it is arguable whether such a transportation system would be more expensive than the dense North American or European style roadway networks serving small, privately owned motor vehicles with one or two occupants each.

Alternative Energy Sources

The alternatives to petroleum all suffer from a common drawback at the present time; they are more expensive as a motor fuel than gasoline or diesel fuel. One way to overcome this disadvantage is through the mechanism of taxation. Several approaches have been proposed: gasoline tax, petroleum tax, carbon tax, tax on heat content or ad valorem tax (simply a surcharge on energy price). These taxes work in various ways to discourage petroleum consumption (either absolutely or in favor of some alternative), to reduce CO_2 emission, or to promote overall energy conservation. What distinguishes these taxes from mere revenue measures is that the tax proceeds would be directed specifically to development of more economically or environmentally beneficial substitutes. The proceeds might be used to carry out research and development of new fuels, to subsidize wider use of petroleum substitutes, or to support more efficient and less polluting forms of transportation (Goldemberg 1987).

TRANSITION: THE NEXT 50 YEARS

Over the next 50 years the opportunities for technological change in transportation become more sweeping. Three main avenues are open: advanced internal combustion engines, new electric propulsion systems, and alternatives to fossil-fuel energy. The choice among these avenues is not mutually exclusive. All are likely to be pursued, and at this point it is unclear which (or which combination) will prove the most beneficial to society and protective of the environment.

Heretofore, the pacing technology in transportation has been the propulsion system. Energy source and supply, although certainly important considerations, have not been the paramount concerns nor the primary factors influencing the direction of vehicle development. This is changing. Combustion of fossil fuel (petroleum or any other form of hydrocarbon) on a large scale is now widely understood to be a major threat to the global climate. It is becoming clear that the long-range demands of human society and economic well-being dictate transition to an energy source that is plentiful to the point of inexhaustibility, affordable by rich and poor nations alike, and environmentally benign. For transportation, the next 50 years will be a period of transition during which scientists and engineers seek a new combination of vehicle technology and energy supply to replace that which has served us for the past century.

Combustion Engines

Automobile manufacturers have been experimenting with alternatives to petroleum as a motor fuel for many years, but with renewed interest since the two sharp oil price increases of the 1970s. The search has turned up many promising technologies; however, they have had little economic success largely because the price of alternatives is high in relation to petroleum, production capacity is lacking, and each possible new fuel dictates some degree of change in engine design and the fuel distribution system. Manufacturers have been reluctant to commit to a particular technology until the question of the future fuel of choice has been resolved.

At this time the field of candidate fuels seems to have narrowed to four, at least for the next 20 years or so: methanol, ethanol, natural gas, and liquefied petroleum gas. With relatively minor changes to conventional engines, all of these fuels could begin to replace gasoline by the end of this century. To obtain the greatest benefit from these substitute fuels, however, new spark-ignition and diesel engines would have to be perfected (Bleviss and Walzer 1990). Other types of engines, including turbines (Brayton cycle), Rankine cycle (steam) engines, and Stirling engines, could also operate on these fuels, in some instances more efficiently than conventional engines. However, the cost, performance (especially at part load), and manufacturability of these alternatives are such that they are unlikely to be competitive with advanced spark-ignition and diesel engines unless there is political intervention to bring them to the market (OTA 1982).

Vehicles powered by any of these substitute fuels could achieve very important reductions in CO_2 emissions in two ways. The fuels themselves are relatively hydrogen-rich and carbon-poor and therefore would yield less CO_2 when burned, and the vehicles would be more fuel-efficient. Offsetting these potential advantages, however, would be penalties in vehicle utility (range, luggage space), performance, maintainability, and refueling convenience (time, complexity) (Table 11.5).

Over the course of 50 years, however, this avenue of development may not be sufficient. The fuel supply itself, although presently abundant, is not renewable, except for ethanol which can be produced from biomass. Large-scale consumption of these fuels to meet the needs of a world motor vehicle

Table 11.5 Comparison of substitutes for gasoline-powered vehicles (adapted from Gray and Alson 1990)

PERFORMANCE CHARACTERISTICS	ALTERNATIVE FUEL OR POWER SOURCE				
	Methanol	Ethanol	Compressed Natural Gas	Liquefied Petroleum Gas	Electricity
Feedstock Size and Diversity	+ +	−	+ +	−	+ +
Environmental Impacts	+ +	+ +	+ +	+ +	+ +
Vehicle Cost	0	0	−	−	− −
Vehicle Utility (range, luggage space)	0	0	− −	0	− −
Vehicle Performance	0/+	0/+	− −	−	− −
Current Fuel Operating Cost (low demand)	−	− −	0	0	0/+
Future Fuel Operating Cost (high demand)	+ +	−	+	0	0/+
Refueling Convenience (time, complexity)	0	0	− −	− −	− −

Relative to Gasoline: + + much better; + somewhat better; 0 similar; − somewhat worse; − − much worse fleet that could number well over 1 billion by 2040 would seriously draw down present reserves and pose a serious threat to the global climate. More important, the reduction of fossil-fuel consumption made possible by the increased energy efficiency of these advanced vehicles could be offset (or perhaps negated) by absolute transportation energy demand that might equal or exceed present levels. In terms of CO_2 production from transportation, the world might be no better off than today.

This line of reasoning argues for a transition path that eventually leads to essentially full replacement of fossil fuels with energy from completely different sources.

Hydrogen

Hydrogen, in the form of water, is one of the most abundant elements on Earth. If a way could be found to produce hydrogen in the quantities needed to sustain transportation demand, and if suitable engines and storage and distribution facilities could be developed, a major and lasting reduction in CO_2 generation could be attained.

Vehicles that use hydrogen have been, and continue to be, tested; however, they are not yet ready for large-scale use. Hydrogen gas has a low energy density, which limits range and payload (Bleviss and Walzer 1990). The energy density problem can be solved by liquefying hydrogen, but this requires additional energy for liquefaction and vacuum-insulated cryogenic containment vessels. Hydrogen may also cause metals to become brittle, posing problems in the design of engines and storage facilities.

Hydrogen is more difficult than petroleum or hydrocarbon fuels to store and distribute in gas or liquid form, but it can be stored easily by chemical binding, e.g., with toluene. The small molecular weight of hydrogen makes containment a major problem. Leakage not only wastes fuel, it also poses safety hazards in storage, handling, and on-board use. An alternative is storage of hydrogen as a chemical compound for release as needed. Metal hydrides, however, are bulky and heavy, which limits their utility as a power source in motor vehicles (C.J. Winter, pers. comm.).

A major impediment to the development of a hydrogen-based transportation system is infrastructure. A completely new fuel production system would be required. The technology to effect this on a worldwide scale would take decades and trillions of dollars to construct. The energy source to produce hydrogen (presumably by electrolysis of water) would have to be something other than fossil fuel (e.g., solar, hydropower, or nuclear) if net reduction of CO_2 discharge is to be achieved. Even if such an energy source were available, there would also be a need for complete revamping of the infrastructure to transport hydrogen from production sites to storage points and then to distribute it to individual vehicles. The capital cost of this infrastructure would be high, perhaps as great as that required to produce hydrogen.

None of these problems is technologically insurmountable in the long run. However, they would take decades to implement; they will be costly; and, of special pertinence here, they will entail a fundamental transformation of vehicle and support system technology. In scale and complexity, such an undertaking would probably surpass that which led to the present petroleum-based transportation system over the past century.

If a hydrogen energy conversion chain that is environmentally clean from end to end can be developed, substantial and permanent reduction of greenhouse gases from the transportation sector can be achieved. The potential is high, but the infrastructure costs are formidable, and the time needed is long—two to three decades or more. Properly handled, hydrogen can meet safety requirements as exacting as those that govern the gasoline storage, transportation, and distribution system today (C.J. Winter, pers. comm.).

Electric Vehicles

A more direct, and probably more energy-efficient approach, would be to use electricity itself for motive power. Electric propulsion has a long history in transportation, and the basic technology is well understood. The chief obstacle to the use of electricity to drive motor vehicles is how to deliver the power on board. In heavy rail, transit, and other fixed-guideway applications, generation of electricity at a central source with delivery to vehicles by power lines is a practical and common solution. For automobile use some form of storage or power generation on board the vehicle itself is needed.

Present battery technology is not fully adequate for this purpose. The ratio of energy storage to weight is low. Storage capacity restricts range. Power delivery drops off steeply as the battery approaches depletion (leading to slow, sluggish performance). Recharging or replacement of the battery pack is time-consuming and cumbersome.

New batteries to replace conventional lead-acid batteries are being developed. The most immediately promising are nickel-cadmium and nickel-iron. Over the longer term, advanced batteries (such as iron-air, zinc-air, zinc-chlorine, sodium-sulfur, and lithium-metal sulfide) have the prospect of providing storage capacity, power density, and recycling capability three to four times greater than the best of those available today (OTA 1982).

A variant of the purely electric vehicle is a hybrid configuration that combines an internal combustion engine with an electric motor to exploit their respective advantages of power and CO_2-free operation. A typical design involves a spark-ignition or diesel engine, a small electric motor, a battery, and a clutch connecting the motor and the engine. Bleviss and Walzer (1990) describe one such hybrid now being evaluated by Volkswagen. Tests indicate that its fuel consumption is 2.4 l/100 km and 25 kilowatt hours of electricity. If the electric charge comes from a source other than fossil fuel, 60% of the CO_2 presently emitted by a conventional vehicle of the same size would be eliminated.

Transportation System Characteristics

Through a combination of these advanced technologies and fuel-efficiency refinements of existing technology, CO_2 emissions from transportation could be cut substantially in the next half century. A study just released by the National Research Council (1990) estimates that by 2050 a 50 to 75% reduction of CO_2 from the transportation sector might be possible. This would be achieved primarily by conversion of motor vehicles (including heavy trucks) to nonpetroleum fuels, significantly increased automotive fuel efficiency, and electrification of rail transport. Aviation, maritime, and pipeline transport could continue to rely on conventional fuels.

LONG-TERM OUTLOOK: THE NEXT 100 YEARS

The world at the close of the 21st century is no more foreseeable than the world of today was from the vantage point of 1890, when the automobile and the airplane were not yet born. What follows therefore is not a forecast or a prediction but a vision of what might be, or perhaps what must be, if we are to sustain a habitable environment on this planet.

By 2090 the transition from fossil energy to noncarbon fuels and electricity must be completed; or, as a minimum, the world must be on a path that will reach this goal soon after. This would be, in effect, a world transportation system relying almost entirely on hydrogen, electricity, and carbon fuels from renewable biomass resources.

The worldwide energy needs at that time, as outlined in the "aggressive response" scenario developed for this workshop (Mintzer, chapter 4), would be roughly three times greater than today. On a per-capita basis, however, the need would be about the same as today (0.6 to 0.7 exajoules per 1 million persons annually). Because of gains in the efficiency of energy use and the switch to noncarbon energy sources, the net environmental burden of CO_2 would be substantially reduced from 4.6 GtC y^{-1} in 1975 to 0.8 Gt y^{-1} in 2075, an 83% reduction (Mintzer, chapter 4).

For this scenario to be realized in the transportation sector, where the proportion of energy use in relation to total world consumption may be as much as one-fourth to one-third more than today, electricity would have to be used extensively both to produce hydrogen fuel for automobiles and aircraft and to propel electric vehicles (chiefly passenger cars).

Where will this electricity come from? Many experts (e.g., NRC 1990; Mintzer chapter 4; Holdren 1990) project that a combination of solar, hydroelectric, and biomass power generation will be able to meet 65 to 75% of the world's needs 100 years from now. Nuclear power might contribute an additional 10 to 20%, and perhaps more depending on the evolution of fission technology and the possible development of fusion reactors.

Nuclear power is still in its infancy. Research and development are being pursued on two fronts. Advanced fission reactors now under study could reduce the cost and environmental hazard and improve the efficiency of power generation. Practical fusion reactors are still many decades away, but 100 years is a very long time in a field that is not yet 50 years old. The advent of large-scale fusion reactors by 2040 could radically alter the energy equation by 2090.

Virtually every aspect of the future technology for reduction of CO_2 emission from transportation is clouded with uncertainty at this time. On theoretical grounds the promise is highly encouraging. Technical and economic practicality, however, are dim at best. In the long run it may be that success depends not so much on engineering as on political factors and our collective commitment to restructuring the world's energy production system. Human ingenuity is great, and a concerted effort over a period as long as 100 years could accomplish much. To achieve a stable long-term equilibrium of the CO_2 cycle will require, above all, political and economic decisions by the nations of the world—the advanced and emerging nations alike—on the steps that must be taken, the sacrifices to be made, and the costs to be borne. In the long run the basic issue is not what we can, but what we will do.

REFERENCES

Bleviss, D.L., and P. Walzer. 1990. Energy for motor vehicles. *Sci. Am.* **263(3)**:103–109.
Bosch GmbH. 1978. Automotive Handbook. 1st English (10th German) edition. Stuttgart.
Davis, G.R. 1990. Energy for planet Earth. *Sci. Am.* **263(3)**:55–62.
DOT (Dept. of Transportation). 1990. National Transportation Strategic Planning Study, 519 pp. Washington, D.C.: U.S. Dept. of Transportation.
Enquête-Kommission. 1990. Technikfolgen-Abschätzung und -Bewertung. Study Commission of the 11th German Bundestag, Bundestagsdrucksache 11/7993. vol. 2 pp. 441–444, 463, and 479. Bonn: Economica Verlag, Verlag C.F. Müller.
FHA (Federal Highway Administration). 1985. 1983–1984 Nationwide Personal Transportation Study, 426 pp. Washington, D.C.: Dept. of Transportation.
Flink, J.J. 1975. The Car Culture, 260 pp. Cambridge, MA: MIT Press.

Fulkerson, W., R.R. Judkins, and M.K. Sanghvi. 1989. Energy technology R&D: what could make a difference? Synthesis Report, Part 1, ORNL-6541/VI, 184 pp. Oak Ridge, TN: Oak Ridge Natl. Lab.

Goldemberg, J. et al. 1987. Energy for a Sustainable World, 219 pp. Washington, D.C.: World Resources Inst.

Gray, C.L., and J.A. Alson. 1989. The case for methanol. *Sci. Am.* **262(5)**:108–114.

Greene, D.I. 1990. Technology and Fuel Efficiency Forum for Applied Research and Public Policy. **5(1)**:23–29.

Hamilton, W. 1980. Electric Automobiles, 425 pp. New York: McGraw Hill.

Hamilton, W. 1982. The future potential of electric and hybrid vehicles. Background paper OTA-BP-E-13. Washington, D.C.: Office of Technology Assessment, U.S. Congress.

Holden, J.P. 1990. Energy in transition. *Sci. Am.* **263(3)**:157–163.

Institute of Transportation Engineers. 1989. A Toolbox for Alleviating Traffic Congestion, 154 pp. Washington, D.C.: GPO.

Lewis, A.L., and W.A. Musciano. 1977. Automobiles of the World, 731 pp. New York: Simon and Schuster.

MVMA (Motor Vehicle Manufacturers Association). 1990a. World Motor Vehicle Data, 352 pp. Detroit: MVMA.

MVMA (Motor Vehicle Manufacturers Association). 1990b. Motor Vehicle Facts and Figures '90, 96 pp. Detroit: MVMA.

NRC (National Research Council). 1990. Confronting Climate Change, 127 pp. Washington, D.C.: Natl. Academy Press.

OTA (Office of Technology Assessment). 1982. Increased automobile fuel efficiency and synthetic fuels, 293 pp. Washington, D.C.: U.S. Congress.

Paaswell, R.E. 1990. Air quality and the transportation community. TR News 148:5–10. Washington, D.C.: Transportation Research Board.

Pisarski, A.E. 1987. Commuting in America, 78 pp. Westport, CT: Eno Foundation.

Seiffert, U., and P. Walzer. 1984. The Future for Automotive Technology, 197 pp. London: Frances Pinter.

White, R.M. 1990. The great climate debate. *Sci. Am.* **263(1)**:36–43.

# 12
Global Options for Reducing Greenhouse-Gas Emissions from Residential and Commercial Buildings

W.B. ASHTON, S.C. McDONALD, and A.K. NICHOLLS
Pacific Northwest Laboratory, Washington, D.C. 20024–2115, U.S.A.

ABSTRACT

A significant portion of current worldwide greenhouse-gas emissions is attributable to energy use in residential and commercial buildings. With expected global population and economic growth trends, buildings will continue to have a major and possibly increasing role in these releases. To combat future emission trends, several recent studies have identified an enormous potential for at least slowing the rate of increase in the growth of greenhouse-gas emissions. For many end-use energy services in buildings, savings as high as 85% of current emissions have been estimated. However, to achieve the bulk of these reductions, some form of enhanced public sector intervention will be necessary.

This paper presents a range of options to reduce greenhouse-gas emissions and discusses both public and private actions necessary to implement them. A wide variety of technical options to reduce energy use in buildings are available through energy service demand restructuring, improved building design, better equipment selection and operation, and fuel-switching. Fortunately, a host of policies exist to implement these options, including economic incentives, regulation, education and information programs, advanced technology research/development/demonstration, and social engineering practices. Choosing the best mix of options and policies to reduce emissions in the buildings sector depends upon the particular situation of each country. Factors such as natural endowments, economic and technological development stages, infrastructure maturity, population growth, and cultural traditions will dictate the most cost-effective set of policies.

Limiting the Greenhouse Effect: Options for Controlling Atmospheric CO_2 Accumulation
Edited by G.I. Pearman published 1992 John Wiley & Sons Ltd

INTRODUCTION

A significant amount of current worldwide greenhouse-gas emissions is attributable to energy use in residential and commercial buildings. With expected population and economic growth trends, buildings will continue to have a significant and possibly increasing impact on these releases. To address these trends, several recent studies have identified an enormous cost-effective potential for reducing greenhouse-gas emissions, or at least slowing the rate of increase, in many sectors, including buildings (see, for example, Carlsmith et al. 1990; Goldemberg et al. 1988; Bevington and Rosenfeld 1990; Fickett et al. 1990). However, these studies also point out that this potential will not be achieved by reliance on market forces alone. In each region of the world, some form of enhanced public sector intervention will be necessary to achieve the bulk of the potential reductions in greenhouse-gas emissions. The central issue for each country is how best to reduce these emissions given the wide range of technical, institutional, and behavioral mechanisms available.

The purpose of this chapter is to present a range of options to reduce greenhouse-gas emissions from buildings and to discuss policies necessary to implement them. It summarizes current building sector emissions and describes measures for reducing these emissions through both public and private actions. The discussion is at an aggregate level and includes both global and long-term perspectives as well as local and near-term issues.

Most greenhouse-gas emissions from buildings occur as a direct result of energy use, particularly combustion of fossil fuels. Accordingly, this chapter concentrates on energy-related emissions, emphasizing CO_2. A significant portion of the emissions of chlorofluorocarbons (CFCs), which are also greenhouse gases, can be considered building-related, and CFCs will continue to be an important contributor to climate forcing into the next century. However, CFCs are not discussed here since international policies already exist to phase out their use.

GLOBAL ENERGY USE IN BUILDINGS AND GREENHOUSE-GAS EMISSIONS

As shown in Table 12.1, the U.S. Environmental Protection Agency (EPA) estimates that the buildings sector energy consumption worldwide is responsible for about 15% of the contribution to global warming potential through greenhouse-gas emissions. Most greenhouse-gas emissions attributable to the buildings sector are a result of burning fossil fuels for space heating, lighting, cooking, and other energy services. These emissions result from the direct combustion of coal, oil, or natural gas and the use of fossil-fuel based electricity. Biomass burning for heating or cooking (organic, biologically based waste such as wood) also contributes to greenhouse-gas emissions, although no net CO_2 emissions arise from this source, given that photosynthetic removal of atmospheric CO_2 offsets the amount of new biomass growth.

The contribution to greenhouse-gas emissions from buildings varies considerably for different regions of the globe. Focusing on energy use, the global portion of buildings energy demand is estimated in Table 12.2 for the Organization for Economic Cooperation and Development (OECD) countries, the centrally planned economies (CPEs) of Asia and Europe (including China and the USSR), and the developing countries.[1] The data are for secondary or delivered energy consumption, both fossil and nonfossil sources, and include both commercially sold and noncommercial (mostly biomass) sources. This table indicates that buildings energy in the OECD is approximately 29% of end-use consumption, while higher consumption

Table 12.1 Global energy use and warming contribution, 1985 (from EPA 1989, p. viii–28); contribution to mean-decadal forcing for the 1980s

	Percent Contribution[1]		
End-Use Sector	Secondary[2] Energy Use	Primary Energy Equivalent	Global Warming Contribution
Energy-Use Sources			
Buildings	29	33	15
Transportation	27	20	20
Industry	44	47	22
Other Sources			
CFCs	—	—	17
Agriculture	—	—	14
Forestry	—	—	9
Industry	—	—	3
Total	100	100	100

[1]Percent contribution shown by column heading.
[2]"Secondary" energy use refers to the energy source delivered to buildings and excludes generation, transmission, and distribution losses associated with electricity use and transmission losses from natural gas use.

---

[1] OECD countries are Australia, Austria, Belgium, Canada, Denmark, Finland, France, Germany, Greece, Iceland, Ireland, Italy, Japan, Luxembourg, the Netherlands, New Zealand, Norway, Portugal, Spain, Sweden, Switzerland, Turkey, United Kingdom, and the United States.

shares occur in the CPEs of Asia and Europe (about 33%) and in developing countries (48%).

Although energy-use patterns in developing countries are not as well understood as in the other two regions, the data in Table 12.2 suggest that fundamental differences in the structure of each region's economy and demography exist. Whereas all regions address residential energy needs of their population, the role of the commercial (or service) sector is quite different in each of these areas. OECD countries have extensive commercial sectors of office and retail buildings, institutions, hospitals and the like, which are large energy users. By contrast, in many CPEs and the developing countries, the commercial sector has not been as well developed.

Determinants of Energy Use in Buildings

A wide range of basic causes are behind the energy consumed in buildings. Figure 12.1 illustrates the major links between certain fundamental determinants, or "Drivers," of energy needs and resulting primary energy use. The intervening sources of energy use are represented as three interacting factors. (The figure indicates that both the fuel-type mix and level of consumption are important features of energy use when considering greenhouse-gas emissions). The first factor affecting building energy use is the *demand for energy services*, such as space conditioning, lighting, and cooking. People do not demand energy directly; instead, their preferences for comfort and convenience create requirements for particular types and levels of services provided by use of energy. Examples of these service demands include the level at which thermostats are set (degree of warmth

Table 12.2 Global energy demand by region for 1985 (from EPA 1989, p. iv-19 [adapted from J. Sathaye et al. 1987, An End-Use Approach to Development of Long-Term Energy Demand Scenarios for Developing Countries, prepared for U.S. EPA. Washington, D.C.; I.M. Mintzer 1988, Projecting Future Energy Demand in Industrialized Countries: An End-Use Oriented Approach, prepared for U.S. EPA. Washington, D.C.])

	End-Use Consumption, Exajoules			
Global Region	Residential/ Commercial	Industrial	Transport	Total
OECD	33.4	42.6	39.7	115.7
Centrally Planned	18.2	29.3	8.4	55.9
Developing	27.0	19.0	10.0	56.0
Total	78.6	90.9	58.1	227.6

demanded), utilization rates for lights (amount and type of illumination demand), and use of powered appliances or office equipment (demand for human labor-saving substitutes). The type and level of service demand are heavily influenced by economics, weather, and local culture, among other factors (SERI 1981; Goldemberg et al. 1988).

Energy-service demands are satisfied through some form of building energy end-use device (or technology application), which is the direct source of energy consumption. The role of end-use devices (e.g., residential gas furnace or compact fluorescent bulb) in energy use can be described in terms of the other two energy use factors: the physical *characteristics of the building stock* (including equipment) and the *operating practices* carried out in using the capital stock. Figure 12.1 indicates that building stock characteristics include items such as the mix of single versus multifamily structures, the thermal integrity of building envelopes, and the conversion efficiency of installed equipment. Conceptually, one might think of these factors as establishing the building capital stock inventory, or the number and type of buildings, *prior* to occupation and use by people. The building stock inventory in turn determines the technical parameters by which energy-service demands might be satisfied. To illustrate, the degree to which warmth can be supplied to residential occupants on a cold winter day is largely a

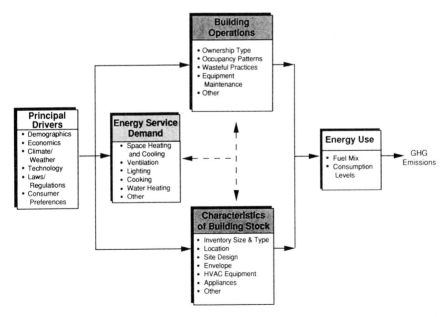

Figure 12.1 Principal determinants of energy use and greenhouse-gas (GHG) emissions from buildings

function of the thermal integrity of the building envelope and efficiency and size of the heating system.

How this heating system is then operated to provide this warmth is also very important in determining energy-use patterns; this factor we deem *building operations*. As Figure 12.1 indicates, the operation of the building will be determined by whether the occupant owns or rents, the frequency of building occupation, how the building capital is maintained, and importantly, by *wasteful practices*. We define wasteful practices as energy use that is not providing an energy service. Examples include thermostats that are not turned down when occupants are on extended leaves from their residence and lights that are left on and which neither provide illumination nor security/safety.

Building characteristics and building operations interact with each other, and both condition the ways in which energy-service demands are established and satisfied. In turn, operations, building characteristics, and energy-service demands are driven by more fundamental determinants, the *principal drivers* of Figure 12.1. These drivers also interact in complex and subtle ways; they represent basic motivators for why and how people use buildings. Figure 12.1 includes the major drivers shown below:

- Demographics—population change, migration, and age composition.
- Economics—economic growth, absolute levels and distribution of wealth, resource prices, interest rates, infrastructure (electrification, fuels availability).
- Climate/Weather—temperature, humidity, rainfall, insolation.
- Technology—availability and characteristics (e.g., operating efficiency, building materials, reliability, longevity, types of equipment).
- Laws/Regulations—contract structure, subsidies, taxes, standards, and building codes.
- Consumer Preferences—cultural, aesthetic preferences.

Although their precise interactions are difficult to delineate even for a single country, these drivers combine to influence the characteristics of the buildings inventory, the energy services demanded from buildings, and the way buildings are operated to meet service demands around the world.

Figure 12.1 indicates interplay between how buildings are physically configured and how people choose to operate them to meet service demand; the end result of this interplay determines energy use and the related levels of greenhouse-gas emissions. This interplay (which can include the principal drivers in Figure 12.1) indicates the presence of certain "aggregate system-level" energy-use influences, which are difficult to measure but are also important in determining ultimate patterns of energy use. To illustrate, consider how aggregate buildings energy use is affected by the process of an aging population moving from colder to warmer climates in a large country, such as the U.S. For the residential sector, the southerly migration could mean closing older homes in the north in favor of construction of new homes, perhaps multifamily dwellings in the south. In the commercial sector, hospitals might be built instead of schools, and perhaps more leisure-related buildings like hotels instead of commercial office buildings. In terms of operation, homes might be occupied more of the day (no one leaving for work) and also kept more relatively comfortable (warmer in winter and cooler in summer).

As a consequence of these changes, demand for space conditioning, lighting, and cooking might all be higher in the residential sector than with a younger, working population. To some extent, these same factors might also affect equipment choice. With greater demand for some energy services, the efficiency of the relevant equipment, and thus its operating cost, becomes proportionately more important and could result in selection of more efficient equipment.

Global Energy Consumption by End Use

Large differences exist among the countries of the world in the relative amount, type, and way energy is used in buildings. For residential buildings, the industrialized countries rely heavily on electricity and gas while CPEs use significant amounts of coal. Biomass is very important to the residential energy requirements of the developing economies, particularly in comparison with the industrialized countries. Patterns of energy use in commercial buildings are more consistent around the globe than for residences, particularly with regard to use of electricity, although district heating is an important thermal energy source in several European and Asian countries (Sathaye et al. 1987; Chandler 1990).

Globally, the main end-use applications for energy consumption in buildings are heating/cooling, cooking, and lighting (AID 1990). The industrialized countries consume most of this energy for space and water heating, supplied primarily by fossil-fuel combustion and, to some extent, electricity; most cooking in industrialized countries is done with gas and electricity. A more detailed example of the demand for energy services in the U.S. is given in Table 12.3. In this case, residential and commercial energy services consumed 27 quadrillion Btu in 1985 (28.5 EJ), or about 36% of total equivalent primary energy for the U.S. The largest component (more than one-third) is for space heating, although the total for all space conditioning moves to more than half (54%) when this figure is combined with air conditioning and ventilation. Lighting accounts for another 15%, hot water heating is 11%, refrigeration is 7%, while the remainder (13%) is divided among all of the other appliances and equipment applications.

Table 12.3 Primary U.S. energy consumption for commercial and residential buildings for 1986 (from DOE 1989b)

	Quadrillion Btu					
	Electricity[2]	Nat. Gas	Oil	Other[3]	Total	%
Residential Sector						
Space heaters	1.81	2.87	1.00	0.39	6.07	39.8
Water heaters	1.61	0.82	0.10	0.06	2.58	16.9
Refrigerators	1.44	—	—	—	1.44	9.4
Lights	1.04	—	—	—	1.04	6.8
Air conditioners	1.08	—	—	—	1.08	7.1
Ranges/oven	0.64	0.21	0.00	0.03	0.88	5.8
Freezers	0.45	—	—	—	0.45	2.9
Other	1.19	0.54	—	—	1.72	11.3
TOTAL Residential	9.27	4.43	1.10	0.48	15.28	100.0
Commercial Sector						
Space heaters	1.00	1.83	0.79	0.18	3.80	32.4
Lights	2.96	—	—	—	2.96	25.2
Air conditioners	0.98	0.12	—	—	1.11	9.4
Ventilation	1.49	—	—	—	1.49	12.7
Water heaters	0.37	0.19	0.08	—	0.64	5.5
Other	1.37	0.24	0.01	0.11	1.73	14.7
TOTAL Commercial	8.17	2.38	0.89	0.29	11.73	100.0
TOTAL Buildings	17.44	6.81	1.98	0.77	27.01	100.0

[1]Totals for residential and commercial consumption from EIA State Energy Data Report (1960-1986). Distribution between end uses based on LBL Residential End-Use Model and ORNL/PNL Commercial End-Use Model. Latest data are available for 1986.
[2]Represents Btu value of primary energy inputs in production of electricity (11,500 Btu/KWh).
[3]For residential: coal and LPG. For commercial: coal, LPG, and motor gasoline (off-road use). Excludes estimated 0.8 quads of energy from wood fuel in residential sector.

In developing countries, most building energy is consumed for cooking, with consumers relying heavily on biomass or kerosene for fuel. In China, about seven-eighths of household energy is from coal, typically for cooking, but with an important role as a space heat source as well (Chandler 1990). A general comparison of the energy-use patterns of southeast Asian developing countries to those of the U.S. is shown in Table 12.4, which illustrates the extremely wide differences in fuel composition (use of biomass, fossil fuels, and electricity) and fuel matching with energy service. For instance, biomass is used extensively for cooking in developing countries; whereas, industrialized countries such as the U.S. cook with natural gas or electricity (see Sathaye et al. 1987; Sathaye and Myers 1985; Leach 1986).

Table 12.4 End-use energy consumption patterns for the residential/commercial sectors (from EPA 1989, p iv-24 [based on data from Sathaye et al. 1987, An End-Use Approach to Development of Long-Term Energy Demand Scenarios for Developing Countries, prepared for U.S. EPA. Washington D.C.; I.M. Mintzer, 1988, Projecting Future Energy Demand in Industrialized Countries: An End-Use Oriented Approach, prepared for U.S. EPA. Washington, D.C.; L. Schipper, pers. comm.])

	(% of Total Energy) Type of Energy			
End Use	Biomass	Fossil Fuels	Electricity	Total
South/southeast Asia				
Heating	0	16	NA	NA
Cooling	0	0	NA	NA
Cooking	75	3	NA	NA
Lighting	0	1	NA	NA
TOTAL	75%	20%	5%	100%
United States				
Heating	<1	59	8	67
Cooling	0	0	6	6
Cooking	0	7	3	10
Lighting	0	0	7	7
Other	0	0	10	10
TOTAL	<1%	66%	34%	100%

TECHNICAL OPTIONS TO REDUCE GREENHOUSE-GAS EMISSIONS

Many authors have catalogued the wide range of technologies relevant to cutting energy use and emissions from buildings (see, for instance, Goldemberg et al. 1988; Fulkerson et al. 1989; Carlsmith et al. 1989; and Cohen 1989). In this chapter the options for using the specific technologies and practices to cut emissions are summarized into the three fundamental generic strategies shown below:

1. restructuring energy service demand;
2. improving the efficiency of energy use (in either design/siting/construction of the building structure or from operating/maintaining efficient equipment);

3. switching to low-carbon fuels (for power generation or end-use applications).

Examples of these strategies are summarized in Tables 12.5 and 12.6 as well as in the remainder of this section.

Restructure Aggregate Building Energy-service Demand

Demand for energy services in the buildings sector can be reduced through a variety of mechanisms, some involving social planning approaches. Incentives can be established to encourage migration to more temperate climates, urbanize using modular housing, build smaller houses and multifamily dwellings, stabilize or reduce population, and better plan for land use (integrated transportation, urban landscaping, district heating or cooling, and co-generation). Moving to more temperate climes would reduce space conditioning requirements, as would reducing the amount of space available per person. Multifamily dwellings use less energy than single family homes; a single family townhouse uses about 25% less energy for heating than a detached house of the same size (Keyes 1980).

Table 12.5 Summary of energy-service demand and supply approaches

Restructing Aggregate Demand
- Demographic
    - Population control
    - Urbanization

- Geographic
    - Movement to more temperate climate

- Land-use planning (buildings, transportation, etc.)
    - Integration of home/work/agricultural needs of community
    - Smaller homes/multifamily homes

Fuel Substitution
- Substituting low-carbon for high-carbon fossil fuels
- Substituting renewable energy for fossil fuels
- Substituting carbon-clean electricity for fossil fuels
- Increasing use of cogeneration and district heating

Table 12.6 Summary of energy-efficiency approaches

Building Design, Siting, and Construction
- Design
    - Maximize use of solar radiation (daylighting, passive solar heating)
    - Minimize load (HVAC system integration)
- Siting
    - Partial burying to take advantage of natural thermal properties of the Earth
    - Southern exposure (in the northern hemisphere)
    - Landscaping
    - Environmental integration
- Construction
    - Increase building albedo (minimize load)
    - Advanced construction materials (phase-change materials, low-E windows)
    - Low-energy construction material (use materials that take less energy to manufacture)

Equipment Selection, Operation, and Maintenance
- Equipment selection
    - High-efficiency HVAC equipment (e.g., air-to-air head exchangers) and other appliances (e.g., ovens, dryers, office equipment)
    - Electrical "peak-smoothing" HVAC equipment (high-efficiency heat pumps, solar "cooling" desiccant systems)
    - Lighting (solid-state tunable ballasts, high-efficiency and compact fluorescents, improved geometry and reflection)
    - Controls (set-back thermostats, occupant sensors, room zoning, "smart" houses and meters)
    - System integration (use waste heat, energy management systems)
- Operation and maintenance
    - Conduct annual "tuneups" to ensure peak performance
    - Inspect and replace air filters
    - Shut off equipment when not in use (plug leaks)
    - Use and store power during off-peak periods
    - Human factors (real time cost feedback — meter on the inside)

Building Design, Siting, and Construction

Once the general demand patterns have been established, design, siting, and construction material and practices can have a significant impact on the overall building load. Buildings can be designed to take advantage of natural lighting by allowing outside light to pass through to the center of the buildings. Building designs can also be integrated with equipment in order to optimize overall system performance. Improving site design by taking advantage of the sun's track, using trees for shading, and employing the insulating properties of the ground can all be used to advantage (see, for example, Rosenfeld and Hafmeister 1990). Finally, building materials and construction practices can help to reduce energy demand. Examples are materials which increase overall building albedo, phase-change materials, and low-emissivity windows. Construction practices would primarily involve sealing the building shell to minimize air exchanges. Both building material use and construction practices are usually the subject of government codes and regulations.

Equipment Selection, Operation, and Maintenance

A wide range of energy-efficient equipment is available for use in both residential and commercial buildings, including energy conversion equipment such as air conditioners, lighting, cooking appliances, furnaces, and ventilation equipment (see Cohen 1989; Rosenfeld and Hafmeister 1990; Fickett et al. 1990; Shepard et al. 1990). The overall efficiency of the equipment is important. Also important is how the equipment interacts with other equipment and with the building. For instance, tightening the building envelope while downsizing space conditioning equipment may also require additional equipment, such as air-to-air heat exchangers, in order to maintain indoor air quality. Just as important to energy use is the manner in which equipment is operated.

Automatic controls, such as sensors that detect the presence of an occupant or set back thermostats, can be used to cut back on space conditioning or lighting needs. However, in terms of importance to energy use, human behavior in operating and maintaining this equipment dominates. Behaviors such as adjusting thermostats, zoning off rooms, turning off equipment and lights, as well as proper maintenance greatly affect energy usage. One approach to managing these human factors is to design in real-time cost-feedback mechanisms which allow users to monitor the cost, and thus, the energy impact of their decisions.

Fuel Substitution

Substitution of low- or no-carbon fuels for high-carbon fuels in the buildings sector can have a direct impact on energy-related building greenhouse-gas emissions. Substitution of natural gas for oil and coal, use of renewables and biomass, electrification (not generated by fossil fuels), and cogeneration and district heating are all possibilities (see Fulkerson et al. 1989). For instance, moderately priced gas stoves can be 5–8 times as efficient as typical woodstoves (which are <10% efficient) in providing heating services; however, the substitution of wood stoves by gas furnaces could well have adverse climate warming impacts. Similarly, compact fluorescent lamps (grid-supplied by an electric generator with 25% overall efficiency) would be on the order of 200 times as fuel efficient as the kerosene lamps used in over three-fourths of Indian households (Goldemberg et al. 1988). These strategies can be applied at many different levels, from individual rooms in buildings up to community installations.

Identifying High-priority Technologies and Practices

A recent study by the National Research Council (NRC 1990) prioritized a number of technical and operational options relevant to buildings on the basis of their near- and long-term potential to cut emissions, especially CO_2. Although directed at U.S. needs, the results shown in Table 12.7 do have relevance for CPEs and developing countries. As can be seen from this table, gas-fired heating technologies hold promise for significant near-term reductions in emissions from both residential and commercial buildings. Also important to both sectors are construction and operation and maintenance practices. Two of the most important targets for the near-term in the commercial sector are equipment controls and high-efficiency lighting.

The relative impacts of some of these technologies in terms of percent reductions in CO_2 emissions attributable to the U.S. buildings sector are shown in Table 12.8. High-efficiency heating equipment leads to the greatest near-term reductions in emissions, followed by lighting. In the long-term, new construction techniques have the largest impact. These measures could result in emissions reductions of nearly 40% in the near-term and nearly 75% in the long-term, although these effects are not strictly additive.

An attempt has been made to estimate the global savings from use of the most energy-efficient equipment (Fisher 1990). According to this study, savings on the order of 75% can be achieved in incandescent lighting (with higher numbers for fluorescent technology) and between 40–70% in space heating and air conditioning if average existing technology is replaced with the best available technology (see Table 12.9). Other researchers have estimated even higher potential savings; work at the Rocky Mountain Institute reports savings which range from 10–40% higher than many of those in Table 12.9, especially for appliances (Shepard et al. 1990). Of course, the actual savings depend on the operating characteristics and the total energy consumed in each application.

Table 12.7 Potential contribution of building technologies/practices to reduce greenhouse-gas emissions in the U.S. (NRC 1990, p. 83).

Technology/Practice	Potential Contribution to Lower Emissions[1]			
	Residential		Commercial	
	Near-Term	Long-Term	Near-Term	Long-Term
Energy Conversion Technologies				
High-efficiency gas heating	3	2	3	2
Advanced heat pumps	1	2	2	2
High-efficiency cooling	1	2	2	3
High-efficiency hot water	3	3	2	3
Solar hot water	1	2	—	—
Commercial refrigeration	—	—	2	2
Solar photovoltaics	—	2	—	2
Cogeneration	—	1	3	2
Energy Storage	—	1	1	2
Building Components and Systems				
Controls	1	2	3	3
Advanced windows	2	3	1	3
High-efficiency lighting	1	3	3	3
Efficient residential appliances	2	3	—	—
Efficient office equipment	—	—	1	3
Insulation	2	2	2	2
Construction materials	—	2	—	2
Design/Practice				
Practice: date, construction, operation and maintenance	3	3	3	3
Community design	1	3	—	3
Design for manufacturing, assembly operation	1	3	1	3

[1] Rating scale: 1 – can make *some* contribution to greenhouse-gas reduction, 2 – can make *considerable* contribution to greenhouse-gas reduction, 3 – can make *substantial* contribution to greenhouse-gas reduction.

POLICIES TO REDUCE GREENHOUSE-GAS EMISSIONS

The options in the previous section have important potential to reduce greenhouse-gas emissions, but only if they are put into place by effective policies and actions. In the near-term, the primary constraints to creating emission improvements are not technological. A vast array of cost-effective

Table 12.8 Potential U.S. CO_2 reductions in the buildings sector (NRC 1990, p. 90).[1]

	Percent Reduction[2] in CO_2	
	Near-Term	Long-Term
Reduce Energy Service Demand		
Retrofit existing homes	2.9	0.9
Advanced construction – new buildings	5.6	24.7
Improved building operating practice	4.5	4.5
Increase Equipment Efficiency		
High-efficiency heating & heat pumps	8.7	17.8
Cogeneration	1.4	6.9
High-efficiency lighting	5.9	5.9
Use Alternate Fuels		
Replace half of heat with gas	3.1	3.1
10% penetration of solar photovoltaics	NA[3]	6.9
Gas/solar water heating	1.7	2.5
Impact of Examples[4]	30.0	60.0
Other Impacts		
25% savings in other areas	7.0	NA[3]
50% savings in other areas	NA[3]	14.0
Total Potential Impact	37.0	74.0

[1] Estimates based on analysis by the Buildings Panel.
[2] Percent reduction from total CO_2 emissions (including those due to primary electric generation) for the residential and commercial sector per unit of service provided (i.e., per household or commercial square foot). Thus, these do not represent reductions from current levels since growth in the number of households and commercial square footage is not included.
[3] Either not applicable or not expected to have impact in time frame shown.
[4] Impacts are not additive. Obvious double counting (e.g., demand reduction and equipment efficiency) has been accounted for; other effects (e.g., cogeneration, solar, photovoltaics, electric) could result in some double counting, especially in the long term.

building technologies and practices exist, particularly to improve energy efficiency. However, a number of barriers to widespread adoption of these technologies have proven difficult to overcome in many countries. This is particularly evident in the industrialized nations, where the barriers to implementation of energy-efficient technologies and construction practices have been fairly extensively studied (see, for example, Carlsmith et al. 1989 and Belzer et al. 1990).

The picture is decidedly different when the focus changes from the near- to the longer-term. While barriers to existing technology are still important factors in any examination of emissions-reduction potential, technology

Table 12.9 Potential global savings from buildings efficiency measures (summarized from Fisher 1990, p. 15, 16, 21 and based on calculations from Environmental Defense Fund 1990).

Building Application	Potential Savings from Replacing:[1]	
	Avg. Existing Unit with Best New Unit[2]	Avg. New Unit with Best New Unit
Residential Sector		
Space heating		
electric	40–60%	40–50%
gas	45–55%	20%
Air conditioning	40–50%	30–40%
Water heating		
electric	50–65%	40–50%
gas	25–40%	25%
Lighting (incandescent)	75%	75%
Refrigeration	40–50%	25–35%
Cooking		
electric	30–40%	20–30%
gas	40%	20%
wood	75%	75%
Commercial/Service Sector		
HVAC	40–70%	30–60%
Lighting		
incandescent	75%	75%
fluorescent	70–85%	60–85%

[1] Savings estimates represent a best-guess estimate of the average savings available on a worldwide basis. Savings potential varies considerably between countries.
[2] Includes savings available from application of retrofit measures to average model in use.

research and development emerges as a more significant factor; the longer the time-frame under consideration, the more this is so. A simple example is illustrative. For technological development ten years out from the present time, it is easy to envision houses equipped with many of the advanced technologies currently in the development stage, technologies such as highly efficient natural gas heat pumps and super-efficient windows. If, however, the focus is thirty years out, it is possible to conceive of houses with technologies that are now only faint glimmers of research ideas, such as houses with dessicant cooling, wind-powered electricity units, micro-co-generation, and phase-change building envelopes.

Barriers to Implementation of Energy-efficient Technologies/Practices

To understand in detail why it can be extremely challenging to encourage the use of energy-efficient technologies and practices in the buildings sector, it would first be necessary to delineate carefully how building and energy markets function and who the important decision-makers are at each stage of the relevant decision-making process. Such a delineation would serve to identify precisely where the "barrier to success" was occurring. While some work has been done in this area in the U.S. and certain European countries, information on many other countries is much less available in the public literature. Despite this drawback, it is possible to generalize from the experience of the U.S., simply because the barriers identified in that country are generic and could easily be applied to a developing country in Africa or a country with a centrally managed economy.

According to the U.S. experience, barriers to adoption of new energy/environmental technologies can be categorized into five general types:

1. Economic barriers—production, market, or financial impediments to new technology investments. Such impediments include distorted price signals; high private rate-of-return/payback requirements; low capital turnover rates; limited access to investment capital (i.e., limited loans for new installations); misplaced economic incentives (e.g., tenant/landlord relationships where the tenant pays for utilities and the landlord pays for building improvements); limitations in delivery infrastructure for new technology.
2. Information barriers—information gaps or inaccuracies regarding building energy consumption, cost-effective savings potential, marginal costs and benefits, risks, reliability, and other important decision criteria for new technology investments. Misperceptions or lack of energy-use information may delay or obscure good decisions on energy system improvements.
3. Institutional barriers—organizational rules, procedures, or policies that block or significantly delay acquisition of new technology. Examples are government fiscal or regulatory practices (such as favoring energy supply over end-use technologies via tax credits) and obsolete codes or standards that discourage construction of energy-efficient homes.
4. Consumer behavior barriers—individual customer practices or tendencies impeding rational purchases of new technology, even when the technology is cost-effective, known and available. Examples of this behavior include general resistance to change, poor customer acceptance or low-priority attitudes that disregard energy-efficiency investments, uncertainty about the future, and perceptions of high risk in new technology investments.
5. Cultural barriers—phenomenon such as language difficulties, social tendencies, religious beliefs, or business practices in a country or region which affect social behavior that drives energy use. For instance,

resistance to the concept of "managed" population growth in some countries is an important barrier to decreasing rates of population increase. In technology transfer between industrialized and developing countries, social or business practices often are serious barriers to use of advanced technology. Even within a single country, the "not-invented-here" syndrome, which occurs when users are reluctant to seek, adapt, and use knowledge or systems that they themselves did not invent or develop, has been a problem between government laboratories developing a new application and industrial firms who manufacture or use the technology.

These barriers can occur throughout the research and development, production, commercialization, acquisition, and end-use phases of technology life cycle. Although many studies have examined these and other barriers, less is known about which ones are the most important and which are most amenable to remedy. Moreover, the role of these barriers in various regions of the world is not well understood.

Policies to Overcome Barriers

Implementing the technical options mentioned above with public policies to overcome barriers is a key step in reducing buildings-based emissions. In general, there are five types of policies available: economic incentives; regulation; education and information; research, development, and demonstration; and social engineering (see EPA 1989; DOE 1989; Koomey and Levine 1989). In all likelihood, actual government policies to reduce greenhouse-gas emissions would blend elements of these approaches to create a policy mix tailored to the specific needs, culture, and barriers of a particular nation.

Economic Incentives

As indicated in Figure 12.1, economics is a major determinant of both the physical characteristics of buildings and of how people "operate" those buildings. Throughout the world, a number of economic signals in the form of energy prices, interest rates, tax considerations, subsidies for energy development, and the like have much to do with the current level of greenhouse-gas emissions. Altering these economic signals could lead to a reduction in greenhouse-gas emissions. The basic strategy here is simply to change the incentive structure and allow the economic actions to do the rest; this attacks greenhouse-gas emissions by encouraging people and institutions to act in appropriate ways.

An obvious candidate for incentives is in the area of energy prices. Prices are very powerful incentives to energy and technology use. If coal is inexpensive and natural gas is relatively costly, coal will be widely used to provide energy services if other things are equal. However, the argument is beginning to be sounded that current market prices do not reflect the full societal costs of energy use, which would include a premium for contributing to the risk of climate change (Goldemberg et al. 1988). One solution to this problem is simply to modify energy prices accordingly to include such environmental considerations. Under this scheme, a tax would be levied on coal relative to natural gas to indicate its higher radiative forcing coefficient. This should create an incentive for natural gas use as opposed to coal, a desired fuel switching outcome. Implementation of concepts such as "time-of-day" energy pricing and the use of marginal prices in general, instead of average prices, would also provide consumers with realistic signals about energy use.

Governments could also subsidize energy consumers who install low- or no-carbon technologies. Perhaps the subsidy could be funded by removing subsidies elsewhere in the economy that contribute to global warming, such as subsidies in South America that encourage wholesale deforestation or subsidies in the U.S. that encourage use of private vehicles (Keyes 1980). Once this incentive was promulgated, consumers would demand the targeted technologies, creating a demand-pull on the market, possibly creating markets for new and attractive technologies.

Providing access to restrictive capital markets is an essential approach for some areas of the world to invest in energy-efficient technologies—especially in many centrally planned and developing countries, where replacing outmoded capital stock and providing for new construction to meet growth must be addressed. Various forms of low-interest loans, credits, or swaps between nations and institutions with capital and those in need of it would provide enormous encouragement for upgrading buildings.

In a somewhat related vein, the concept of Integrated Resource Planning (sometimes also called Least Cost Utility Planning), which combines both pricing and demand management, is beginning to gain acceptance in the U.S. In one promising approach, utilities subsidize customers to replace existing lighting with more efficient equipment; the parties share the savings (Fickett et al. 1990).

New Markets or Market Instruments

Promoting the idea that high initial costs for new technology can translate into much lower operating costs would greatly help the introduction of efficient or low-carbon technologies. This can be accomplished by encouraging new approaches to marketing energy services like heating, cooling, and lighting instead of simply selling energy supplies (see Reddy and Goldemberg

1990). In the U.S., private energy service companies have begun to appear and many electric utilities are starting to market energy efficiency to their customers (Fickett et al. 1990). Use of new instruments like "fee-bates" are producing renewed interest in efficiency; in this scheme, buildings that use more than the average electricity per square foot would be charged a high utility hook-up fee by comparison with efficient buildings which would get a rebate (Bevington and Rosenfeld 1990).

Regulation

A regulatory policy compels people and institutions to act in a prescribed manner to achieve certain ends, such as to cut harmful atmospheric releases. In a number of countries, model codes or standards which recognize energy/environmental implications of construction practices have been developed and are starting to be implemented. In the U.S., builders in California have begun using a set of prescribed computer-based formulas for optimizing design and construction of houses. This approach often leads to houses which are both more energy efficient and cheaper to build. About two-thirds of new California houses are now designed this way (NRC 1990).

Minimum appliance efficiency standards are also becoming more common, at least in the industrialized countries. Again, to use the U.S. as an example, the federal government has established minimum efficiency standards for a range of equipment types, including fossil-fuel furnaces, refrigerators, lighting ballasts and so forth (Bevington and Rosenfeld 1990).

A very important option is to revise current utility regulations so that utilities simply do not have an incentive to sell more energy, which decreases incentives for efficient equipment. For example, the government could mandate that utilities must design and then implement demand-side management plans to reduce the need for new generating capacity. The government could establish legislation making it incumbent upon utilities to exhaust all demand-side efficiency options before allowing new capacity to come on line.

Education and Information

Better knowledge of effective practices in managing energy needs is a cornerstone to reducing emissions for all regions of the world. Many producers and consumers must be taught the implications of inefficient and emissions-intensive energy use on the global environment. Widely disseminated education and training through public education, effective publications, and advertising can provide information to the user. Private sector producers can be encouraged to label energy-related products.

An example of an information-based policy in the residential sector would be to have some sort of home energy rating system to indicate energy use and associated costs for prospective real estate owners, along with a "House Doctor" mechanism to evaluate homes and recommend changes. Consumers could then compare various homes on the margin of energy efficiency, information that may not have previously been available. Another example would be for the government to establish regional technology testing centers that would provide credible, measured data on equipment performance to consumers free-of-charge (see Koomey and Levine 1989).

Research, Development, and Demonstration

One way of reducing the risk of climate change is to focus on the technologies that provide needed energy services. A research, development, and demonstration policy aims to improve the energy efficiency of technologies using conventional fuels and to deliver low- or no-carbon technologies to end-users. Put simply, research, development, and demonstration policies aim to push out society's technology frontier, and although industrialized countries have such programs, many developing countries do not. Governments can pursue research, development, and demonstration with a wide range of approaches. The argument for government action in research, development, and demonstration as opposed to reliance on private financing, is that, except in the very largest companies, the building industry in many countries is too highly fragmented and too driven by fluctuating business cycles to conduct enough private research and development (see DOE 1989; Belzer et al. 1990).

It is important to note that research, development, and demonstration are more than just the means of developing altogether new technologies; they are also processes of developing and demonstrating technologies that are currently technically feasible, but that require advancements to become economic, reliable, or safe. Many of these buildings research and development needs have been summarized in several recent studies, some of which also cover long-term technology concepts (see Fulkerson et al. 1989; DOE 1989; NRC 1990).

Social Engineering Policies

Finally, a government may set policies that aim to alter the rate of population growth, migration, or urbanization within a country of region. Such policies would very likely have goals that complement the reduction in greenhouse-gas emissions, such as quality of life or health, and would not be focused exclusively on climate change. The policies might be direct and mandatory, as with China's edicts on population growth, or might attempt to attain the objectives in a more oblique manner. As an example of the latter, a government concerned with high fertility rates might pursue a policy aimed at increasing the wealth of individual families, because wealth and low fertility rates are very highly correlated (Keyfitz 1989).

CONCLUSION—THE NEED FOR COMPLEMENTARY POLICIES

Attention to creating complementary strategies, which form consistent packages of actions, is a highly important matter for greenhouse-gas emission reduction in the buildings sector. Reliance on pure market pricing strategies (e.g., where increased energy costs encourage conservation) alone is a limited approach, particularly in areas where capital is difficult to obtain. The high first-cost sensitivity of both residential and commercial consumers can only be overcome with joint approaches like investment incentives along with research and development to help cut costs. Many desirable codes, standards, and regulations will be difficult to implement without also providing resource incentives that make it easy for producers and consumers to meet the requirements.

Several studies of consumer behavior show that selective financial incentives, when combined with information and training programs to meet regulated standards, can be effective beyond expectations, enhancing compliance with standards, leading to early adoption of standards, and promoting beneficial innovations not covered by the standards. The resulting energy efficiency performance can be substantially better than that achieved through compliance with minimum standards (see, for example, Vine and Harris 1988). More study is needed in virtually all regions of the world to understand the best ways for combining policies to achieve the needed complementarity.

ACKNOWLEDGEMENTS

This work was supported by the U.S. Department of Energy under contract DE-AC06-76RLO 1830. Pacific Northwest Laboratory is operated for the U.S. Dept. of Energy by the Battelle Memorial Institute.

REFERENCES

AID (Agency for International Development). July 1990. U.S. Congress. Greenhouse gas emissions and the developing countries: strategic options and the U.S. AID response. Report to U.S. Congress. Washington, D.C.: U.S. AID.

Belzer, D., et al. 1990. Policies to reduce energy use and greenhous gas emissions in the commercial sector. Working paper. Richland, WA: Pacific Northwest Lab.

Bevington, R., and A.H. Rosenfeld. 1990. Energy for buildings and homes. *Sci. Am.* **263(3)**:76–86.

Carlsmith, R.G., et al. 1989. Energy efficiency: how far can we go? ORNL/TM-11441. Oak Ridge, TN.: Oak Ridge Natl. Lab.

Chandler, W.U. 1990. Carbon emissions control strategies: case studies in international cooperation. Washington, D.C.: World Wildlife Fund and the Conservation Foundation.

Cohen, S.D. 1989. Energy efficiency technologies for residential and commercial buildings. Energy Analysis Prog. Berkeley, CA: Lawrence Berkeley Lab.

DOE (Dept. of Energy). 1986. Analysis and technology transfer annual report, DOE/CH/0016-HZ. Washington, D.C.: GPO.

DOE (Dept. of Energy). 1989. A compendium of options for government policy to encourage private sector responses to potential climate change. Vol. 2. DOE/EH-0103. Report to Congress. Washington, D.C.: Office of Env. Analysis.

EPA (Environmental Protection Agency). 1989. Policy options for stabilizing global climate. Draft report to U.S. Congress. Washington, D.C.: EPA Office of Policy, Planning, and Evaluation.

Fickett, A.P., C.W. Gellings, and A.B. Lovins. 1990. Efficient use of electricity. *Sci. Am.* **263(3)**:64–74.

Fisher, D., ed. 1990. Options for reducing greenhouse gas emissions. Draft report. Stockholm: The Stockholm Env. Inst.

Fulkerson, W., et. al. 1989. Energy technology R&D: what could make a difference? Vol 2 (1):End-Use Technology. ORNL-6541/V2/P1. Oak Ridge, TN: Oak Ridge Natl. Lab.

Goldemberg, J., et al. 1988. Energy for a Sustainable World. New Delhi: Wiley Eastern Ltd.

Keyes, D.L. 1980. The influence of energy on future patterns of urban development. In: The Prospective City: Economic, Population, Energy and Environmental Development, ed. A.P. Soloman. Cambridge, MA: MIT Press.

Keyfitz, N. 1989. The growing human population. *Sci. Am.* **261(3)**:119–126.

Koomey, J., and M. Levine. 1989. Policies to increase energy efficiency in buildings and appliances. Energy Analysis Prog. Berkeley, CA: Lawrence Berkeley Lab.

Leach, G. 1986. Household Energy in South Asia. London: Intl. Inst. for Env. and Development.

NRC. (National Research Council). 1990. Confronting climate change: strategies for energy research and development DOE/EH/89027P-H1. Washington, D.C.: Natl. Research Council.

Reddy, A.K., and J. Goldemberg. 1990. Energy for developing countries. *Sci. Am.* **263(3)**:111–118.

Rosenfeld, A.H., and D. Hafmeister. 1989. Energy-efficient buildings. *Sci. Am.*

Sathaye, J., A. Ghirardi, and L. Schipper. 1987. Energy demand in developing countries: a sectoral analysis of recent trends. *Ann. Rev. Energy* **12**:253–281.

Sathaye, J., and S. Meyers. 1985. Energy use in cities of the developing countries. *Ann. Rev. Energy* 1985.

SERI. (Solar Energy Research Institute). 1981. A New Prosperity: Building a Sustainable Energy Future. Andover, MA: Brick House Publ.

Shepard, M., et al. 1990. The state of the art appliances. COMPETITEK Report. Old Snow Mass, CO: Rocky Mountain Inst.

Vine, E., and J. Harris. 1988. Planning for an energy-efficient future: the experience with implementing energy conservation programs for new residential and commerical buildings (vol. 1 & 2). LBL-25525 and 25526. Berkeley, CA: Lawrence Berkeley Lab.

13

Potentials to Reduce Greenhouse-Gas Emissions by Rational Energy Use and Structural Changes

E.K. JOCHEM
Fraunhofer Institute for Systems and Innovation Research, Breslauer Straße 48, 7500 Karlsruhe 1, F.R. Germany

ABSTRACT

Many energy economists expect a further increase in energy demand, even in industrialized countries, and, hence, a substantial shift to natural gas, nuclear power, and renewables to avoid climatic changes due to energy-related greenhouse gases. I question the judgement that feasible improvements in energy efficiency are limited to 30 to 40%. I argue that potentials of rational energy use higher than 80% can be achieved on the basis of theoretical considerations in the long term by improving the exergy (available useful energy) efficiency (which is today less than 10%) and by decreasing the level of useful energy by reduced losses, substitution of energy-intensive processes, applying new materials, and intensified recycling of energy-intensive materials.

Because of the unbalanced perception of the future potentials of rational energy use and of energy conversion technologies, it is very likely that the high potentials for reducing the greenhouse-gas emissions by improved energy efficiency will be underestimated. The large potential for economically or ecologically justified efficiency improvement in energy use is hindered from being fully realized by obstacles and market imperfections. Efficiency improvements are mostly group specific and have to be matched with a suitable mix of energy policy instruments.

I envision a similar tendency to underestimate the potential of structural changes in the economy of industrialized countries to less energy-intensive production and expect a yearly average decrease of 1% of total primary energy intensity for industrialized countries in the future due to structural changes (without the structural effects due to increasing net imports of energy-intensive products from countries such as Canada and Australia).

Limiting the Greenhouse Effect: Options for Controlling Atmospheric CO_2 Accumulation
Edited by G.I. Pearman © 1992 John Wiley & Sons Ltd

INTRODUCTION

Many member countries of the Organization for Economic Cooperation and Development (OECD) have decreased their primary energy intensity, defined as the ratio of total primary energy required to gross domestic product, by 1.5 to 2.7% per year between 1973 and 1988. The structural changes within the economy towards less energy-intensive products and services and the trend towards products with higher shares of value added account for 10 to 25% of this improvement. Of the changes, however, 75 to 90% are caused by improved energy efficiency in the technical sense (Morovic et al. 1989).

The point has been made by several individuals that the new "energy source" of rational energy use will be more or less exhausted within the next 20 to 30 years. The arguments are that most of the economic energy-saving potentials have been realized within the past 20 years and that one should not expect substantial efficiency improvements over and above today's potential of some 30%. The slowing of the decline of energy intensity since 1985 in North America and Japan seems to justify this argument (see Figure 13.1). Therefore, if a continuing increase in energy demand is not to cause climatic changes due to greenhouse-gas emissions, a substantial shift to renewables, gas, and nuclear power seems to be unavoidable.

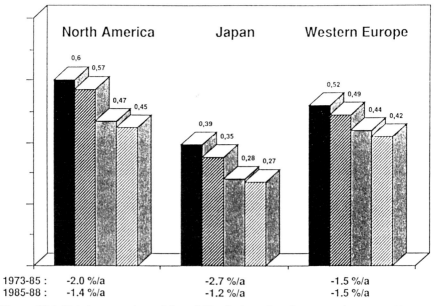

Figure 13.1 Total energy intensities of North America, Japan, and western Europe from 1973 to 1988 in tons of oil equivalent (toe) 1985 U.S. $ gross domestic product (from IEA 1989; own calculations)

The declining trend of energy intensity changes since the mid-1980s, however, is attributed to several factors which do not indicate exhausted potentials of rational energy use (Jochem and Morovic 1988):

- Government energy conservation activities boomed following 1977 in many OECD countries and subsequently diminished to a rather low level (comparable to the early 1970s) in most OECD countries by 1986/1987, when the prices of oil and natural gas fell below the 1974 level in real terms. This policy has been called "procyclical energy demand policy."
- Budgetary arguments have been put forward by politicians and administrations to justify reduced efforts to improve energy efficiency.
- More important was the reaction of the energy consumers to the oil- and gas-price drop in 1986/1987 and the fact that additional coal, oil, and gas then became available on the world energy market and that overcapacities of electricity generation grew because of unrealistically high demand projections made in the 1970s (IEA 1987).
- Finally, structural effects in the manufacturing industry due to the upswing of the business cycle in the mid 1980s (i.e., increased production of energy-intensive basic materials) as well as an increased growth of private income, which in turn increased the demand for bigger cars, air transport, and other energy-intensive forms of consumption, contributed to the smaller decline or stagnation of energy intensity in western Europe (see Figure 13.2) or elsewhere.

One of the consequences of the assumption that energy conservation potentials are likely to become exhausted is a rather high energy demand projection, for instance that of the World Energy Conference of 1989. The per capita energy demand of western industrialized countries is expected to stabilize at 4 tons of oil equivalent (toe) per capita or even increase by 13% up to 4.7 toe per capita by 2020. Even more surprising is the inconsistency: the less efficient eastern industrialized countries are expected to increase their energy consumption by between 28 and 45%, up to a level of 5.4 to 6.2 toe per capita by 2020. If the growing energy demand of the developing countries is added, global primary energy demand is expected to increase by some 75% between 1985 and 2020. These projections of the World Energy Conference of 1989, resulting in a yearly 1.2% increase of CO_2 emissions, strongly contradict the goal of the Toronto Conference of 1988, where climatologists argued that a 30% global reduction of CO_2 and other greenhouse-gas emissions has to be reached by 2020 in order to limit the increase in average global temperature to 2°C within the 21st century (see Figure 13.3). There is no doubt that substitution of carbon-rich fossil fuels by gas, renewables, and nuclear energy could not reverse the trend towards increasing CO_2 emissions in the case of such high energy demand growth

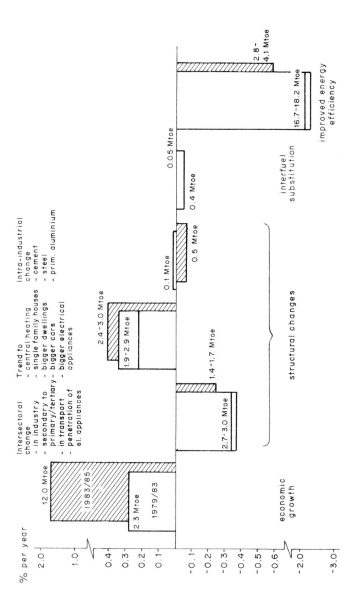

Figure 13.2 Average annual changes (given in %) in the final energy consumption due to different influencing factors for the European Community (EUR-12) for the period 1979 to 1985

because the share of renewables is too small and the acceptance of nuclear power and the easy availability of natural gas are limited.

THE ROLE OF RATIONAL ENERGY USE

Proved Technical and Economic Potentials for Rational Energy Use: 40% in the Next 20 to 30 Years

The known technical potentials for improved energy efficiency in the transformation sector and the final energy sectors as well as at the useful energy stage will amount to at least 40% in western industrialized countries (see values in Table 13.1). Technical potentials for rational energy use are defined as those energy savings which could be achieved using today's available technical solutions and knowledge if applied 100% in the sectors considered. Therefore, one may question whether a 40% improvement of energy efficiency does in fact represent the upper limit that we can expect in the next 30 to 40 years (Becht and van Soest 1989).

Figure 13.4 shows the energy flow scheme of the F.R. Germany. The relation between useful energy and primary energy consumption, i.e., the overall efficiency, is at present 0.33. As one has to expect some losses in any energy conversion process, the technical potential may be on the order of 50%. However, of course, only a part of this doubling of energy efficiency

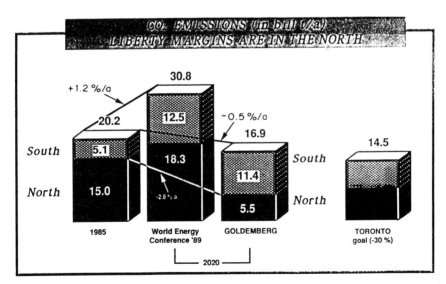

Figure 13.3 Projected CO_2 emission of two energy demand projections and the goal of Toronto, 1985–2020

Table 13.1 Technical potentials for the rational energy use in F.R. Germany in % (from Jochem and Schaefer 1990; Johansson et al. 1989; Maier and Angerer 1988)

Sector and Energy Use 1987 Final Energy	(in PJ)	Technical Potential in %	Remarks
Space Heat	(2370)		
in existing buildings		70 to 90	major contribution by increased insulation, without active solar energy
in new buildings		70 to 80	as compared to existing building codes
Hot Water Production		10 to 50	depending on the type of warm water generation system
Electric Appliances	(250)		
refrigerators		60	data of potentials as compared to average consumption of new
freezers		60 to 70	electric appliances (without
washing machines		30 to 40	substitution of electric heat generation)
dryers		50	
dishwashers		30	
Vehicles	(1990)		
cars	(1230)	50 to 60	data of potentials compared to present new vehicles
buses, trucks		15 to 25	high potential in public transport
electric vehicles		15 to 25	
airplanes		50 to 60	
Commercial/Public			
area 1	(500)	40 to 50	high share of process heat
area 2	(795)	50 to 70	high share of space heat
Industry	(2200)		
basic materials			
– fuels	(1326)	15 to 20	only technical improvements,
– electricity	(358)	ca. 10	changes due to structural changes not included
investment goods			
– fuels	(181)	15 to 20	
– electricity	(115)	15 to 20	
consumer goods			
– fuels	(150)	40 to 45	high fuel savings in the glass
– electricity	(66)	ca. 10	and textile industry
food industry			
– fuels	(126)	25 to 30	
– electricity	(30)	ca. 10	

Table 13.1 Continued

Sector and Energy Use 1987 Final Energy	(in PJ)	Technical Potential in %	Remarks
Transformation Sector			
refineries		20 to 25	equal potential for fuels and electricity
condensation power plants			
– existing stock		0 to 15	savings are compensated by increased environmental protection
– new combined gas/vapor plants		20 to 30	gas use in the case of high potential
cogeneration		ca. 15	
Weighted Average		40 to 45	

will be economically achievable within the next few decades. This seems to support the argument of many energy economists that the feasible savings are limited to around 30 to 40% and that after a period of some 30 to 40 years starting in the early 1970s, the energy markets still have to expect a continuing rise in energy demand in highly industrialized countries (the "moving staircase effect").

In this chapter I question this opinion and argue, from a technological point of view, that energy efficiency potentials are higher than 80% in the

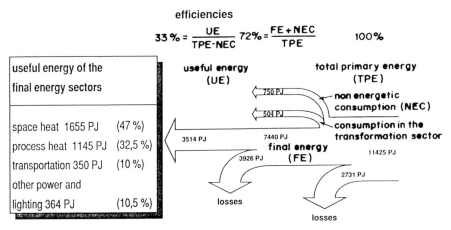

Figure 13.4 Energy flow scheme of primary final and useful energy for the F.R. Germany, 1988 (from RWE 1990)

long term. I consider three aspects: the exergy (the available useful energy), the level of useful energy, and the demand for energy services.

Future Energy Savings by Improved Exergy Efficiency

Energy efficiency does not consider the work capability (or availability) of energy use. From this exergetic point of view, energy use is still very inefficient although energy efficiency may have high values above 85%. For instance, the heat of a chemical reaction of burning fossil fuels can produce temperatures above 1000°C, yet in most instances it is simply used to heat water between 40 and 90°C. This process thus has an exergetic efficiency of less than 10%. This low exergetic efficiency is not only restricted to space heating and hot water production, it is also the case in many industrial processes (Riekert 1980). The exergetic efficiency of steel mills is on the order of 25%, of paper production 5%, of cement production 20%, and of propane-propylene rectification 5% (Morris et al. 1988). The overall exergetic efficiency, i.e., the relation between availability of useful energy and the availability of primary energy, is estimated for industrialized countries to be on the order of 10% (Fredriksson 1984; Gyftopoulos and Widmer 1982).

If this low efficiency is to be improved, it will be necessary to use energy with small losses of availability or only to apply heat at small temperature differences (Kashiwagi 1989). For example:

- use the available energy of burning fossil fuels via gas turbines or combustion engines before the generated heat is used;
- try to avoid heat losses by using heat exchangers and vapor compressors;
- try to use heat transformers or heat pumps to increase temperature levels of waste heat in order to use the "upgraded" heat for useful purposes (Alefeld 1988).

The chemical industry in F.R. Germany, for instance, has increased production since 1973 by more than 40%, whereas its fuel consumption has been reduced during this period. This success is due (apart from structural changes to less energy-intensive products) to an extensive use of waste heat with high temperatures in other processes with lower temperatures (Körner 1988). Theoretical calculations on how much energy could be saved by the industry in F.R. Germany if all industrial processes were optimized according to exergetic considerations resulted in an energy saving potential of 50% (Groscurth et al. 1988).

Of course, this maximum theoretical potential will never be economically realized. Even high proportions of this potential presume long-term structural changes of industrial areas and a connection by district heat grids with surrounding residential and commercial areas. This does not mean, however, that the realistic future energy-saving potential of industry is smaller than 50%.

Future Energy Savings by Reducing Specific Useful Energy Demand

The discussion about energetic and exergetic efficiency does not consider one major possibility for energy conservation: the reduction of specific useful energy demand, which is often taken as a constant of nature. Useful energy demand depends on the level of technical know-how and energy prices, on habits of decision-making, and habits of consumption. Today's level of specific useful energy demand (and useful exergy demand) reflects nothing but:

- the state of the art of industrial production of today; this has been developed over the past 50 to 100 years, over a period when real energy prices were continuously decreasing until the early 1970s;
- the traditional construction techniques (of buildings) and "modern" consumption patterns; increased incomes rendered consumption of energy-intensive products possible and facilitated mobility, particularly by cars and airplanes; increased incomes also converted the cyclical mode of the preindustrial economy to a linear production and "ex and hopp" consumption economy with huge streams of "waste."

The level of specific useful energy demand can be reduced by innumerable technological changes without reducing the energy services provided by energy use or impairing comfort. A few examples may demonstrate these almost unconverted possibilities:

- The quality of insulation and airtightness determines the amount of useful energy in buildings or refrigerators and freezers. Swedish low-energy houses only need 10 to 20% of the heat per square meter which is used on average in residential buildings in F.R. Germany (Feist 1987). A cold-storage depot or a refrigerator could be operated by outdoor air in the winter in zones of moderate climate.
- Catalysts, enzymes, and new materials will render possible the substitution of many energy-intensive processes (Riekert 1980). High-energy demand of activation of chemical reactions, high pressures, and temperatures of processes may be rendered unnecessary by new catalysts or biotechnological processes. Membrane processes will only use a small percentage of the useful energy which is needed today in thermal separation processes (Strathmann 1989). The production of iron, which today involves energy-intensive sintering and coke making, will be switched to the new coal

metallurgy with substantial energy savings. The energy-intensive rolling-mill operation of steel making will be substituted by continuous thin slab casting (Graf et al. 1987). Lasers will reduce the specific energy demand of metal cutting, and inductive electric processes will save energy in thermal surface treatment.
- New materials for cutting edges will improve surface quality, hence avoiding several machine operations. New compound plastic materials or foamed metals will substitute metals or reduce their specific consumption, thus inducing less energy demand in manufacturing and (because of smaller specific weight) in their use in vehicles and moving machine parts.
- Even a resubstitution of energy-intensive materials by wood, natural fibers, and natural raw materials for chemicals is expected in a few fields of application (Shamel and Chow 1988).
- The possibilities for recycling of energy-intensive materials are still numerous today. In the F.R. Germany the share of recycled material in the consumption of steel is 60%, of lead 54%, of paper 50%, of copper 40%, of aluminium 35%, of zinc 25% and of plastics only 7%. When such energy-intensive materials are recycled, energy savings are on the order of 50 to 80% compared with the production from primary raw materials.
- New concepts for vehicles, in particular for cars and airplanes, which could be realized due to new materials, microelectronics, and new thermo- and aerodynamic knowledge open up further potentials for saving useful energy (Horton and Compton 1984). New communication systems will contribute to a reduction of some 10% in passenger transport and a small part of paper use in offices.

The theoretical potential for energy savings due to these factors (substitution of processes and materials, avoiding losses of heat and power, and closure of open material cycles) has not been studied systematically until today. My preliminary estimates and calculations conclude that the theoretical potential of useful energy savings is at least 60% in industrialized countries, with a high share of space heat demand comparable to the F.R. Germany. The 60% includes the savings of useful energy in space heating, cryotechnics, the manufacturing industry, transport, and the transformation sector.

Finally, the specific consumption of energy-intensive materials could be reduced by improved material properties or changes in construction, or by increasing the life cycles of products by improved protection against corrosion or improved possibilities for repairs and maintenance.

If we summarize the findings of the sections above, the theoretical potential of rational energy use is at least 80%. The speed of realization of these potentials, which could only be estimated by basic technological considerations, depends, of course, on technological progress as well as on other important factors, such as the development of energy prices and transportation and environmental policies. The realization of an 80% theoretical energy-saving potential may take more than 100 years because life cycles of buildings and industrial areas may be too long and a resource-conscious behavior of a postindustrial society may only develop over several generations.

FUTURE ENERGY SAVINGS BY REDUCING ENERGY SERVICES IN OVERINDUSTRIALIZED COUNTRIES: A GLIMPSE OF A TABU

The contradiction between the increasing CO_2 emissions expected by the World Energy Conference in 1989 and the maximum CO_2 emission rate acceptable for a moderate change of the global climate according to the Toronto Conference in 1988 is so severe that even the use of existing potentials of energy conservation by improved efficiencies, solar energy use, and interfossil-fuel substitution may not be enough to avoid unacceptable changes of climate.

If we assume this restricted future, one might consider under what circumstances and to what extent a reduction of energy services would be acceptable. Each kilowatt of useful energy is used to satisfy a demand for an energy service, e.g., a building at a given temperature, the transport of a person or goods from point A to B in a given time, or the construction of a wall out of concrete. This demand for energy services is (again) not a "natural" constant but is the result of a complex sociopolitical process (habits, marketing, personal income, available leisure time, etc). In general, energy services per capita grew even in highly industrialized countries during the last decade, when primary energy consumption remained constant, and they are still likely to increase slightly in the next few decades.

There are examples, however, which demonstrate that a reduction in energy services, or "comfort" so to speak, does not necessarily lead to a reduction in well-being. For instance, building temperatures of one or two degrees colder during the winter (or warmer in the summer) may be acceptable (and even healthier!), or perhaps a small percentage of western Europeans will change their shopping habits in the next few years and will walk or take a bicycle instead of a car. In the view of the developing countries, the OECD countries are oversaturated with energy services. It is precisely this "oversaturation," they say, that is leading to the problems of global climate change (average yearly per capita CO_2 emissions in less-developed countries [LDCs]: 1.2 t and in industrialized countries: 15 t). This judgement may be considered harsh today by many Europeans or

North Americans, but it is quite possible that in several decades one will look back in dismay at an energy-wasting "fashion" which proved itself to be socially and environmentally unacceptable (Hohmeyer 1988). Western society is imprisoned by this "fashion" and it is therefore impossible to estimate how much energy services in the industrialized countries could be reduced by changes in the lifestyles of their inhabitants. First moderate estimates, made in the studies for the Enquête-Kommision of the F.R. German Bundestag, lie between 4 and 7% (Enquête-Kommission 1990; see also Table 13.2). In addition, structural effects of a postindustrial society towards less energy-intensive consumption could indirectly support these possible reductions of energy demand.

THE ROLE OF AN ENERGY CONSERVATION POLICY

There is little doubt that the potential to act and to reduce energy-related greenhouse-gas emissions is similar for new energy sources as well as for efficiency improvements over the next several decades. However, looking

Table 13.2 Examples of energy-conscious behavior and reduced energy services in the F. R. Germany, 2005.

Energy-conscious Behavior/Reduction of Energy Services	Related Energy Consumption[1] in PJ	Reduced Final Energy in PJ
1. 1 to 2°C decrease of room temperature in the residential and nonresidential sectors (6% per °C)	1500	90 – 180
2. Reduction of heated area while away and in very cold periods (10% of residential sector with 20% savings)	1000	20 – 40
3. Reduction of 10 to 20% in warm water consumption; open air drying instead of electric dryer in 5 to 10% of households	150	15 – 30
4. Reduction of car use and air trips (10 to 15% of a total of 750 billion person-km)	1180	120 – 180
5. Appropriate turnoff of machines, illumination at night, red lights, etc. less use of wrapping material (1 to 3%)	2500	25 – 75
Total	6770	270 to 505 (4% to 7.5%)

[1] consumption of the energy policy scenario 2005

back at the past or at today's power structures, it is very likely that the high potentials of rational energy use will be overlooked or judged as "purely theoretical," i.e., not "feasible." Of course, it is not easy to explore the potentials of rational energy use. It is not the few, large national energy supply companies or engineering companies that need to be examined, rather millions of energy consumers who will need to decide upon new investments and organizational measures. The heterogeneity and diversity of energy consumers and manufacturers causes a low perception of the high potentials of energy efficiency.

Obstacles and Market Imperfections

Perfect market performance would produce an optimal economic allocation of resources for producing and using energy, including an appropriate effort on efficient energy use. In practice, however, many obstacles and market imperfections, outlined below, prevent profitable energy-saving potentials from being fully realized (IEA 1987; Jochem and Gruber 1990).

Lack of Knowledge, Know-how, and Technical Skills in Many Final Energy Sectors

One of the main obstacles to efficient energy use is inadequate information and knowledge. Above all, private households and car drivers, small- and medium-sized companies, and small public administrations do not have enough knowledge about possibilities for energy saving or enough technical skills. Lack of information and knowledge does not only exist on the side of energy consumers but also on the side of architects, consulting engineers, and installers (Stern 1985). These groups exert remarkable influence on the investment decisions of builders, small- and medium-sized companies, and public authorities. The construction industry and many medium-sized firms face the same problem as small companies: managers are completely occupied with day-to-day business and can only engage themselves in the most immediately important tasks.

Lack of Access to Capital and Historically or Socially Grown Investment Patterns

Difficulties in raising funds for energy-efficiency investments are often obstacles for the same groups of energy consumers mentioned above. In some cases, owned capital is scarce and additional credits are expensive. In other cases, private homeowners or small companies do not want to utilize credit options. Especially when the interest rate is high, firms and private households prefer to accept higher current costs and the risk of rising energy prices instead of taking a credit (ex post-adaption).

In most small- and medium-sized companies, all investments (except those for infrastructure) are mostly decided on payback periods rather than on internal rates of return calculations. If the lifespan of energy-saving investments is greater than the useful life of production plants and machinery, the entrepreneur expects, consciously or unconsciously, a higher profitability of energy-saving investments. For small- and medium-sized local government units, lack of finance is a severe constraint. Funds for investment are restricted and many communities with high rates of unemployment are highly indebted nowadays, e.g., in F.R. Germany. In addition, energy-saving investments mostly remain invisible and do not contribute to politicians' public profiles.

Separation of Expenditure and Benefit

The owner of a building or energy-consuming equipment is not always identical with the user. Therefore, two problems exist (IEA 1987):

- For several reasons (e.g., lack of market transparency, difference in the financial and operating public authority of a public building) the owner does not receive the complete payoff of energy-saving investments, so he may make his decision in favor of a more profitable investment.
- When making energy-saving investments, if he is allowed to do so, the user runs the risk of not getting the payoff for the whole life cycle of the investment if he terminates the contract earlier.

These obstacles becomes more and more relevant for efficient energy use, not only in the case of space heating but also with respect to air conditioning, ventilation, cooling, and lighting equipment in leased buildings and appliances. This is also important in the public sector where buildings such as schools, sport halls, hospitals, or leased office buildings have different owners and users, or where buildings have been financed by state or federal sources but are operated by local governments.

Disparity of Profitability Expectations of Energy Supply and Energy Demand

This situation becomes even more important when considering the fact that there is a disparity in the rate-of-return expectations between energy consumers and energy suppliers. According to available information, energy supply companies in most OECD countries request nominal internal rates of return of 8 to 20% after taxes for major supply projects (IEA 1987;

Chesshire 1986). For energy conservation investments, energy consumers consciously request payback periods between one and five years, which are equivalent to nominal internal rates-of-return of about 15% to more than 50%. Decisions of homeowners often do not include life cycle costing, i.e., they tend to consider investment costs only and this may lead to an implicit rate-of-return above 100%. This disparity in the rate-of-return expectations between energy consumers and energy suppliers leads to a bias in favor of supply investments. Preliminary estimates suggest that the effect of this intersectoral disparity of profitability expections is at least a 10 to 15% distortion of energy-saving investments.

The Impact of Electricity and Gas Price Structures on Efficient Energy Use

The structure of gas and electricity tariffs for small consumers and the load-independent energy charge are important for energy conservation. As tariff structures are usually designed in two parts to reflect both the potential to get a certain amount of capacity at any given time and the delivered energy, the capacity charge plays an important role in profitability calculations. With regard to the residential sector, for instance, the share of the fixed standing charge for electricity varies between 53 and 94% among 20 different utilities in OECD member countries. This difference implies a 75% increase in the profitablity of electricity saving investments in the residential sector. The effect on profitability may be on the same order of magnitude in the industrial sector in those cases where electricity saving investments do not reduce capacity demand, such as inverters of electric engines. In addition, costs of electricity production and distribution related to peak demand or gas storage are not reflected in time-of-use or seasonal rates in most OECD countries in the residential and small consumer sectors.

Legal and Administrative Obstacles

There are legal and administrative obstacles to the realization of energy-saving potential in almost all energy sectors. They are mostly country-specific and often originate from periods before 1973, i.e., from periods with declining energy prices and no threat of global warming. Examples are:

- Social housing with upper limits for rents restrict capital-intensive solutions for refurbishing multifamily buildings or for the installation of highly efficient electrical appliances.
- Price-dependent taxes on electric light bulbs discriminate against high-efficiency bulbs.

- In public budget planning, the budgets for operating costs are often treated quite separately and not simultaneously with the investment budget.
- German legislation does not allow the construction of oil- or gas-fired cogeneration plants of a capacity above 10 MW, whereas the coal-fired plants only become profitable above 40 MW.

The obstacles and market imperfections described above do not cover all aspects of the question, why existing economic potentials for efficient energy use, and hence for reducing energy-related greenhouse gases at no cost, are not fully realized. But even the facts described above demonstrate that energy demand policy could act today without any risk with regard to scientific uncertainties about the severeness of the greenhouse effects ("no regrets policy").

Strategic Policies for Reducing Greenhouse-gas Emissions by Improved Energy Efficiency: National Policy Aspects

The obstacles to achieving the potential energy savings, and hence emission reductions, are closely related with each other and have typical patterns in the different energy consumer groups. These group-specific patterns of obstacles have to be matched by similar patterns of measures of energy policy (see Figure 13.5). Individually designed instruments, such as provision of information, training, grants or energy taxes, may produce very poor improvements in energy efficiency; this has been observed in many sectors and OECD countries (e.g., Gruber et al. 1982; Bonaiti 1989). On the other hand, integrated energy demand policies which have considered the interdependence of regulations, consulting, training programs, and financial aids have been very successful (e.g., the Swedish and Danish residential programs since 1978).

I expect that instruments of energy demand policy should be initiated by governments, as well as by companies, utilities, and industrial associations. Whoever undertakes certain measures to improve energy efficiency in a given group will normally look for a "central measure" whose impact may dominate other measures designed to alleviate other obstacles. This strategic question may be answered positively in the case of mass products and of basic innovations in small- and medium-sized firms. The answer will be rather negative in cases of individual technical solutions (e.g., refurbishing of buildings, energy savings in industrial plants).

The design of a policy package for energy demand should reflect the obstacles and market imperfections of a certain energy consumer group because all existing obstacles have to be reduced or removed at a given time.

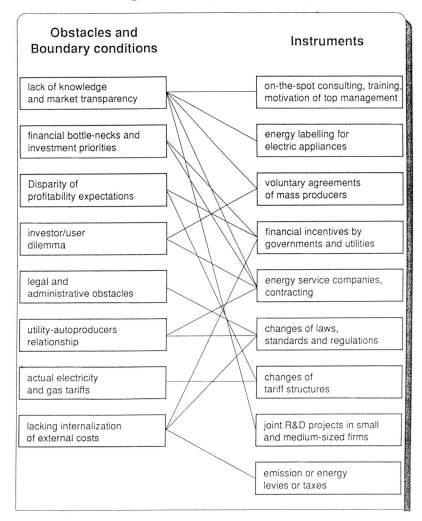

Figure 13.5 Scheme of the interrelationships between market imperfections of efficient energy use and policy measures to alleviate them

A good example is the refurbishing of buildings in the residential sector. Residential buildings consume about 20% of the total final energy in the European Common (EC) Market countries. Refurbishing of buildings is mostly an individual event and the obstacles are numerous; therefore, a package of instruments may include:

- Advanced education and training of architects, planners, and craftsmen necessary to improve the diffusion of knowledge and technical skills.

- Information and education of landlords and homeowners needed to improve their knowledge and understanding (particularly the substitution of energy cost by capital cost).
- This causes a need for professional advisors to perform audits and give practicable recommendations.
- These energy audits may be considered too costly by the landlords or homeowners; a subsidy for energy audits, therefore, has proved to be cost-effective (Smith-Hansen 1988).
- The subsidizing of investments may be bound to a registered energy consultant and a formal heat survey report.
- The government may create a market for these services by a regulation that demands a formal heat survey in the case of a changeover of tenants in dwellings or a change of ownership of houses and buildings.
- An investment subsidy scheme for specified groups of homeowners or multifamily buildings may be justified to overcome financial bottlenecks. However, the cost-effectiveness of this instrument has often been overestimated (Gruber et al. 1982).
- An economically justified minimum thickness of insulation may be secured by new building codes which also cover the refurbishing of buildings.
- In addition, new rules for the pricing of electricity and gas could abolish fixed standing charges for these energies as they do not contribute to the economics of energy-saving measures.
- Flexible import taxes on oil may avoid irritation of homeowners and landlords in periods of falling oil prices.

The Danish and Swedish energy-saving programs in the residential sector do have this multi-measure character. The results of these policies are impressive: between 1972 and 1987, annual consumption in space heating dropped from 1.26 GJ/m^2 (or by 45%) in Denmark and by 35% during the past 10 years in Sweden (Smith-Hansen 1988).

The question of the extent to which obstacles and market imperfections hinder energy-saving potentials from being realized cannot be answered on precise scientific evidence. Looking at recent policy studies, however, it is expected that energy conservation policy will play a major role in reducing greenhouse-gas emissions by improved energy efficiency. For example, according to most recent studies on the final energy demand between 1987 and 2005 for F.R. Germany (see Figure 13.6; Enquête-Kommission 1990):

- energy savings are expected to slow down the growth of energy consumption by 1% annually (see Fig. 13.6: "laissez-faire" scenario compared with "frozen efficiency" trend);
- additional improvements in energy efficiency of 20%, or 1.2% per year,

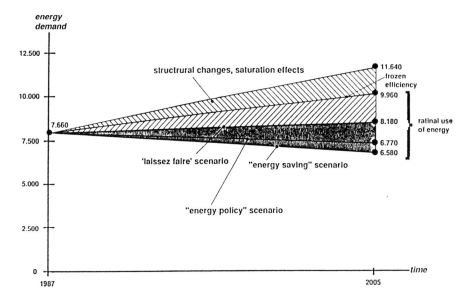

Figure 13.6 Effects of structural changes and rational use of energy on energy demand of final energy sectors in the F.R. Germany 1987 to 2005

have been identified as a result of a dedicated energy conservation and transportation policy (see "energy-saving" scenario).

These results demonstrate the importance of both the high potentials of rational energy use and the impact of policy. Instead of a 7% increase in final energy demand between 1987 and 2005, energy conservation and transportation policy could reverse this trend and reduce final energy consumption, and hence CO_2 emissions, by almost 1,100 PJ or 14% between 1987 and 2005 (see Fig. 13.6). Similar results have been predicted for the period between 1985 and 2010 by Becht and van Soest (1989) for the Netherlands, where a 2% efficiency improvement is expected to reflect the difference between laissez-faire development and a sustainable energy policy. The possibilities for reducing greenhouse-gas emissions seem to be high and achievable if policy considers energy efficiency with great care.

International Policy Aspects

A significant emission reduction potential is available at low or negative resource cost. The Energy and Industry Subgroup of the IPCC (IPCC 1991) estimates in its reference scenario that "around 20% of global emissions projected for 2020 are primarily attributable to the accelerated implementation of energy efficiency and conservation measures." But achieving a 20% reduction from current emission levels would require major changes in global energy markets, plans, infrastructure, and intervention by governments on the national and international level. A global policy is required to meet the challenge; however, many uncertainties remain about the impacts and trade-offs of this political response: macroeconomic and second-round effects, changing social and environmental costs and benefits (which often cannot be monetarized).

Three countries (Norway, the Netherlands, and Sweden) have formally adopted policies to limit greenhouse-gas emissions. To what extent they will be able to pursue such a policy in the international context will be affected by:

- Impacts on trade and international competitiveness may be associated with energy or CO_2 taxes or surcharges, energy efficiency regulations, and emission targets.
- National policies of the Netherlands may be hampered by EC regulation and harmonization activities.
- Energy-intensive production facilities may move from countries with high energy taxes and environmental control standards to countries with low energy taxes and lower standards.
- National governments may be inclined to rely on high "reference" or "status quo" projections for national energy demand (with unrealistically high expectations of economic growth and low assumptions for structural changes, saturation effects and energy efficiency improvements) in order to start from a "safe" bargaining position when it comes to an international convention.

These potential conflicts or reactions demonstrate the necessity for internationally organized response strategies. There are many good reasons why the EC and OECD countries should go ahead and take up their responsibilities for reducing the disproportionately high per capita emissions of greenhouse gases. This might be achieved if:

- The rich industrialized countries should try to internalize the social costs of today and future generations by a long-term and steady increase in energy or CO_2 taxes and by implementation of strict standards of efficiency for equipment, appliances, and buildings.
- The EC commission should develop high sensitivity as to where EC regulations or harmonization could help to promote rational energy use, instead of building up new obstacles.
- Low energy taxes and lower environmental emission standards should be revisited under the aspect of moving energy-intensive industries.

- Encouragement for accelerated implementation of energy efficiency will need to be supported by enhanced research and development if momentum is to be maintained. Former and recent policy declarations stress the dominant role of rational energy use among the technical options to reduce CO_2 emissions. These statements have been in strict contradiction to the government research and development budgets in all OECD countries ever since the data were collected in the early 1970s. Government research and development budgets for energy efficiency have grown smaller by one order of magnitude, which seems to be a typical indication that rational energy use has a weak lobby, in contrast to coal, breeder or fusion technology (see Figure 13.7).
- The western industrialized countries should generate or intensify programs to increase the flow of capital and environmentally sound technology to

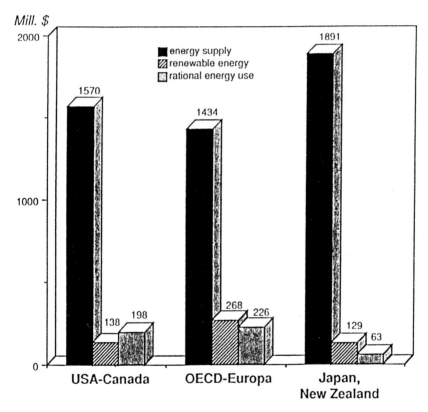

Figure 13.7 Government research and development budgets of OECD regions for energy supply (fossil and nuclear fuels), renewables, and energy efficiency in 1988 in million U.S. dollars (from IEA 1989)

developing east European countries, including technical assistance and professional training.
• Close cooperation among energy economists of major energy-consuming countries is necessary, to discuss recent national energy demand projections and the underlying assumptions regarding major determining factors such as economic growth, structural changes, and energy prices.

Although industrialized countries have to lead the way towards a sustainable energy future, different interests exist among them: differences in domestic energy resources and supply industries as well as differences in climate, economic structures, international competitiveness, and national energy policies. These differences will call for highly skilled politicians, scientists, and economic leaders who will look for solutions of policy options from a national and international point of view and with the eyes of future generations.

REFERENCES

Alefeld, G. 1988. Problems with the exergy concept (or the missing second law). *Newsletter of IEA Heat Pump Center* **6(3)**:19–23.
Becht, H.Y., and J.P. van Soest. 1989. Energy conservation for a long-term, substainable energy policy. Delft, Netherlands: Centrum voor energiebesparing.
Bonaiti, J.P. 1989. Local Conservation Programmes in France: 1979 – 1986. Methods of Diffusion of Technical and Social Innovation. Manuscript. Grenoble: Inst. Energy Policy and Economics.
Chesshire, J. 1986. An Energy Efficient Future: A Strategy for the UK. Brighton: SPRU, University of Sussex.
DOE (Department of Energy). 1982. Energy Conservation Investment in Industry. Energy Paper Nr. 50. Laondon: HMSO.
Enquête-Kommission. 1990. Preventive Measures to Protect the Earth's Atmosphere, 11th German Bundestag, Bundestagsdrucksache 11/8030. Bonn: Deutsche Bundestag.
Enquête-Kommission. 1991. Preventive Measures to Protect the Earth's Atmosphere. Third report, vol. 2. Bonn: Univ.-Buchdruckerei.
Feist, W. 1987. Niedrig-Energiehäuser in Dänemark und Schweden. Darmstadt: IWU.
Fredriksson, R. 1984. The technological potential for more efficient use of energy in industry. In: STU-Report, pp.7–10. Stockholm: STU.
Graf, H., H. Rüdiger, K.-H. Schubert, and K. Takahama. 1987. Die verschiedenen Rohstahlerschmelzungstechniken, die Sekundärmetallurgie und Stahlgießverfahren. *Stahl und Eisen* **107(22)**:1021–1028.
Groscurth, H.-M., R. Kümmel, and W. van Gool. 1988. Thermodynamic Limits to Energy Optimization. Manuscript. Universität Würzburg and Utrecht.
Gruber, E., F. Garnreiter, E. Jochem, U. Kuntze, W. Mannsbart, F. Meyer-Krahmer, and S. Overkott. 1982. Evaluation of Energy Conservation Programmes in the EC Countries. Karlsruhe: ISI.

Gyftopoulos, E.P., and Th.F. Widmer. 1982. Cost-effective waste energy utilization. *Ann. Rev. Energy* **7**:293–327.
Hohmeyer, O. 1988. Social Costs of Energy. Berlin: Springer.
Horton, E.J., and W.D. Compton. 1984. Technological trends in automobiles. *Science* **225**:587–593.
IEA (International Energy Agency). 1987. Energy Conservation in IEA Countries. Paris: OECD.
IEA (International Energy Agency). 1989. Energy Policies and Programmes of IEA. Paris: OECD.
IPCC (Intergovernmental Panel on Global Climate Change). 1991. Energy and Industry Subgroup Report. Executive Summary. Washington D.C.: Island Press.
Jochem, E., and E. Gruber. 1990. Obstacles to rational electricity use and measures to alleviate them. *Energy Policy*, pp.340–350. Surrey: **18(4)**.
Jochem, E., and T. Morovic. 1988. Energy use patterns in Common Market countries. *Ann. Rev. Energy* **13**:131–157.
Jochem, E., and H. Schaefer. Reduction of Emissions by Rational Energy Use (in German). Bonn: Economica Verlag GmbH, in press.
Johansson, Th.B., B. Bodlund, and R.H. Williams. 1989. Electricity-Efficient End-Use and New Generation Technologies and Their Planning Implications. Lund: Lund University.
Kashiwagi, T. 1989. Present status and future prospects of energy utilization technology in Japan for greenhouse gas mitigation. Proceedings of the IEA/OECD expert seminar on energy technologies for reducing emissions for greenhouse gases. Paris: OECD.
Körner, H. 1988. Optimaler Energieeinsatz in der Chemischen Industrie. *Chem. -Ing. -Tech.* **60(7)**:511–518.
Maier, W., and G. Angerer. 1988. Rationelle Energieverwendung durch neue Technologien in der Industrie. 2 vols. Köln: TÜV Rheinland.
Morovic, T., G. Gerritse, G. Jaeckel, E. Jochem, W. Mannsbart, H. Poppke, and B. Witt. 1989. Energy Conservation Indicators II. Berlin: Springer.
Morris, D.R., F.R. Steward, and J. Szargut. 1988. Energy Analysis of Energy Production Processes. Fredericton, N.B. Canada: University of New Brunswick.
Riekert, L. 1980. Energieumwandlung in chemischen Verfahren, Ber. Bunsenges. *Phys. Chem.* **84**:946–973.
RWE Anwendungstechnik. 1990. Energieflußbild der Bundesrepublik Deutschland (1990). Essen: RWE.
Shamel, R.E., and J.J. Chow. 1988. Biotech's potential impact on the chemical industry. *Bio/Technol.* **6(6)**:681–682.
Smith-Hansen, O. 1988. Energy efficiency policy – the Danish case. Rockwool A/S Denmark. Discussion paper at the Intl. Conference of IAEE in Luxembourg.
Stern, P.C., ed. 1985. Energy Efficiency in Buildings: Behavioral Issues. Washington D.C.: Natl. Academy Press.
Strathmann, H. 1989. Stand der Membrantechnik und ihre wirtschaftliche Bedeutung. *Swiss Biotech.* **7(1)**:13–25.

14
Least-Cost Climatic Stabilization

A.B. LOVINS and L.H. LOVINS
Rocky Mountain Institute, Old Snowmass, CO 81654–9199, U.S.A.

ABSTRACT

The Earth's climate can almost certainly be stabilized *not at a cost but at a profit*, taking no account of the avoided cost of adapting, or failing to adapt, to possible climate change. Because it is generally cheaper today to save fuel than to burn fuel, energy efficiency can avoid the resulting CO_2 emissions at a strongly negative net internal cost. Thus energy efficiency (plus some cost-effective renewable sources) can avoid more than half of global warming while saving a global sum ultimately approaching $1 trillion per year. Virtually all the rest of global warming can be avoided by other measures (e.g., sustainable farming and forestry practices plus CFC displacement) whose net internal cost appears to range from slightly positive to significantly negative.

This potential for abating global warming at a profit, hence largely via market mechanisms, is first reviewed from an international engineering/economic perspective, assembling diverse empirical data on new technologies that apply to radiatively active gases both individually and in previously overlooked synergistic combinations. Applicable, often innovative, ways to implement that technical potential are then explored in the context of OECD, formerly planned, and developing countries. Finally, the policy implications of recent developments in both technology and implementation are discussed and contrasted with fallacious econometric assumptions that abatement must be very costly or it would already have been done.

INTRODUCTION: STABILIZING GLOBAL CLIMATE SAVES MONEY

The threat of serious, unpredictable, and probably irreversible changes in the Earth's climate has moved from conjecture to suspicion to near-certainty

---
Essay submitted to the Mitchell Prize competition, 15 October 1990.
Copyright © 1990 by Rocky Mountain Institute. All rights reserved. The research and logistical support of the RMI staff, especially Rick Heede and Catherine Henze, and the essential contributions of many colleagues around the world, are gratefully acknowledged.
Limiting the Greenhouse Effect: Options for Controlling Atmospheric CO_2 Accumulation
Edited by G.I. Pearman published 1992 John Wiley & Sons Ltd

(Houghton et al. 1990). Denial is now confined to the uninformed (Brookes 1989). Yet the threat's cause continues to be widely misunderstood even by many experts on its mechanisms.

Global warming is not a natural result of normal, optimal economic activity. Rather, it is an artifact of the economically inefficient use of resources, especially energy. Advanced technologies for resource efficiency, and proven ways to implement them, can now support present or greatly expanded worldwide economic activity while stabilizing global climate *and saving money*. New resource-saving techniques (chiefly in energy, farming, and forestry) generally work better *and cost less* than present methods that destabilize the Earth's climate.

In short, even based on energy-efficiency assessments by such organizations as the Electric Power Research Institute (the utility industry's think-tank) and leading U.S. National Laboratories, most of the best ways known today to abate climatic change are:

- not costly but profitable;
- not hostile but vital to global equity, development, prosperity, and security; and
- reliant not on dirigiste regulatory intervention but on the intelligent application of market forces.

Newer analyses summarized here reveal even bigger and cheaper energy-saving potentials, as well as innovative ways to abate other sources of global warming at unexpectedly low costs. These findings imply that most global warming can be abated not at roughly zero net cost, as two government-sponsored analyses have recently found, but at *negative* net cost, without ascribing any value to the abatement itself.

However, technical and implementation options, obstacles, and strategies vary widely with culture and geography. This paper therefore identifies opportunities to use resources more efficiently, describes concrete successes in capturing those opportunities, and outlines an agenda for systematically harnessing them. This is discussed for three main types of societies: the industrialized Organization for Economic Development (OECD) countries, the USSR and its former satellites, and developing countries. The discussion highlights needed interactions (e.g., issues of trade, technology transfer, and emulation) and possible synergisms between these three regions.

MAJOR ABATEMENT TERMS

Over a century's time horizon, about half of global warming in the 1980s[1] was caused (Houghton et al. 1990) by burning fossil fuel, which produces

---

[1] Counting contributions by gases projected to be released in 2025 rather than in the 1980s (IPCC 1990a, Figs. 2 & 3), or integrating their effects over a longer period (Lashof and Ahuja

carbon dioxide (CO_2) and monoxide (CO), nitrous oxide (N_2O), fugitive methane (CH_4), and ozone (O_3). Another one-fourth or more was driven by unsustainable farming and forestry practices, which produce biotic CO_2 and CH_4 and nitrogen-cycle N_2O. Virtually all the rest was driven by the release of halons and chlorofluorocarbons (CFCs), whose production (but not use) will be phased out during the 1990s to protect the stratospheric O_3 layer. Among the ~57% of current worldwide contributions to global warming estimated by Environmental Protection Agency to arise directly from energy use (EPA 1989, p. VII–28), 20% is ascribed to transportation, 22% to industry, and 15% to buildings. The sizeable uncertainties in all these figures are immaterial for the purposes of this paper, since, as will be shown, major abatements are available and cost-effective for each gas and in each application.

Stabilizing atmospheric concentrations of heat-trapping gases at current levels will require that their present rates of emission be reduced by different amounts, because different gases remain in the air for different periods and react to form different products. These reductions are believed (Houghton et al. 1990) to be more than 60%[2] for CO_2, 15–20% for CH_4, 70–80% for N_2O, 70–85% for the most important CFCs[3], and 40–50% for HCFC-22. These reductions, too, are not as interchangeable in practice as they might seem mathematically (Krause et al. 1990). Yet reductions generally larger than these—large enough to return the atmosphere to a composition likely to entail no climatic change (*id*)—will be shown below to be very cheap, free, or better than free.[4]

Specifically, this paper will show that demonstrated technologies and implementation methods can:

- save most of the fossil fuel now burned, at a cost far below that of the fuel itself, making this abatement cost less than zero;
- change soil from a carbon source to a carbon sink (and incidentally reduce related emissions of CH_4 and N_2O too) at a net cost around zero or less; and
- displace CFCs (and often their proposed hydrohalocarbon substitutes) at a net cost close to zero, though this cost is irrelevant since the substitution

---

1990), would strengthen this paper's conclusions, which are most detailed for energy efficiency. Those conclusions hold independently of such assumptions, since they rest on the demonstrated potential to reduce dramatically *all three* source terms (energy, agroforestry, and CFCs).
[2] Simulations suggest 50–80% (EPA 1989).
[3] EPA (1989) gives 75–100% for CFC-11 and -12.
[4] This paper does not consider several recently proposed innovations that could allegedly capture and sequester CO_2 from electric generation and other combustion processes at relatively low costs (Hendriks et al. 1990; Wolsky and Brooks 1989). If such processes turn out to work, that is "icing on the cake." On the early data available, however, while probably cheaper than global warming, these processes appear to abate CO_2 at a cost several to many times that of the proven energy-efficiency options described below.

is already required by international treaty to abate stratospheric O_3 depletion.

The cost of the main global warming abatements therefore ranges, broadly speaking, from strongly negative to roughly zero to irrelevant—with policy implications discussed in the CONCLUSIONS.

TECHNOLOGICAL AND ECONOMIC OPTIONS

Energy Efficiency

Removing a 75-watt incandescent lamp and screwing into the same socket, a 15-watt compact fluorescent lamp will provide the same amount of light for 13 times as long, yet save enough coal-fired electricity over its lifetime to keep about a ton[5] of CO_2 out of the air (plus 8 kg of SO_x and various other pollutants). If the quintupled-efficiency lamp saves oil-fired electricity, as it would in many developing countries, it can save more than enough oil to drive a standard 20-mi/gal car for a thousand miles, or to drive the most efficient prototype car across the U.S. and back. If it saves nuclear electricity, it avoids making a half-curie of long-lived wastes and two-fifths of a ton-TNT-equivalent of plutonium. Yet far from costing extra, the lamp generates tens of dollars' net wealth—it saves tens of dollars more than it costs—because it displaces replacement lamps, installation labor, and utility fuel. It also defers hundreds of dollars' utility investment (Lovins 1990).[6]

This is an example—one of the costlier ones that could be given—of the proposition that today it is generally cheaper to save than to burn fuel. The CO_2 and other pollution avoided by substituting efficiency for fuel is thus avoided not at a cost but at a profit.

Most of the best energy-saving technologies are less than a year old, especially the superefficient lights, motors, appliances, and other end-use devices that save electricity. Saving electricity gives the most climatic leverage, because it takes 3–4 units of fuel (in socialist and developing countries, often 5–6 units) to generate and deliver a single unit of electricity,

---

[5] All tons (t) in this paper are metric; all miles (mi) and gallons (gal) are U.S.; and all costs, unless otherwise noted, are in 1986 U.S. $. Costs of saved energy are levelized at a 5%/y real discount rate using the Lawrence Berkeley Laboratory methodology (Lovins 1988, pp. 11ff), whereby the marginal cost of buying, installing, and maintaining the efficient device is divided by its discounted stream of lifetime energy savings. Standard metric prefixes are used (M = 10^6, G = 10^9, T = 10^{12}, P = 10^{15}, E = 10^{18}), along with mostly metric units (e.g., 1 ha = 2.4 acre).

[6] These calculations (Lovins 1990) assume that each kWh saved directly by the lamp also saves 0.36 kWh of net space-conditioning energy, as it would in a typical U.S. commercial-sector building (Lovins and Sardinsky 1988; Shepard et al. 1990).

so saving that unit displaces many units of fuel, mainly coal, at the power plant. Power plants burn a third of the world's fuel and emit a third of the resulting CO_2, as well as a third of the NO_x and two-thirds of the SO_x, both of which also contribute to global warming (Krause et al. 1990)—a little directly, and more by degrading forests and other ecosystems that otherwise store carbon. Electricity is also by far the costliest form of energy[7], so it is the most lucrative kind of energy to save. Saving electricity saves much capital: in the mid-1980s the U.S. spent as much private capital and public subsidy (Heede et al. 1985) expanding its electric supply, about $60 billion per year, as it invested in all durable-goods manufacturing industries. Moreover, a fourth of the world's development capital goes to electrification, and about five times as much such capital is projected to be needed in the 1990s as is likely to be available. This ~$80 billion annual shortfall (Churchill 1989; Reddy and Goldemberg 1990) may imperil proposed development. For these reasons, this discussion emphasizes the frequently undervalued opportunities to save electricity.

Electricity

Many utilities still think that only ~10–20% of the electricity used can be cost-effectively saved. However, a recent reassessment by the Electric Power Research Institute found a potential, mainly cost-effective, to save 24–44% of U.S. electricity within this decade, not counting a further 9–15% already in utilities' demand forecasts or program plans (EPRI 1990; Fickett et al. 1990). The California Energy Commission has similarly identified a potential to save electricity 2.5%/y faster than projected load growth (CEC 1990, Figs. 3–1 and C–4). As will be shown below, such electrical savings, and analogous nonelectrical energy savings, can save enough money to pay for most *non*energy kinds of global-warming abatement. Most of this electricity-saving potential is untapped: for the non-Communist world during 1973–87, oil intensity fell by 32%, but non-oil intensity by only 1%.

Analyzing technologies even a few years old, however, can make potential savings seem much smaller and costlier (Lovins 1980). Today's best electricity-saving technologies can save twice as much as five years ago, but at only a third the real cost. Still more detailed assessments[8] of these new

---

[7] Average 1989 U.S. retail electricity at 6.44¢/kWh is equivalent in heat content to oil at $110/bbl.
[8] These are presented chiefly in the COMPETITEK Hardware Reports cited below, believed to be the most thorough and up-to-date such information available. They are currently supplied to, and undisputed by, more than 170 member organizations, including 50+ utilities and a similar number of government agencies, in 30 countries. The first three such reports (Lovins and Sardinsky 1988; Lovins et al. 1989; Shepard et al. 1990) total 1,268 single-spaced pages documented with more than 3,000 notes. The findings of the three remaining Hardware Reports are already known in sufficient detail through earlier analyses (e.g., Lovins 1986, 1986a, 1988) and from many other authors' studies, some of the most important of which are cited below.

opportunities, based on measured cost and performance data, thus reveal that full retrofit of U.S. buildings and equipment with today's most efficient commercially available end-use technologies would deliver unchanged or improved services while saving far more electricity, and at far lower cost, than previously supposed. This makes it possible to abate a large fraction of global warming—enough, it appears, to stabilize the Earth's climate—at negative net cost.

The modern U.S. electric-efficiency potential includes saving:

- half of motor-system (or a fourth of total) electricity through 35 motor, control, drivetrain, and electric-supply improvements collectively paying back in ~16 months (Lovins et al. 1989; Fickett et al. 1990)[9]—a key opportunity in reindustrializing countries like the USSR, whose motors already use 61% of all its electricity (R.V. Orlov, pers. comm.; Makarov et al. 1988);
- 80–92% of lighting electricity (or a fourth of total electricity including net space-conditioning effects) at a net cost less than zero, because much of the lighting equipment more than pays for itself by costing less to maintain (Lovins and Sardinsky 1988; Piette et al. 1989);
- a sixth of total electricity through numerous design improvements to household appliances, commercial refrigeration and cooking, and office equipment—where the potential saving exceeds 90% at roughly zero or negative cost (Shepard et al. 1990);
- two-thirds of water-heating electricity through eight simple improvements (insulation, high-performance showerheads, etc.) (Lovins 1986);
- most of the electricity used for space-heating and -cooling, through both mechanical-equipment retrofits and improved building shells[10], including "superwindows" that can now insulate 2–4 times as well as triple glazing but cost about the same (Rosenfeld and Hafemeister 1989; Bevington and Rosenfeld 1990; Lovins 1986a, 1988);
- three-fourths of all electricity used in typical U.S. houses and commercial

---

[9] These savings are so cheap because one needs to pay for only seven of the 35 motor-system savings. The other 28 are free byproducts of those seven, reducing their average cost below 0.5¢/kWh. To capture such big, cheap savings, however, requires not just new technologies but also new thinking—whole-system engineering with meticulous attention to detail. Improvements to or beyond the machine driven by the motor are not included here, but can often save about half the remaining energy (e.g., Johansson et al. 1983).

[10] And through ligher-colored paving and building surfaces and smarter landscaping, especially urban forestry: direct shading, evapotranspiration, and reduction in the mesoscale urban "heat island" effect will enable a half-million trees in Sacramento, for example, to save 15 peak MW and 35 GW–h/y in the tenth year of a program costing 2¢/kWh (SMUD 1990; Akbari et al. 1988). In Sacramento, simulations show savings up to 14% in peak and 19% in annual cooling energy just from whitewashing buildings, and up to 35% and 62% from all measures to modify urban albedo (to average values of 40% overall and 90% for houses) (Taha et al. 1988).

buildings at respective retrofit costs of 1.6¢/kWh and −0.3¢/kWh (Lovins 1988);
- about three-fourths of total U.S. electricity at a net cost averaging about 0.6¢/kWh (Fig. 14.1)—several times cheaper than just *operating* a typical coal or nuclear power plant, even if building it cost nothing.[11] Of course, considerably more could be saved at less than long-run marginal cost, which is at least tenfold higher, and higher still when externalities are included.[12]

Potential savings appear to be only slightly smaller and costlier in the most efficient countries than in the United States. Detailed studies have found a potential to save half of Swedish electricity at an average cost of 1.3¢/kWh (Bodlund et al. 1989), half the electricity in Danish buildings at 0.6¢/kWh or three-fourths at 1.3¢/kWh (Nørgård 1989), and 80% in West German households with a 2.6-year payback (Feist 1987). To be sure, Europeans do (for example) light their offices less intensively, and turn the lights off more, than Americans do, but that does not affect the *percentage* savings available in the lighting energy that *is* used—a function only of the lighting technology itself, which is quite similar in both places.

Abundant observational evidence confirms that the potential savings in socialist and developing countries (Goldemberg et al. 1988) are much larger and (at world equipment prices) cheaper than in OECD. Differences in what electricity is used for between industrialized and developing or capitalist and socialist countries are surprisingly small (*id*; Reddy and Goldemberg 1990; Lovins 1979), and major savings are available in essentially every significant application. The feasibility of major electric savings is confirmed by comparisons at all scales: the micro-scale of individual technologies (e.g., Shepard et al. 1990), sectoral intensities[13] (e.g., Kahane 1986), and aggregate intensities.[14] It therefore seems reasonable, and probably conservative, to

---

[11] Fickett et al. (1990) contrast RMI's and EPRI's supply curves, but much of the difference shown is methodological, not substantive: EPRI's supply curve excludes the 9–15% savings already predicted or planned, counts only savings achievable by 2000 rather than long-term, shows only values near the lower end of a range of uncertainty spanning 18 percentage points, ignores saved lighting maintenance costs, and counts drivepower savings three times smaller and at least eight times costlier than agreed to in the accompanying text. Remaining differences arise from the modernity, disaggregation, and thoroughness of characterization of the technologies analyzed by RMI. Disaggregation alone—counting many small savings as well as a few big ones—can roughly double the quantity of savings.

[12] An exhaustive compilation (Ottinger et al. 1990) found that external costs due to SO_x, NO_x, CO_2, and particulate emissions total about 5.8¢ for coal-fired generation, 2.7¢ oil-fired, 1.0¢ gas-fired, and 2.5¢ nuclear, all per busbar kWh.

[13] For example, Kahane (1986) found that in the car, paper, and cement industries, electricity per ton of product was falling in Japan but rising in the U.S.

[14] For example, International Energy Agency statistics show that the Japanese GNP in 1986 was 36% less electricity-intensive than the American, and this gap was projected by those governments to widen to 45% in 2000.

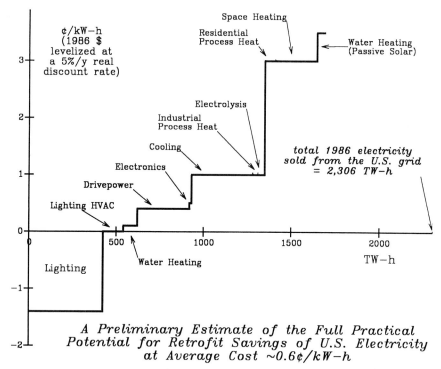

Figure 14.1 Supply curve of the full technical potential to save U.S. electricity by retrofitting the best commercially available 1989 end-use technologies wherever they fit in the 1986 stock of buildings and equipment. The vertical axis is levelized marginal cost (1986 ¢/kWh delivered, 5%/y real discount rate). Costs are negative if the efficient equipment's saved maintenance costs exceed its installed capital costs. The horizontal axis is cumulative potential saving corrected fo interactions. Measured cost and performance data are summarized for about a thousand technologies, condensed into end-use blocks. Fuel-switching, lifestyle changes, load management, further technological progress, and some technical options are excluded. How much of the potential shown is actually captured is a policy variable, but many utilities have in fact captured 70 +−90 +% of particular efficiency markets in a few months or years through skillful marketing, suggesting that most of the potential shown could actually be captured over a few decades. Note that savings totalling around 50% have a net internal average cost of zero, and that new-construction savings would be larger and much cheaper (often negative-cost) than the retrofit savings shown. For comparison, utilities such as Bonneville Power Administration and Wisconsin Power & Light Co. report empirical total costs of 0.5¢/kWh for saving business customers' electricity. The "electronics" saving has turned out in more recent analyses to be larger and cheaper than shown (Shepard et al. 1990, Ch. 6), and the drivepower saving, to be probably twice as large as shown (Lovins et al. 1989, Fickett et al. 1990), although saving more drivepower energy leaves less cooling energy available to be saved.

treat the U.S. values as a surrogate for the global average of potential electrical savings' fractional quantity and average cost.

Oil

The potential for saving oil with today's best demonstrated technologies is also large and cheap. Unlike electricity, about half of the needed technologies are not yet on the market, though they could be within a few years. There are large potential savings in transportation (~44% of world oil use), industrial heat (~12%), building heat (~14%), electric generation (~10%), and feedstocks (~14%). The rest of the oil is used or lost in refineries (G. Davis 1990).

Transportation

Personal mobility, the most familiar and pervasive use of oil, accounts for about two-thirds of OECD transportation energy use (EPA 1989, p. VII–37) and offers some of the most dramatic savings. To start with, a U.S. DOE study (Difiglio et al. 1989) describes 15 proven, readily available improvements in car design. These, plus two more equally straightforward ones (Ledbetter and Ross 1990), can maintain average 1987 U.S. new-car size, ride, and acceleration at 33.7 actual mi/gal (7.0 l/100 km). That is 39% less fuel-intensive than the average new 1987 U.S. car, and 69% more efficient than the present fleet. The measures' average cost is 53¢/gal saved (14¢/l). This result is conservative:[15] at least seven attractive cars with actual performance over 42 mi/gal (55 EPA-rated mi/gal) are already on the U.S. market. Savings ~72% as large are achievable in light trucks at about half the unit cost (Ledbetter and Ross 1989, Table 10).

A further doubling or tripling of car efficiency has been demonstrated by ten automakers[16] whose prototypes have achieved average on-road efficiencies of 67–138 mi/gal (1.7–3.5 l/100 km)— ~4–7 times the present OECD fleet or ~2½–5 times the low-powered USSR/eastern European fleet (Chandler et al. 1990). The Toyota AXV, for example, carries 4–5 passengers at EPA ratings

---

[15] E.g., a drag coefficient of 0.3 is assumed, but 0.12 is readily achievable and consistent with attractive appearance (0.20 for vans) (P.B. MacCrady, pers. comm.); the Ford Sable and several other models already get 0.3, Renault's Vesta prototype, 0.186, and the experimental Ford Probe v,0.137. A curb weight of 2,490 lb (1,132 kg), only 10% below the recent U.S. average of ~2,800 lb, is also assumed, but is ~2.4 times as heavy as some 4- to 5-passenger prototypes. Indeed, a U.S. car fleet averaging 2,000 lb and hence 50 mi/gal could be achieved by materials substitution *alone* (Flemings et al. 1980); each 200 lb reduction improves fuel economy by ~5% (Bleviss and Walzer 1990).

[16] Ten examples from eight companies are discussed by Bleviss 1988 and 1988a and by Goldemberg et al. 1988. More recent varieties reported in the trade press, from Audi, Citroen, GM, Fiat, Honda, Nissan, Volkswagen, and others, support the same conclusions.

of 89 mi/gal city and 110 highway, while Renault's 4-passenger Vesta 2 prototype was tested in 1987 at 101 and 146 mi/gal. Two prototypes—a 71–mpg Volvo and a 92–mpg Peugeot—are said to cost about the *same* to build as ordinarily inefficient cars of comparable size.[17] Thus such radical efficiency improvements appear to be far cheaper than paper studies had predicted by simply extrapolating the cost of far smaller incremental savings. Though each prototype has individual design peculiarities, collectively they prove that cars more than three times as efficient as the world fleet can be at least as comfortable, peppy, safe[18], and low in emissions as today's typical new OECD cars. Comparable opportunities apply also to light trucks (Bleviss and Walzer 1990).

Even this dramatic efficiency gain results only from incremental progress, not basic redesign. For example, one could choose crushable-metal-foam bodies (which would be extremely crashworthy at a curb weight of ~750 lb (~340 kg), series-hybrid drives, switched-reluctance motors integrated into the axles at zero marginal weight, power-electronic regenerative braking (eliminating the hydraulic system), variable-selectivity windows, and other innovations. Systematically combined, such features to be able to yield a very safe, peppy station wagon averaging upwards of 150 mi/gal (1.6 l/100 km). Other major innovations show additional promise, e.g., membrane oxygen enrichment of air intake, variable-geometry turbochargers, direct-injection diesels, ceramic engines, perhaps new two-stroke engines, etc. (Bleviss 1988a).

There is also the option of a more diverse mix of car sizes. Minicars (no more than $4\frac{1}{2} \times 10\frac{1}{2}'$ and 0.55 l displacement) recently held a fifth of the Japanese domestic car market. General Motors, too, has invested over $50

---

[17] This is largely because their ~1200–1400 lb curb weight (~550–640 kg) requires extensive use of plastics and composites, so large, complex assemblies can then be molded as a unit and snapped together. Not having to make and assemble many small parts, and being bettter able to design for easier assembly, saves more money than the molding dies and the more exotic materials cost. The leftover money pays for better areodynamics, smarter chip controls, etc., and the total net marginal costs is about zero. This negative cost of weight reduction is consistent with consultancy data recently provided to Mark Ledbetter (pers. comm., 8 October 1990), and with Chrysler's finding (*Automotive News* 1986) that a largely plastic/composite car could cut a steel car's part count by 75% and production cost by up to 60%. Earlier incremental calculations (e.g., von Hippel and Levi 1983) of a $500 marginal cost to achieve 71 mpg have therefore proven, as hoped, to be overly conservative. Moreover, plastics and composites can improve safety and cut maintenance costs, although they require careful design for recyclability.
[18] Today, some light cars are among the safest and some heavy ones are among the most dangerous. This proves that for safe momentum transfer in a crash, design and materials are far more important than mass. Some of the prototypes, such as the Volvo LCP, are designed for survival in a 35-mph head-on crash—a 36% higher energy-dissipation capability than the U.S. 30-mph standard. Energy-absorbing materials and body designs can thus ensure that ultralight cars are safer than today's—without taking credit for their greater maneuverability, faster acceleration (because they are so light, despite their ~20–50 hp engines), and shorter stopping distance.

million in successfully developing a 1- or 2-passenger, ~0.75-liter "Lean Machine" rated at ~150–200 mi/gal. Said to be safer than a normal car (because of its "bouncy" composite materials and its maneuverability), and occupying less than half the driving or parking width of a normal car (Sobey 1988), it is licensed to Opel and Suzuki but stalled by regulatory ambiguities.

Similarly, the minivehicles popular in many developing countries can be much improved. Taiwan is now doing this with motorcycles and scooters. An unwelcome development, in contrast, is the widespread improvization of carrying passengers on adapted tractors, especially in China and India: a typical Chinese tractor carrying 1 ton is estimated by the World Bank to use 75% more fuel per mile than a fully loaded 4-ton truck, and such tractors used 27% of China's diesel fuel around the mid–1980s (EPA 1989, p. VII–58). Efficient alternatives are clearly important.

The world's half-billion motor vehicles—twelve times the 1945 fleet, and now consuming nearly half the world's oil—has grown by about 5%/y for two decades (Bleviss and Walzer 1990). If this growth continues, we'll run out of land and air long before we run out of oil. Transportation alternatives are an essential supplement to light-vehicle efficiency, because urban highway congestion handicaps even the most efficient cars. California's congestion, for example, is frequently severe on the urban roads that carry 40% of traffic. It costs $17 billion/y in wasted fuel and lost time (360 million person-h/y), and is expected to triple in the 1990s (CEC 1990, p. 7–13). Congestion is even worse in many developing countries. The main technical options for alleviating it include:

- Improved road design, signalling, signage, and controls (right turn on red, computer-controlled traffic lights, freeway entry-ramp flow controls, etc.). In-car computers linked to computer-driven transmitters that suggest time-minimizing route changes are already under test in Berlin. Automatic proximity-control systems could safely pack 2–3 times as many cars per lane-mile and "save up to 20% of the fuel consumed" (Bleviss and Walzer 1990).
- Symmetrical treatment of competing transportation modes. A thorough study of 32 cities on four continents (Newman and Kenworthy 1989) found that after controlling for per-capita income and other variables, vehicle-miles travelled per capita in Australian and western European cities and in Tokyo are respectively 85%, ~45%, and ~25% of the U.S. average—correlated with mass-transit shares of 8%, 32%, 63%, and 4% respectively. The American cities' high gasoline intensity was fundamentally due to their overprovision of roads and of downtown parking as apparently free goods, while other modes had to pay far more of their own costs. In California, for example, cars pay only ~10% of

their full costs through taxes and tolls, while mass transit pays ~20–25% through fares (CEC 1990, p. 7–9).
- Coordinated land-use/transport development. A mile of travel by mass transit is commonly assumed to displace a mile of travel by car, implying that mass transit has limited effect and uninviting economics (though almost always severalfold cheaper than roadbuilding: Goldstein et al. 1990, App.B). Recent studies, however, indicate that the one-for-one assumption is misleading, because "the availability and usage of transit services also changes the location of trip origins and destinations in a way that reduces the need to travel by car, and reduces the distance of travel required by [most] . . . people who will continue to drive their cars" (Goldstein et al. 1990).[19] That study found a nearly twofold difference in vehicle-miles per comparable, income-normalized household in two nearby California communities, one with and one without light-rail service. Each mile of mass-transit travel displaced *ten* miles of car travel. Less thorough studies elsewhere have found a 4–5:1 ratio, essentially independent of cultural factors (Newman and Kenworthy 1989, pp. 77–100). These results are empirical, reflecting *actual* transit ridership, not market potential. Because doubling residential density reduces vehicle-miles per household by 25–30% (Newman and Kenworthy 1989), doubling a region's population through infill would increase car travel by only 40–50%, rather than the 100% expected from sprawl. The effect of transit corridors on commercial density, too, is even greater than on residential density, enabling many more errands to be done per trip. Capturing these benefits requires zoning that encourages infill and highly mixed land-use but discourages sprawl.[20] It also requires careful coordination of land-use and of parking with transport planning, and frequent, fast, safe transit. Such variables are far more important to per-capita gasoline use than gasoline prices, incomes, and vehicle efficiencies (*id.*).
- Ridesharing in private vehicles, and vanpooling, long organized by employers but paid for by riders, work well in some U.S. cities, encouraged

---

[19] This should come as no surprise, since the U.S. Interstate Highway System has proven to be the most important determinant of land-use in this century: as night satellite photographs reveal, an estimated ~95% of Americans live in counties on or adjacent to interstates. Of course, the roads were built between cities and through or near towns to start with, but they have since accreted most new "greenfield" developments too.
[20] Newman and Kenworthy (1989) found that fuel use rises steeply and nonlinearly as density drops too low to support satisfactory transit service. Urban per-capita gasoline use worldwide falls into roughly three classes (Newman and Hogan 1987): car cities (<30 persons/ha, typically North American), public transport cities (30–130, typically European), and walking/cycling cities (>130, typically Asian). Australian residents have claimed to be as satisfied with their lifestyle in high- as in low-density suburbs even if they might not originally have preferred the former (Duxbury et al. 1988).

Least-Cost Climatic Stabilization

by dedicated carpool lanes and other incentives. In the U.S., where the average car carries only 1.7 people, full 4–passenger carpooling would save 45% of all gasoline (Bleviss and Walzer 1990). During 1973–88 alone, California's vanpool and rideshare programs saved 2.5 billion vehicle-miles and 156 million gallons of fuel, or over 5 million tons of carbon (CEC 1988). Another option, voluntary carsharing among 4–6 households, is proving quite successful in West Germany. And a Belgian innovation—a national hitchhikers' club with a variety of cost- and risk-reducing features—could be widely emulated.

- Telecommuting via electronic media (Shepard et al. 1990, pp. 432–436) is now the main workstyle of some ten million Americans and growing rapidly. It saves money, time, stress, unhappiness, and pollution (Washington State Energy Office 1990). A related Swedish experiment is fitting one car per subway train with computers, modems, FAXes, etc. to making commuting time more productive.
- Offering safe and convenient bicycle (and pedestrian) lanes or paths, and coordinating with public transit (so bikes can be taken on trains and buses or rented at stations), enables bikes to carry 9% of all Dutch commuter traffic, and "in some cities, they account for more than 40% of all passenger trips" (Bleviss and Walzer 1990). In contrast, although 54% of working Americans work within five miles of home, only 3% bike to work. The scope for overcoming obstacles to biking is enormous: all U.S. biking is currently estimated to displace more than 14 billion car trips per year, saving the marginal portion of ~$6 billion in differential costs, but the gasoline displaced is currently less than 1% of total usage (Calwell et al. 1990, p. 26).
- Some capital-short developing countries have devised cheap, highly effective transit designs. Jaime Lerner, for example, developed unsubsidized, 10¢/ride commuter bus systems in Rio and Curiciba, Brazil, some with one-minute intervals. His onstreet "boarding pods" and special door designs nearly trebled density to 12–18,000 passengers per corridor-hour (J. Lerner, pers. comm.). Coordinated land-use policy and a three-tier bus system (including dedicated radial express lines) gave Curiciba "one of the highest rates of motor vehicle ownership and one of the lowest rates of fuel consumption per vehicle in Brazil"—because most car-owners prefer mass transit for routine city travel (Bleviss and Walzer 1990).

Heavy transportation can save considerable energy too. For example, commercial jet aircraft efficiency is about twice as high in today's 757/767/MD9–80 aircraft as in the older aircraft in U.S. fleet (Lovins and Lovins 1981), and that fleet in turn is about twice as efficient as that of, say, Aeroflot. Further savings of ~50% have been demonstrated in the

Boeing 7J7 and of ~40% in the McDonnell Douglas MD–91/92.[21] New methods of drag reduction (Vaughan 1988) can save 20–40% of fuel. Just GE's unducted propfan engine, with a bypass ratio of 36,[22] uses 40% less fuel than the 727's JD8D–17 engines (Kavanagh 1990), at a 25% ($1 million/engine) price premium. Against its nearest competitor, its saved energy[23] costs only 19¢/gal or 5¢/l (a 14–21%/y rate of return @ $1/gal); against the fleet, it looks about twice that good. It is therefore slightly too costly for cash-strapped airlines at pre-Saddam Hussein fuel prices, but extremely attractive at social discount rates, or at long-run replacement fuel costs, or counting externalities like global warming.[24] Aircraft can also benefit from further improvements (still not begun in many countries) in computerized operations management, fuel-load minimization, idle reduction, flightpath optimization with improved weather monitoring, more frequent aircraft washing, weight paring[25], etc. In all, EPA (1989, p. VII–52, emphasis added) has identified improvements that "could reduce fuel use per passenger mile *to less than one-third of the current [U.S.] average . . .* "

Even in the U.S. with its relatively modern railway equipment, potential energy savings of ~25% were estimated (Ephraim 1984) before Caterpillar introduced a diesel locomotive with doubled efficiency (*Fortune* 1986). Savings estimated at ~80% (NRC 1990, p. 73; Sobey 1988) are also available by substituting rail for long freight hauls by truck, using e.g., GM's Roadrailer, which converts in seconds between a semitrailer and a railcar, and pays back in a few years just from lower demurrage and enabling just-

---

[21] The latter, described by Hene (1989), meets or exceeds all current or anticipated noise rules, sharply reduces pollution, and has less interior noise, than any other commercial airliner. Both aircraft were fully developed but shelved in 1989 for lack of a market in those cheap-fuel, cash-short days.

[22] Vs. ~6 for conventional engines and 10 for GE's latest large conventional engine, the GE90.

[23] Levelized at 5%/y real over a 15-y operating life, assuming a 16.8% weighted saving from initial fuel consumption of 1.424 milion gal/engine-y, typical of a narrowbody two-engine aircraft flying 3,200 h/y (Kavanaugh 1990, p. 8). Unfortunately, the >40% drop in fuel use per passenger-mile in the U.S. during 1970–80 and the >50% decline in real jet-fuel price during 1980–87 greatly reduced airlines' economic incentive to save more fuel (EPA 1988, p. VII–52). However, at this writing, the Mideast crisis's doubling of fuel prices has increased U.S. airlines' average fuel share of total costs from ~15% to nearly 25%.

[24] The contribution to Ottinger et al. (1990) by P. Chernick & E. Caverhill found the local and regional CO_2 externality alone to be worth 1.1¢/lb, equivalent to 7.5¢/gal or about an eighth of 1989 jet-fuel prices; adding SO_x, NO_x, and particulates increased this by fivefold. Soviet estimates *excluding* CO_2 are even higher, at over 60% of the 1989 U.S. jet-fuel price (Chizhov and Styrekovich 1985). Chernick & Caverhill further estimate the external cost of U.S. oil imports at $2.26/million BTU, or 52% of the 1989 jet-fuel price. Thus counting the externalities would more than double the price, making new aircraft like the 7J7 or MD91/92, or engines like GE's unducted fan, immediately attractive even at the airlines' high discount rates.

[25] At mid-October 1990 prices, according to a 12-13 October 1990 CNN Headline News "Dollars and Cents" feature, it costs the average U.S. airline ~$22.50/y in fuel to carry one extra can of soda onboard.

in-time inventorying. Substituting Japanese- or French-style (or more advanced) high-speed trains for long car or short air trips could likewise save ~80% and ~87% of travel energy, respectively, at current European efficiencies (Parson 1990), and similar fractions of travel cost. New proprietary developments in unusually simple and cost-effective magnetic-levitation trains show particular promise. Simply electrifying main rail lines, in countries with efficient utility systems, can save tens of percent of primary fuel compared with typical OECD diesels (EPA 1989, p. VII–61).

Heavy trucks can directly save ~60% of their energy, with paybacks of probably a few years, through turbocharged and adiabatic (uncooled) low-friction engines, improved controls and transmissions, better tires and aerodynamics, exhaust heat recovery, regenerative braking, improved payloads and payload-to-capacity matching, and reduced empty backhauls through better shipping management (Lovins et al. 1981; Samuels 1981). Savings of 40% have already been prototyped (*Automotive News* 1983), and savings of 50% have been found to have reasonable cost with present technology (Goldemberg et al. 1988). Most of these techniques also apply to buses, and many apply to agricultural traction, the need for which can be further reduced by improved cultural practices (*infra*).

Even after the doubling of ships' energy efficiency per ton-mile during 1973–83, considerable further savings remain available (Spyrou 1988) from improved propellers, engines, and hydrodynamics, antifouling paints, heat recovery, and (in some cases) modern versions of sails. Just the same engine innovations identified for trucks may achieve 30–40% savings (EPA 1989, p. VII–50).

Transportation equipment stays efficient only with proper maintenance and operation. Much of China's truck energy intensity (twice that of the U.S. [Chandler et al. 1990]), and similarly in many developing countries, is due to poor maintenance. Nor is OECD adequate in this respect: about 1 mi/gal, or ~5% of the fuel used by the U.S. car fleet, could be saved simply by proper tire inflation. Road quality, too, is a key determinant of vehicle efficiency and life, especially in the USSR and eastern Europe. Poor roads, while not discouraging vehicle ownership (EPA 1989, p. VII–59), subtly nudge designers toward tank-like designs—in effect, substituting extra fuel (to haul around the extra weight, drive more slowly, and stop often) for pothole repairs.

Low-temperature Heat

Buildings can use the same improvements that save electricity in water- and space-heating to save oil (or gas fungible for oil[26]). New options include

---

[26] This discussion assumes such fungibility. This is reasonable on a timescale of several decades—sufficient to achieve flexibility in refinery product-slate allocation—but in the short

furnaces up to 97% efficient (while also saving >90% of fan energy), superwindows that gain net winter heat even facing away from the Equator, ventilation heat recovery, and cost-effective ways to insulate or "outsulate" a wide range of existing buildings. Even with 1979 technologies, a major government study found that careful retrofits could save 50% or 75% of U.S. space heat at average costs of $10/bbl and $20/bbl respectively—severalfold cheaper than heating oil (SERI 1981). Technological progress since (Rosenfeld and Hafemeister 1988; Bevington and Rosenfeld 1990; Shepard et al. 1990) has cut these costs by probably half.[27] EPA considers a 75% reduction in households' total energy intensity achievable by 2025 (EPA 1989, p. VII–6).

A compilation (Rosenfeld et al. 1990) prepared for a National Academy of Sciences study, which is discussed below, adopted a midrange finding that presently commercial technologies could save 45% of the electricity (EPRI 1990) and 50% of the direct fuel used by U.S. buildings in 1989. Those savings would respectively save $37 billion and $20 billion a year more than they would cost. An additional $4.3 billion per year could be saved by cost-effective fuel-switching. The carbon avoided would thus total 232 million tons a year, or a sixth of total U.S. emissions, at a net cost of *minus* $61 billion per year (or –$263 per ton of carbon). The paper also summarized eight other studies, several of which documented much larger and cheaper savings than those adopted.

High-temperature Heat

Most of the oil used for industrial process heat, being fungible for gas, could be replaced by less carbon-intensive natural gas saved in buildings—and by far less of it. U.S. industry reduced its primary energy intensity, nearly all by saving process heat, by 30% during 1977–1985. Similar savings continue today, chiefly through improved insulation, heat recovery, controls, and process design (Ross and Steinmeyer 1990): computerized process simulation and controls, and substitution of membrane and other nonthermal processes for distillation, offer especially important opportunities still largely untapped. Numerous conversations with industrial energy managers confirm that many tens of percent more industrial energy remain to be saved, at typical paybacks often around two years, even in the most efficient countries, where many firms have already cut energy intensity in half since 1973. Swedish industry in the mid-1970s, for example, was a third more energy-

---

term, saved gas may mainly displace residual oil currently in surplus, rather than scarcer light products.

[27] Even against low gas prices, and with a relatively new building stock much of which was built under modern standards, the California Energy Commission has found it cost-effective to save half of the natural gas used in existing households (CEC 1990, Fig. 3–1).

efficient than U.S. industry despite having a more energy-intensive product mix (Schipper and Lichtenberg 1976), but ~50% of its 1975 energy intensity could still be cost-effectively saved by using the best ~1980 technologies, or ~60–65% by using the best technologies entering the market around 1982 (Johansson et al. 1983). Both opportunities continue to expand: leading European chemical firms are privately reporting *typical* savings around 70% from pinch technology (thermodynamic process-design optimization) and better catalysts.

Further large savings are available from long-term redesign and coordination of industrial systems to cascade industrial process heat through successively lower temperatures on a regional scale. Using heat pumps, cogeneration (with heat transmission up to 50 km), and heat exchangers (up to 25 km), ~25% of industrial energy could thereby be saved in West Germany, 30% in the U.S., and 45% in The Netherlands and Japan. Much of this potential appears cost-effective. Probably all of it is at long-run marginal social cost (Groscurth and Kümmel 1989).

These are technical improvements only. But the rapid "dematerialization" of the industrial economies (Larson et al. 1986; Herman et al. 1989) has reduced industrial energy intensity in the U.S. and western Europe nearly as much as improved energy efficiency has (Lovins et al. 1981). U.S. steel consumption per real dollar of GNP, for example, is now below its *1860* level and falling. Worldwide, raw-material use per unit of industrial output has fallen by at least 60% since 1990, and this decline is accelerating so quickly that Japan's intensity fell by 40% just during 1973–84 (Colombo 1988). As will be noted below, the scope for future compositional energy savings is especially large in the USSR and similar economies distorted by excessive output of primary materials that are largely wasted (Chandler et al. 1990).

Furthermore, reductions in the throughput or resources needed to maintain a given stock of material goods represent an additional revolution just beginning (Lovins et al. 1981). These reductions involve recycling, reuse, remanufacture, scrap recovery, minimum-materials design (often by computer), near-net-shape procesing, increased product lifetime, and substituting elegantly frugal materials (such as optical fibers for copper cables, reducing their tonnage by $97\frac{1}{2}$% and their manufacturing energy by 95% [Colombo 1988]) or processes (such as ambient-temperature biological enzymatic catalysis for chemical-engineering pressure-cookers[28]). Recycling alone typically saves about half of materials-processing energy, so "[t]he potential energy savings are staggering" (Ross and Steinmeyer 1990, p. 96).

---

[28] For example, if we were as smart as chickens (as Ernie Robertson points out), we would know how to make eggshell at ambient temperature, rather than calcining limestone at ~1250°C into Portland cement that's several times weaker.

This is especially true in a garbage-rich, landfill-poor country like the United States, which, for example, throws away enough aluminium to rebuild its commercial aircraft fleet every three months, even though recycling aluminium takes only ~5% as much energy as making it from virgin ore. Collectively, these materials-policy options can probably reduce long-run industrial energy intensity per unit of maintained stock (not throughput) *by an order of magnitude*. If this were combined with technical gains in process efficiency, little industrial energy use would remain: industrial energy use smaller than today's could support a worldwide western European material standard of living (Lovins et al. 1981; Goldemberg et al. 1985).

Electric-utility Hydrocarbon Fuels

Another kind of oil saving comes via electricity.[29] The potential to save electric utilities' small remaining oil input by substituting other forms of generation has been exaggerated (Lovins and Lovins 1989, pp. 25–26). Yet that oil and the larger amount of gas still used in thermal power plants were together equivalent to 13% of all oil burned in the U.S. in 1989. Nearly twice that much electricity could be saved by lighting retrofits alone, at negative net cost (Piette et al. 1989; Lovins and Sardinsky 1988). Globally, oil-and-gas use in power plants is equivalent to ~22% of total oil use (G. Davis 1990), but lighting plus other cheap electrical savings can clearly displace far more than 22% of electricity at negative net cost (Fig. 14.1).

Miscellaneous Oil Uses

The oil and gas used as feedstocks (10% and 3% of their respective total U.S. consumption in 1986, 14% and 7% globally) are subject to unknown but probably substantial savings. These mainly involve more efficient petrochemical processes (Ross and Steinmeyer 1990), internalization of solid-waste disposal costs (leading, as in Europe, to high plastics recycling rates, improved product design and longevity, agricultural reform [*infra*], and lower use of disposable packaging), and reduced but more durable highway construction leading to lower asphalt requirements (and less fuel burned to make cement). In addition, at least half of the ~2% of oil used to propel the other 98% through pipelines would be saved by the above

---

[29] Besides the scope for saving power-plant fuel by saving electricity, small but useful amounts of oil and gas can be directly replaced by electrotechnologies that are cheaper in certain applications, but they will increase industrial electric use only a few percent as much as improved electric efficiency decreases it (EPRI 1990).

measures, and analogously for refinery fuel and losses (~6%) and for gas compressor energy (~3%).

Total Oil-saving Potential

The combined potential to save oil by these means in the United States, shown in Figure 14.2, is ~80% at an average cost below $3/bbl (plus a further 20% of leftover saved gas at ~$10/bbl-equivalent). Qualitative

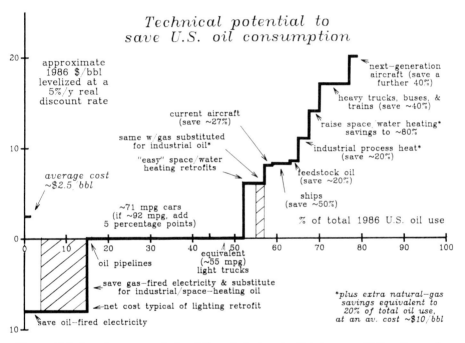

Figure 14.2 Supply curve of the full technical potential to save U.S. oil use by retrofitting or substituting the best demonstrated 1988 end-use technologies. The vertical axis is levelized marginal cost (1986 $/barrel delivered, 5%/y real discount rate). The horizontal axis is cumulative potential saving (% of total 1986 U.S. end-use) corrected for interactions. Shaded areas represent savings of natural gas that then displaces oil used to heat buildings or industrial processes. The cost and performance data are empirical (Lovins and Lovins 1989); costs above $10 bbl are quite uncertain, but this has little effect on the result. No lifestyle changes or intermodal transport shifts are assumed. The curve reflects many conservatisms: e.g., omission of any light-vehicle improvements whose marginal cost exceeds zero, and translating the negative-cost lighting retrofits (which save the oil- and gas-fired electricity) directly into equivalent $/bbl without taking credit fr the value of the fuel displaced. The overall uncertainty appears to be ~10 percentage points in total quantity and <2x in average cost.

evidence that potential oil savings and costs are comparable in other OECD countries include:

- the similar efficiences of new light vehicles (nearly 30 mi/gal for cars) throughout OECD;[30]
- the cost-effectiveness of large additional industrial and building heat savings even in such efficient countries as Sweden (Johansson et al. 1983; Bodlund et al. 1989), West Germany (Lovins et al. 1981; Feist 1987), and Denmark (Nørgård 1979), and hence even more so in less efficient countries; and
- the virtual irrelevance of differences in oil end-use structure, because such large savings are available in each end-use.

These considerations apply *a fortiori* in the other two world regions, since they are even less efficient than OECD.

Aggregate energy intensities per unit of economic output are typically 2–3 times as high in socialist and developing countries as in OECD (Chandler et al. 1990; Goldemberg et al. 1988). Both this fact and the field observations reported universally in the literature suggest that if all countries became as energy-efficient as OECD countries should be, the potential percentage savings would be even larger in socialist[31] and developing countries than in OECD, and the costs correspondingly lower (Goldemberg et al. 1988). As such countries build or rebuild their infrastructure, too, more opportunities will arise in new construction, and fewer in retrofit, than in OECD. This will furher increase savings and reduce costs. One can therefore conclude that most of the oil now used in the world can be saved at an average cost far below mid-1990 world oil prices—perhaps an order of magnitude below.

Other Energy

Natural gas, natural-gas liquids (NGL), and coal used for process or building heat or for feedstocks are subject to the same categories of savings just described, and can be saved with similar effectiveness and cost. This is

---

[30] New cars are a few mi/gal less efficient in the U.S. than in western Europe and Japan, but this effect is immaterial compared with the potential improvements, partly because it is diluted by those areas' half-as-big transportation share of oil use, meaning that more oil is used in industry and buildings. The thermal efficiency of buildings in such countries as Germany, Britain, and Japan is particularly low.

[31] Their efficiency analyses so far (e.g., Chandler et al. 1990) show savings of only about a sixth for the Soviet Union by 2030 (or a third including changes in output structure). Extensive discussions there support our and other observers' belief that this is not because of a lower actual potential—quite the contrary—but only because Soviet analysts have not yet become familiar enough with disaggregated analyses, modern Western technology, and market mechanisms to apply these opportunities to their own difficult situation.

especially true in heating applications, where oil, gas, and NGL are used essentially interchangeably and in nearly identical technologies. In broad outline, therefore, no additional treatment of these other fuels is necessary. There are two exceptions: (1) Seven-eighths of Chinese household energy is from coal, nominally for cooking. Although its use is officially forbidden for space-heating, despite indoor winter temperatures often below freezing in many provinces (Chandler et al. 1990), cooking coal nonetheless contributes precious heat. Better-insulated houses would thus improve comfort more than they would save coal in that instance: only in combination with gas (e.g., biogas) cooking will they displace much coal. (2) Much Soviet and eastern European steel is still produced in open hearths, which are half as efficient as basic-oxygen plants that are themselves no longer state-of-the-art. The USSR, by far the world's largest steelmaker but consuming at least two-thirds more energy than Japan to make each ton of steel (EPA 1989, p. VII–107), can thus save large amounts of coal by this improvement alone. Continuous casting and more advanced processes can save even more (Eketorp 1989).

Unsustainably harvested wood and dung burned for fuel release biotic carbon from soil to air. This contributes ~23% of global CO_2 emissions and hence increases poor countries' 1987 share of those emissions from 19% (from fossil fuels) to 42% (Reddy and Goldemberg 1990). Such fuels also cause erosion, deforestation, loss of soil fertility, blindness among women and children, and many other social ills. A combination of end-use efficiency,[32] reforestation, cascaded fuel-switching, and social changes (chiefly related to the role of women) is needed to address this complex problem (Goldemberg et al. 1988). Among the desirable substitutions here is LPG and biogas for electric and kerosene lighting and cooking (*id.*); a good gas stove can also be 5–8 times as efficient as a traditional <10%-efficient woodstove (Goldemberg et al. 1988). Similarly, a compact fluorescent lamp driven by 25%-efficient electric generator and grid is some 200 times as fuel-efficient as the kerosene lamps used in ~80% of Indian households (*id.*).

Such substitutions are only part of a complex chain of successive fuel-switching (Goldemberg et al. 1988, pp. 255–273) needed to address simultaneously the fuelwood *and* oil problems of countries like India. The main steps are: replace household kerosene, wood, and dung with biogas (thereby producing more and better fertilizer too); use sustainably grown gasified fuelwood and a little biogas to run old diesel pumpsets, new

---

[32] Comprising not only efficient stoves (Baldwin et al. 1985; Baldwin 1987) but probably also efficient pots, perhaps with double walls and lids, enhanced heat transfer, and internal hot-gas paths—a concept that has as yet received no serious engineering attention for developing-country use.

pumpsets (if not photovoltaic), and short-haul freight; use the electricity saved from pumpsets to electrify all homes and replace kerosene for lighting; desubsidize kerosene and diesel fuel; and shift long-haul freight from trucks back to revitalized railways. End-use efficiency is the key throughout.

Conversion and Distribution Efficiency

More efficient use of delivered energy is only part of the energy savings available. Major savings are also available in converting and distributing primary fuels. Salient opportunities that could reduce global fossil-carbon emissions by a third or more at negative net cost include:

- improving maintenance and operational techniques at most developing countries' thermal power plants, where efficiencies below 20%, output several times lower than nameplate ratings, and poor availability are endemic;
- substituting 42+%–efficient combined-cycle gas turbines, or better still, >50%–efficient intercooled steam-injected gas turbines (Williams and Larson 1989), for 25– to 35%–efficient classical thermal power plants (a ~50+% CO_2 reduction), at a fourth the marginal capital cost and lead time of scrubbed coal plants—an especially attractive opportunity in the gas-rich Soviet Union (Chandler et al. 1990) and when integrated with efficient biomass gasifiers (Larson and Williams 1988; Larson et al. 1989);
- substituting low- for high-carbon fuels (e.g., natural gas emits half as much carbon per unit energy as coal);
- expanding economic power wheeling by using Japanese advances in power electronics and mainly Soviet advances in control theory to raise grid capacity;
- saving several percent of electric energy in OECD and up to ten times that in socialist and developing countries through advanced distribution management and metering at negative cost;
- reducing the major losses of natural gas (~8%, Arbatov 1990[33]) and

---

[33] Makarov and Bashmakov (1990) put the losses at ~2%, but Arbatov's higher estimate, or something close to it, has been informally confirmed by other knowledgeable Soviet experts. Arbatov found 50 Gm³/y of losses (mostly CH_4) due to leakage and ruptures, excluding 17 Gm³/y in extraction, out of 800 Gm³/y of total production, 80 burned for compression, and 30 of associated gas burned (not vented). The losses are very nonuniform in time and space. In contrast (Abrahamson 1989), U.S. production losses are ~0.13%, ~0.54% is "lost and unaccounted for" in interstate pipelines, and corresponding losses in retail distribution are a highly variable 1–6%, averaging 2–3%. These figures are all too high, and should be greatly reduced to save both money and fire-danger, as well as to reduce global warming. Abrahamson (1990) further cites direct CH_4 measurements in ambient urban air consistent with urban leakage of ~2.9–5.9% of U.S. natural gas consumption, some perhaps from storing and using coal.

district heating energy (half of two-thirds, K.S. Demirchian, pers. comm.) in the Soviet grids, and the even more dismaying loss of much of the *delivered* district-heating energy (which heats about three-fourths of all buildings) owing to the lack of operable controls, let alone meters, for each office or apartment, making open windows serve as thermostats[34];
- extending advanced Scandinavian district-heating technology to cold countries not yet taking advantage of it (i.e., most of them) wherever superinsulation retrofits aren't cheaper;
- displacing electric space and water heating with gas or passive-solar techniques, as some North American electric utilities already pay customers to do;
- recovering about half of gas-pipeline compression energy with a city-gate turboexpander/generator (a U.S. opportunity to generate probably on the order of 10 GW at costs below 1¢/kWh);
- eliminating gas flaring, which accounted for nearly 1% of 1986 fossil carbon releases (Marland et al. 1989), by using the gas as a feedstock or to fuel steam-injected gas turbines;
- making industrial cogeneration—a common and lucrative practice in West Germany—the universal practice in all countries that can use both the electricity and its process-heat coproduct; and
- using simple kiln improvements to double, or more, the efficiency of traditional charcoal-making in developing countries (EPA 1989, p. VII-140).

Renewable Energy Sources

Many renewable energy sources are already, or are rapidly becoming, competitive without subsidy. Where some do not quite compete yet, their margin of disadvantage is usually less than the external costs they avoid (Ottinger et al. 1990; SERI 1990). The speed of progress and the variety of options showing great promise of further improvement are both most encouraging (Weinberg and Williams 1990; SERI 1990).[35] Integrating

---

[34] Similarly, Makarov (Chandler et al. 1990) states that "metering and individual control of residential heating and hot-water systems, coupled with repairs and improved management of district heating systems, could permit [Polish] housing space to double with only 15% growth in energy consumption." This implies a 43% efficiency gain. Jászay (1990) estimates a corresponding 30% sectoral saving potential in Hungarian buildings.
[35] For example, apparently competitive solar-thermal-electric technologies now include not only the Luz trough-concentrator technology mentioned by Weinberg and Williams 1990, but also the Sunpower/Cummins combination of an Ericsson engine with a solar dish, now believed to cost ~5¢/busbar kWh with cheap dishes (W. Beale, pers. comm.) and perhaps less with very cheap ones like Solar Steam's. Yet only a few years ago, solar-thermal-electric technologies looked unpromising.

different types of renewable power sources, in different places, tends to provide needed storage automatically (*id.*), or at least more cheaply than with nonrenewable central stations (Lovins and Lovins 1982, pp. 268–270 & Apps. 1–3; Lovins 1978). Even the costlier electric technologies, like photovoltaics, are already competitive in remote sites (Weinberg and Williams 1990) or those requiring high power quality and reliability.[36] Progress is similarly rapid in bringing down the costs and raising the conversion efficiency of both thermal and biochemical processes for converting biomass to liquid fuels (*id.*, SERI 1990). And new developments in solar process heat[37] promise to provide it economically even in unfavorable climates.

The combined potential of the rich menu of renewable sources is very large (*id.*) and its growth potential quite rapid—contrary to the impression that can be created by dividing renewable energy into many small pieces and discussing only one at a time. In the most detailed official study to date in the U.S., for example, five National Laboratories (SERI 1990) found that fair competition plus restored research and development priority[38], or proper counting of avoided environmental costs, could increase competitive renewable electric output from 363 TW–h/y (363 billion kilowatt-hours per year) in 1988[39] to 1,573 TW–h/y in 2020. The 2020 figure is 60% of all 1989 U.S. electricity sold, and hence is enough to run an expanded but efficient economy without fossil-fueled or nuclear power stations. The 2030 renewable electric potential was found to be 94% larger still, giving that conclusion a large safety margin. All the technologies assumed would compete with assumed 2030 prices of 6¢/kWh baseload (1988 $), 9¢ intermediate, and 15¢ peaking. The same study also found that cost-effective electric *and* nonelectric renewable sources could together supply 44 EJ/y (41 quadrillion BTU/y) in 2030, equivalent to half of present total U.S. demand. A vibrant and much expanded economy using energy in a way that saves money would need no more than that. Yet 2030 is about the retirement date of a standard power station ordered today; and the Interlaboratory Study's results, based on midrange expectations of economic performance and artificial constraints

---

[36] The Federal Aviation Administration, for example, is converting hundreds of ground avionics stations to PV power even where there is already grid power to the site, because cleaning up and backing up the grid power costs more than starting with an isolated source.

[37] Especially David Mills's development, at the University of Sydney, of semiconductor-sandwich surfaces which should soon be able to absorb visible light 85–90 times as well as they emit infrared (pers. comm., December 1989 & October 1990). Such a surface in a hard vacuum, if it is sufficiently heat-resistant, can be calculated to yield heat at high enough temperatures for most industrial processes, even on cloudy winter days at high latitudes.

[38] Real Federal renewable-energy research, design and development (RD&D) funding fell by 89% during 1979–89.

[39] Excluding nonelectric sources, such as at least 2.7 EJ/y of direct biofuels: renewable supply of all kinds was probably not the cited ~8% but rather ~11–12% of total U.S. primary supply, and the fastest-growing part, outpaced only by savings (Lovins and Lovins 1989).

on intermittent sources like windpower (Sørensen 1979), do not represent "an upper limit on the potential contribution of renewables" (*id.* p. ix).

In many countries, including the U.S., the combination of sustainably grown biofuels[40] with modest efficiency gains (chiefly doubled light-vehicle efficiency) could *eliminate* the need for fossil fuels both for light vehicles[41] and for power plants. The same was shown a decade ago to be true in each region of the globe, using only renewable sources then already available and cost-effective on the long-run margin. Indeed, the same study, for the German government, found that such sources could provide essentially all the energy needed, at levels of end-use and energy-system efficiency available and cost-effective in 1980, to sustain a 1975 West German standard of living *throughout* a long-run world with a population of 8 billion (Lovins et al. 1981). Another study, though twofold more conservative, reached broadly similar conclusions about the combined potential of efficiency and renewables (Goldemberg et al. 1988), consistent with a decade's shorter-term analyses for the U.S. (SERI 1981, 1990).

Two specific foreign investigations of efficiency-plus-renewables strategies also merit emphasis:

- A detailed analysis by the Swedish State Power Board (Bodlund et al. 1989) found that doubled electric end-use efficiency (costing 78% less than marginal supply), plus fuel-switching to natural gas and wood, plus environmental dispatch[42], could together support 54% growth in Swedish real GNP during 1987–2010 and handle the voter-mandated phaseout by 2010 of the nuclear half of the country's electric generation, yet at the same time *reduce* the heat and power sector's CO_2 output by one-third and *reduce* the cost of electrical services nearly $1 billion per year. (This reduction arises because efficiency saves more money than fuel-switching and environmental dispatch cost.) This result is especially striking because Sweden is arguably the world's most energy-efficient country (in aggregate or in many details: e.g., Schipper and Lichtenberg 1976) to start with,

---

[40] I.e., those whose production can be indefinitely repeated because it depletes nothing. If such biomass were not burned, it would rot or be eaten by respiring animals and release its carbon anyhow; the issue is only whether that carbon release is taken up again, promptly and in equal measure, by new photosynthesis.

[41] Pure-electric cars are not considered here. They appear unlikely in principle to compete in cost, range, and performance with efficient fueled cars (including those which convert the fuel to electricity with an onboard motor-generator or fuel cell, the "series hybrid" concept). They also do not reduce global warming if powered by anything like the present utility fuel mix (DeLuchi et al. 1988). For the same reasons of economics and an actual worsening of global warming, coal synfuels are not considered either (*id.*). Compressed or liquefied natural gas can modestly reduce CO_2/vehicle-mile, but far less than efficiency and biofuel options, and with some drawbacks, so they are best considered a transitional niche fuel.

[42] I.e., operating most of the stations that emit the least carbon, and vice versa, by including externalities in economic dispatch.

with a heavily industrialized economy and a severe climate. Any other country should therefore be able to do better.
- At the same time, a study for the Indian state of Karnataka analyzed the combination of several end-use efficiency measures (but far from a comprehensive list), small hydro dams, bagasse cogeneration, biogas/producer gas, a small amount of natural gas, and solar water heaters. This combination would achieve far greater and earlier development progress than the utility committee's plan, since rejected, yet would use three-fifths less electricity to do so. It would also cost a third as much and emit only 1/200th as much fossil-fuel CO_2 (Reddy and Goldemberg 1990). This is encouraging too, since India already emits ~5% of global carbon (*id.*) and projects this fraction, assuming the traditional coal-based strategy, to increase enormously.

These two analyses are especially interesting when considered together, because between them they scope essentially the full global range of energy intensity and efficiency, technology, climate, wealth, income distribution disparities, and social conditions. Yet both find that the money saved by efficiency more than pays for the renewables, yielding a net profit on the whole carbon-displacement package in the energy sector.

Reports commissioned by the governments of Australia (Greene 1990) and Canada (DPA Group 1989) similarly found that national CO_2 emission reductions of ~20% via energy efficiency would be highly profitable. A 36% Australian energy saving from projected 2005 levels, reducing forecast fossil-fuel CO_2 emissions by 19%, would produce net nonenvironmental savings, in today's Australian dollars, of $6.5 billion *per year* by 2005: each $5 invested in efficiency could save $15 worth of new energy supplies and 1 ton of CO_2. Similarly, the Canadian report found a cumulative net saving of $100 billion through 2005 (present-valued Canadian dollars) from a 20% CO_2 cut compared to present emissions. And a private analysis for California has detailed potential savings of 26% from projected levels in 2000 and of 54% in 2010, both at negative net cost (Calwell et al. 1990).

Farming and Forestry

As noted earlier, nearly all greenhouse gases not related to energy use or CFCs arise from unsustainable farming and forestry practices. These emissions include ~46% of anthropogenic CH_4, from livestock-gut fermentation and rice-paddies, rising to 68% if biomass burning[43] is included; ~57% of CO,

---

[43] Much biomass is burned, often unsustainably, for fuelwood, slash-and-burn shifting cultivation, disposing of crop residues, and clearing forests to extend cultivation. The last three of these terms probably releases ~1.4–2.9 GtC/y, *vs.* ~1.6 from burning wood and dung for fuel (Krause et al. 1990, p. I.3–15n8). Burning inevitably emits *non*-CO_2 trace gases whether the carbon is recycled or not—i.e., independently of whether the carbon release is compensated

from forest clearing, chiefly for agriculture (Newell et al. 1989) and fuelwood; ~52% of anthropogenic NO_x and ~15% of anthropogenic N_2O, from biomass burning; a further ~33% and ~18% of anthropogenic N_2O from cultivating and fertilizing natural soils[44], respectively; and about a fourth of all CO_2, from deforestation, desertification, and simplification of terrestrial ecosystems including farmland (Krause et al. 1990, Ch. 3).

Ecological simplification in its myriad forms is less visible, but no less important to global warming, than clearcutting American forests or burning Brazilian forests. Just the loss of *above*-ground biomass and diversity, assuming no loss of soil carbon, means that replacing an old-growth Pacific Northwest forest with a young one reduces its total carbon inventory by two- to threefold (Harmon et al. 1990).[45] But typically at least as much terrestrial biomass is belowground as aboveground—and in temperate farmland, some *20–30 times* as much (Krause et al. 1990, p. I.3–32[46]). This invisible but enormous carbon stock, typically upwards of 100 metric tons of carbon per hectare (tC/ha), is at risk of mobilization into the air if insensitive practices defeat living systems' carbon fixation. In essence, turning (for example) prairie into corn and beans, and substituting synthetic for natural nutrient cycles, puts a huge standing biomass of soil bacteria, fungi, and other biota out of work. They then tend to lose interest, die, oxidize or rot, and return their carbon to the air.

At the same time, soil erosion, still endemic throughout most farmlands, transports soil organisms and other soil organic constituents ("finely pulverized young coal") into riverbeds and deltas, where they decay into CH_4—a greenhouse gas many times as potent as CO_2 (Houghton et al. 1990). Reduced soil fertility via erosion, biotic simplification, compaction, or poisoning requires ever greater inputs of agrichemicals, notably nitrogen fertilizers, whose production consumes ~2% of industrial energy (Ross and Steinmeyer 1990) and whose use drives N_2O emissions. Other well-known agricultural problems include:

• burgeoning pest resistance—the world loses more of its crops to pests

---

by re- or afforestation or by other biotic carbon sinks. How the carbon is harvested will of course affect the ecosystem's ability to sustain such compensation (*id.*, p. I.3–15).

[44] Forcing the nitrogen cycle boosts the yield from side-reactions whereby denitrifying bacteria in the soil produce N_2O from nitrate and nitrifying bacteria produce N_2O from ammonium. Cultivation also appears to increase microbial N_2O emissions even without fertilizer, and nitrate runoff into surface- and groundwaters appears to result in increased N_2O emissions (Krause et al. 1990, p. I.3–20).

[45] Hence Oregon has estimated that cutting of old-growth forest is responsible for ~17% of the state's total carbon emissions (Oregon Department of Energy 1990).

[46] Contrary to the conservative assumption made by Harmon et al. (1990), the data cited by Krause et al. (1990) do show a 10% soil-carbon loss when the natural forest becomes managed.

now than before the pesticide revolution—and health problems, especially among fieldworkers;
- rapidly growing OECD demand for food free of chemical contamination (Wall St. J. 1989)[47];
- crops' narrowing genetic base as diverse native stocks are inexorably lost to habitat destruction and seed-bank neglect;
- problems of water quality and quantity;
- many farmers' marginal profitability as their revenues immediately flow back to input suppliers; and
- the distressing spectacle of simultaneous food surpluses and famines.

These trends all confirm the need for "a major overhaul of current agricultural production methods" (Krause et al. 1990, p. I.3–14). Such an overhaul would be essential to achieving *adequate, acceptable, and sustainable* food and fiber output even if global warming were not of concern. Achieving it will be difficult because of the rapid loss of rural culture and traditional ecological knowledge as farmers vanish into cities (Jackson 1980): every year's delay loses more of those irreplaceable human resources.

Livestock

Just as saving electricity reduces CO_2 emissions disproportionately by displacing severalfold or manyfold more fuel, so affecting the numbers and rearing of livestock—which convert ~3–20+ units of grain to one unit of meat—can disproportionately help to protect existing forest-, farm-, and rangeland while reducing emissions of CO_2, N_2O, and CH_4. Such high-priority actions include (Krause et al. 1990; EPA 1989):

- reducing OECD dairy output to match demand[48];
- desubsidizing livestock production, especially for cattle, which emit ~72% of all livestock CH_4 (Crutzen et al. 1986): many dairy and beef cattle would not be grown without large subsidies, especially in OECD (Soden 1988);
- reforming beef grading and distribution, particularly in the U.S., to reduce the inefficient conversion of costly, topsoil-intensive grains to discarded fat (Browning 1990);

---

[47] For example, a 1986 National Institutes of Health study found that *every* U.S.-registered fungicide is a known carcinogen (E.N. Davies, pers. comm.). A ballot initiative mandating major reductions in the use of the more dangerous pesticides is pending in California, and the Dutch Parliament is shortly expected to pass a law requiring 25% reductions by 1993 and 50% by 2000 (*id.*).
[48] Dairy cows produce extra methane because they are fed about three times maintenance level (Krause et al. 1990, p. I.3–16).

- regulating or taxing methane emissions from manure so as to encourage its conversion to biogas for useful combustion;
- improving livestock breeding, especially in developing countries, to increase meat or milk output per animal, consistent with other important qualities and with humane practices;
- shifting meat consumption to less feed- and methane-intensive animals[49] and to aquaculture (preferably integrated with agriculture, a highly flexible and productive approach that may also help cut rice-paddy CH_4); and
- developing, if possible, alternative feed, fodder, and rumen flora that minimize CH_4 output[50].

Many of these livestock options would have important side-benefits. For example, many OECD cattle herds are fed, at conversion ratios of 8:1 or worse, with grain from developing countries. The western European herd consumes two-thirds of the domstic grain crop, yet still imports >40% of its grain from developing countries (Krause et al. 1990, p. I.3–19). OECD consumption of this inefficient form and large amount of feedlot beef is thus "directly related to starvation in the poor countries of the world." If OECD countries replaced part of their feedlot beef consumption with range beef and lamb, white meats, aquaculture, marine fish, or vegetable proteins, then Central America would feel less pressure to convert rainforest to pasture. Many developing countries would free up arable land. There could be less displacement of the rural poor onto marginal land, and renewed emphasis on traditional food crops rather than on export cash crops. Above all, this one action could save enough grain, if properly distributed, to feed the world's half-billion hungry people (*id.*).

Low-input Sustainable Agriculture

Organic farming techniques that are already rapidly spreading in OECD for economic, health, and environmental reasons (Wall St. J. 1989) can simultaneously reduce biotic CO_2, N_2O, and CH_4 emissions directly from farmland, and indeed can *reverse* the CO_2 emissions. These techniques can

---

[49] For example, shifting half of beef consumption to pork and poultry would maintain dairy output and total meat consumption while reducing methane emissions by ~40%—about twice the stabilizing CH_4 reduction (Krause et al. 1990, p. I.3–18). In OECD, the market is already shifting in this way, largely because of health concerns. Ultralean, organic range beef (which grazes only on natural grasslands and is not grain-fed), which alleviates those concerns and can cost less, may also produce less methane than equivalent feedlot beef (EPA 1989, p. VII–270).
[50] EPA is encouraged about this option and believes that the resulting productivity increases would often yield a significant net profit: as J.S. Hoffman of EPA put it, "This creates a very economic picture for methane [abatement]" (Stevens 1990). (1990a) apparently concurs. Validating field experiments are now underway.

and often do use standard farm machinery, but require it less often[51], and can work well on any scale. They substitute natural for synthetic nutrients (e.g., legumes for synthetic nitrogen), mulches and cover crops for bare ground, natural predators and rotations in a polyculture for biocides in a monoculture, and nature's wisdom for humans' cleverness. They integrate livestock with crops, and garden and tree crops (*infra*) with field crops. They maintain often tens and sometimes hundreds of cultivars instead of just one or a few. In Asia, they draw on a particularly rich tradition of integrating many kinds of production—vegetables, fish, rice, pigs, ducks, etc.—in a sophisticated quasi-ecosystem that efficiently recycles its own nutrients.

Green Revolution seeds and artificial fertilizers are often assumed to be essential to grow enough food in land-short developing countries. Yet diverse African field studies have demonstrated that "ecoagriculture," which substitutes good husbandry and local seed for these purchased inputs, yields nearly as much maize, sorghum, etc. in the short term. The small yield difference probably narrows the time "[i]n view of the accelerated degradation of soils that usually accompanies chemical agriculture." Such results "suggest that regenerative farming could be greatly expanded both in industrialized and developing countries without negative consequences for the goal of increasing Third World agricultural yields. On the contrary, without [such] a conversion . . . the loss of arable land, notably in the tropics, threatens to accelerate out of control. . ." (Krause et al. 1990, pp. I.3–23 & 24).

Whether in OECD or developing countries, ordinary organic farming practices modelled on complex ecosystems generally produce comparable or slightly lower yields than chemical farming but at much lower costs. They therefore produce *comparable or higher farm profits* (NRC 1989)—without counting the considerable premium many buyers are willing to pay for food free of unwelcome biocide, hormone, and antibiotic residues (*Wall St. J.* 1989). The organic practices' economic advantage is clear in large commercial operations over a wide range of crops, climates, and soil types (NRC 1989). That advantage tends to increase at family-farm scale, which brings further social benefits (Jackson et al. 1984). Similar economic benefits have been found in many hundreds of diverse U.S. and West German farms (Brody 1985; Bechmann 1987; Bossell et al. 1986).

---

[51] In California (CEC 1990), ~3% of energy is used directly in agriculture. Of the one-third of that used for irrigation, ~40% can readily be saved through simple and highly cost-effective water-efficiency measures—thereby saving electricity (coal-fired on the margin), since pumping water is the largest single use of electricity in the state. The further ~22% of agricultural energy use for synthetic pesticides and fertilizers would be virtually eliminated by organic techniques, while the ~24% for traction could be cut in half (*id.*) through the reduced need for field operations. In all, therefore, organic farming would save nearly half of California's agricultural energy use (Calwell et al. 1990, p. vi).

Little is yet known about CH_4 and N_2O cycles, so it is only a plausible hypothesis, not yet a certainty, that the reduced tillage and fertilization that accompany profitable organic farming, together with reductions in the burning of biomass and fossil fuels, will suffice to eliminate most of the ~35% of total N_2O that is released by human activity. Yet even if N_2O reductions from organic fertilizers turned out to be less than hoped, the CO_2 benefits would still be large, because CO_2 can be absorbed by building up organic matter in soil humus through the gradual accumulation of a richly diverse soil biota. Today, in both OECD and developing countries, and reportedly in eastern Europe and the USSR too, soil loss, and especially the physical loss or biological impoverishment (hence carbon depletion) of humus, is far outpacing soil and humus formation and enrichment. But successful conversions to organic practices, chiefly in the U.S. and West Germany, have demonstrated that after a few year's reequilibration, these carbon losses can be not only eliminated but reversed.

Not just forests, then, but also farmland can be changed from carbon sources to carbon sinks. For example, an ordinarily impoverished soil in the U.S. cornbelt could plausibly start at 2% organic content of ~1% C. A decade's organic practices—corn/alfalfa or /clover rotations, manure and green-manure use, and integrated pest management—could raise the organic content by a conservative 0.02%/y[52], thereby adding ~4.8 tC/ha over the decade (Holmberg 1988). Doing this on the United States' 50 million hectares (MHa) of farmland would offset the annual combustion of ~10% of current U.S. gasoline use per year. Thus a very efficient U.S. car fleet (~5–7 times as efficient as now), getting a substantial part of its fuel from sustainably grown biomass, would emit only as much fossil carbon as the farmland would reabsorb into soil humus (*id.*). Based on organic-farming comparisons by the National Research Council (1990) and others (e.g., McKinney 1987), this carbon sink could be achieved at zero or negative net internal cost. Specifically, each hectare of sustainably grown corn or other fuel feedstock could, for example, produce 600 gal of anhydrous ethanol,

---

[52] For example, Holmberg (1988) cites Herman Warsaw, a successful (370 bu/acre in 1985) organic corn farmer who in the past 30 y has increased organic content of his soil from ~$3\frac{1}{2}$% to ~8% in the top 3" and from 1% to 3% at 1' depth. The average increase in the top foot of topsoil was ~4% (~2% in carbon terms). He believes he has now learned how to achieve the same improvement in five years. We assume half that speed. This is probably quite conservative: for example, Holmberg also cites Steve Pavich's 1% carbon gain profitably achieved in 12 y and probably reproducible in half the time in dry Arizona and California soils, and USDA/Beltsville test plots' achievement of 0.2–0.6%/y carbon gains through light (40 t/ha-y) applications of compost and manure. The carbon uptake assumed here is 45× slower than Beltsville's two-year achievement at a 160 t/ha-y manure application rate.

fix enough soil carbon to offset the combustion of 200 gal of gasoline (Holmberg 1988)[53], and increase the farmer's profits.

There are also many techniques for substantially reducing the use of nitrogen fertilizer (Krause et al. 1990, p. I.3–20) within the context of predominant OECD farming practice or of lower-input but still not truly organic modifications. Most of these techniques are cost-effective, too, because they reduce chemical and application costs and nitrate-runoff pollution without cutting yields. In many developing countries, too, additional measures to reduce CH_4 emissions are available and desirable: e.g., biogas-digester preconditioning of rice-paddy fertilizer, improved dryland rice options, and reducing in-paddy anerobic fermentation of rice residues.[54] Substantial reductions in N_2O and CH_4 releases can undoubtedly be obtained by the simpler of these management techniques at costs on the order of $3–30 per ton of carbon-in-CO_2-equivalent. On closer examination, side-benefits, such as saving fertilizer and reducing runoff through more precise application, may well turn out to pay for some important abatement measures.

Sustainable Forestry

The needs and opportunities in forestry are strikingly analogous to those in farming. The two are directly linked, chiefly by ways to reduce agriculture pressure on forests and by opportunities for agroforestry—applying agricultural traditions to tree crops. The former options include (EPA 1989; Krause et al. 1990):

- reducing the area and increasing the fallow period of slash-and-burn to sustainable levels;
- replacing slash-and-burn with sustainable techniques (often proven by indigenous cultures), including agroforestry using native or cultivated tree-crops or both;
- using trees felled during land-clearing for timber and biofuel;
- using crop wastes not as direct fuel but rather for composing, mulch, efficiently burned low-leakage biogas, or gasification/steam-injected-gas-turbine cogeneration (Larson et al. 1989);
- controlling artificial burns;
- adopting low-input/organic farming and forestry techniques (the former because they reduce pressure for land-clearing);

---

[53] This is not to say that corn is necessarily the best feedstock nor ethanol the best biofuel, but the conclusion holds for other examples too.
[54] Perhaps by frequent pond-switching between rice and aquaculture, a phenomenally productive traditional South Asian technique in which fish graze the rice stubble, or by using the rice straw more widely for roofing or as a biogas feedstock.

- improving developing countries' farm productivity in order to reduce land needs per person; and
- reducing feedlot if not total beef consumption (*supra*).

Agroforestry (Leach and Mearns 1988) is especially applicable to developing countries. Projects in many African conditions have demonstrated profitable 40–90% crop-yield gains while providing surplus woodfuel (*id.*). Such practices can also permit the beneficial substitution of organic for artificial fertilizer and other agrichemicals (Krause et al. 1990, p. I.3–23ff). Some tree crops can yield not only wood and food but also oils, resins, or terpenes (*id.*, p. I.3–41) that are directly usable as fuel, especially in diesel tractors and pumpsets. The oils can be combined with dirty, wet alcohols in a simple solar catalytic reactor to yield superior diesel fuels such as methyl and ethyl esters.

Nonagricultural ways to relieve pressure on existing forests (Krause et al. 1990, p. I.3–40) involve energy and materials policy: e.g., recycling forest products[55] and "stretching" their effect (as by honeycomb structures, *infra*), substituting electric efficiency for tropical-forest hydroelectric projects, designing frame structures for minimum timber waste, improving the protection and hence the lifetime of outdoor structural wood, and wringing far more work from biofuels (e.g., Baldwin et al. 1985; Larson et al. 1989; Goldemberg et al. 1988).

Additional forest-protection measures include taking better care of existing forests, harvesting more thoughtfully, shelterbelts, livestock exclusion (e.g., with photovoltaic-powered electric fences), regenerating degraded forests, and promoting recreation and ecotourism to create a supportive constituency. These actions should supplement silvi/agri/aquacultural integration, reforesting surplus OECD farmland and degraded drylands, planting fuelwood around developing-country cities and along roads, and urban forestry. Together, such forestry practices could probably sequester a maximum of ~1.3 billion tons of carbon per year (GtC/y)—enough, with the concomitant stabilization of existing carbon pools, to offset one-fourth of worldwide 1985 fossil carbon releases (Krause et al. 1990, p. I.3–49). This broader menu of options yields nearly twice the 0.7 GtC/y net sink calculated by EPA (1990, p. VII–7). Much of the difference is practices like agroforestry that generally exhibit negative costs.

Some of the most promising and profitable new forests, too, can be in cities. Urban tree-planting programs are an especially cheap carbon sink

---

[55] "The average OECD person consumes about as much wood in the form of paper as the average Third World person consumes in the form of fuelwood" (Krause et al. 1990, p. I.3–43). Producing a ton of paper takes ~3/4 ton oil-equivalent of energy (Herman et al. 1989). If that energy were biomass, as the majority of it is in the U.S. forest products industry, it could still be used instead to make liquid biofuels for sale to replace oil.

because one tree planted in a typical U.S. city sequesters or avoids ~*10–14 times* as much carbon release as if it were planted in a forest where it could not also save space-cooling energy (Akbari et al. 1988). Urban forests and woodbelts can go far to relieve developing countries' fuelwood shortages and urban sprawl. The biomass produced by urban trees, even if not systematically and densely planted, can be substantial: Los Angeles County alone sends ~3,600–7,300 t/d of pure, separated tree material to landfills, not counting mixed truckloads. That ~1 GW of currently wasted (and costly-to-dispose-of) thermal energy is equivalent, at 70% conversion efficiency, to 0.5 million gal/d of gasoline—enough to drive a 60-mi/gal car >10 mi/d for every household in the County.

Urban forestry is also consistent with urban agriculture—long practiced in western Europe and in China, where it provides 85+% of urban vegetables (100% in Beijing and Shanghai), plus large amounts of meat and treecrops (Wade 1981). Urban farming in turn further reduces greenhouse-gas emission from centralized agriculture, saves energy otherwise needed to process and transport food, and improves nutrition, esthetics, community structure, and urban culture. Even the most crowded cities can farm on rooftops. With superwindows and air-to-air heat exchangers, climate is no obstacle: Rocky Mountain Institute's 99+%-passive-solar headquarters grows bananas with no heating system despite outdoor temperatures as low as $-44°C$ ($-47°F$).[56]

In considering these supplements to conventional forestry, however, it could be overly sanguine to suppose that forestry in its present form will be able to sustain its vital carbon-sequestering role. Most discussions of CO_2 abatement through fiber-production forestry emphasize *planting trees*, often of specific, fast-growing, genetically engineered kinds. But planting trees is very different—often by severalfold in carbon inventory (Harmon et al. 1990)—from maintaining a diverse forest that changes at its own pace. Much "modern" forestry repeats the ecological errors of monocultural, chemical-driven agriculture, treating trees like rows of giant corn—short-rotation (annual) monoculture instead of long-rotation (perennial) polyculture (Jackson et al. 1984). This is as true with *Pinus radiata* in New Zealand or southern pine in the U.S. as with eucalyptus in developing countries, where >40% of all new hardwood plantings are now eucalypts (Krause et al. 1990, p. I.3–46).

Much modern forestry rests on mechanistic assumptions that appear from historic evidence to be ecologically unsound and unsustainable (Plochmann 1968; Cramer 1984; Maser 1988). Clearly what is needed is to sustain and

---

[56] The 373-m^2 building, at 2,165 m in a 4,900 C°-d/y climate, also saves half of normal water use, 99–100% of water-heating energy, and 90+% of household electricity (reducing the lights-and-applicances bill to $5 a month @ 7¢/kWh). Its marginal cost for all these savings, $16/$m^2$, paid back in 10 months with 1983 technology, and would pay back faster today, mainly because windows can now insulate twice as well (<0.5 W/m^2K or >R-11) at nearly the same cost.

increase both the quantity *and* the ecological quality (diversity, health, cycle time, resilience) of existing and new forests. This will require forest managers—just like farmers—to think like ecologists, not accountants. Currently, forestry economics is being questioned chiefly on an accounting basis (the U.S. Forest Service, for example, is the world's largest socialized roadbuilder), not on fundamental grounds. But new questions are emerging, and new answers will follow. In this decade, researchers may discover whether in forestry, as in farming, ecologically sound practices are also the most profitable kind, and whether, as some forest scientists are starting to suspect (Maser 1988), a forest is worth more as a going (growing) concern than in liquidation.

Unfortunately, the kinds of re- and afforestation that might prove not to be very productive or sustainable in the long run are the kinds currently considered in global-warming economic analyses. By normal forestry-economics standards, the fast-rotation plantings now dominating forestry practice appear profitable: that is why they are so widely practiced. Even massive tree-planting programs have been found in several analyses, including a U.S. Forest Service/ex-Council of Economic Advisors report to USEPA, to sequester carbon at low costs, typically on the order of \$10/tC (ICF 1990, p. 39), despite taking no credit for the discounted revenues from ultimate timber harvest, nor for other benefits meanwhile.[57] Unpublished cost estimates circulating within a National Academy of Science study (discussed below) are reportedly in a similar range. A study for Pacific Power suggests that timber revenues could in fact repay the cost of the reforestation about twice over (Reichmuth and Robison 1989); if this proves true, as the apparent profitability of current forestry activities suggests, then these canonical ~\$10/tC abatement costs should be negative.

Consistent with this, EPA (1990, p. VII-7) considers reforestation "one of the most cost-effective technical options for reducing CO_2 and other gases." To the extent, at least, that greater forest growth involves agroforestry, urban forestry, and other ecologically sensitive techniques, the average cost of sequestering carbon in trees should be even lower.

CFC Substitution

The projected cost of the CFC/halon production phaseout now required by the London Amendments to the Montreal Protocol has declined by roughly half over the past two years through closer scrutiny and ingenious technological innovations. In some instances, including refrigeration with reoptimized design, the substitute may actually *improve* performance

---

[57] These include, e.g., erosion and flood control, groundwater recharge, runoff holdup (stretching reservoir capacity), fish and wildlife enhancement, esthetic value, and recreation.

(Shepard et al. 1990, p. 62). In others, notably the cleaning of printed-circuit boards with terpene derived from orange peels, the substitute works better and costs *less*. Although ~$135 billion worth of equipment in the U.S. alone uses CFCs, the need is expected to be met by a combination of ~29% efficiency/maintenance/recovery/recycling/reclamation, 32% substitution by not-in-kind replacements, and 39% "drop-in" (or nearly so) replacements (Manzer 1990).

According to USEPA analyses (EPA 1989a, ICF 1990), late-1989 data indicated that a U.S. phaseout by 2000 would cost ~$1.3 billion, or ~$2.4/tC–equivalent, at a 6%/y real discount rate. This figure is the sum of many disparate terms, all subject to technological change that tends to reduce costs. New technologies, like the advanced thermal insulation mentioned next, are reducing the total cost quickly enough that it may before long become slightly negative.

There is no obvious reason abatement should cost more in other countries, given access to OECD technologies: quite the contrary, since a third of global CFC use (Krause et al. 1990, p. I.3–3) is for aerosol propellants long ago cheaply displaced in the U.S. The U.S. accounts for ~29% of global CFC consumption (Turiel and Levine 1989). The global abatement cost appears, therefore, to be ~$4½ billion, and is continuing to decline with further technological development. Since CFCs in the 1980s accounted for ~24% of global warming (Houghton et al. 1990), $4½ bilion is equivalent to <$2 per ton of carbon-equivalent in CO_2.

Multigas Abatements

There are important opportunities to abate two or more greenhouse gas emissions simultaneously, both at a net profit. A few examples illustrate the diversity of such options:

- Five advanced classes of thermal insulation, along with other design refinements, can save 90+% of the electricity used by refrigerators and freezers (Shepard et al. 1990, pp. 44–60)—the biggest users of electricity in households lacking electric space and water heating, in countries as diverse as the U.S. and Brazil (Reddy and Goldemberg 1990). This saving alone can avoid burning roughly enough coal each year to fill up the refrigerator. These insulations also substitute a vacuum for the CFCs normally used to fill plastic foam; eliminate most of the refrigerant inventory (currently CFCs); and can accommodate the modest efficiency losses, if any, caused by switching to non-CFC refrigerants. Best of all, certain advanced insulations can make the appliance's walls thinner, because they insulate up to twelve times as well as CFC-filled plastic foam. The resulting increase in interior volume may be worth about enough to pay for the insulation (Shepard et al. 1990, pp. 59–60).

- Landfills emit ~30–70 Mt/y of uncontrolled methane, about a sixth of total methane from human activities (Krause et al. 1990, p. I.3–11). Capturing this gas and using it as fuel—ideally for cogeneration—both prevents its emission *and* displaces a larger amount of CO_2 otherwise emitted by a coal- or oil-fired power plant or boiler. ICF (1990, pp. 43–45) calculates that 75% recovery just from U.S. landfills holding >0.9 million tons of waste would burn ~55% of U.S. landfill methane at a cost roughly half the market value of the electricity. At least 123 U.S. landfills already recover methane as fuel (EPA 1989, p. VII–191), but far more do not yet. Capturing and burning coal-bed methane (*id.*) looks similarly profitable (or at least breakeven) and makes mines safer. Converting livestock or human manure into biogas, whether on a commercial or a village scale, and using the biogas to displace fossil fuel or unsustainably grown firewood (Goldemberg et al. 1988), also appears economic with modern technologies available at many scales. Interestingly, since ~20% of U.S. methane emissions can be captured just from landfill, coal, and natural-gas leakage, and there are apparently attractive agricultural abatement options too, there is a growing consensus that the ~15–20% CH_4 abatement needed to stabilize this gas's heat-trapping will prove costless or profitable.
- Recycling paper, or composting food and garden wastes, reduces landfill CH_4 output; saves the CO_2 and NO_x otherwise emitted when fuel is burned to produce and transport those materials; and saves money. Compost can also displace synthetic fertilizer, whose manufacture releases CO_2 and whose use released N_2O, and by improving tilth, can help the soil to retain water, saving irrigation pumping energy and hence CO_2. If part of a locally based agriculture, compost can help to substitute fresh food for perishable food refrigerated with CFCs and fossil-fueled power plants (~9 GW in the U.S.: Shepard et al. 1990, p. 115). Local agriculture can also avoid burning oil to transport food; the average molecule of American food has been estimated to travel ~1,200 miles before it's eaten.
- Native building materials such as adobe, caliche, mudbricks, etc. can displace CO_2-intensive production and transportation of cement. Sustainably grown timber or bamboo incorporated into buildings can also temporarily sequester carbon (Krause et al. 1990, p. I.3–14). Where timber is scarce, however, pressed-wood/paper-honeycomb materials can reduce tree use per building by up to ~90% at negative net cost, incidentally increasing the building's energy efficiency too (D. Hartwell, pers. comm.). Improved energy infrastructure, especially for more efficient use of fuelwood, can also divert large amounts of wood from fuel to fiber use (Leach and Mearns 1988; Goldemberg et al. 1988; Baldwin et al. 1985).

- Efficient motor vehicles can cost-effectively and simultaneously reduce emissions of CO_2, CO, O_3, N_2O, NO_x, SO_x, hydrocarbons, and other radiatively active gases or photochemical products such as peroxyacetyl nitrate (Krause et al. 1990). The reduced air pollution, especially O_3 (Newell et al. 1989), can also reduce forest death and other vegetative damage, maintaining more and healthier trees as carbon sinks. On a microdesign scale, more efficient car air conditions, or reduced cooling loads due to such improvements as lighter-colored paint or spectrally selective glass, can reduce the inventories and leakage of CFCs, save compressor operating energy (hence CO_2), and save CO_2 in all driving by transporting a smaller, lighter-weight compressor.
- Electrical savings that displace new hydroelectric dams can achieve their fuel-saving goal but preserve the carbon inventories in the impoundment area's above- and below-ground biota, rather than emitting them both as CO_2 when the area is cleared and, even worse, as CH_4 after flooding converts it to an anerobic swamp. Since the electric savings are cost-effective (cheaper than the dam or a thermal power station), both these abatements cost less than zero.

Though not thoroughly catalogued or characterized, such multigas abatements will generally reduce the total cost of abating global warming by providing multiple benefits for single expenditures. Omitting them from supply-curve analyses is thus a conservatism. It may well be a significant one.

Supply Curves for Abatement

In 1989, the Amsterdam office of McKinsey & Co. prepared for the Dutch government one of the first attempts at a supply curve for global-warming abatements (D. Six, pers. comm.). It was explicitly illustrative and incomplete, but heuristically valuable. Since then, an increasing body of ever more detailed and empirically grounded evidence has taught two lessons: that using supply curves to relate the marginal quantities and marginal costs of abatement is a useful way to gain understanding of policy options, and that closer scrutiny tends to raise the quantities and lower the costs. For example, the data from one such compilation (ICF 1990[58]) of diverse government and industry studies of the U.S. potential for abating

---

[58] Plotted from Tables 10–15; some utility-sector data in Table 14 differ from those in Appendix H, although their total differs little.

Least-Cost Climatic Stabilization 389

emissions of CO_2, nonbiotic CH_4, and CFCs in 2010, are plotted in Fig. 14.3.

This curve's basic structure is arrestingly simple: a long flat section in the middle at roughly zero cost (the cheap CFC abatements, landfill and coalbed methane capture, and reforestation), plus "tails" of essentially equal area at both ends. These "tails" comprise chiefly ~25% energy savings at negative net cost on the left, and the costlier kinds of renewable energy and industrial fuel-switching at positive cost on the right (i.e, costing respectively less and more than competing fossil fuels). The calculated potential reduction from this quite incomplete list of measures[59] totals 1.72 billion metric tons of carbon per year (GtC/y), or ~52% of the base-case emissions projected for that year. This large abatement is significant for two reasons:

- some major options, such as sustainable farming, forestry other than standard reforestation, most industrial and heavy-transport savings, and other trace gases were not counted; and
- the net private internal cost of halving the U.S. contribution to global warming is roughly zero.

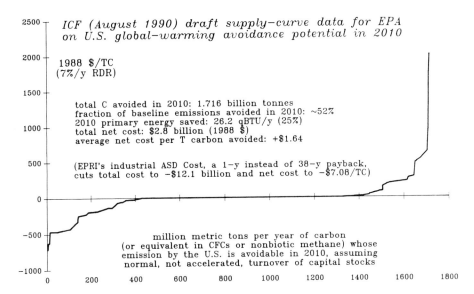

Figure 14.3

---

[59] As a small example, only half of the industrial motor-system retrofit potential was considered, and that half—all from adjustable-speed drives (ASDs)—was assigned a cost ~38 times EPRI's value (Fickett et al. 1990) or ~15× RMI's (Lovins et al. 1989). Correcting this apparent error changes the net cost of abating a ton of carbon-equivalent emission from +$1.6 to −$7.

This approximately symmetrical and costless arrangement appears to be qualitatively consistent with the findings of the Mitigation Subpanel of a National Academy of Sciences study currently nearing completion.[60] The Subpanel has reportedly analyzed, on conservative assumptions, a much wider range of technical measures (some of which the Subpanel felt might entail modest lifestyle changes). The Subpanel therefore found a larger abatement potential. The measures on which the Subpanel has reached consensus could collectively abate a reported ~85% of 1990 U.S. contributions to global warming[61] at an average net cost ~$6/tC—again, quite affordable. The net cost was also reportedly calculated to equal zero at a global-warming abatement of ~71%—typical of levels considered likely to be needed to stabilize the climate.

The point is not whether either of these sets of figures is exactly right— both inevitably reflect many uncertainties—but rather that sound public policy requires an open process to identify all the parts of such supply curves and harness the public's imagination and ingenuity in refining and achieving each. It should be possible to reach near-agreement about the numbers, or at least to understand the origins of residual disagreements.

As a small initial contribution to that goal, how might the additional opportunities described earlier in this paper change the conclusions of the ICF or NAS analyses? Qualitatively, these extra options would clearly have two effects: reducing global warming even more, and converting those studies' low net cost to a substantially negative net cost, whether for the United States or worldwide. The rough magnitude of some of these changes can be estimated as follows:

- The Academy reportedly assumed industrial electricity-saving potential equivalent to one-third of the savings that RMI has demonstrated (Lovins et al. 1989) and EPRI concurred with (Fickett et al. 1990) for motor systems *alone*. The RMI/EPRI-agreed cost of such savings is also an order of magnitude lower. Ignoring the substantial non-motor electricity-

---

[60] The Panel on Policy Implications of Greenhouse Warming of the Academy's Committee on Science, Engineering, and Public Policy, whose report to be published around the end of 1990. The Chairman is Governor Dan Evans, a distinguished engineer and former Chair of the Northwest Power Planning Council. It appears that the report will be a valuable contribution to public policy formation. Its main shortcomings are likely to include outdated energy-efficiency potential, inadequate attention to less conventional forestry options and to organic agriculture, omission of multigas abatements, and probably an excessively restrictive and short-term view of renewable energy potential (SERI 1990). Improving the analysis in any of these respects should raise the calculated abatement and lower its cost (*infra*).
[61] Assuming these were equivalent to 2.24 GtC/y in 1988 (ICF 1990, p. 2, based on Houghton et al. 1990), adding a further 2% as a surrogate for 1988–90 growth in emissions. Of course, the fractional abatement would be smaller—perhaps around half, consistent with the ICF data in Figure Three—compared to the presumably larger emissions by the time the abatement were actually implemented.

saving possibilities, just substituting the RMI/EPRI motor findings would therefore triple the saving and cut its cost (i.e., increase its net present-valued financial saving) by about $60 billion. Considering industrial fuel savings more fully, and adding industrial energy savings from the leaner materials flows described earlier, would also yield major gains. So would substituting the gas saved in industrial and building heat for more carbon-intensive fuels.

- EPRI's assumed potential for electricity savings in buildings was presented to the NAS panel as 45% at an average cost of 2.5¢/kWh. Yet detailed, empirically based, and undisputed retrofit analyses for Arkansas, a slightly more difficult climate than the U.S. average, found a retrofit potential to save 77% in houses at 1.6¢/kWh and 74% in commercial buildings at −0.3¢/kWh (Lovins 1989). More recent developments (Shepard et al. 1990) would support even more favorable results, but just the Arkansas results would raise the EPRI/NAS building-electricity carbon saving by two-thirds and nearly double its net financial saving.
- The 50% fuel saving assumed in buildings is smaller and costlier than the potential found by a major Federal study a decade ago (SERI 1981), and improvements in the building stock meanwhile do not seem to account fully for the difference. Such key developments as superwindows, which have captured a large share of the U.S. insulated-glass market in the past few years, were not considered. Interestingly, the Arkansas house-retrofit analysis just cited found that gas savings of 60% would result at *no* extra cost as a free byproduct of the 77% electric savings—or far more, at modest cost, if the gas appliances were also made more efficient. In an era when skilled practitioners can retrofit superinsulation that virtually eliminates space-heating even in cold climates, and retrofit savings of two-thirds of water-heating fuel are straightforward, a 50% fuel-saving potential is clearly outdated.

A simple thought-experiment illustrates the importance of giving the energy-efficiency potential the most searching and up-to-date possible scrutiny. In 1989, Americans paid $453 billion at retail for commercial fuels and power (Rosenfeld et al. 1990, 1989 $). The U.S. energy system is quite competitive in most respects.[64] Competitively providing the world's commercial energy would probably cost, shorn of taxes and subsidies, not far from the same amount scaled up for consumption, or on the order of one and two-thirds trillion dollars per year. (We use this very rough estimate

---

[64] Direct Federal subsidies reduced the apparent total energy bill by ~10% in FY1984 (Heede et al. 1985), but became smaller in 1986, so the overall distortion is probably lower now, though it may be more unevenly allocated between competing options.

of an internal shadow price because actually quoted prices often do not reflect actual costs.)

Now recall that just the direct electricity and oil end-use efficiency potentials described earlier would save upwards of three-fourths of that energy, and at typical late-1980s energy prices, would repay their cost in a few years. The savings would be slightly smaller and costlier in the most efficient OECD countries, but substantially bigger and cheaper in other countries. The U.S. potential would thus be a reasonable-to-conservative global average. The present-valued cost of such large savings was shown above to be on the order of a tenth of present energy prices. Subtracting that tenth from the gross savings leaves a potential long-run net saving of money[65] of at least $1 trillion a year from fully implementing the efficiency opportunities described.

That is about as big as the global military budget. The money now wasted on inefficiently used energy is certainly needed for more productive purposes than wasting energy. But more importantly for the purposes of this paper, the negative-cost energy savings that yield this ~$1-trillion-a-year of net wealth creation represent probably the cheapest increment of global-warming abatement. Cheap energy efficiency can reduce global warming by about a third—*and free about a trillion saved energy dollars per year to pay for other abatements, with money left over*. Recent official assessments show abatements in the ~50–70% range at roughly zero net cost despite assuming energy savings several times smaller and costlier than those demonstrated above. But those studies slighted the modern energy-efficiency potential (and some other opportunities too). Properly counting that potential is thus bound to abate even more global warming, and to change the total net cost of abating global warming from about zero to a value robustly less than zero.

It is also important to note that a trillion saved dollars per year is so much money that it can buy a great deal of additional abatement even at relatively high unit costs. For example, many of the costlier ways to achieve the last pieces of abatement required for climatic stabilization might cost (say) $165 per ton of carbon. If so, then the net money saved by the energy efficiency could buy six billion tons of carbon abatement per year—about equivalent to today's entire global output of CO_2 from all combustion processes. Since most of that fossil-fuel CO_2 would already have been abated by the energy efficiency itself, and since most other kinds of abatement (biotic carbon, CFCs, and most CH_4 and N_2O) appear, as noted above, to be relatively inexpensive, it is not easy to construct an optimized menu of abatements that could stabilize the climate *without* having a considerable

---

[65] Long-run because the sunk capital costs of the energy system cannot be saved in the short run, but re-incurring those costs—rolling over retiring capacity into replacement capacity—can still be avoided by achieving durable, reliable, long-term efficiency gains.

amount of money left over. In other words, although the amount of money that energy efficiency can save is not precisely known, it is clearly large enough to ensure a financial surplus despite the considerable uncertainties about what non-energy abatements may cost.

Opportunity Cost Requires "Best Buys First"

There is, however, one caveat requiring emphasis. To abate global warming promptly with finite resources, *it is vital to choose the best buys first*. This is because of "opportunity cost"—the impossibility of using the same money to buy two different things at the same time.

If, for example, you spend a dollar on a costly source of electricity, such as nuclear power or photovoltaics, then you'll get relatively little electricity for that dollar—that's what "expensive" means. You'll therefore be able by such means to displace little coal-burning in power stations. But if you use the dollar to buy a very cheap option instead, such as superefficient lights or motors, then the resulting bounty of electricity could displace a lot of coal. Therefore, whenever you spend the dollar on a costly option *instead of* on a cheap one, you'll unnecessarily release into the air the extra carbon that would not have been released had you bought the cheapest option first. That is why, for example, nuclear power makes global warming worse:[66] it emits less carbon per dollar than coal-fired plants, but many times more than efficiency (Keepin and Kats 1988 and 1988a; Lovins 1989c).

This is not an academic point; it is at the crux of essential policy choices. Investors must understand which options are cheapest, hence most profitable, and policymakers must avoid or amend regulatory structures that divorce these two attributes. Even in societies where capital is allocated by planning rather than by markets, the planners must be able and eager to determine the best buys and then buy them. *Any other sequence of investments prolongs and enlarges climatic risk.* We therefore turn next to how the least-cost

---

[66] Algebraically, if K is the carbon intensity of existing coal-fired plants (in tC/kWh), CC_n is the levelized cost of marginal nuclear kWh, and C_e is the levelized cost of a marginal saved kWh (both in $/kWh), then $K([C_n/C_e] - 1)$ tons of extra carbon are released per kWh made in a new nuclear plant instead of saved by improved end-use efficiency. This assumes simple fungibility of dollars between the two investments. Recent experience of the U.S. utility industry, however, suggest that matters are actually worse: when utilities overinvested in capacity by ~$200 billion, many, seeking to recover sunk costs, turned their efficiency departments into surplus-power marketers, making power-plant dollars not just a neutral complement to but a direct enemy of efficiency dollars. Thus EPRI estimated a few years ago that today's "strategic marketing" programs will directly result, by 2000, in some 36 GW of new *on*peak demand—about two-thirds of the savings projected from the industry's efficiency programs. Whenever nuclear investments, therefore, mean not only foregoing an efficiency investment but deliberately seeking to boost electric demand, more carbon will be released than C_n/C_e would indicate.

investment sequence can actually be discovered, bought, and successfully marketed and delivered in diverse societies.

IMPLEMENTATION TECHNIQUES

The foregoing discussion has highlighted both commonalities and differences between the technical options available to the world's three main regions. In all regions, for example, energy efficiency is an urgent priority, though for somewhat different reasons: for example, high energy efficiency in industrialized countries is vital to global climatic protection and is economically valuable for those countries, while high efficiency in the other two regions is currently less important for the world but economically vital to their own development (which would in turn raise their CO_2 output to the majority of the global total). Moreover, non-OECD countries now building infrastructure and stocks of consumer goods have a chance to build in efficiency from scratch, often as a natural and inexpensive feature of investments they are making anyhow, whereas OECD faces the daunting task of retrofitting trillions of dollars' worth of obsolete buildings and equipment.

Most instructive, however, are the differences in implementation strategy between the regions. Many of these have been extensively treated in a huge literature, so we seek here only to summarize main points that may previously have been overlooked or underrated.

OECD Countries

The OECD countries emitted nearly three-fifths of 1950–86 fossil CO_2 (Krause et al. 1990, p. I.1–9) and cause the lion's share of global warming today. They have such large economies largely because they have vigorous markets[67] with high innovation rates and rapid discovery, application, learning, and corrective feedback in most sectors. Most OECD countries also have a powerful public consensus for environmental protection. That is why, for example, government policy is committed to major CO_2 reductions in West Germany, Denmark, and the Australian State of

---

[67] Centrally planned exceptions, such as the French electric sector, are likely to prove short-lived when 1992 European economic integration unleashes new competitive pressures and transparent-pricing requirements. Increasing North American integration under Canadian and (soon) Mexican free-trade agreements will similarly hone competition. Some economies, such as Sweden's and Japan's, and in some respects the EEC itself, are more properly regarded as mixed, with an overlay of sophisticated public-sector planning and coordinating apparatus. Yet the vitality of the market sector, the diversity of the public institutions within OECD, and the force of market and political accountability in most OECD countries have permitted as impressive a range of initiatives to flower in the mixed as in the most *laissez-faire* economies.

Victoria[68], to CO_2 stabilization (variously defined) in Britain, Sweden, Canada, and Japan, and to stabilization followed by reduction in The Netherlands and Norway.[69]

This region is ideal for adopting and adapting the best worldwide experience of market mechanisms for capturing profitable global-warming abatements. Favorable conditions include widespread acceptance of the polluter-pays principle[70]; availability of generally sound statistical data and reporting systems; widespread sophistication in industrial organization to meet financial objectives; a large population of skilled entrepreneurs; and the world's highest mobility of labor, capital, and (most importantly) information.

Although most western European "Greens" traditionally distrust markets and the private sector, there are clear signs that many are starting to appreciate the power of properly structured and informed markets[71], hence the importance of *making markets in avoided depletion and abated pollution* (Lovins 1989). The basic concept is simple: economic *glasnost*—prices that tell the truth—can scarcely achieve efficient behavior without a market where the buyers and sellers of technological solutions can meet and do business. My potential loss from a carbon tax must be convertible to your potential profit from selling me a more efficient lamp. If I can't buy that lamp, my only response is a behavioral change (using a dimmer lamp or using it less often), which is relatively weak and impermanent.

Ways to make markets in saved energy and water, developed at Rocky Mountain Institute and elsewhere, are now entering widespread use both in the United States and abroad. Based on successful early experience, they show promise of accelerating energy efficiency and appear applicable also to other key ways to abate global warming, as described next.

---

[68] The German proposal, whose implementation is being worked out, is to reduce CO_2 emissions by 25% of 1987 levels by 2005. A stricter proposal, with very large long-term reduction targets, is reportedly to be recommended by the Enquête-Kommission of the Bundestag, which has produced two excellent reports on the subject. Victoria has also adopted the 25%-reduction Toronto goal. Denmark's plan calls for 20% reductions by 2005 and 50% by 2030.
[69] There is no analogous U.S. policy, apart from some commendable state initiatives (Calwell et al. 1990, p. 12); on the contrary, in both global warming and ozone protection, a strong consensus in the rest of OECD was openly thwarted by U.S. representatives during 1989–90.
[70] Finland and The Netherlands introduced carbon taxes in early 1990, Sweden will do so in January 1991, and Germany is weighing one.
[71] Acknowledging this trend, in Royal Dutch/Shell's latest Group Planning scenarios, the "Sustainable World" scenario was somewhat *dirigiste* as first drafted, but was changed to rely far more on market forces—logically enough, since its options cost less.

Energy

Electric Utilities

More than half of Americans can already get financing from electric utilities for electricity-saving equipment, in the form of concessionary loans, gifts[72], rebates, or leases (Fickett et al. 1990; Lovins and Shepard 1988). Such financing is essential because customers typically want to get their energy-saving investments back within a couple of years, whereas if they don't become efficient and the utility builds a power station instead, *its* technical and financial strengths enable it to accept a payback period closer to 20 years (*id.*). This roughly tenfold "payback gap" is rational to both parties, but societally, it causes a severe misallocation (~\$60 billion a year in the U.S. in the mid-1980s) by effectively diluting price signals tenfold. Utility financing helps to close this gap by reducing customer's implicit >60%/y real discount rate approximately to utilities' ~5–6%/y (*id.*).

Utility financing of efficiency enables all supply- and demand-side options to compete on a "level playing field." Such competition, through either a planning or a market process, is now mandatory in the ~43 states with a "least-cost utility planning," or best-buys-first, policy (Moskovitz 1989). Such financing works so well that if all Americans saved electricity at the same speed and cost at which the ~10 million people served by Southern California Edison Company *did* save electricity during 1983–85, chiefly financed by SCE's rebates, then national forecast needs for power supplies a decade ahead would decrease by ~40 GW/y, at a total cost to the utility of ~1% of the cost of new power stations (Lovins 1989a; Fickett et al. 1990).

This sort of service-delivery, engineering-driven model of how utilities promote customers' efficiency gains retains an important place in dealing with specific market failures, such as the split incentives between builders and buyers or landlords and tenants. (Government performance standards and labels are also a key part of the policy toolkit in such situations.[73]) But a complementary approach is now starting to supplement and may even supplant the "we-will-wrap-your-water-heater" philosophy. Rather than merely marketing "negawatts," many utilities are also starting to make markets *in* negawatts: to make saved electricity into a fungible commodity

---

[72] Southern California Edison Company, for example, has given away more than a million compact fluorescent lamps, because that was cheaper than operating the company's existing power plants.
[73] California's Title 24 building standards alone, now saving the state's citizens ~\$1 billion worth of electricity every year, are a good model of how to combine performance standards with prescriptive options to reduce hassle to the builder.

subject to competitive bidding, arbitrage, derivative instruments, secondary markets, etc. For example, some utilities are

- buying back savings from customers by paying "generic rebates" per kWh or peak kW saved—including rebates for beating government standards or for scrapping old equipment;
- in eight states, operating "all-source bidding" in which all ways to make or save electricity compete in open auction and the utility takes the low bids, which are generally for efficiency[74];
- starting to buy saved electricity from other utilities—a form of arbitrage on the difference between the cost of supply and efficiency;
- considering making spot, future, and options markets in saved electricity;
- exploring ways to broker saved electricity between customers, rewarding any customers who goes "bounty-hunting" by correcting inefficiencies anywhere in the system[75];
- selling electric efficiency in *other* utilities' territories (Puget Power Company, for example, sells electricity in one state and efficiency in nine states); and
- in seven states, experimenting with sliding-scale hookup fees for new buildings—"feebates" whereby the builder either pays a fee or gets a rebate when the building is connected to the grid (which and how big depends on the building's efficiency).[76] Feebates can offer major economic advantages to all the parties, can generate tens of thousands of dollars' net wealth per U.S. house so built, and are readily coupled with efficiency labelling.

In addition, gas utilities can make money selling *electric* efficiency, thereby changing the behavior of buildings in ways that also help them open up new gas markets (*id.*). Electric utilities can also sell gas efficiency, and both should be rewarded for selling either. Wisconsin's utility regulators have even ordered that state's utilities to help customers switch to any competing energy form that costs less.

---

[74] Maine, largely through such bidding, also raised its private, mainly renewable, share of power generation from 2% in 1984 to 20% in 1989 to 30% in 1991 (based on 1990 construction), according to former Maine Public Utilities Commission Chairman David Moskovitz.

[75] This has already been done with saved water (Menke and Woodwell 1990, p. 21). All the other mechanisms described here are also being applied to water efficiency, and many appear useful for other resources or for services such as transport (Lovins 1989).

[76] Formally, the "feebate" should satisfy three boundary conditions: revenue-neutrality (the fees pay for the rebates); fees for inefficient buildings are based on long-run marginal costs including externalities; and rebates exceed the builder's marginal cost of achieving the efficiency. The slopes and intercepts can be adjusted annually as needed to maintain these conditions. When construction of inefficient buildings has been driven off the market, one can declare victory and stop.

Rapid experimentation in these and other market-making methods has been facilitated by the great diversity of the U.S. electric utility system: ~3,500 utilities of all shapes and sizes in ~50 major and hundreds of minor regulatory jurisdictions. Results so far are encouraging. A few years ago, some utilities had captured ~70–90+% of particular efficiency micromarkets, mainly difficult ones (residential shell retrofits), in only one or two years[77]. In 1990, greater marketing experience (Lovins and Shepard 1988a) has enabled, for example, New England Electric System to capture 90% of a small-commercial pilot market in two months, and Pacific Gas and Electric to capture 25% of its entire new-commercial-construction market in three months.[78]

Such entrepreneurship is being encouraged by a nationwide agreement in principle among U.S. utility regulators[79] to change the rules of price formation so as to ensure that utilities' cheapest options are also their most profitable ones. Although many ways to do this are available (Moskovitz 1989), the most common is to decouple utilities' profits from their sales and then let them keep as extra profit part of what they save their customers. Under a new policy approved in summer 1990, for example, Pacific Gas and Electric Company will be allowed to keep 15% of certain savings— adding ~$40–50 million to its 1991 profits—but the customers are better off getting 85% of an actual, prompt saving than getting all of nothing. At this writing, four states have approved such reforms and another 19 are doing so. The previous regulatory scheme rewarded utilities for selling more electricity (or gas) and penalize them for selling less. Similar perverse incentives still exist in many countries, despite the diversity of utility structures, and can be corrected by similar means.

Several states' least-cost planning comparisons have also begun to credit efficiency and renewables, or penalize their competitors, with shadow prices reflecting externalities (Ottinger et al. 1990). The correct externality values are not exactly known, but they certainly exceed zero, so in adopting a nominal figure, the New York Public Service Commission's Chairman, Peter Bradford, notes that it is better to be approximately right than precisely wrong. This approach, too, is attracting considerable interest in Europe.

These regulatory moves toward simulating efficient market outcomes have accelerated the already rapid shift in U.S. utilities' culture and mission, away from selling more kilowatt-hours and toward the profitable production of customer satisfaction (Fickett et al. 1990). About a third of U.S. utilities

---

[77] E.g., The Hood River County experiment in Oregon, and several Iowa municipal utilities' load-management promotions.

[78] Data from presentations at the 1990 COMPETITEK Members' Forum (5–8 September) and from John Fox of PGandE (pers. comm., 6 October 1990).

[79] Unanimously approved by the National Association of Regulatory Utility Commissioners in November 1989.

Least-Cost Climatic Stabilization 399

have already made this transition. Selling more efficiency may reduce their electric sales and revenues, but their costs go down more; and under the new rules, they can keep part of the difference, making money on margin instead of volume. Such a utility can indeed make money in six ways: it saves operating, construction, and replacement costs, plus associated risks and externalities; under the new U.S. Clean Air Act its fuel savings will be able to generate tradeable emission rights (currently for acid gas, but extendible by future legislation to fossil carbon); as soon as its regulators reform their ratemaking rules, it will be specifically rewarded for efficient behavior; and it can earn a spread on financing customers' efficiency improvements.[80]

Implementing Oil Efficiency

Analogous concepts are starting to enter the oil-efficiency market. Most importantly, on 30 August 1990, the "Drive+" feebate proposal passed the California Legislature with overwhelming bipartisan support[81] (Levenson and Gordon 1990). This bill would enact a revenue-neutral, open-ended sales-tax adjustment based on both fuel efficiency (measured as CO_2 emissions per mile) and smog-forming emissions. Buyers of dirty, inefficient cars would pay two fees; buyers of clean, efficient cars would get two rebates. By influencing car choice directly, feebates overcome the ~6:1[82] dilution of gasoline prices by the other costs of owning and running a car. Similar proposals are pending in Iowa and Massachusetts, are being drafted in other states, and have been proposed nationally.

This rapidly spreading idea has gotten several of the makers of superefficient prototype cars seriously interested in entering the U.S. market immediately, in order to maximize their share of the early adopters' market. Interestingly, General Motors did not oppose Drive+, reportedly because the company prefers market-oriented feebates to the standards or other direct regulations which might otherwise be imposed if the "Big Green" ballot initiative passes in November 1990.[83]

Governor Deukmejian vetoed Drive+ on 30 September 1990, but it

---

[80] Arbitrageurs get rich on spreads of a fraction of percent, but the difference in discount rates between utilities and customers is more like a thousand percent.
[81] By 61–11 votes in the Assembly and 31–2 in the Senate.
[82] In the United States at pre-Gulf-crisis prices; typically the ratio is nearer ~3–4:1 in other OECD countries with higher motor-fuel taxes.
[83] This is understandable: the Corporate Average Fuel Economy (CAFE) standards passed by Congress in 1975 were "at least twice as important as market trends in fuel prices, and may have completely replaced fuel price trends as a base for long-range planning about [new-car efficiency]" (Greene 1989), but they are an administrative headache for automakers, who have to juggle production and inventory continuously, almost ona daily basis, according to the efficiency of the models customers are buying.

seems bound to become law in 1991, since both the candidates to succeed him endorse it. Drive+ is then likely to launch a powerful national trend. A useful early refinement, too, would be "accelerated scrappage": basing rebates for efficient new cars on the *difference* in efficiency between the new car and the old one that is scrapped (if it is worse than a certain level); drivers who scrapped a functioning car and didn't replace it would get a bounty on presenting a death certificate that it had been duly recycled. By offering a far higher price than dealers do for trade-ins, the state would put a premium on getting the least efficient, most polluting cars off the road soonest. This incentive would greatly accelerate the energy and pollution savings. It would also enable poor people, to whom the worst cars tend to trickle down, to afford to buy a highly efficient new car that they could then afford to run.

Such feebates have wide application. They could spread from cars, light trucks, and buildings to appliances, aircraft, heavy road vehicles, etc.[84] In each case, they would transfer wealth from those whose inefficient choices impose large external costs on society (global warming, acid rain, oil-important dependence, etc.) to those who save such costs. Their self-financing makes them politically attractive, as the California Legislature's vote confirms. They entail three straightforward steps: set a target level of efficiency; charge an open-ended variable fee for all new devices that are worse; and rebate the fees (less administrative costs), also on an open-ended variable basis, to all new devices that are better (Rosenfeld et al. 1990a).

Feebates can be usefully supplemented by three further policy innovations:[85]

- In many EEC countries, and some others like New Zealand, much urban commuting—often half or more—is in company-owned cars, provided as a perquisite because this form of compensation is taxed less than equivalent salary (or not at all). This tax dodge is often meant to inflate domestic car sales for the benefit of domestic automakers. It contributes disproportionately, however, to congestion and hence to fuel waste and pollution. Since removing even a small number of cars from a crowded highway can markedly relieve its congestion, eliminating tax breaks for company cars should be a high priority.
- Greater symmetry between modes of transport requires that cars pay more of the cost of providing their infrastructure. Singapore, for example,

---

[84] Draft recommendations by the California Legislature's Joint Committee on Energy Regulation and the Environment (12 September 1990) include feebates for buildings *and*, on a pilot basis, appliances.

[85] A possible fourth, a novel "pay-as-you-drive" way to charge for car insurance, is starting to receive scrutiny too (El-Gasseir 1990).

limits traffic by a no-exceptions daily tax on downtown driving. In an even more interesting system being introduced in Stockholm, downtown residents who wish to drive their cars during a given month must buy a permit which also serves as their free pass to the regional public transit system for that month. Then they have it, so they might as well use it.

- "Golden carrot" rebates are designed to elicit the production of specific energy-saving products that are cost-effective but are not yet brought to market because cautious or undercapitalized manufacturers are unwilling to risk retooling costs for uncertain sales. For example, utilities from San Diego to Vancouver may soon join together to pay, say, a $300 rebate for each of the first 10,000 or 100,000 refrigerators sold in their territories that beat the 1993 Federal efficiency standard by at least 50%.[86] Such incremental efficiency would require one or more of the advanced insulating materials mentioned above to be put into mass production. If the refrigerators are not sold, no rebate is paid, so the utility is at no risk of not getting the desired savings; and once that many *are* sold, the manufacturing hurdle has been leapt, the rebates can be discontinued, and continuing sales will then yield far larger savings. A larger "platinum carrot" can then be offered for the next incremental advance, and so on until cost-effective opportunities are exhausted. This proposal was originally developed for North American electric utilities and appliances—and is now being considered by European and South Pacific utilities too—but there is no obvious reason why it could not be offered by governments and apply to other products such as light vehicles, in addition to or in lieu of feebates.

So far, less progress has been made in fundamentally changing mission and culture in free-market oil companies than in regulated, franchise-monopoly utility companies. But there are strong reasons for private oil companies, too, to promote customers' energy efficiency (Lovins and Lovins 1989):

- Under long-run competitive equilibrium conditions (ignoring fluctuations as war or peace breaks out in the Middle East), the major oil companies expect fairly flat long-term real prices, implying little prospect for large upstream or downstream rents. The major rent still largely uncaptured is the spread between the cost of lifting and of saving barrels.
- With long-run real oil prices fluctuating between, say, ~$12/bbl and

---

[86] This scheme is promoted by A.H. Rosenfeld (Lawrence Berkeley Laboratory) and R. Cavanagh (Natural Resources Defense Council, San Francisco) and has already been accepted by the main utilities involved.

~$25/bbl, it is easier to make one's margins selling "negabarrels" that cost $5 to produce than barrels costing $15.
• Selling a variable mixture of fuel and efficiency can be used to hedge risks in supply-side markets (*id.*).
• Efficiency can fundamentally reduce long-term price volatility (*id.*), benefiting a capital-intensive industry.
• Efficiency promotes global development, which is good for the oil business, and deflects Persian Gulf war, acid rain, and global warming, which are bad for it.

For these and similar reasons, several major and independent oil companies are starting to express interest in a significant commitment of effort and resources to marketing efficiency. Some of these firms are, in essence, very large, technically oriented banks, and are starting to realize that they can get better at selling financial products, such as leasing efficient end-use technologies.

Generic Issues

The main cultural obstacles to this transition, as in many utilities, are changes in mindsets: selling services rather than commodities, working on both sides of the customer's meter or the vendor's pump, and getting used to doing fewer big things and more smaller things. This difference of unit scale is perhaps the most uncomfortable, because traditional energy-supply systems are ~5–8 orders of magnitude larger than end-uses. But such large scale is not technically or logically necessary, and often it is not economic either. The well-known economies of scale in engineering and manufacturing certain energy systems can easily be swamped by even a small subset of the dozens of known *dis*economies of scale[87]. There is now abundant empirical evidence that minimizing whole-system cost generally entails matching scale, at least roughly, between supply and end-use. This does not mean that everything should be small: it would be nearly as silly to run a huge smelter with thousands of little wind machines as to heat millions of houses with a fast breeder reactor. But it does mean that making supply systems the right size for the job usually makes them cheaper.

Oil and gas companies are starting to compete in electricity markets by promoting smaller-scale technologies such as packaged gas-fired cogeneration plants, steam-injected gas turbines, and combined-cycle retrofits of classical

---

[87] These arise from, e.g., higher distribution costs and losses, reduced unit availability, increased reserve requirements, longer lead times (hence greater risks of cost escalation, technical or political obsolescence, or mistimed demand forecasts), higher ratio of onsite fabrication to factory mass-production, more difficult maintenance, greater awkwardness of using waste heat, etc.— ~50 identified mechanisms in all (Lovins and Lovins 1982, App.1).

combusion turbines. By some estimates, private additions to U.S. generating capacity exceeded utility additions in 1990 (Borr 1990), and 27 states that recently ran supply-side auctions were offered, on average, eight times as much private generation as they wanted (Blair 1990).[88] Fuel vendors are becoming significant players in this competition. Gas companies diversifying downstream now routinely note that building and running a gas/combined-cycle power plant undercuts just the *running* cost of a typical nuclear plant. As oil and gas companies increasingly bundle both electric and gas efficiency with downstream applications of their fuels, they become involved with customer scale and start to think more like customers.

Another obstacle being slowly overcome is the pervasive asymmetry in public policy, long dominated by special interests subjected to scant performance accountability. Public policy in almost all countries has for decades been overwhelmingly biased toward supply over efficiency, depletables over renewables, electricity over heat and liquids, and centralization over appropriate scale. There are reasons for this, but they are certainly not economic reasons. They are the same reasons that efficiency got 2.8% and renewables 3.1% of USDOE's FY1988 proposed civilian RD&D budget, *vs.* 12.4% for fusion, 16.6% for fossil fuels, and 11.1% directly for fission; or that U.S. fission, after decades' devoted effort, continues to receive strong policy support despite having missed its cost target by an order of magnitude, while renewables, which quickly met or bettered *their* cost targets and now provide twice as much energy (Lovins and Lovins 1989), continue to be dismissed as futuristic or impractical. Too often, the balance of official effort between competing options is like the old recipe for elephant-and-rabbit stew: one elephant, one rabbit. But in the U.S., the EEC, and many developing countries, the pendulum is starting to swing towards economic rationality, if only because there is no longer the capital to misallocate.

These frustrating, though gradually resolving, problems must not obscure the major gains already made. During 1979–86, for example, the United States got more than seven times as much new energy from savings[89] than from all net expansions of supply, and more new supply from sun, wind, water, and wood than from oil, gas, coal, and uranium (Lovins and Lovins 1989). By 1986, U.S. CO_2 emissions were *one-third lower* than they would

---

[88] Many, but not all, of these proposals could be relied upon to yield actual, reliable capacity if accepted. Opinions differ on the "real" fraction. Many utilities have also found private cogeneration to be more reliable than their own central plants.

[89] As indicated in aggregate by improvements in primary energy consumption per unit real GNP—a crude and sometimes misleading measure, but useful shorthand in most cases. About 65–75% of that improvement is generally considered to be due to technical gains in energy efficiency, nearly all the rest to changes in composition of output, and only a few percent to behavioral change.

have been at 1973 efficiency levels; the average new car alone expelled almost a ton less carbon per year; annual energy bills were ~$150 billion lower; and annual oil-and-gas savings were three-fifths as large as OPEC's capacity (Rosenfeld et al 1990a, p. 4).

During 1977–85, the United States increased its oil productivity four-fifths faster than it had to in order to match both economic growth and declining domestic oil output. By 1986, the annual savings, chiefly in oil and gas, were providing two-fifths more energy than the entire domestic oil industry, which had taken a century to build (Lovins and Lovins 1989). Oil, however, has dwindling reserves, rising costs, and falling output, whereas efficiency has expanding reserves, falling costs, and rising output.

By 1989, the United States was getting 91% as much annual primary energy from post-1973 savings as from *all* oil, domestic and imported: the 1986–88 hiatus in efficiency gains was history, and the previously steady downward trend of energy intensity had resumed its pace of several percent per year. Some other OECD countries have done even better. During 1973–88, while energy intensity declined 26% in the U.S., it fell 30% in Japan. In the 1980s, the countries, like Japan and West Germany, with the highest energy efficiencies have proven among the toughest economic competitors—and are now redoubling their efficiency efforts as they start to discover the new technological opportunities.

The task now for OECD is twofold: to accelerate these historic efficiency gains by harnessing today's far more powerful and cost-effective technologies, e.g., by promoting superefficient cars through feebates with accelerated scrappage; and to extend to electricity the rapid, consistent efficiency gains obtained during 1973–86 for direct fuels. In principle, it should be possible to save electricity as least as quickly as oil, because

- far more electricity than oil is used in highly concentrated applications[90] and in standardized commodity-like devices installed in relatively few places[91];
- there is far more economic and environmental incentive to save electricity than oil;
- electric applications do not have psychological complications analogous to those of cars, especially in the United States;
- service quality is more likely to be markedly improved by saving electricity than by saving oil; and
- for electricity, unlike oil, a skilled engineering and financial institution,

---

[90] About a million industrial motors >125 hp use ~12% of all U.S. electricity (Lovins et al. 1989, pp. 28–29 & 39); most of this is probably used by ~10^{3-4} motors rated at 10^{3-4}hp.
[91] For example, the ~1.5 billion 2×4' fluorescent lamp fixtures installed mainly in large U.S. office buildings.

with a relationship with every customer, stands ready to deliver efficiency programs and can easily be rewarded for doing so.

The best utility programs confirm this thesis: Southern California Edison's 1983–85 program mentioned above reduced the decade-ahead forecast of peak load by the equivalent of $8\frac{1}{2}\%$ of the then-current peak load *per year*, at average program costs of a few tenths of a cent per kWh saved (Fickett et al. 1990). Taken together, many U.S. and some foreign utilities' experience, especially during 1989–90, now extensively confirms that such rapid, reliable, cost-effective electric savings are possible. As usual, the limiting factor in rapidly propagating such success is the number of skilled practitioners. Major initiatives to expand recruitment, training, cross-pollination, and career development opportunities are therefore emerging, and merit reinforcement.

One last category of policy initiative requires emphasis: incentives within the public sector. Washington State, for example, splits the money saved by energy efficiency improvements to government buildings into three unequal parts: part returns to the General Fund, part is used to reward everyone responsible for achieving the saving (by supplementing institutional budgets or paying personal cash bonuses), and part finances more efficiency, thus bootstrapping successively longer-payback investments without requiring recourse to the capital budget. California already returns half its dollar savings to the institution achieving the efficiency, and proposes to earmark the other half for a revolving loan fund to achieve more savings throughout state government. Such mechanisms, plus careful tracking of energy costs so that responsibility for reducing them can be assigned, are the beginning of sound energy management. Without such exemplary leadership by governments at all levels, citizens will take calls for efficiency less seriously.

Farming and Forestry

More is known about the mechanisms of transition to sustainable energy systems than to sustainable farming and forestry, but the latter seems analogous in principle and seem to offer scope for market mechanisms such as feebates. Sweden, for example, has long taxed agrichemicals and rebated the proceeds to help farmers make the transition to organic techniques. Iowa has a similar agrichemical tax to finance groundwater protection. Fees on synthetic nitrogen fertilizers could be rebated to manures, green manures, or legumes. Fees on logging could be rebated to tree-planting or (better) to forest protection, especially old-growth.

Many agrichemicals are already costly enough, and engender enough health among farmers, that little more incentive is needed not to use them. What is lacking is transitional advice, reassurance, and financing: most farmers do not have practical knowledge of sound alternatives, and have too little financial safety-margin (or lender flexibility) to undertake the perceived risk of trying anything new. But in the U.S. Canada, Germany, Japan, and other OECD countries, lively networks of farmers are emerging to help match successful conversion case-studies with farmers facing similar challenges. Expanding and endorsing such clearinghouse activities should be a high government priority.

Alert agrichemical companies are already starting to plan their own transition. One of us (ABL) has recently been told by senior planners at three major agrichemical manufacturers that they all plan to get out of the business; the only question is how soon, how gracefully, and what to do instead. Like utilities that have revised their product from electricity to end-use services, these agrichemical firms have revised their mission from selling chemical commodities to helping farmers grow nourishing food. Like utilities, too, such firms often have financial resources, marketing skills, and technical capabilities that will be important for the agricultural transition. They are becoming interested in providing such assets as sophisticated soil-test kits (conveniently testing the health and diversity of soil biota is still in its infancy), mineral amendments, targeted predators and other integrated-pest-management tools, adaptable native seeds, technical advice, farm-management software, and transitional financing and risk insurance. Together, such elements could make an attractive package for a harried farmer who wants to change but is deterred by "hassle factor," novelty, and perceived risk.

It is is also important to make markets in carbon sequestered or not emitted ("negations" of carbon). Applied Energy Services, Inc., an Arlington, Virginia, firm, is planting trees in Guatemala to offset the output of its new coal-fired generation plant and funding the planting by a voluntary <5% surcharge on the plant's output. As utilities or private market-makers start to broker such carbon offsets to and from utilities and industrial fuel users, farmers and foresters should be able to bid to provide carbon-absorbing services. For example, it should be straightforward to make a market in forest- or prairie-preservation rights, which would certify that at least a certain carbon inventory or density will be maintained for a given period. Such a market would, for example, add value to farmers' decisions to enhance humus through organic practices, or to foresters' decisions to lengthen cycle times or preserve old-growth forest. Analogies already exist: in Southern California, cogeneration deals were made in 1989 in which one of the partners' in-kind equity contributions was reduced smog formation, assessed at the day's market value as quoted by the local pollution-reduction broker (a new profession created by EPA's "bubble concept" for air-pollution offsets).

In fact, a market in carbon offsets could in effect transfer some of the large amounts of money saved by electric efficiency into financing the transition to sustainable farming and forestry. Electric utilities, especially those burning coal, would have to reduce their carbon emissions, or purchase "offset rights" representing extra carbon sequestered in trees or soil biota. (Utilities in the parts of the U.S., Germany, and Britain that burn the most coal tend to be the least interested in energy efficiency, so they would need to buy the most offsets.) Until the utilities got tired of this income transfer, it would provide a timely injection of capital to help launch a serious farming and forestry transition—and thereafter, a spur to the utilities to get serious about their own demand-side opportunities. Such a capital flow could easily exceed $10 billion per year in the United States alone. That is not large compared with the $175-billion-a-year electric bill, but would be a godsend to cash-short farmers and small-scale foresters unable to finance fundamental changes through traditional lenders.

It is hard to estimate the attainable speed of reforms to make farming and forestry more sustainable and carbon-conserving. It is not even known how quickly organic farming is spreading in the United States. Anecdotal evidence suggests it is far faster than anyone had expected (*Wall St. J.* 1989), and informal reports from many regions in 1989–1990 have indicated that demand for organic produce is often tens of times larger than supply. Now that organic farming has finally received an official economic endorsement (NRC 1989), official definitions and standards are emerging, and most areas and types of operations can find a successful example of conversion relatively close by; many previously skeptical farmers are starting to consider a transition as a more serious near-term option. Extension agents from many parts of the United States report overwhelming demand for transitional counsel. In time, the supply of the needed information and risk capital will catch up. Bundling the global-warming benefits with their other, more familiar benefits of sustainable practices will tend to attract more capital, reduce perceived risk, and hence speed the transition.

Some encouragement may be drawn from the speed with which other farming changes have lately been adopted: low- or no-till herbicide-based cultivation, land set-aside programs, and hedging in commodity futures markets. U.S. farmers responded eagerly enough to financial set-aside incentives to cause spot shortages of some crops. Similar incentives for conversion—analogous to electric utilities' loans, gifts, rebates, and leases for efficiency—could bear similar fruit.[92] If government agricultural departments assigned a tenth as high a priority to helping free farms of their chemical dependence as law-enforcement departments do for citizens, and mounted

---

[92] Just the major energy benefits in reducing heat-island effects could motivate utilities to fund urban forestry, including agroforestry, on a large scale, as Sacramento's municipal utility is already starting to do (*supra*).

a major campaign to renew the old arts of soil conservation and tilth improvement, there is every reason to think that the time is ripe and many farmers could make suprisingly rapid changes.

CFC Substitution

Mandatory production phaseouts and rapidly increasing taxes are already a fact of life for CFCs. Less mature, however, are mechanisms to recover, store, and destroy the large *existing* inventories of CFCs. A few utilities that already pay customers to scrap old, inefficient refrigerators and freezers are starting to integrate CFC recovery, usually by hiring specialist appliance-recycling firms (Shepard et al. 1990, pp. 95–96). The City of Palo Alto, California, is also considering collecting all CFC-containing products found in local landfills and recovering the CFC for reuse or disposal (Turiel and Levine 1989, p. 197). CFC recovery from air conditioners in both inservice and scraped cars is also a rapidly growing business.

The recoverable CFC inventories currently in circulation, or sitting in (and leaking from) scrapped cars and appliances, are important for both global warming and ozone depletion. A typical 18-ft^3 U.S. refrigerator/freezer contains ~0.9 kg of CFC-11 in the foam insulation and ~0.23 kg of CFC-12 refrigerant.[93] These CFCs are ~14,000–20,000 as heat-trapping per molecule as CO_2 (Krause et al. 1990, p. I.1–10). Their potency, and the high policy priority therefore accorded to their replacement, suggests that "offsets" for CFCs would also be worth marketing—permitting their continued *use* (even though their production may meanwhile have been phased out) so long as an equally potent quantity of CFCs is removed from the environment (Turiel and Levine 1989, pp.197–198). Properly done, such "tradeable use rights," analogous to the EPA's "bubble concept" for conventional air pollution, could cap the effective prices of CFC substitutes, reduce energy demand, and smooth the transition to CFC substitutes (*id.*).

Formerly Centrally Planned Economies (FCPES)

The Soviet Union emits 15% of the world's fossil carbon from ~12% of world economic output (calculated by Soviet methodology: Makarov and Bashmakov 1990). The USSR and its former satellites, which are similarly or more carbon-intensive, are engaged in an historic transition of extraordinary dimensions. These countries' great opportunities to help abate global warming can be neither assessed nor achieved without wrenching structural,

---

[93] Technology to recover the former may be available in Germany, but information seems scarce and R&D appears to be a high priority.

political, and economic changes that are only just beginning. These changes include:

- Major reallocation of national resources, especially in the USSR, from military to civilian production within the context of lessening tensions, European partial demilitarization, increased popular control over governmental military adventures (this is needed in the U.S. as well as in the USSR), and a widespread commitment to a global security regime that makes others more secure, not less (Harvey et al. 1990).
- Radical economic reform based largely on market principles, truthful prices, and integration into the world economy, including convertible currency and fair opportunities for foreign ownership and joint ventures.
- Equally radical social and political reform reinvigorating productivity, initiative, and personal responsibility, and hence requiring comprehensive educational renewal.
- Economic structuring markedly reducing the relative role of primary materials production, enhancing the nascent service sector, favoring smaller enterprises, decentralizing much overcentralized production, introducing competition to monopolies (including electric utilities), and creating a working distribution system that now scarcely exists at all.

This is a tall order; yet there is no way out but through. As the economist P.G. Bunich (now an advisor to President Yeltsin of the Russian Republic) remarked in early 1989,

> Here we are, 280 million Soviets gathered together on a vast beach and wading together into the surf. We're at that awkward point where it's too deep to wade but not yet deep enough to swim, so we're losing our footing—and anyway, none of us know how to swim. But, by God, some of us will figure out how to swim, and those who figure out first will teach the rest. Of course, we'll have to make some lifejackets, but only a few million, not 280 million.

That historic transition comes at just the time when the extremely energy-intensive Soviet/eastern European economies are broken and need fixing. Poland, for example, has about the same per-capita energy use and carbon emissions as Austria, but only a fourth the per-capita GNP (Sitícki et al. 1990). Hungary has about the same per-capita energy use as Japan, but one-fifth to one-fourth the per-capita GDP (Jászay 1990). The Soviet Union is one-third to two-thirds more energy-intensive than the United States, and 2–4 times more than the most efficient countries such as Sweden (with a similar climate) and Japan.

However, significant progress is being made. Hungary got 38% of its 1970–88 increase in energy services from efficiency (31%) and structural change (7%), and can both save much more and considerably expand biomass growth and use (*id.*). Although Poland has so far sustained substantial efficiency gains only in transportation, further cost-effective structural and efficiency improvements could even hold long-term per-capita energy use constant despite 2–3%/y income growth (Sitnícki et al. 1990). The Soviet Union's heavy use of natural gas contributed to a 21% drop in its CO_2/primary-energy ratio, which fell by 22% during 1970–84 (about the same as in western Europe) compared with only 1% in the United States (Makarov and Bashmakov 1990).

These high intensities have four especially important aspects:

- Energy prices are heavily subsidized (Chandler et al. 1990). A senior Soviet colleague who set up a cooperative selling energy-efficiency services in Moscow said, "You might suppose this activity is three times less cost-effective than in the West, because in the Soviet Union we sell energy for about a third of its production cost. But we're also about three times less efficient, so it works out the same." Energy subsidies in Poland in 1987 (before the "big bang" price reforms) were ~49% of the delivered price of coal, 83% of gas, and 27% of electricity (Sitnícki et al. 1990). For eastern Europe as a whole, "Until recently, electricity, coal and natural gas were priced at one-fourth, one-half and four-fifths of world market levels" respectively (Chandler et al. 1990). Yet "it is not enough simply to get the prices right: prices must also matter. Making prices matter means not permitting enterprises to pass the cost of energy waste on to consumers" without risking bankruptcy for their uncompetitiveness (*id.*). Absent such fundamental reform, based on understanding that prices are *information*, not a social entitlement, the Eastern economies are simply, as one Soviet scholar put it, "vast machines for eating resources."
- Decades of central planning by empire-building bureaucracy ministries rewarded for setting ever-larger quantitative targets have left the Eastern economies exceptionally overbuilt in primary materials industries. An astonishing 70% of Hungary's energy consumption goes to raw materials production, which provides only 15% of national output (Jászay 1990). Similarly, by producing less unnecessary material and using it more efficiently, market reforms could save a sixth of Soviet energy and half of Czech steel production—even more in chemicals and nonferrous metals (Chandler et al. 1990). This is especially good news because more than two-thirds of Soviet fossil-carbon output is from the industrial and energy sectors (Makarov and Bashmakov 1990).
- Distortions of production, blockage and compartmentalization of information, administered prices, and mandatory allocation of goods and services have led to serious weaknesses in some sectors, notably electronics, that are vital to energy efficiency. That is why "only 25–30% of the cost

of [Soviet] energy efficiency . . . [is the cost of] implementing measures at the point of use. The remaining 70–75% results from the expense and difficulty of expanding domestic production of energy-efficient equipment and materials" (Makarov and Bashmakov 1990). Price distortions make inferior domestic equipment look only half as costly as foreign versions, but those cannot be imported without a convertible ruble, soft credits from abroad, or less restricted joint ventures.

- As both Western and Soviet analyses independently showed in 1983[94], energy efficiency is by far the most critical technical measure for the success of *perestroika*, because it produces a double benefit: it frees scarce resources (capital[95], hard currency, technical skills, etc.) to modernize industry and agriculture, and it frees the saved oil and gas for export to earn hard currency to buy technologies for the same purpose. For these reasons, they estimated, each 1% improvement in aggregate energy productivity (especially in electricity) may increase national output by several percent, and bring even more important qualitative improvements. Because saved energy is fungible for hydrocarbons that can be sold for hard currency, Soviet energy savings are properly denominated not in kopecks per gigajoule but in dollars per barrel.

These circumstances require a combination of three approaches: *economic reforms, structural shifts in the composition of output, and strong improvement in end-use efficiency.* The alternative, on standard projections, is to double Soviet fossil-carbon output by the 2020s (Makarov and Bashmakov 1990), and Polish a decade earlier (Sitnícki et al. 1990). Such projections may be too high because the growth is self-limiting: Makarov and Bashmakov (1990) state flatly that "The Soviet economy can no longer sustain continued growth in energy consumption and the corresponding demand for increasing energy production. If current trends continue, capital and other resources will be required in amounts so large as to preclude the possibility of realizing any but the Pessimistic Variation of the Base Case Scenario." But the prospect conjured up by this kind of involuntary grinding to a halt is not pleasant either.

Besides economic rationalization, structural changes, and end-use

---

[94] By Royal Dutch/Shell's Group Planning in London and Corresponding Member V.A. Gelovani et al. at the USSR Academy of Sciences' Institute for Systems Studies in Moscow.
[95] Nearly a fifth of Soviet investment goes to fossil-fuel extraction (not counting the electric sector) (Chandler et al. 1990). The same fraction went to Polish coal-mining *alone* in 1980, almost double the 1970 share, while gas and oil extraction's share of investment went from <1% of total industrial investment in 1980 to 39% in 1986. These trends squeezed out nonenergy industrial investment, which fell from 74% of total industrial investment in 1970 to 61% in 1986 (Sitnícki et al. 1990). That is partly why Sitnícki advocates halting *all* energy-supply investment until the demand side is set in order. In Hungary, likewise, "investment in energy supply grew to 40% of all industrial investment by 1986" (Jászay 1990).

efficiency, some eastern European countries would benefit greatly from raising fuel quality by substituting mainly natural gas. This is most true for Poland: the world's fourth largest extractor of coal, 75% coal-fired throughout its economy, and the user of a third of eastern Europe's 20 EJ/y of primary energy (Chandler et al. 1990). Most eastern European countries have major energy supply problems—Hungary imports half its energy, Soviet fuel reserves are shifting inexorably from high-grade western to remote, costly, and often low-grade Siberian and Far Eastern sites[96], etc. But Poland epitomizes the most acute supply problems. Severe air and water pollution are contributing to economic shrinkage (Chandler et al. 1990), coal-mining consumes a fifth of all steel (up >150% since 1978) and nearly a tenth of grid electricity, the average depth of mines is increasing 2–4%/y, more difficult mining conditions are cutting labor productivity in the worst mines to a sixth of the British or West German norm, social and administrative costs are high and rising, and land is so scarce that some mines "transport waste rock and coal washing refuse as far as 80 km for disposal" (Sitníckiet al. 1990). Coal exports for hard currency must cease in the 1990s in order to fill domestic needs as coal quality and accessibility decline. In any event, the economic benefits of the exports have been illusory because they greatly speeded the shift from high- to low-quality coal (high sulfur, high ash, high cost, more global warming).

Such conditions offer unusual leverage in abating global warming, because the costs and the environmental impacts of burning such poor fuels and inflated in four ways: more fuel must be burned to yield a given amount of primary energy, and most of that energy is then wasted by inefficient conversion, delivery, and end-use. Conversely, end-use efficiency improvements bring benefits which are amplified manyfold by avoiding system losses. Any such improvements, in a country like Poland, will push the most intractable coal-mining problems further into the future, buying more time to substitute gas which (efficiently used) might bridge to renewables.

In principle, the Soviet Union has three advantages in promoting energy efficiency. One is its gifted scientists and technologists. Soviet achievements in such areas as control theory, materials coproduction, materials science (diamond films, ceramics, supermagnets, composites, etc.), and mathematical physics can not only help meet domestic needs but also compete in world markets. It is now becoming common, with good reason, for the computer literature to speculate that later in this decade, the combination of Taiwanese hardware and Soviet software may make a strong market showing. Thus

---

[96] In recent years, Soviet coal-mining's shift to the East has resulted in mining more coal but getting less energy out of it, because grade is falling faster than tonnage rises. Some Eastern coals are of such poor quality that no one has figured out a reliable way to burn them (OTA 1981).

the USSR has far more to sell than its unparalleled storehouse of raw materials. So far, however, Soviet technical prowess has not been mobilized to advance energy efficiency in any fundamental way, and indeed there is not yet any really detailed and modern (by Western standards) efficiency research or analysis in the country, though there is plenty of talent to apply to it.

Second, Soviet energy-using hardware is highly standardized. Lighting retrofits in the U.S. or western Europe are complicated by a fragmented industry distributing thousands of types of fixtures from hundreds of manufacturers. In the Soviet Union, only 10–20 types of fixtures are in general use. For that matter, somewhere in the files of GOSPLAN (if one knows which numbers are real and which are fake) one can ostensibly find a complete record of the entire Soviet capital stock, because its production and shipment were planned. No Western country has a comparable "paper trail": American analysts, for example, know less about their stock of electric motors, which use more primary energy than highway vehicles, than they know about moonrocks.

Third, what is left of the Soviet command economy—which has been largely retained in the strictly monopolistic utility/electrotechnical sector—may still be useful in shifting production towards end-use efficiency. The design of every electricity-using device manufactured by the State, for example, must be approved by a single engineer in Leningrad—everything from toasters to giant motors, lamps to computers. He likes efficiency, but his authority is circumscribed. For example, a small lighting institute on the Volga has been trying for years to get permission to pilot-produce some improved compact fluorescent lamps (none are made in the Soviet Union). But the Ministry of Energy and Electrification has never approved the request: the Ministry's job is to build and run power plants, and there is no demand-side institution with any force as a counterweight. There are recent signs that the Ministry is starting to think more about efficiency—having only $3\frac{1}{2}$–5% more generating capacity than expected peak load concentrates the mind wonderfully—but such change of mission will probably be slow. Happily, the view of the leading Soviet climatologist and many of his Academy colleagues—that global warming would probably, on the whole, be good for the Soviet Union—is apparently giving way to a realization that the risks of mispredicting are too great (M. Budyko and G. Golitsyn, pers. comm.). Yet there is still a very limited understanding of energy efficiency within the old Ministries.[97]

---

[97] For example, Budyko and others have misinterpreted analyses by Makarov et al. as indicating that major gains in Soviet energy efficiency would cost the entire capital investment available. What the studies said was that such gains could accompany a complete, from-the-ground-up reconstruction of the whole Soviet economy and infrastructure—which of course would cost that much, but would yield far more joint benefits than just reducing global warming.

The Soviet/eastern European energy problem is part of a nearly infinite onion containing layer upon layer of challenging social, political, and economic problems. Peeling that onion will be slow and difficult. Soviet experts have suggested, however, some of the specific initiatives that are most needed. Paraphrasing Makarov and Bashmakov (1990), these include:

- International lenders' soft credits for importing energy-saving equipment.
- Joint ventures to make such equipment in the Soviet Union.[98]
- Joint exploration of efficiency opportunities and implementation methods.
- Public education stressing "that energy savings is the principal, and most economically effective, means of solving many global problems of world energy development."
- Collaboration to make nuclear plants safer (a major concern for Western utilities whose operations are hostage to the next Soviet accident) and cheaper. (In our view, and that of many Soviet colleagues, such cooperation may be worthwhile but is very unlikely to lead to a Soviet nuclear revival: as in most other countries, the economic and political obstacles are too daunting.)
- Systematic reduction of the natural gas system's methane leaks, which increase the Soviet contribution to global warming by ~8% (*id.*) to ~26% (Arbatov 1990).
- Expanded cooperation in renewable energy development—an area of much ingenious Soviet design—and in trapping and isolating CO_2.

Our own experience suggests the need to add seven major items to that list:

- A high initial priority should be given to superwindows—the coating technology exists in the Soviet military sector but has not yet been transferred to the civilian sector—and to superinsulated modular house/apartment construction. It would be silly to try to relieve the housing shortage, let alone meet emergency housing needs in e.g., Armenian-earthquake reconstruction, using the same poorly insulated, shoddily built, seismically hazardous technologies whose use is still so widespread.
- Soviet analysts are prone to suppose that building up consumer goods and the service sector will require rapid growth in the electric share of end-use energy, accounting for half the projected rise in CO_2 output (Makarov and Basmakov 1990, p. 5). Instead, there is good reason to

---

[98] Poland has introduced and now provides a three-year tax holiday for such ventures, along with measures to encourage reinvestment domestically. Those measures include low-interest credits and 50% taxable-income deductibility for investments in modernization—100% if they are for environmental protection (Sitnícki et al. 1990).

believe that the electricity required could come from larger-than-expected savings both in the new equipment itself and elsewhere, especially in industrial drivepower. But those efficiency opportunties are revealed only by very detailed and up-to-date analysis, so they are not yet well understood by most Soviet experts. A joint USSR Academy of Sciences/RMI book now in preparation may help in this regard.

- As is now starting, it is important that skilled western and east Asian firms feel able to participate for mutual benefit in major Soviet oil and gas projects. Potentially very large reserves of both will require Western technology to find an extract. Such new reserves plus high end-use and conversion efficiency could probably "completely cover" Soviet domestic energy needs from high-grade resources—hydrocarbons could cover ~80% of energy supply—for many decades to come (Chandler et al. 1990). Such new reserves would only be squandered in the USSR, however, and unaffordable to eastern Europe, without major reforms in price, output structure, conversion efficiency, and end-use efficiency (*id.*).

- To help achieve those reforms, neighboring countries in both Asia and western Europe could and should provide a massive infusion of capital and technology specifically focused on Soviet and eastern European end-use efficiency. That is among the cheapest ways in the world to abate global warming, because efficiency is so low to start with and the fuel displaced on the margin is mainly low-grade coal (or high-grade, low-carbon gas that can in turn displace more coal). The improved efficiency, preferably using equipment made by joint ventures in the USSR, would free up oil and gas for resale to the efficiency providers. Those fuels, especially the gas, would displace German coal, reducing global warming even more, and earn hard currency to pay for the efficiency technologies and financing. That payback would be very rapid, leaving most of the oil-and-gas revenue stream available to buy other technologies for modernization outside the energy sector. Similarly, Western aeroturbine manufacturers could joint-venture with their Soviet counterparts in producing steam-injected and combined-cycle gas turbines—a good use of the military engine production capacity now being partly demobilized. Those turbines, with their modularity and very short lead times, could help quickly to relieve Soviet and eastern European power shortages while also yielding major long-term gas savings.

- A formal mechanism to provide up-front, pump-priming credits for Western investments in Eastern energy efficiency would be through the sale of carbon offsets from East to West. Since, for example, West Germany, a carbon-intensive country, is considering introducing a carbon tax, the introduction of German offsets without territorial limits could readily lead to a flow of hard currency eastwards to pay for carbon savings that can initially be achieved at lower cost ore cheaply in the

East than in the West. That hard currency would then be recycled westwards to pay for German or other world-market efficiency technology installed in the East to fulfill the carbon-abatement contract previously sold. Sale of the subsequently saved gas to the West, and its use to offset still more carbon, could be part of the same deal.[99]

- Japanese industry's remarkable training programs for energy-efficiency managers should be exported to the Soviet Union. Those managers' attention to detail is unrivalled elsewhere in OECD, and has much to teach all other countries. In principle, Japanese industry might obtain carbon offsets through fuel savings achieved in Soviet industry, yielding another mutually beneficial currency flow to be coordinated with increasing oil-and-gas collaboration in the Soviet Far East.

- A *sine qua non* for Soviet restructuring is continued cooperation by the NATO countries, especially the United States, in rapid disarmament and demilitarization of both sides' economies (Makarov and Bashmakov 1990). This will require redefining not only military force structures and missions but also strategic doctrine and the whole concept of what security is and how to obtain it at least cost (Harvey et al. 1990). Recently U.S.-Soviet cooperation in the Persian Gulf crisis is an encouraging step in this direction. Rapid, effective, comprehensive Western help with Eastern energy efficiency is a critical component of Western security interests: the alternative is not only an unstable meterological climate but also political instability, stagnation, or worse. NATO has for many years sponsored leading international seminars on technical aspects of energy efficiency, and may be able and willing, given high-level encouragement, to turn its energies in this direction.

In short, the United States and the Soviet Union, which between them release about half the world's fossil carbon, are not only natural technical and financial partners in helping to abate global warming; their intricate security embrace demands such cooperation, mobilizing the best talent in both countries.

This partnership should not stop with energy. Both countries have much to learn from each other's experience of sustainable farming and forestry—rather than continuing misguided efforts to sell American-style chemical

---

[99] This would be timely, since much German coal will become uncompetitive after 1992 anyway, when its "Kohlepfennig" subsidy becomes legally unsupportable. This could offer an opportunity to make lemons into lemonade. Ideally, such offset arrangements should be brokered in markets that include derivative instruments with a range of maturities, so that rather than locking themselves into contracts lasting far longer than the timescales of installation and technological evolution, the parties would retain some flexibility to keep hunting for better buys as factor costs and gas and carbon values all shift. As noted below, Krause et al. (1990) suggest such periodic carbon auctions.

farming to a country whose internal obstacles have so far happily hampered, at least outside Kazakhstan, its efforts at chemical self-contamination of its vast lands. Early citizen-organized farmer-to-farmer exchanges have proven such an eye-opener to both sides that they clearly merit expansion: it will be difficult to elicit enough entrepreneurship to break up the state farms unless more Soviet farmers realize what is possible. Further, a large proportion of Soviet vegetables, fruit, meat, and dairy products come from the tiny fraction of farmland privately farmed with some semblance of real economic incentives. Yet the tools used on those plots are often medieval. Important productivity gains could probably be had by as simple a means as transferring to Soviet artisans and cooperatives the technologies of advanced gardening tools (from such firms as Smith and Hawken) that could do so much to ease the lot and expand the output of those private plots. Superwindows could in time further provide more fresh winter produce by making indoor farming possible in major cities like Moscow.

Carbon offsets are especially important also to preserving the fragile carbon inventories of the Soviet Union's boreal forests, now being plundered by overbuilt pulp-mills at home and in Scandinavia. Similar opportunities exist elsewhere in the parts of eastern Europe not yet severely damaged by acid rain: Hungary, for example, "possesses rich biomass production potential and could serve as a testbed for new biomass utilization and sequestration technology," involving, e.g., "increasing the humus in soil from 2 to 3 percent and . . . forest area from 17 to 24%" (Jászay 1990, p. 9). Among the comparative advantages of the Eastern cultures in this context is their often intact rural culture and knowledge of how the land works—assets long since all but destroyed in the overurbanized West (Jackson et al. 1984). There may be more ecological wisdom to be learned from many a Byelorussian or Bulgarian countryperson than from a Western agronomist.

Developing Countries

For all their historic handicaps, most of the peoples of the Soviet Union and eastern Europe represent industrial cultures with a long and distinguished history of sophisticated technical achievements. Comparable native talents, however, have had a very different historic pattern of expression in developing countries. While their achievements in many arts, sciences, and ways of living have often been extraordinary, often predating analogous Western progress by many centuries (Arab and Chinese astronomy, Chinese medicine, Sri Lankan irrigation, and Polynesian navigation come immediately to mind), they have not in general had the opportunity to exploit cheap resources, usually taken from poorer countries or bought at competitively depressed prices, with Western manipulative technologies to produce "modern" infrastructutre, sell manufactures at monopoly rents, and create, at least for themselves, widespread material wealth. Therein lies their great opportunity today *not to seek to tread that same development path, but to proceed more wisely.*

If developing countries try to repeat the mistakes of OECD, they will never develop. The cheap resources are dwindling; the force-fed monopolistic markets are going fast; the direction of the global economy no longer supports colonists' former ability to buy cheap and sell dear. Rather, one now increasingly buys raw materials at monopoly rents and sells manufactured goods at competitively depressed prices.

The collision between old cultures, new technologies, and perennial aspirations is perilous, both for the little peoples for whom a major mistake may be the end of the line, and for the great peoples whose ultimate weight in the world is so ponderous: a billion Chinese times anything is a big number, and before long, on present trends, India will have more people than China. This is not to say there are no hopeful signs: in the 1980s, the Chinese economy did apparently grow twice as fast as coal consumption (which represents three-fourths of energy use)—a commendable ~4.7%/y annual decline in aggregate energy intensity (Chandler et al. 1990). Yet China is still at least three times as energy-intensive as Japan, and a plausible scenario for 1990–2025 (*id.*) is widely assumed to end up with at least 1.4 billion Chinese, with quadrupled per-capita income (rising to a third of the current U.S. level), and with total energy and coal consumption tripling to nearly the present U.S. level. If anything like that actually happens (let alone whatever happens after that), then OECD will need efficiency even more to help stabilize the earth's climate.

Preventing Chinese energy use from tripling under such a 5%/y-GNP-growth scenario requires far more detailed and comprehensive efficiency efforts. Yet without such efforts, the growth may not occur. Consider, for example, the sad story of Chinese refrigerators. When the government decided people should have them, more than a hundred factories were built, and the fraction of Beijing households owning one rose from 2% to 62% in six years. But through inattention, they were built to an inefficient design. China now needs billions of unavailable dollars' worth of new power plants to run those inefficient refrigerators. An effort at development instead created crippling shortages of both power and capital. The officials to whom this was pointed out said the error would not, if they could help it, be repeated: it had taught them that China can afford to develop only by making energy, water, and other kinds of resource efficiency not just an add-on program but *the very cornerstone of the development process.* Otherwise, as is already true even in such a fundamentally wealthy country as the USSR (*supra*), the waste of resources will require so much and so costly supply-side infrastructure that too little money will be left to build the things that were to *use* those resources. As a wise homebuilder once put it, "If you can't afford to do it right the first time, how come you can afford to do it twice?"

Recall, too, that the average poor country[100] derives nearly three times less economic output per unit of primary energy used (commercial and traditional) than the average rich country—which in turn can cost-effectively at least quadruple *its* energy productivity. These two facts together imply that if poor countries leapfrogged over the mistakes of the rich countries, they could in principle expand their economies *roughly tenfold* without incrasing their energy use at all—while the rich countries could in principle sustain or improve *their* standard of living while using several times less energy than now. (Principle and practice probably differ here. Because of practical constraints in getting organized, the former pace of savings seems less plausible than the latter.) Ultimately, both groups would meet in the middle, and there would be enough energy for all.

Lack of energy or fuel, however, is not the problem. In Reddy and Goldemberg's masterly summary (1990):

> If current trends persist, in about 20 years the developing countries will consume as much energy as the industrialized countries do now. Yet their standard of living will lag even farther behind than it does today. This failure of development is not the result of a simple lack of energy, as is widely supposed.
> Rather, the problem is that the energy is neither efficiently nor equitably consumed. If today's most energy-efficient technologies were adopted in developing countries, then only about one kilowatt per capita used continuously—roughly 10 percent more than is consumed now—would be sufficient to raise the average standard of living to the level enjoyed by western Europeans in the 1970s.

This discussion therefore seeks to supplement the large literature on energy and development—much of it compactly summarized elsewhere (Goldemberg et al. 1988; Lovins et al. 1981)—with a few observations on how to "do it right the first time" and improve the prospects for a "leapfrog strategy."

Marketing for Diversity

First, action requires understanding of choices, and "Consumers who do not obtain their energy efficiently fall into three categories: the ignorant, the poor and the indifferent" (Reddy and Goldemberg 1990). They continue:

---

[100] Most poor countries are in fact "dual societies," consisting of small islands of affluence in vast oceans of poverty. The elite minorities and the poor masses differ so much in their incomes, needs, aspirations and ways of life that, for all practical purposes, they live in two separate worlds" (Reddy and Goldemberg 1990). For that reason, this discussion focuses mainly on the needs of poor *people* within poor countries.

The first [group] consists of people who do not know, for example, that cooking with LPG is more efficient than cooking with kerosene. They can be educated to become more energy efficient.

The second category consists of those who do not have the capital to buy more efficient appliances—An Indian maid may know that her employers spend less money cooking with LPG than she does with kerosene, yet she may not be able to switch because LPG stoves are about 20 times more expensive than kerosene stoves.

This is the position, albeit in a far starker context, of the Western householder who is deterred by the high cost of a compact fluorescent lamp. The remedy is analogous:

. . . utilities or other agencies should help finance the purchase of efficient equipment with a loan that can be recovered through monthly energy payments. Alternatively, a utility can lease energy-efficient equipment. A consumer's savings in energy expenditures can exceed the expenses of loan repayments and new energy bills. In principle, this method of converting initial costs into operating expenses can be extended to commercial and industrial customers as well, thereby improving efficiency and modernizing equipment at the same time.

This is what an RMI-cosponsored pilot project seeks to do with compact fluorescent lamps in Bombay (Gadgil and Sastry 1990). The lamps would be leased with an always-favorable cashflow to the user. The lamps would save the utility at least six times their total cost. There are two bonuses: the power saved in the heavily subsidized household sector, where it is sold at a loss, can then be resold to businesses which pay full rates, so the utility converts a loss to a profit; and in a city where 37+% of the evening peak load is from lighting, the saving should help to prevent the evening crashing of the grid, as well as improving reliability in some adjacent Western-grid states now short of capacity. Even such simple measures can have a profound economic effect: by one expert estimate, giving away such lamps throughout a very poor country like Haiti might raise the average household's disposable income by as much as one-fifth, because so much of the sparse cash economy goes directly or indirectly for electricity, mainly for lighting.

The third category of consumers consists of those with little incentive to raise their energy efficiency because their energy costs are so small or because the costs are almost unaffected by efficiency changes . . . Enticing those customers . . . will depend on intervention at higher levels [e.g., via government] . . . efficiency standards.

Bombay has such people too—especially affluent householders who can afford tubular fluorescent lamps three times as efficient as the poor user's incandescents. Marketing to the affluent group may require emphasizing the compact fluorescent lamp's longer life and esthetic superiority (better color, no flicker, no hum, nicer shape).

Institutional Parallels

These three categories and remedies are not unique to developing countries: they have exact parallels in OECD, and for that matter in the USSR and eastern Europe. Parallel, too, is the problem that most "utilities, financial institutions and governments" lack "methods for converting the initial cost of efficient systems into an operating expense." Further parallel is the issue of scale:

> Spending $2 billion on end-use improvements is much more complicated [than spending $2 billion on a single power plant]. If each efficiency measure costs between $2,000 and $20 million, then between 100 and a million subprojects are involved. Organizing so many diverse activities is difficult. . .

—especially in a nonmarket economy. And equally parallel is the almost universal bad habit of using energy consumption, or (worse) its rate of growth, as a measure of progress rather than as an indicator of inefficiency in meeting social goals with elegant frugality. Until energy planners start by asking what the energy is *for*, and how much of it, of what kind, at what scale, from what source, will do that task in the cheapest way, the outcomes will be far from rational or even affordable. Until energy planners appreciate that it is ten times as important to eliminate the "payback gap" as to get the prices right, their exhortations simply to desubsidize energy prices will continue to prove inadequate.

In developing and ex-socialist countries, while seeking to harness or mimic market forces, it is especially important to remember that markets do not in general produce justice, equity, or sustainability. They were not meant to. Equity requires political and ethical instruments, and an appreciation that, as Reddy's field experience has taught him, "If you look after the needs of the poorest, everything else will look after itself."

Lending Institutions

One of the chief obstacles to sound energy-for-development policies is the World Bank's and other multinational lenders' persistent lack of interest and insight (Van Domelen 1989). This is partly because while reorganizing a few years ago, the Bank laid off most of its technical staff and retained chiefly economists who don't understand the engineering and tend toward not-invented-here reflexes. But it's also because the system of rewards within such institutions, as in commercial banks, creates incentives to make bigger loans, not smaller ones, and to maximize loan volume in a way that a small, centralized staff can do only in large chunks. The Bank and its peer organizations also lack "templates" for successful Third World energy-efficiency projects. These structural adjustments within the lending organizations will be made only when their major supporters insist that lending follow the same cost-minimizing, all-things-considered rules that their own utilities are already institutionalizing—i.e., that energy financing be least-cost (*id.*). The Bank's ~$2 billion a year worth of power-plant investments would practically all fail such a test.

To encourage this major shift of emphasis, the Bank and its regional counterparts may wish to experiment with becoming a carbon-offset broker. For example, American, German, and Japanese coal-fired utilities and industries might choose to fund exceptionally cheap Third World carbon-abatement projects—agroforestry, lighter-colored buildings and pavements, Curiciba-style bus systems, lighting and motor retrofits, etc.—*through the Bank* via carbon-offset contracts. (They could even train the relevant Bank and host-country staff in the implementation techniques those modern utilities have developed, and perhaps learn too about the often smarter ideas that resource-short hosts tend to develop out of necessity.) This is at first blush an adventurous financial concept, but in structure it is no odder than a debt-for-nature swap. Another alternative, suggested by Reddy and Goldemberg (1990), is that carbon taxes, collected chiefly from industrialized countries under the OECD "polluter pays" principle, be earmarked for energy- and other carbon-saving projects in developing countries. Certainly OECD's historic responsibility for much of the cause of global warming justifies no less.

It has been argued (Krause et al. 1990, pp. I.5–7 & –8) that even if tradeable emission rights have an initial allocation based on an equitable per-capita formula rather than one which "grandfathers" OECD's historically high emissions, such melancholy experiences as the Third World debt crisis suggest "that approaches based mainly on market exchange will not work among nations as structurally different as developing and industrialized countries." In today's political climate, that reservation may be valid (although, as with the Berlin Wall, the unlikely can happen). But as a limited component of a far broader implementation strategy, tradeable rights do make good sense (*id.*). After all, in societal terms they are allocating not costs but benefits, and the only question is how to allocate the benefits to give all parties the necessary incentives to act. Since abating global warming is better than free, *the object is not to figure out how to share sacrifices for the common good, but rather how to help individuals, firms and nations behave in their economic self-interest.* To this end, Krause et al. (*id.*, p. I.5–18ff) have further proposed an ingenious mechanism with some potential advantages—a Superfund-like international Climate Protection Fund combined with carbon-reduction auctions.

International lending, both bi- and multilateral, badly needs to be restructured "from support for specific projects (such as building dams) to support for goal-oriented programs (such as lighting more homes)" (Reddy and Goldemberg 1990). Another worthy goal would be to support, as in eastern Europe and the USSR, joint ventures to produce efficient equipment locally. Many utility officials in developing countries, for example, say that their countries have refrigerator factories—but are unaware that those factories could joint-venture with, or license from, a Danish firm to make refrigerators that look and cost about the same but use ~80–90% less electricity (Shepard et al. 1990). It appears that such industrial marriage-brokerage is not on lenders' agenda. It should be. It is indeed extraordinary that some developing countries, like Brazil, manufacture for export to rich countries certain appliances more efficient than are available for sale to their own people.

In some instances, it may be possible to package an advantageous three-way swap: e.g., using a western European compact-fluorescent lamp technology, plus rare-earth phosphors from Soviet minerals, manufactured in an Indian free-trade zone (using rupees as a bridge between guilders and rubles), for sale in both socialist and developing countries if not to the West as well. Currently, however, such high-leverage opportunities apparently are not on lenders' radar screens at all.

Critical to these shifts of emphasis are education, training, and institution-building. Interesting programs are starting to emerge in countries like Brazil, Ghana, India, Thailand, and Tunisia. Their university and government efforts could be nurtured into regional centers of excellence in energy-efficiency technology, field implementation, and policy. A significant private initiative to build such centers' capabilities for regional outreach, research, and training is under serious discussion. Trained people from *within* each culture are essential: only they can understand behavioral issues and novel kinds of mistakes that would baffle a Western expert.[101] A "Negawatt University" network that induced a frew smart graduate students to devote their careers to energy efficiency could make a big difference. Today, a few such nascent centers exist (e.g., Bangalore, Bangkok, Berkeley, Genève,

---

[101] We recently heard of a major tourist hotel, near the Equator, that installed a 250-ton chiller to attract affluent visitors. It never worked: the building seemed to get hotter in the summer and colder in the winter. No expert even from the capital could figure out what was wrong with it. The manager started asking every engineer who came to stay to take a look. Finally, one Dutch engineer instantly spotted the problem: the chiller's chilled-water output was being pumped straight into tecooling tower.

Grenoble, Lund, Lyngby, Princeton, São Paulo, Sydney), but they need support.

Overarching all these issues, even education, is the power of example. Americans, for example, cannot preach that others should protect their forests as we clearcut our own in Alaska, /Hawaii, and the other 48 states. We cannot preach the virtues of population control, or energy efficiency, or sustainable rural economies, as we officially erode their foundations at home. But the power of a positive example can be even stronger than the power of a negative example. Acting from our highest traditions has moved the world before, and can again.

The almost-conventional agenda mentioned so far is essential. But several more steps, seldom discussed, seem warranted too. The central one is a major effort to inventory, then shift, the efficiency of energy-using equipment in international commerce.

Technology Transfer—Positive and Negative

In Osage, Iowa (population 3,800), the municipal utility has worked with local vendors to make good equipment easy to get and bad equipment hard to get. The hardware stores have a good selection of compact fluorescents, but may keep incandescent lamps out back where you have to ask for them. The lumber yards don't stock 2×4" studs for poorly insulated frame walls, nor ordinary double glazing—"Sorry, sir, that's obsolete—special order, it'll take a month"—but instead sell wider studs and superwindows, and explain their superiority to help make the sale.[102]

Nobody does that in global commerce. From Bangkok to Cairo, apparently cheap Taiwanese ballasts are being installed in fluorescent lighting fixtures. Those ballasts are often made from unrefined recycled-scrap-copper wire with such high resistivity that its overheating causes most of Taipei's house fires. Seemingly cheap Taiwanese, Czech, and other motors, gearboxes, and the like are also common, and often so inefficient that if run at more than a modest fraction of their rated power, they burn up. Although apparently no one has yet done a formal survey of this component market, nor of the equipment that incorporates such components, many anecdotes leave a strong impression that the world is awash in very inefficient end-use devices

---

[102] This is a small part of a series of initiatives whereby Osage's residential energy savings have enabled the utility to prepay all its debt, build up a large interest-bearing surplus, cut its rates five times in five years (to half the average Iowa level), thereby attract two big factories to town, and keep recirculating in the local economy more than $1,000 per household per year. This plugging of unnecessary dollar leaks has made the local economy noticeably more prosperous than in neighboring towns. The same principle, elaborated in RMI's Economic Renewal Project, applies to villages, states, and countries.

marketed to people with limited technical understanding, no independent information sources, and almost infinite discount rates.

Making and selling those devices, however, is not such an innocent pastime. The precious capital, especially foreign exchange, consumed in trying to build and fuel power plants and other energy-supply systems to run such inefficient equipment cannot be used to buy vaccines, to provide clean drinking water and family planning, to plant trees, to teach women to read. The overloaded power supply is unreliable and cannot run a productive industry, imposing huge backup-generator costs. Too little electricity is available for direct needs such as lighting (further hampering literacy) and pumping water. Tracing back such interlinked opportunity costs suggests that those ballasts and motors may ultimately create about as much human misery as the drug traffic. This "negative technology transfer" needs to be put squarely on the international agenda.

We do not know what it would take to get bad equipment off the world market, or at least to stigmatize it as the menace to development that it represents; but some countries have ideas. In Tunisia, for example, the national energy office has achieved an exemplary but little-known decoupling of economic activity from energy use, chiefly by developing nearly two dozen national minimum performance standards for basic appliances, lights, motors, and vehicles. You can buy a variety of cars, but they've all been bid, bought, and imported in bulk by the government on the basis of their fuel efficiency. Inefficient models are deliberately excluded. The saved capital is then available for what people really need—a prerequisite to the sensible investment approach that has paid off so richly in Costa Rica.[103]

Another possible analogy is the recent UNEP convention on international trade in hazardous wastes, requiring the "informed consent" of the recipient. Recipients of electricity-wasting equipment could surely be better informed—say, by labelling with life-cycle costs, and a list of alternatives, perhaps put out by the U.N. itself.

An international convention banning trade in energy-guzzlers, as in endangered species, may be too much to hope for, but perhaps Customs authorities, in countries from Bermuda to India where they play a key role in trade regulation[104], could be encouraged to attach tariffs varying inversely

---

[103] Research at Shell and by the late human-rights barrister Paul Sieghart has confirmed the basic premise of basic-human-needs strategies by showing three nearly perfect correlates of conventionally measured development success: absence of subsidies to basic commodities, adherence to basic human rights, and the health and education of the people ten years earlier. (The Costa Rican electric utility, incidentally, runs a lottery for everyone who uses less electricity this month than last month; but their opportunities to use less are currently limited to behavioral change, since they can't yet buy very efficient equipment and receive no financial help in doing so.)

[104] But often a perverse one, e.g., in India, where they charge far higher duties on energy-saving than on energy-supplying imports.

with energy efficiency (or even positive-or-negative duties structured like feebates). Devoting to import regulation of energy efficiency a small fraction of the engineering skill that now goes into designing power plants could pay big dividends. One can even imagine multicountry consortia that bulk-buy energy-using commodities (lights, motors, windows, etc.), and maintain a common secretariat that tracks and helps to get the best buys. One way or another, it is important to make it cheaper and more politically attractive for both buyers and sellers to deal in efficient than in inefficient equipment. The industrialized countries, too, have a major but unacknowledged responsibility in this matter: they typically set a bad example in what *they* buy, and they often export their most inefficient and obsolete equipment to boot—like deregistered pesticides all over again.

Another issue meriting exploration is the ways in which intellectual property rights may collide with transfer of energy-saving technology to countries that need them but cannot afford the royalties. Where such problems occur, a possibly good use for carbon-tax revenues or carbon-offset sales revenues might be to pay those royalties—a sort of *pro bono* fund. It may also be possible to organize a form of nonmonetary recognition, perhaps by UNEP, for inventors and firms who waive royalties in such circumstances. There may be helpful analogies, too, with the ways in which intellectual property rights were handled when Green Revolution crop strains were developed for use by developing countries.

The need for further research, demonstration, development, and outreach—for there is already much good news to report—hardly needs further emphasis. But it is important to stress a major hole in most current research agendas: the integrative design of future energy systems that integrate efficiency with renewables. This includes the redesign of industrial processes to match solar heat more conveniently, as those processes were in the past successively redesigned to fit the characteristics of wood, coal, oil, gas, and electricity. It also embraces efforts to anticipate bottlenecks, problems, and technical gaps as renewables and efficiency come to play major roles in energy-service supply in each sector.

Reddy and Goldemberg (1990), asking whether the needed transformation of energy policies can take place in the next few decades, conclude that aside from actions by the international bodies charged with, but largely ignoring, this responsibility, "the best hope for change lies in a convergence of interests." Industrialized countries need to protect the environment and ensure sustainable development; the environmental movement is joining hands worldwide; advocates for the poor, and increasingly the disenfranchised themselves, are crying out for policies that work better and cost less. The realization is spreading around the world that high levels of energy efficiency, and farming and forestry practices that treat nature as model and mentor, are not hostile but vital to a stable climate, a healthful environment, sustainable development, social justice, and a liveable world.

This convergence will be seen and used more clearly for what it is when everyone, especially we in the industrialized world, better understand what so often obscures our view of poor countries' energy and development needs. It is not our regions' differences, important though they are, but their similarities.

ADDITIONAL BENEFITS

Energy efficiency and sustainable farming and forestry practices, and the other ways described above to abate global warming and make money simultaneously, all have additional benefits which presumably could be expressed as monetary values. Counting those values would make their sources look even cheaper. The following list of omitted joint benefits is illustrative, not exhaustive.

Environmental Protection

Energy efficiency and renewable energy sources largely eliminate the environmental impacts of mining, transporting, and burning the fuel they displace. Reduced CO_2 is only one such benefit. Others include avoided SO_x, NO_x, O_3, hydrocarbons, particulates, and other air pollutants; despoiled land; acid mine drainage; oil spills; and impaired esthetic, wilderness, and widlife values.

A simple example suffices. Rather than raising electric bills to clean up dirty coal-fired power stations, a utility can help its customers to get superefficient lights, motors, appliances, etc., so they need less electricity to do the same tasks. The utility will then burn somewhat less coal and emit less sulfur (ideally backing out the dirtiest plants first), but mainly the utility will save a lot of money, because efficiency costs less than coal. That saved operating cost can then be used partly to clean up the remaining plants, partly to make electricity cheaper, and partly to reward the utility's shareholders. Similar principles permit the negative-cost abatement of urban smog by such measures as superefficient cars (Lovins 1989b). In either case, not only the health of people and other living things (and the longevity of cultural monuments and natural artifacts) will benefit; forests and other ecosystems, being less degraded, will also be better able to sequester carbon.

Sustainable agriculture and forestry, too, do not just reduce biotic CO_2, CH_4, and N_2O emissions; they also help to preserve topsoil, genes (biodiversity), water, fuels, farms, and farmers. They control floods, reducing the siltation of dams and of navigable waterways. (Dams that last longer displace more fossil-fuel CO_2 from thermal power plants.) Sustainable practices increase the habitat and population of land, aquatic, and marine wildlife. They preserve or create diverse and beautiful landscapes. They protect rural culture—an important cultural "anchor"—and help to reverse rural depopulation, a key contributor to urban problems. They reduce or eliminate dependence on biocides, whose manufacture accounts for a substantial fraction of all toxic-waste generation, and reduce associated problems ranging from occupational exposure and drift in application to runoff and groundwater contamination after application. They similarly reduce fertilizer contamination of ground- and surface waters, hence eutrophication, nitrate toxicity in drinking water, etc. They also restore the wholesomeness of food now contaminated by biocides, hormones, and antibiotics, probably thereby benefiting public health.

Moreover, such sustainable practices reduce crop losses to pests by rebuilding predator stocks and diversity; may modestly raise net farm or forest income, and certainly make that income stream less vulnerable to weather, pests, crop prices, and other uncontrollable variables; and should reduce dependence on government subsidies. They keep farming, in short, as profitable as now or (in most cases) more so (NRC 1989), and far more consistently so. They thereby reduce strains on the rural banking system, reliance on commodity brokers, and risks arising from commodity speculation. They set a good example for agricultural evolution abroad. They foster a land ethic and the practice of stewardship. And they make farming more diverse, interesting, and appealing to the young.

In sum, informal estimates at EPA's Pollution Prevention Office suggest that most—perhaps around 90%—of the problems EPA deals with could be displaced at negative cost, just by energy efficiency and by sustainable farming and forestry. That is a pleasant byproduct of abating global warming at a profit.

Security

Security—freedom from fear or privation or attack—may be achieved by other and cheaper means than armed might and threats of violence (Harvey et al. 1990). Security comprises access to reliable and affordable necessities of life (water, food, shelter, health, and healthful environment, a sustainable economy, a legitimate system of government, certain cultural and spiritual assets). It also requires some combination of conflict prevention, conflict resolution, and defense—preferably nonprovocative—that is predictably able to defeat aggression (*id.*). Resource efficiency in all its forms is an essential element of both providing needed resources and forestalling conflict over resources.

Energy/security links take many forms, and the list grows with awareness that climatic stability and biotic productivity are essential elements of global security. High on anyone's security list are equitably providing the energy needed for sustainable development; reducing domestic energy vulnerability (Lovins and Lovins 1982); abating the spread of nuclear weapons (Lovins and Lovins 1981); and avoiding conflict over fuel-rich areas like the Persian Gulf.

At this writing, the United States and other nations are putting their youths in tanks[105] in the Middle East because the same youths weren't put in efficient cars. A three-mile-per-gallon improvement in the ~18–mi/gal U.S. household vehicle fleet would displace the mid-1990 rate of oil imports from Iraq and Kuwait; a 10 mi/gal improvement would displace all U.S. oil imports from the Persian Gulf. If the military cost of current actions in the Gulf were somehow internalized in the oil price, rather than socialized through taxes and deficits, it would be interesting to how well the oil would sell at probably not far from $200 per barrel.

This paper has described cost-effective efficiency improvements in U.S. oil use that could potentially displace the United States' oil imports from the Persian Gulf roughly seven times over. Similar opportunities, differing only in detail, are available to other oil-importing countries, even the most efficient. It is bad enough to pay in dollars the cost of continuing to ignore such opportunities. It is tragic to pay it in blood. Given that the Middle East is uniquely rich in oil, and full of diverse peoples who have fought each other for millenia, there is wisdom in at least making the oil under their disputed territories as irrelevant to the world's continued peace and prosperity as modern technology permits.

Equity

An original and pragmatic treatment of global warming (Krause et al. 1990) has proposed that the long-term fossil-fuel "budget" of ~300 billion tons of carbon (GtC) believe to be consistent with a probably tolerable rate of climatic change be split evenly between developing and industrialized countries. This would "push industrialized countries to fully mobilize their technological, financial, and organizational capacities for phasing out fossil fuels without creating infeasible goals" (*id.*, p. I–6.9). It would also leave developing countries more leeway to cope with their presumably lower adaptive capacity, greater needs, and greater population momentum. All regions together, under such a budget, might find it a reasonable milestone to return *global* carbon emissions to ~1985 release rates by ~2005—by

---

[105] An M-1 tank gets 0.56 miles per gallon. An oil-fueled aircraft carrier gets 17 feet per gallon.

which time the 20% reduction target set at Toronto should have been achieved by the industrialized countries. This seems a reasonable trajectory, and is certainly fairer than most. But one must also consider the "micro-equity" features of the specific tools proposed. In that respect, the strategy proposed both by Krause et al. (1990) and here seems attractive.

One of the best features of the efficiency-and-renewables energy strategy is that many of the technologies are "vernacular," able to be locally made with fairly common skills, and the renewable energy itself is equitably available to all. Sunlight is indeed most abundant where most of the world's poorest people live (Sørensen 1979). In no part of the world within the polar circles is freely delivered renewable energy inadequate to support a good life indefinitely and economically using present technologies (*id.*, Reddy and Goldemberg 1988; Lovins et al. 1981).

Much the same is true of sustainable farming and forestry practices. By relying chiefly on natural processes and assets, these practices minimize dependence on inputs that must be bought or brought from afar. In both cases, future generations' rights seem to be far better protected than now.

Resilience

Life is full of surprises. Energy analysts who owe their careers to singular events in 1973 and 1979 cheerfully go on to assume a surprise-free future. It will not be like that at all. There will even be surprises of kinds nobody has thought of yet (just as some people unfamiliar with the history of climatic science suppose global warming is). However, the fundamental principles of resilient design, borrowed chiefly from biology and engineering (Lovins and Lovins 1982), are completely consistent with the energy and agri/silvicultural strategies proposed here. More diverse, dispersed, renewable, and above all efficient energy systems can make major interruptions impossible in principle (*id.*). Growing green things in a way duly respectful of several billion years' design experience is the best way anyone knows to ensure that the earth will keep on handing down that experience.

Buying Time

In an earlier analysis of least-cost climatic stabilization (Lovins et al. 1981), two colleagues and we noted that if the terrible exponential arithmetic of burning more and more fuel, faster and faster, were simply reversed—if the amount of carbon released each year steadily *shrank*—then the "tail" of "global warming commitment" would soon become so slender that its length would be unimportant matter. A very long period would then be safely available for displacing the last remnants of fossil-fuel use. That simple idea remains valid.[106] In a more subtle sense, however, the time-buying value of techniques like energy efficiency is greater than meets the eye. Efficiency buys not just money and avoided pollution but also *time*—the most precious and least substitutable resource.

An anecdote is useful here. Around 1984, Royal Dutch/Shell Group planners foresaw the 1986 oil-price crash, and warned that a deepwater North Sea oilfield called Kittiwake would have to be brought in at 40% below the planned cost, because by the time the field opened, it would be possible to sell the oil only for $12 a barrel, not $20. The engineers, who had been sweating over one-percent cost cuts, were aghast. But offered the alternative of being fired and leaving the oil where it was, they cut the cost by 40% in about a year.

It turned out they had previously been asked the wrong question: how to bring on fields as fast as possible with cost no object, rather than how to do it cheaply even if it took longer. Asked the new question, they came up with completely different technological answers. (In how many other situations have we gotten the wrong answer by asking that same wrong question?) But the key result was that the new technology made oil that used to cost $30 to extract into oil that costs only ~$18 to extract: the whole oil supply curve therefore flattened out. This in turn postponed depletion, and this in turn bought time in which to develop and deploy still better technologies, on both the demand side and the supply side, which broaden the range of choice and which reinforce each other by buying still more time, together pushing depletion far into the future and facilitating a graceful transition to renewables.

Especially in view of the start-up obstacles to achieving major efficiency gains in the non-OECD regions, time-buying strengthens the case for strong and rapid efficiency gains in the countries (OECD) best equipped to achieve them. If that might be "overachievement" relative to some theoretical goal of equitable sharing, nobody should mind: on the evidence presented above, maximizing the size and speed of energy savings is likely only to bring larger and earlier economic *benefits*.

---

[106] Perhaps by coincidence, the rough and illustrative estimates presented in 1981, at a time when the role of trace gases other than CO_2 was poorly understood, are surprisingly close to those now emerging from today's far more sophisticated analyses. Indeed, Krause et al.'s rough timeline for a 300-GtC global carbon budget (1990)—a 20% reduction in global carbon emissions between about 2005 and 2015, 50% around 2025–2035, and 75% before the middle of the next century—is "roughly equivalent to the efficiency-plus-renewables scenario of Lovins et al. (1981)" (Krause 1990). The 450-GtC budget associated with more sanguine climatological assumptions is close to the Goldemberg et al. (1988) scenario, "modified to include significant renewables penetration" (Krause 1990). Krause concludes (*id.*): "The former scenario can be seen as representing the limit of practical logistic feasibility, while the latter scenario can be seen as marking the limit of climatic acceptability.***The range of climate-stabilizing energy scenarios is circumscribed by" those two scenarios.

CONCLUSIONS

This paper has rebutted ten prevalent myths (in italics) about abating global warming:

- *Greater scientific certainty should precede action.* The uncertainties about global warming and its potential consequences are substantial, interesting, and likely to cut both ways. But they are also irrelevant to policy, because virtually all the actions needed to abate global warming (if it does turn out to be a real problem) should be taken anyway to save money. These "no-regrets" actions are about enough to solve the problem if it does exist, and are highly advantageous even if it doesn't. The problem with global warming isn't decision-making under uncertainty; it's realizing that in this instance, uncertainty doesn't matter.
- *The issue is whether to buy a "climatic insurance policy" analogous to fire insurance or to defense expenditures (a major investment mobilizing most of the country's scientific and technological resources, and meant to forestall or respond to unlikely but potentially catastrophic threats to national security).* The "insurance" analogy is partly valid, because delaying action until obvious climatic changes are unambiguously underway makes abatement too little, too late, and too costly—just like trying to install a sprinkler system in a hotel that's currently on fire, or build military forces while you're already under attack, or buy collision insurance after you've crashed your car. Abating global warming will require significant efforts affecting large stocks of people and capital over long periods and with long lead times, so waiting too long will certainly raise cost, difficulty, and risk of failure (Schneider 1989). But the analogy breaks down if, as was shown above, the real choice is not balancing uncertain future benefits against daunting present costs, but rather making the investment as wisely and quickly as possible in order to achieve both the uncertain future benefits *and* the guaranteed financial *savings*. Any insurance "premium" is actually negative: the actions that can stabilize global climate will save money anyway, without counting the avoided costs of trying, or failing, to adapt, to possible climatic change. This "insurance" is unquestionably a good buy.
- *Abating global warming would be costly.* Distinguished econometricians have claimed that just achieving the Toronto interim target of cutting CO_2 emissions by 20%—roughly a third of the reduction probably required for climatic stabilization (Houghton et al. 1990)—would cost the United States alone on the order of $200 billion per year (B. Davis 1990; Nordhaus 1989; Manne and Richels 1990; Passell 1989). Such calculations are wrong by at least an order of magnitude (e.g., Williams 1990; Zimmerman 1990). Worse, their high-cost conclusion is a bald *assumption*

masquerading as a fact (W. Nordhaus, pers. comm.). The econometric analysis merely asks how high energy prices would need to be, based on historic price elasticities of demand (typically from decades ago), to reduce fossil-fuel use by x%, then counts those higher prices (or their equilibrium econometric effects) as the cost of abatement. This approach ignores the compelling empirical evidence that saving most of the fuel now used is cheaper than even its short-run marginal cost, and hence is profitable rather than costly. The econometricians thus have the amount about right but the sign wrong: using modern energy-efficient techniques to achieve the Toronto target would not cost but *save* the U.S. on the order of $200 billion a year.[107] These techniques did not exist at the time of the behavior described by the historic price elasticities: those elasticities summarize how people used to behave under conditions that no longer hold. Indeed, cost-minimizing energy policy—if not derailed by the blunder of treating future energy needs as fate instead of choice—will seek to change those conditions as much as possible.

- *Abating global warming would drastically curtail American and similar lifestyles, and would mean less comfort, mobility, etc.* Nothing could be further from the truth. The fuel-saving technologies that can stabilize global climate while saving money actually provide unchanged services: showers as hot and tingly as now, beer as cold, rooms as brightly lit, torque as strong and reliable, homes as cozy in the winter and cool in the summer, cars as peppy, safe, and comfortable, etc. The quality of these and other services can often be not just sustained but substantially improved by substituting superior engineering for brute force, brains for therm: e.g., with efficient lighting equipment, you get the same amount of light, but it looks better and you see better. The same is broadly true of sustainable agriculture and silviculture, which provide comparable yields with superior quality, resilience, human health, and (generally) profitability.

- *If such cost-effective abatements were available, they would already have been bought.* This is reminiscent of the econometrician who, asked by his mannerly granddaughter whether she could pick up a $20 bill she'd just noticed lying on the sidewalk, replied "No, my dear, don't bother: if it

---

[107] Rosenfeld et al. (1990) note that commercial direct fuels cost Americans $283 billion and electricity $175 billion in 1989. Saving (for illustration) a fifth of each at average prices would save $92 billion a year, but the costliest sources would in fact be displaced first, and long-run marginal costs generally exceed present prices. Together, these effects probably at least double the value of the savings. A similar result could be obtained by a conservative method using longer-term savings potentials: saving about two-thirds of the direct fuels (an understated and rough composite of the potential for all sectors) and, in the short run, only one-fourth (utilities' average fuel-cost share) of 75% of the electricity, would total ~$221 billion. More sophisticated calculations are of course possible, disaggregated by sector, fuel, region, timing, etc., but not very useful.

were real, someone would have picked it up already" (M. Gell-Mann, pers. comm.). The striking disequilibrium between how much energy efficiency is now available and worth buying and how much has already been bought arises from distinctive, well-understood market failures that leave cheap efficiency seriously underbought at present prices. (For example, consumers have poor access to information and to mature mechanisms for conveniently delivering integrated packages of modern technologies. Discount rates are about tenfold higher for buying efficiency than supply, thus diluting price signals. Many energy utilities misunderstand their business and want to increase their sales—even though reducing their sales would increase their profits by decreasing their costs even more. Perverse regulatory signals often reward inefficient and penalize efficient behavior. Markets in saved energy are sparse or absent. And present market signals, omitting externalities that may be as big as the apparent fuel prices, make consumers indifferent to whether they buy, for instance, a 20- or a 60-mile-per-gallon car, since both cost about the same per mile to own and drive.) Solutions exist for each of these market failures. These solutions have been proven in market economies and are rapidly emerging in a wide range of other societies, so there is an ample range of effective policy instruments to choose from. The technical and implementation options—the everyday work of energy-efficiency practitioners—are mostly unknown, however, to those econometricians who lie awake nights worrying about whether what works in practice can possibly work in theory.

- *Abatements would be so costly and disagreeable that they could only be achieved by draconian, authoritarian government mandates incompatible with democracy.* On the contrary, the abatements described above are so profitable and attractive that they can be largely if not wholly achieved by existing institutions, within the present framework of free choice and free enterprise. Planners unaware of market-driven alternatives seem anxious to set up new bureaucracies to tell people how to live. Many bizarre schemes have been suggested for substituting dirigisme for markets, penury for development, risks for rewards, and costs for profits. This paper seeks to provide an antidote to this perversion of economic rationality.
- *Combating global warming requires tough tradeoffs—swapping one kind of pollution or risk for another.* Abating global warming by resource efficiency can simultaneously reduce or eliminate many other hazards too—oil-security risks, nuclear proliferation, utilities' planning and financial risks, declining farm and forest yields, etc.—without creating new ones.
- *Available means of abatement, singly or combined, will be too small and too slow, so global warming is inevitable and we must start trying to adapt*

to it. This counsel of despair is misguided. To be sure, some significant degree of climatic change or increased climatic volatility in some places may already be unavoidable if the more sensitive models prove valid (Houghton et al. 1990; Krause et al. 1990), or if greater climatological or ecological understanding continues to bring unpleasant surprises. A modest degree of adaptation may therefore be prudent if not inevitable: e.g., planning coastal developments to accommodate some sea-level rise and water projects to tolerate shifts in rainfall, or reversing the narrowing of crops' and forests' genetic bases. Nonetheless, the techniques described here, if their benefits are properly understood, show promise of such rapid and widespread deployment that most of the harm projected in today's best models could almost certainly be avoided. Many abatement measures also have the valuable side-effect of increasing resilience in the fact of whatever climatic change may nonetheless occur.

- *Abating global warming would lock developing countries into abject poverty, or at least prevent their achieving their legitimate aspirations—even though most global warming so far has been caused by the industrialized countries.* On the contrary, the abatement options discussed above are not merely compatible with but essential to affordable and sustainable global development and increased equity.
- *Policymakers already know what their options are and haven't chosen those described here, so either the policymakers are stupid or the options don't work.* Many policymakers suppose that abatement must be slow, small, costly, inconvenient, and nasty—not because that's true, but simply because they don't know any better. The difficulty, we suspect, may be the one economist Ken Boulding described: that a hierarchy is "an ordered arrangement of wastebaskets designed to prevent information from reaching the executive." The options described above are available, demonstrated, and often in widespread and successful use. Many, however, are so new that they are not yet widely known even to technical experts, and will take many years to filter up to decisionmakers through normal channels. What is needed, therefore, is better and faster technology transfer to the policymakers. This paper is hoped to contribute to that effort.

REFERENCES

Abrahamson, D. 1989. Relative Greenhouse Heating from The Use of Fuel Oil and the Use of Natural Gas. Minneapolis, MN.: Hubert H. Humphrey Inst. of Public Affairs, Univ. of Minnesota.

Akbari, H. et al. 1988. The impact of summer heat islands on cooling energy consumption and global CO_2 concentration. LBL-25179. Berkeley: Lawrence Berkeley Lab.

Arbatov, A.A. 1990. Deputy Chairman, Commission for the Study of Production Forces and Natural Resources, USSR Academy of Sciences, Moscow, pers. comm. to U. Fritsche, Ökoinstitut, Darmstadt.
Automotive News. 1983. Cat's 3306B makes it big in the real world. 7 Nov.
Automotive News. 1986. Chrysler genesis project studies composite vehicles. p. 36, 5 May.
Baldwin, S.F. 1987. Biomasss Stoves: Engineering Design, Development and Dissemination. Arlington, VA.: VITA.
Baldwin, S.F., H. Geller, G. Dutt, and N.H. Ravindranath. 1985. Improved woodburning stoves: signs of success. Ambio **14(4–5)**:280–287.
Bechmann, A. 1987. Landbau-Wende: Gesunde Landwirtschaft—Gesunde Ernährung. Frankfurt/M.: S. Fischer.
Bevington, R. and A.H. Rosenfeld. 1990. Energy for buildings and homes. *Sci. Am.* Sept.: 76–86.
Blair, P. 1990. Presentation to Aspen Institute for Humanistic Studies Energy Committee, August.
Bleviss, D.L. 1988. Saving fuel. *Technol. Rev.* Nov./Dec.: 47–54.
Bleviss, D.L. 1988a. The New Oil Crisis and Fuel Economy Technologies. Westport, CT.: Quorum Books.
Bleviss, D.S. and P. Walzer. 1990. Energy for motor vehicles. *Sci. Am.* **263**:102–109.
Bodlund, B. et al. 1989. The challenge of choices: technology options for the Swedish electricity sector. In: Electricity, ed. T.B. Johansson, B. Bodlund, and R.H. Williams, pp. 883–947. Lund: Lund Univ. Press.
Borré, P. 1990. Presentation to Aspen Institute for Humanistic Studies Energy Committee, August.
Bossell, H. et al. 1986. Technologiefolgeabschätzungen für die Landwirtschaftliche Produktion. Hannover: Inst. für Systemanalyse und Prognose.
Brody, J.E. 1985. Organic farming moves into the mainstream. *New York Times*, p. 20, 8 October.
Brookes, W.T. 1989. The Global Warming Panic. *Forbes* Dec.: 96–100.
Browning, W. 1990. Steaks and Mistakes. unpublished MS in course of revision for publication by Rocky Mountain Institute.
Calwell, C., A. Edwards, G. Gladstein, and L. Lee. 1990. Clearing the Air. San Francisco: Natural Resources Defense Council.
CEC (California Energy Commission). 1988. 1988 Conservation Report. Sacramento, CA. October.
CEC 1990: 1990 Conservation Report (Staff Draft), Sacramento, CA. August.
Chandler, W.U. 1989. Control of Carbon Emissions in the Soviet Union, Poland, and Hungary. IPCC draft expert report, 15 January.
Chandler, W.U., ed. 1990. Carbon Emissions Control Strategies: Case Studies in International Cooperation. Conservation Foundation, Washington, D.C., in press.
Chandler, W.U., A.A. Makarov, and D. Zhou. 1990. Energy for the Soviet Union, eastern Europe and China. *Sci. Am.* **263**:120–127.
Chizhov, N. and M.A. Styrekovich. 1985. Ecological advances of natural gas over other fossil fuels. In: The Methane Age, ed. T.H. Lee et al., pp. 155–161. Dordrecht: Kluwer and Laxenburg, Austria: ILASA.
Churchill, A. 1989. Digest of the 14th World Energy Conference. Montreal. Washington, D.C.: World Bank. Discussion at p. 120.
Columbo, U. 1988. The technology revolution and the restructuring of the global economy. In: Globalization of Technology: International Perspectives, eds. J.H. Muroyama and H.G. Stever, pp. 23–31. Washington, D.C.: Nat. Acad. Pr.

Cramer, H.H. 1984. On the predisposition to disorders of middle European forests. *Pflanzenschutz-Nachrichten* **37**:98–207.
Crutzen, P.J. et al. 1986. Methane production by domestic animals, wild ruminants, other herbivorous fauna and humans. *Tellus* **28B**:271–284.
Davis, B. 1990. Bid to slow global warming could cost U.S. $200 billion a year, Bush aide says. *Wall St. J.* p. B4, 16 April.
Davis, G. 1990. Energy for planet Earth. *Sci. Am.* **263**:54–62.
DeLuchi, M.A., R.A. Johnston, and D. Sperling. 1988. Transportation fuels and the greenhouse effect. *Transportation Research Record* **1175**:33–44.
Difiglio, C., K.G. Duleep, and D.L. Greene. 1989. Cost effectiveness of future fuel economy improvements. *J. En.* **2(1)**:65–83.
DPA Group. 1989. Study on the Reduction of Energy-related Greenhouse Gas Emissions, report to Federal-Provincial Task Force on Global Warming. Toronto.
Duxbury, M.L., S.D. Neville, R. Campbell, and P.W.G. Newman. 1988. Mixed Land Use and Residential Satisfaction: An Evaluation. Transport Research Paper 2/88, Perth, Western Australia: Envir. Sci., Murdoch Univ.
Eketorp, S. 1989. Electrotechnologies and steelmaking. In: Electricity, ed. T.B. Johansson, B. Bodlund, and R.H. Williams, pp. 261–296. Lund: Lund Univ. Press.
El-Gasseir, M.M. 1990. The Potential Benefits and Workability of Pay-As-You-Drive Automobile Insurance. 8 June testimony to California Energy Commission, Docket #89–CR–90.
EPA. 1989. Policy options for stabilizing global climate, draft Report to Congress, D.A. Lashof and D. Tirpak, eds., 2 vols., February.
EPA. 1989a. Costs and Benefits of Phasing Out Production of CFCs and Halons in the United States. Office of Air and Radiation, 3 November review draft.
Ephraim, M. Jr. 1984. Locomotive energy-related technology: recent improvements and future prospects. In: ed. B.C. Houser. Rail Vehicle Energy Design Considerations. Amsterdam: Elsevier.
EPRI (Electric Power Research Institute). 1990. Efficient Electricity Use: Estimates of Maximum Energy Savings. CU-6746. Palo Alto CA.: EPRI.
Feist, W. 1987. Stromsparpotentiale bei den privaten Haushalten in der Bundesrepublik Deutschland. Darmstadt: Inst. Wohnen und Umwelt.
Fickett, A.P., C.W. Gellings, and A.B. Lovins. 1990. Efficient use of electricity. *Sci. Am.* **263**:64–74.
Flemings, M.C., et al. 1980. Materials substitution and development for the light weight, energy efficient automobile, report to U.S. Congress's Office of Technology Assessment. 8 February.
Fortune. 1986. Updating the diesel locomotive, p. 52, 3 March.
Gadgil, A.J., and M.A. Sastry. 1990. The BELLE Project: Bombay Efficient Lighting Large-Scale Experiment. Overview: An Innovative Method for Reducing Peak Electric Demand. LBL/RMI.
Goldemberg, J., et al. 1988. Energy for a Sustainable World. New Dehli: Wiley Eastern.
Goldstein, D.B., J.W. Holtzclaw, and W.B. Davis. 1990. NRDC/Sierra Club Testimony for Conservation Report Hearing on Transportation Issues. California Energy Commission Docket #89–CR–90, 23 April. San Francisco: Natural Resources Defense Council.
Greene, D.L. 1989. CAFE or Price? An Analysis of the Effects of Federal Fuel Economy Regulations and Gasoline Prices on New Car MPG, 1978–89. Oak Ridge National Laboratory/USDOE Office of Policy, Planning and Analysis, 10 May draft.

Greene, Deni. 1990. A Greenhouse Energy Strategy: Sustainable Energy Development for Australia, report to DASETT, Canberra, by Deni Greene Consulting Services, Melbourne.
Groscurth, H.-M., and R. Kümmel. 1989. The Cost of Energy Optimization: A Thermoeconomic Analysis of National Energy Systems. Physikal. Inst. der Univ. Würzburg, 16 May.
Harmon, M.E., W.K. Ferrell, and J.F. Franklin. 1990. Effects on carbon storage of conversion of old-growth forests to young forests. *Science* **247**:699–702.
Harvey, T.H., M. Shuman, and D. Arbess. 1990. Reclaiming Security: Beyond the Controlled Arms Race. Rocky Mountain Inst. New York: Hill and Wang.
Heede, H.R., R.E. Morgan, and S. Ridley. 1985. The Hidden Costs of Energy: How Taxpayers Subsidize Energy Development. Center for Renewable Resources, Washington, D.C. (Summarized in H.R. Heede and A.B. Lovins. Hiding the true costs of energy sources. *Wall St. J.* p. 28, 17 Sept. 1985.)
Hendriks, C.A., K. Blok, and W.C. Turkenburg. 1990. Technology and cost of recovery and storage of carbon dioxide from an integrated combined cycle plant. *Appl. En.*, in press.
Henne, P.A. 1989. MD-90 Transport Aircraft Design, Long Beach, CA.: Douglas Aircraft Co., McDonnell Douglas Corp. Am. Inst. of Aeronautics and Astronautics Aircraft Design, Systems and Operations Conf. Seattle, WA.
Herman, R., S.A. Ardekani, and J.H. Ausbel. 1989. Dematerialization. In: Technology and Environment, pp. 50–69. Washington, D.C.: Nat. Acad. Pr.
Holmberg, W. 1988. Farm Ethanol—Time for a National Policy. Washington, D.C.: Information Resources, Inc.
Houghton, J.T., G.J. Jenkins, and J.J. Ephraums, ed. 1990. Climate Change. The IPCC Scientific Assesment. Cambridge: Cambridge Univ. Press.
ICF. 1990. Prelimary Technology Cost Estimates of Measures Available to Reduce U.S. Greenhouse Gas Emissions by 2010, draft report to EPA, August.
IPCC. 1990a. *Report from Working Group III*, Policymakers' Summary, WMO/UNEP, June.
Jackson, W. 1980. New Roots for Agriculture. San Francisco: Friends of the Earth, and Salina, KS: The Land Institute.
Jackson, W., W. Berry, and B. Colman. 1984. Meeting the Expectations of the Land. San Francisco: North Point.
Jászay, T. 1990. Carbon Dioxide Emissions Control in Hungary: Case Study to the Year 2030. PNL-SA-18248. Richland, WA: Battelle Pacific Northwest Lab.
Johansson, T.B., et al. 1983. I Stället for Kärnkraft: Energi År 2000. DsI 1983:18. Stockholm: Industridepartementet, summarized in *Science* **219**:355–361 (1983).
Kahane, A. 1986. Industrial Electrification: Case Studies of Four Industries. Steel, Paper, Cement and Motor Vehicles in the United States, Japan and France, LBL report to EPRI, available from author at PL/12, Shell Centre, London SE1 7NA.
Kavanaugh, M. 1990. Fuel economies available from ultrahigh bypass (UHB) jet engines. memo to P. Schwengels (EPA), 10 March, Att. D. ICF, *op. cit. supra*.
Keepin, W.N., and G. Kats. 1988. Greenhouse warming: Comparative analysis of nuclear and efficiency abatement strategies. *En. Pol.* **16(6)**:538–561, and calculational supplement, "Greenhouse Warming: A Rationale for Nuclear Power?," available from RMI.
Keepin, W.N., and G. Kats. 1988a. Global Warning [*sic*]. *Science* **241**:1027.
Krause, F. 1990. Required Speeds of Fossil Fuel Phase-Out under a 2 Degree C Global Warming Limit. In: Energy Technologies for Reducing Emissions of

Greenhouse Gases, pp. 2:102–109. OECD/IEA Experts' Seminar, Paris, 12–14 April 1989, Organization for Economic Cooperation and Development, Paris 1989.

Krause, F., W. Bach, and J. Koomey. 1989. Energy Policy in the Greenhouse, report to Dutch Ministry of Housing, Physical Planning and Environment, International Project for Sustainable Energy Paths. El Cerrito, CA, September, vol. 1, summarized in Krause (1990).

Larson, E.D., M.H. Ross, and R.H. Williams. 1986. Beyond the era of materials. *Sci. Am.* June: 34–41.

Larson, E.D., P. Svenningsson, and I. Bjerle. 1989. Biomass Gasification for Gas Turbine Power Generation. In: Electricy, ed. T.B. Johansson, B. Bodlund, and R.H. Williams, pp. 697–739. Lund: Lund Univ. Pr.

Larson, E.D., and R.H. Williams. 1988. Biomass-fired Steam-injected Gas Turbine Cogeneration. Procs. 1988 ASME Cogen-Turbo Sympos. Montreaux, 30 Aug.–1 Sept.

Lashof, D.A., and D.R. Ahuja. 1990. Relative global warming potentials of greenhouse gas emissions. *Nature* **344**:529–531.

Leach, G., and R. Mearns. 1988. Beyond the Woodfuel Crisis: People, Land, and Trees in Africa. London: Earthscan.

Ledbetter, M., and M. Ross. 1989. Supply Curves of Conserved Energy for Automobiles. Att. B to ICF, *op. cit. supra.*

Ledbetter, M., and M. Ross. 1990. A supply curve of conserved energy for automobiles. Proc. 25th Intersoc. Energy Convers. Eng. Conf. (Reno, NV, 12–17 August). New York: Am. Inst. Chem. Engs.

Levenson, L., and D. Gordon. 1990. Drive+: promoting cleaner and more fuel efficient motor vehicles through a self-financing system of state sales tax incentives. *J. Policy Analysis & Mgt.* **9(3)**:409–415.

Lovins, A.B. 1976. Energy strategy: The road not taken? *Foreign Affairs* **55(1)**:65–96.

Lovins, A.B. 1978. Soft Energy Technologies. *Ann. Rev. En.* **3**:477ff.

Lovins, A.B. 1979. Re-examining the nature of the ECE energy problem. 1978 U.N. Economic Commission for Europe paper. *En. Pol.* **7**:178.

Lovins, A.B. 1980. Economically efficient energy futures. In: Interactions of Energy and Climate, ed. W. Bach et al. Dordrecht: Reidel.

Lovins, A.B. 1986. The State of the Art: Water Heating. Old Snowmass, CO: Rocky Mountain Inst./COMPETITEK.

Lovins, A.B. 1986a. The State of the Art: Space Cooling. Preliminary Edition, RMI/COMPETITEK.

Lovins, A.B. 1988. Negawatts for Arkansas, vol. 1, RMI #U88–30.

Lovins, A.B. 1989. Making Markets in Resource Efficiency. RMI Publication #E89–27.

Lovins, A.B. 1989a. End-Use/Least-Cost Investment Strategies. paper #2.3.1, World Energy Conference, Montreal, September.

Lovins, A.B. 1989b. Abating air pollution at negative cost via energy efficiency. *J. Air & Waste Mgt. Assoc.* **39(11)**:1432–1435.

Lovins, A.B. 1989c. Energy Options. *Science* **243**:12.

Lovins, A.B. 1990. Factsheet: How a Compact Fluorescent Lamp Saves a Ton of CO_2. Old Snowmass, CO: Rocky Mountain Inst. #E90–5. (*cf.* Gadgil, A.J., and A.H. Rosenfeld, Conserving Energy With Compact Fluorescent Lamps. Berkeley: Lawrence Berkeley Lab. 1 May 1990.)

Lovins, A.B. and L.H. 1981. Energy War: Breaking the Nuclear Link. New York: Harper and Row Colophon. Expanding A.B. and L.H. Lovins and L. Ross.

Nuclear power and nuclear bombs. *Foreign Affairs* **58**:1136ff (1980), **59**:172ff (1980).
Lovins, A.B. and L.H. 1982. Brittle Power: Energy Strategy for National Security, report to USDOD Defense Civil Preparedness Agency/FEMA/CEQ, Brick House, Andover, MA., summarized in "Reducing Vulnerability: The Energy Jugular," in R.J. Woolsey, ed., *Nuclear Arms: Ethics, Strategy, Politics*, Inst. for Contemporary Studies, San Francisco 1983, and The fragility of domestic energy, *Atlantic*, Nov. 1983, pp. 118–126.
Lovins, A.B. and L.H. 1989. Drill rigs and battleships are the answer! (But what was the question?). In: The Oil Market in the 1990s, ed. R.G. Reed, III, and F. Fesharaki, 83–138. Boulder, CO.: Westview.
Lovins, A.B. and L.H., F. Krause, and W. Bach. 1981. Least-Cost Energy: Solving the CO_2 Problem. Andover, MA.: Brick House. 2nd printing, Old Snowmass, CO: Rocky Mountain Inst. 1989; summarized as: Energy, economics, and climate, *Clim. Change* **4**:217–220 (1982); see also *id.* **5**:105ff (1983).
Lovins, A.B. and R. Sardinsky. 1988. The State of the Art: Lighting. Rocky Mountain Inst./COMPETITEK, Old Snowmass, CO.
Lovins, A.B., and M. Shepard. 1988. Financing Electric End-Use Efficiency. RMI/COMPETITEK Implementation Paper #1, and "Update," 1989.
Lovins, A.B., and M. Shepard. 1988a. Customer Behavior and Information Programs. RMI/COMPETITEK Implementation Paper #2.
Lovins, A.B., et al. 1989. The State of the Art: Drivepower. RMI/COMPETITEK.
Makarov, A.A., et al. 1988. Otsyenka parkov dvigatyeleiy v narodnom khozyaistvye SSSR i SShA. Moscow: Inst. of Energy Economics, USSR Academy of Sciences.
Makarov, A.A., and I.A. Bashmakov. 1990. The Soviet Union: A Strategy of Energy Development with Minimum Emission of Greenhouse Gases. PNL-SA-18094. Richland, WA: Battelle Pacific Northwest Lab.
Manne, A.S., and R.G. Richels. 1990. Global CO_2 Emission Reductions—the Impacts of Rising Energy Costs. EPRI and *En. J.*, in press; see also "CO_2 Emission Limits: An Economic Cost Analysis for the U.S.A.," EPRI and *En.J.*, April.
Manzer, L.E. 1990. The CFC-ozone issue: progress on the development of alternatives to CFCs. *Science* **249**:31–35.
Marland, G., et al. 1989. Estimates of CO_2 Emissions from Fossil Fuel Burning and Cement Manufacturing. ORNL/CDIAC-25, Oak Ridge, TN: Oak Ridge Nat. Lab.
Maser, C. 1988. The Redesigned Forest. San Pedro, CA: R. and E. Miles.
McKinney, T.R. 1987. Comparison of Organic and Conventional Agriculture: A Literature Review. RMI Publication #A87–28.
Menke, K., and J. Woodwell. 1990. Water Productivity and Development: Strategies for More Efficient Use. RMI Publication #W90–10; see also the fuller EPA-sponsored RMI water-efficiency implementation handbook, in press, 1990.
Moskovitz, D. 1989. Profits and Progress Through Least-Cost Planning. Washington, D.C.: Nat. Ass. of Regulatory Utility Commissioners.
Newell, R.E., H.G. Reichle, Jr., and W. Seiler. 1989. Carbon monoxide and the burning earth. *Sci. Am.* Oct. 82–88.
Newman, P.W.G., and T.L.F. Hogan. 1987. Urban Density and Transport: A simple model based on three city types. Transport Research Paper 1/87. Perth, Western Australia: Envir. Sci., Murdoch Univ.
Newman, P.W.G., and J.A. Kenworthy. 1989. Cities and Automobile Dependence: A Sourcebook. Aldershot, Hants., England: Gower Technical.

Nørgård, J.S. 1979. Husholdninger og Energi. København: Polyteknisk Forlag.
Nørgård, J.S. 1989. Low electricity appliances—options for the future. In: Electricity, ed. T.B. Johansson, B. Bodlund, and R.H. Williams, pp. 125–172. Lund: Lund Univ. Press.
Nordhaus, W. 1989. The Economics of the Greenhouse Effect. New Haven: Yale Univ. and presentation at AAAS meeting, New Orleans, 18 February 1990.
NRC (National Research Council). 1989. Alernative Agriculture. Washington, D.C.: National Academy Press.
NRC. 1990. Confronting Climate Change. Washington, D.C.: National Academy Press.
Oregon Department of Energy. 1990. Oregon Task Force on Global Warming: Report to the Governor and Legislature, June, p. 1–9.
OTA (Office of Technology Assessment, U.S. Congress). 1981. Technology and Soviet Energy Availability. Washington, D.C.
Ottinger, R., et al. 1990. Environmental costs of electricity. Pace Univ. Law School report to USDOE and NYSERDA, June.
Parson, E. 1990. The transport sector and global warming. Global Envir. Policy Proj. Disc. Paper, Harvard Univ. at p. 80.
Passell, P. 1990. Cure for Greenhouse Effect: The Costs Will Be Staggering. *N.Y. Times*, pp. 1 & 10, 19 November.
Piette, M., F. Krause, and R. Verderber. 1989. Technology Assessment: Energy-Efficient Commercial Lighting. LBL-27032. Berkeley: Lawrence Berkeley Lab.
Plochmann, R. 1968. Forestry in the Federal Republic of Germany. Corvallis, OR: Hill Family Found. Series, Oregon State Univ. School of Forestry.
Reddy, A.K.N., and J. Goldemberg. 1990. Energy for the developing world. *Sci. Am.* Sept.:110–118.
Reichmuth, H. and D. Robison. PE. 1989. Carbon dioxide offsets. Portland, OR.: Lambert Engineering/Pacific Power.
Rosenfeld, A.H., C. Atkinson, J. Koomey, A. Meier, and R.A. Mowris. 1990. A compilation of supply curves of conserved energy for U.S. buildings. Paper for Mitigation Subpanel, Panel on Policy Implications of Greenhouse Warming, U.S. Natl. Acad. of Sci.
Rosenfeld, A.H., and D. Hafemeister. 1988. Energy-efficient Buildings. *Sci. Am.* Apr.:78–85.
Rosenfeld, A.H., and A. Meier. 1990. Energy Efficiency vs. Global Warming: A No-Regrets Policy. Berkeley: Lawrence Berkeley Lab.
Rosenfeld, A.H., R.J. Mowris, and J. Koomey. 1990a. Policies to Improve Energy Efficiency, Budget Greenhosue Gases, and Answer the Threat of Global Climate Change. 15–20 February AAAS meeting, New Orleans, 26 January.
Ross, M.H., and D. Steinmeyer. 1990. Energy for industry. *Sci. Am.* Sept.: 88–98.
Rowberg, R.E. 1990. Energy Demand and Carbon Dioxide Production. 90–204 SPR, U.S. Congressional Research Service, Library of Congress, Washington, D.C., 17 April.
Samuels, G. 1981. Transportation Energy Requirements to the Year 2010. ORNL-5745. Oak Ridge, TN: Oak Ridge Nat. Lab.
Schipper, L., and A.J. Lichtenberg. 1976. Efficient energy use and well-being: the Swedish example. *Science* **194**:1001–1013.
Schneider, S. 1989. Global Warming: Are We Entering the Greenhouse Century? New York: Vintage.
SERI (Solar Energy Research Institute). 1981. A New Prosperity: Building a Sustainable Energy Future. Andover, MA.: Brick House.

SERI. 1990. The Potential of Renewable Energy. Interlaboratory White Paper. SERI/TP-260-3674. Golden, CO.
Shepard, M., et al. 1990. The State of the Art: Appliances. RMI/COMPETITEK.
Sitnícki, S., K. Budzinski, J. Juda, J. Michna, and A. Szpilewicz. 1990. Poland: Opportunities for Carbon Emission Control. Richland, WA: Battelle Pacific Northwest Lab.
SMUD (Sacramento Municipal Utility District). 1990. Shade Tree Program ("Trees Are Cool"). With Sacramento Tree Foundation.
Sobey, A.J. 1988. Energy use in transportation: 2000 and beyond. In: Summary of Presentations at a Workshop on Energy Efficiency and Structural Change: Implications for the Greenhouse Problem, ed. S. Meyers. Berkeley: Lawrence Berkeley Lab.
Soden, K. 1988. U.S. Farm Subsidies. RMI Publication #A88-22.
Sørensen, B. 1979. Renewable Energy. New York: Academic Pr.
Spyrou, A. 1988. Energy and Ships: The Effects of the Energy Problem on the Design and Efficient Operation of Future Merchant Ships. London: Lloyds of London Press.
Stevens, W.K. 1989. Methane from guts of livestock is new focus in global warming. *N.Y. Times*, 21 November p. C4.
Taha, H., H. Akbari, A.H. Rosenfeld, and J. Huang. 1988. Residential cooling loads and the urban heat island—the effects of Albedo. *Building and Envir.* **23(4)**:271–293.
Turiel, I., and M.D. Levine. 1989. Energy-efficient refrigeration and the reduction of chlorofluorocarbon use. *Ann. Rev. En.* **14**:173–204.
Van Domelen, J. 1989. Power To Spare. Washington, D.C.: Conservation Found.
Vaughan, C. 1988. Saving fuel in flight. *Sci. News* **134**:266–269.
von Hippel, F., and B.V. Levi. 1983. Automotive fuel efficiency: the opportunity and weakness of existing market incentives. *Res. and Cons.* **10**:103–124.
Wade, E. 1981. Fertile cities. *Development Forum*. Sept./Dec.
Wall Street Journal. 1989. Big firms get high on organic farming. p. B1, 21 March, and Back to the future: a movement to farm without chemicals makes surprising strides, p. A1, 11 May.
Washington State Energy Office. 1990. Telecommuting: An Alternative Route to Work. Olympia, WA.
Weinberg, C.J., and R.H. Williams. 1990. Energy from the Sun. *Sci. Am.* Sept.:147–155.
Williams, R.H. 1990. Low-cost strategies for coping with CO_2 emission limits. *En. J.* **11(3)**:in press.
Williams, R.H., and E.D. Larson. 1989. Expanding roles for gas turbines in power generation. In: Electricity, ed. T.B. Johansson, B. Bodlund, and R.H. Williams. pp. 509–554. Lund: Lund Univ. Press.
Wolsky, A.M., and C. Brooks. 1989. Recovering CO_2 from large stationary combustors, pp. 2:179–185 in OECD/IEA, *op. cit. supra*
Zimmerman, M.B. 1990. Assessing the Costs of Climate Change Policies: The Uses and Limits of Models. Alliance to Save Energy, Washington, D.C., 10 April.

15
Energy Saving in the U.S. and Other Wealthy Countries: Can the Momentum Be Maintained?

L. SCHIPPER
International Energy Studies, Energy Analysis Program, Energy and Environment Division, Lawrence Berkeley Laboratory, Berkeley, CA. 94720 U.S.A.

ABSTRACT

In this brief chapter, I examine trends in key energy-use sectors, commenting on recent changes that may signal a slowdown in the trend towards improved energy-use efficiency. The wealthy industrialized countries reduced their energy use through greater efficiency by nearly 32 Exajoules, or 16 million barrels per day of oil equivalent between 1972 and 1985, a savings of close to 30%. These savings played a crucial role in forcing down world oil prices. However, the pace of these improvements in energy use has begun to slow.

Also examined will be some of the differences between countries in efficiency. I will show that differences in the efficiency of energy use among countries account for close to one-half of the gap between the ratio of energy use to gross domestic product in the U.S. and that of a European country or Japan. However, differences in the structure of manufacturing output, distance traveled, and the total area of homes and service-sector buildings account for at least half of this "gap." Moreover, both changes in efficiency and change in these other factors account for changes in energy use in industrialized countries since 1973. Hence the ratio of energy use to gross domestic product is a somewhat misleading tool for measuring improvements in energy efficiency, and a poor indicator by which to compare the efficiency of energy use of different countries.

INTRODUCTION

Recently, many scientists have voiced concern that increased human-made production of so-called greenhouse gases, such as CO_2, methane, or chlorofluorohydrocarbons (CFCs), may have lasting and possibly costly effects on the Earth's climate (Schneider 1989; Lashof and Tirpak 1990). This concern has led to interest in reducing the emissions of these gases, particularly through more efficient use of fossil fuels, wherever more efficient fuel use saves money.

Indeed, the wealthy industrialized countries reduced their energy use through greater efficiency by nearly 32 Exajoules (EJ), or 16 million barrels per day of oil equivalent (Schipper 1987), between 1972 and 1985 after oil prices increased twice. These savings played a crucial role in forcing down world oil prices. However, the pace of these improvements in energy use has begun to slow (Schipper and Ketoff 1989). This brief paper examines trends in key energy-use sectors, commenting on recent changes that may signal a slowdown in energy efficiency.

MEASURING STICKS OF ENERGY EFFICIENCY

How can progress towards more efficient energy use be measured? The ratio of energy to gross domestic product (GDP) is often used as a measure of the differences in energy efficiency among economies (e.g., IEA 1989). Schipper and Lichtenberg (1976), however, found that this comparison was misleading as an indicator of energy efficiency performance. Moreover, changes over time in that indicator were only vaguely related to changes in individual energy intensities. Further, Schipper, Howarth, and Wilson (1990b) found that while the energy/GDP ratio in Norway fell 30% between 1973 and 1986, sectoral energy intensities (energy use per unit of residential heating, passenger or freight transport, services GDP, or manufacturing output) were constant or increased. Clearly, use of the energy/GDP ratio can give very misleading signals about changes in energy-use efficiency. In addition, one striking finding is that differences in sectoral energy intensities among various countries are almost always significantly smaller than the differences in energy/GDP ratios among countries. Therefore, we reject this indicator for our analysis, focusing instead on measures of energy intensity from each subsector of each economy.

SUMMARY OF FINDINGS FROM THE UNITED STATES AND OTHER COUNTRIES

We recently reviewed changes in U.S. energy-use intensity since the early 1970s (Schipper, Howarth, and Geller 1990a). The principal findings of that review may be summarized as follows:

- Aggregate energy intensities in the U.S. residential, services, manufacturing, freight, and passenger transportation sectors, adjusted for changes in the level and structure of sectoral activity, fell by an average of 23% between 1973 and 1987. Adjusted primary energy intensities fell by an average of 19%. Since the U.S. energy/GDP ratio fell by 31.8% for delivered energy and 26.3% for primary energy over this period, this analysis suggests that about three-quarters of the decline in the energy/GDP ratio was induced by improved energy efficiency, while the remainder was caused by structural change and fuel substitution.
- Actual energy use for the five sectors surveyed in detail was 51.8 EJ in 1987, or 71.4 EJ including electricity generation and transmission losses. Taking into account changes in the level and structure of energy-use activities, the efficiency improvements described above translate into savings of 15.6 EJ of delivered energy or 17.7 EJ of primary energy.
- The largest reductions in energy intensities occurred for automobile and air travel, home heating, and fuel use in the manufacturing and service sectors. The energy intensity of truck freight, in contrast, actually increased. A decline in load factors and a rise in the importance of light trucks for personal transportation together limited the decline in the system intensity of private vehicles, measured in energy/passenger-km, to only 13%.
- Changes in the aggregate of sectoral activity and the structure of activity within each sector had offsetting impacts on U.S. energy use. The sheer increase in sectoral activity would have boosted delivered energy use by 33% and primary energy use by 35%. However, the activity levels of the freight, passenger transportation, and residential sectors lagged behind GDP growth while the proportion of GDP generated in the manufacturing sector remained relatively constant. Only service-sector output grew more rapidly than GDP. Structural change reduced manufacturing energy use but increased energy use in the residential, freight, and passenger transportation sectors because the mix of activities became more energy-intensive. Overall, structural change within sectors increased U.S. energy delivered by only 1.2% and increased primary energy use by 5%. Thus the increase in activity had a major impact on each sector's energy use, but the impact of structural change on each sector's energy use was in most sectors small.

A recent slowdown in the improvement of U.S. energy efficiency has manifested itself in almost every sector except manufacturing. This slowdown represents a market plateau, not the confrontation with thermodynamic or technological limits. Public policies could restore some of the interest in raising the efficiency of energy use.

How do these global results compare with those from other countries?

Unfortunately few studies of this scope have been performed; however, analyses of F.R. Germany (Schipper 1988) and, more recently, of Norway (Schipper, Howarth, and Wilson 1990b) indicated that manufacturing energy intensity declined more in Germany than in the U.S., but less in Norway as compared to the U.S. Transportation energy intensity increased in both of the European countries, but decreased in the U.S. Residential heating intensity fell more in the U.S. than in the Germany, because the low penetration of central heating in Germany in 1973 (less than 55%) increased to over 70% by 1987, pushing up heating energy use significantly. These results are not inconsistent with those from a study of the members of the European Economic Community, which analyzed changes between 1979 and 1985 (Jochem et al. 1989).

The study of these three countries showed that only part of the change in the ratio of energy use to GDP resulted from improvements in energy-use efficiency. Indeed, the increase in energy-intensive activities in both Germany and Norway, particularly driving and heating, was far greater than in the U.S. and offset energy savings in other sectors in the European countries. Moreover, some of the differences in efficiency among these countries decreased because the efficiency of automobiles in the U.S. improved more rapidly than in other countries. In all three countries, however, a significant energy-saving potential remains.

Is energy use in the U.S. less efficient than in other countries? Using the other countries analyzed as guides, as well as the individual sectoral studies described below, important estimates of how U.S. energy efficiency compares with that of other countries can be made. American cars, homes, appliances, and services buildings still use 20–33% more energy per unit of activity than do those in other industrialized nations. American industries use 10–25% more energy per unit of activity. Thus while improvements in the U.S. have kept pace with improvements elsewhere, and even narrowed the gap considerably for space heating and driving, significant differences between energy efficiency in the U.S. and other countries still exist. However, these differences are much smaller than the differences in energy/GDP ratios imply.

BRIEF SECTORAL COMPARISONS

Presently, the International Energy Studies (IES) group at the Lawrence Berkeley Laboratory is analyzing the intensity of energy use in the major consuming sectors (passenger transport, freight, manufacturing, households, and the services sector) of major countries in Europe, as well as in Japan, the USSR, and important energy-consuming developing countries. In this section I focus on results of IES studies of energy use in the major developed countries.

Transportation

The transportation sector accounts for 20–35% of final energy use in Organization for Economic Cooperation and Development (OECD) countries. In spite of oil prices shocks, transportation energy use in most OECD countries was considerably higher in 1987 than in 1972. Total travel, in passenger-km per capita, increased; travel per unit of GDP also increased for most countries (Figure 15.1). The main reason was the growth in the number of cars. Car ownership increased more rapidly in Japan and Europe than in the U.S. (a consequence of higher U.S. ownership prior to 1973). Air travel increased its share of total travel, and rail and bus transit lost part of its share to air travel and the automobile in almost every country. As a result of these shifts towards the automobile and air travel, the aggregate energy intensity of passenger travel increased more in other countries than in the U.S. Still, the U.S. travel structure is some 33–50% more energy-intensive than it is in other countries: Americans have larger cars and travel more in them (and in the air) than do Europeans or Japanese. One important finding often overlooked, however, is that automobiles dominate passenger transportation and energy use in Europe as well as in North America, and their share is growing rapidly in Japan. Thus future energy savings in this sector must focus on automobiles.

U.S. freight patterns are different than those in other countries. The volume of freight, relative to GDP, is higher in the U.S. than in virtually

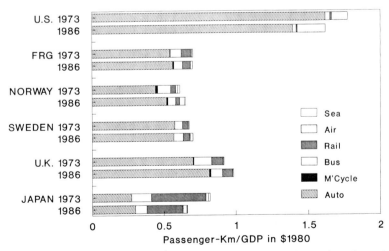

Figure 15.1 The ratio of passenger travel and gross domestic product for selected countries and modes of travel

every other industrialized country (the USSR is an important exception).[1] However, the share of energy-intensive truck freight in the U.S. (20%) is lower than in most other countries and has remained steady while increasing elsewhere. This means improving the efficiency of trucks will be an important step in reducing future energy use for freight.

Overall, the level of passenger travel, relative to GDP, declined in the U.S., but increased in most other countries. Freight volume, relative to GDP, fell in almost all industrialized countries. These structural changes narrowed the gap between the U.S. and other countries, but the U.S. still remains one of the most transportation-intensive countries.

The energy intensities in transportation behaved in a mixed way. The intensity of air travel fell everywhere, principally because the same new, fuel-efficient aircraft, manufactured by Boeing, McDonnell Douglas, and Airbus Industrie entered almost all the fleets of the industrial nations. The fuel intensities of automobiles behaved in a different way. The sales-weighted fuel intensity of new U.S. automobiles declined dramatically, as noted above; that for light trucks also declined, but from a higher initial level. Although the share of light trucks increased to over 20% of personal vehicles in the U.S., the figure for combined new-vehicle fuel intensity still fell more than 45%. The same measure for new automobiles in most other countries showed very little change, declining by 20% at most, as Figure 15.2 shows.[2] When the fleet fuel intensities for these countries are displayed (Figure 15.3), the results are different. Intensity in the U.S. fell by 30%, in the other countries, by 10% at most. The disparity between new-car-test economy and on-the-road performance in Japan and Europe was greater than in the U.S., both because of the inaccuracy or the paucity of sales-weighted new-car-intensity figures in Japan and Europe and because the increase in congestion and city driving was so great. By 1987, the fuel intensity of new cars in the U.S. was only slightly higher than that of the wealthiest countries in Europe, but the U.S. car/light truck fleet fuel intensity was still 30–40% greater than intensities in Europe or Japan. Thus the gap between per capita travel in the U.S. and other countries as well as the differences in the role of the automobile shrunk, while the U.S. narrowed the differences with European or Japanese fuel intensity. One major

---

[1] One reason for this is the size of the U.S. and the importance of shipments of bulk materials and energy over long distances. Another is that much of the trade that could be considered international (by sea) between central Europe, Scandinavia, or Japan is not counted in domestic freight activity. The same activities take place by domestic routes within the U.S.

[2] Data taken from national sources: L'Agence pour La Matrise d'energie (France); Verkehr in Zahlen (German Ministry of Transport, Bonn, FRG); Transportraadet and National Energy Board (Sweden); Transport Oekonomisk Institute (Norway); Ministry of International Trade and Industry (Japan); Agip SPa (Italy); and Schipper et al. 1990a (U.S.). See also Schipper et al. 1992.

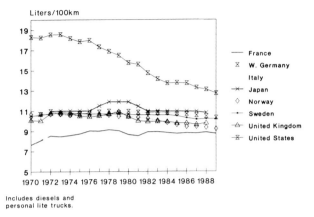

Figure 15.2 Recent changes in automobile fuel ecomony (new car fleet averages) for selected countries

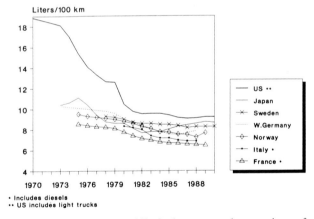

Figure 15.3 Recent changes in automobile fuel economy (comparison of on-the-road fleet averages, including diesels and common personal light trucks) for selected countries

difference between U.S. energy use and that in other countries shrunk considerably.

What are the prospects for continued energy savings in transportation? The world's airlines are still adding newer, more efficient aircraft, which will reduce average energy use per kilometer of travel. However, both

Boeing and McDonnell Douglas have postponed plans to introduce a very fuel-efficient prop-fan engine, a consequence of low oil prices and high interest rates according to experts at one of these companies. The slowdown in improvements in new-car efficiency, coupled with the small improvements in fleets (outside of North America) is more worrisome as the role of the automobile increases. While many highly fuel-efficient prototypes exist, the challenge of the next decade appears to be the improvement in more conventional, i.e., family-type cars that dominate present markets by improving motors, drive trains, reducing drag, and reducing weight through design and new materials (Mellde et al. 1989; Ross 1989; Sperling and DeLuchi 1989). The present fleets in Europe average about 9 l/100 km (26 miles per U.S. gallon [MPG]), those in North America about 12 l/100 km (19–20 MPG). These could be improved to about 5.5 l/100 km (about 40 MPG) at less than a 10% increase in vehicle cost (Ross 1989). Equally important, however, is the reduction of emissions from all vehicles, as well as the improvement of traffic, which plays a key role in future fuel economy.

Manufacturing

Manufacturing accounts for between 25 and 35% of final energy use in most countries. Important changes took place in the role of manufacturing in overall energy consumption in most OECD countries between 1972 and 1985 (Howarth and Schipper 1991). The most transparent structural change in manufacturing energy use was the reduction in raw materials production per unit of GDP. Figure 15.4 shows the situation for steel in 1973 and 1986, for example, which is indicative of that for cement as well. Note that the U.S. produces somewhat less steel and other raw materials per unit of GDP than do other countries, particularly Japan, Sweden, and F.R. Germany. Indeed, the share of U.S. manufacturing output in six major energy-intensive industries[3] (16.1% in 1985 excluding refining) is lower than the shares of most other countries (23% in 1985–1987 in Japan, 20% in the five European countries [EU-5], France, F.R. Germany, Norway, Sweden, and the U.K. in Figure 15.5). Thus if the U.S produced the same mix of output as do these other countries, manufacturing energy use in the U.S. would be higher than it actually is.

Changes in the role of these industries since 1973 had a measurable impact on manufacturing energy use in all countries. Overall, the shift away from production of raw materials (measured in value-added) alone reduced energy use for manufacturing by 7% in the EU-5, by 12% in Japan, and by 15% in the U.S., as shown by the plots of structural change in Figure 15.5.

---

[3] Iron and Steel (ISIC 371), Non-Ferrous Metals (342), Basic Chemicals (351), Petroleum Refining (354,4), Stone-Glass-Clay (36), and Paper and Pulp (341)).

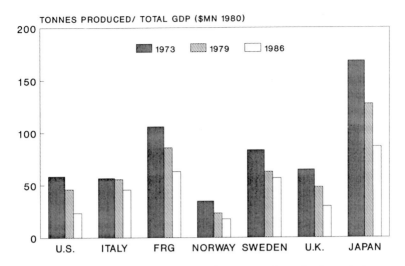

Figure 15.4 Recent changes in energy-intensive raw steel production relative to gross domestic product (GDP) for selected countries. GDP is given at 1980 purchasing power parities (MN=million)

Structural change had a larger impact on manufacturing energy use in the U.S. than in the other countries. Corrected for this structural change (i.e., holding the 1973 mix of industries constant), energy use per unit of value-added fell by 30% in the U.S., by 38% in Japan, and by 27% in the EU-5, as shown by the plots of intensity in Figure 15.5. Note that the intensity indices fell significantly more than the structural indices, i.e., intensity changes caused more energy saving than structure changes. The share of oil in all countries fell markedly, and the intensity of oil use fell by more than 66% everywhere. The overall U.S. record, measured as reduced energy intensity of manufacturing, lies between that of Japan (greatest decline) and Europe (least decline). While energy use per unit of monetary output is not always a reliable indicator of changes in efficiency, available data show that energy use per unit of physical output for most of the key raw materials fell in all countries.

The improvements in energy efficiency in manufacturing since 1973 represent only an acceleration of a long-term trend. Indeed, few observers find any slowdown in the improvement of efficiency occurring after the oil price crash, because the increase in output from basic industries that followed the recovery in the mid-1980s stimulated turnover of old inefficient plants and innovation to save all resources. Thus, it is widely believed that industry will continue to save energy.

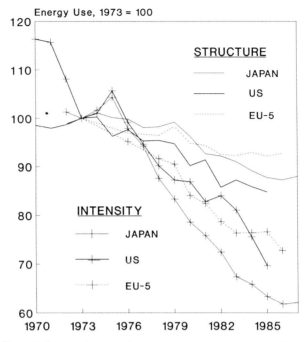

Figure 15.5 Recent changes in manufacturing energy use expressed as the impact of changes in sectoral intensity or the mix (structural) of sectoral output for selected OECD countries (1973 energy use = 100)

Residential Sector

The residential sector accounts for nearly 25% of final energy use in most OECD countries (Schipper, Ketoff, and Kahane 1985a; Ketoff and Schipper 1990). In spite of two sharp hikes in energy prices, standards of comfort and convenience increased in all OECD countries. Differences in standards also narrowed: House area per capita and appliance ownership grew more rapidly in Europe and Japan than in the U.S. The penetration of central heating increased to over 70% of homes in central Europe, and water heating reached nearly every home. Still, Americans enjoy 50–80% more house area per person than Europeans and more (or larger) major appliances as well.

Fuel choices and fuel shares in the U.S. evolved in line with the pattern in Europe: the choice of oil as heating fuel fell drastically and that of liquid propane gas declined as well (Schipper and Ketoff 1985). Substitutes were principally gas and wood in existing homes in the U.S., and gas, district heat, or electricity in Europe. In new homes, gas and electricity filled the gap left by oil. Electricity made important gains in water heating and cooking in both the U.S. and in most European countries. Two countries where oil maintained an important share of final consumption were Germany (45%) and Japan (65%).

The U.S. space-heating intensity is above average, with that of Scandinavian countries lying below, and those of Britain, Holland, and F.R. Germany higher, as Figure 15.6 shows (calculated from IES data from each country; Schipper et al. 1985).[4] The energy intensity of heating (in $kJ/m^2/deg$ day of useful energy, i.e., the energy delivered to the heating system minus losses in combustion of liquid, gaseous, and solid fuels) fell somewhat more in the U.S. than in most other countries. The decline in Britain, F.R. Germany, France, and Holland understates the real savings in space-heating energy since 1973 in these countries, however. This is because the penetration of central heating rose from under 50% to over 70% in these countries in the period considered, a change that would have boosted space heating energy use/home by some 33% in these countries. Intensity in Norway and Japan, not shown in Figure 15.6, is difficult to compare with those illustrated because although homes in these two countries have little central heating (circulating hot water or hot air systems), heating intensity grew steadily as more heat was supplied to more rooms.

There were several components of these improvements (Schipper, Ketoff, and Kahane 1985a). Rapid reduction in comfort in oil-heated homes in most countries, including the U.S., led the decline. Occupants of gas- and electrically-heated homes also cut back on comfort. Gradually, some or much of this short-term savings were supplemented by efficiency improvements throughout. The efficiency of heating equipment improved; in the U.S., the penetration of heat pumps rose higher than in every country except Japan, although in Japan these are now used to complement kerosene heaters. Improvements in the thermal integrity of new homes in most European countries, as implied by stiffer building code requirements (Wilson et al. 1989) or levels actually insulated in new homes (Schipper, Meyers, and Kelly 1985) contributed to a significant decline in heat requirements for new homes. The thermal resistance of attics virtually doubled in all countries, but the improvements in insulation in walls were uneven. The insulation levels now required in walls of new homes built in Europe are twice the pre-1973 values, but improvements in U.S. were much less (Meyers 1987). These factors resulted in the space-heating intensity, in the most recent homes in the U.S. and Europe, dropping by 30–50% compared with that

---

[4] The intensity for Japan (not shown) lies far below those displayed because heating is only intermittent and by room. Note that Japan has only 1975 Degree Days (base 18C); the U.S. has 2599 (Meyers 1987), while the European countries we consider range from 2450 (France), 2800 (Britain), 3017 (Holland), 3113 (F.R. Germany), 3316 (Denmark), to over 4000 for Sweden and Norway. The average for the European countries, weighted by population, is about 15% greater than the U.S. population weighted average.

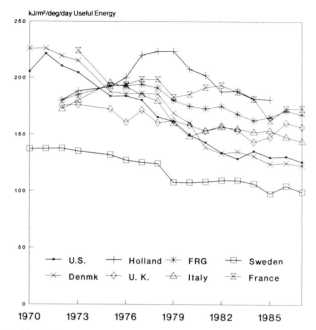

Figure 15.6 Residential space heating intensity (excluding boiler losses and including effects of home area and climate) since 1970 for selected OECD countries

in representative homes built before 1975. The intensity of U.S. residential space heating fell far enough between 1973 and 1987 to be nearly comparable to the average for European countries; the significantly larger size of U.S. homes by comparison would permit economies of scale, so U.S. values should be lower than those in Europe for a given thermal integrity. Considering all factors, U.S. residential space-heating intensity is about 25% more energy intensive than that in Scandinavia but close to the level of central Europe.

We examined electric appliances and other home electricity uses in one dozen OECD countries (Schipper, Ketoff, Hawk, and Meyers 1987; Schipper and Hawk 1991). We found that appliance efficiency improved in all countries over the period under consideration, with the U.S. improvements about average. Figure 15.7 summarizes the developments for the U.S. We have displayed electricity use per year for a variety of appliances in the 1973 and 1987 stock, as well as average values for those sold today. Note the decline in average consumption as well as the lower consumption of new models. (Appliances sold in other countries show similar progress.) More efficient appliances have reduced household electricity use for appliances (40–60% of total household electricity use) by 10–20% in almost

Energy Saving in Wealthy Countries 455

every country. After these improvements are counted, however, new U.S. appliances still consume about 20–30% more electricity, per unit of service, than do those in Europe.

In conclusion, structural differences betwen the U.S. and Europe narrowed significantly between 1973 and 1987. Still, if the residential sectors of European countries (and Japan) had U.S. characteristics, residential energy use in Europe (Japan) would be nearly 50% (>100%) larger than it is. The intensity of heating and electric appliances in the U.S. fell significantly between 1973 and 1987, but U.S. homes and appliances still require roughly 20–25% more energy, per unit of service, than do those in Europe.

The outlook for improvements in energy use in the residential sector is mixed. In each country there is still a large fraction of homes with poorly insulated walls and heat-leaky windows. Even in Sweden and Norway, improvements to existing homes make sense. The insulation levels in new homes could be improved somewhat as well. New appliances could be improved significantly through adoption of the energy-saving features already found in some models of every appliance sold (Geller 1988a; Schipper and Hawk 1991). All of these improvements would be economically advantageous if introduced.

Figure 15.8, for example, illustrates the potential for electricity saving in key electric appliances sold in Sweden. The first three levels of consumption for each appliance reflect progress in the average appliance in the stock. Next, models are shown models with the highest and lowest yearly consumption of those typical in size and features of those sold in 1987.

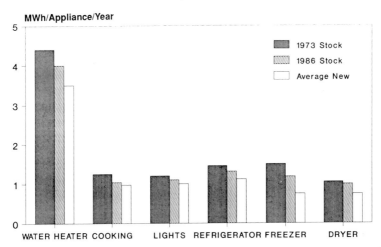

Figure 15.7 Annual unit consumption of electricity by domestic appliances in the U.S. for existing stock and new equipment

Finally, we show values for a low-energy refrigerator sold in Denmark, designed by a team led by Prof. Joergen Noergaard of the Technical University of Lyngby (see the review in Schipper and Hawk 1992), as well as Lawrence Berkeley Laboratory's projections for refrigerator-freezers and freezers. If consumers pay no attention to electricity use, they will, on average, buy models that use about 25% less electricity than the average of the stock. If they focus on models using the lowest level of electricity, usage per appliance will fall by 50%. Also, if consumers would send a strong signal about electricity efficiency to manufacturers, then 66–75% of present-day use/appliance could be cut from household electricity bills.

Unfortunately, present-day markets provide little stimulus to homeowners or occupants to make these improvements because of a variety of market and non-market barriers. Put simply, consumers have very short time horizons, ignoring efficiency except where the payoff is very rapid, i.e., in two years or less. A combination of standards and other incentives is probably necessary to accelerate the technical progress that will lower the energy needs of comfort and convenience.

Services Sector

The services sector accounts for only 15% of delivered energy in most countries, but a higher (and growing) share of electricity. International comparisons of energy use in the services sector are difficult. Nevertheless, available data do make some partial comparisons meaningful (Schipper et

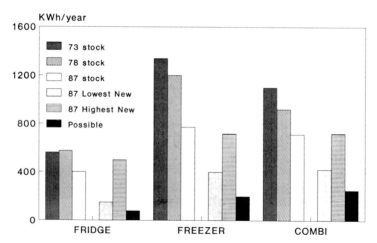

Figure 15.8 Annual appliance-specific consumption of electricity in refrigerators in Sweden (sizes: fridge (refrigerators), 150–200 l; freezer, 250–300 l; combi (combined refrigerator/freezer), 350–375 l)

al. 1986). The share of GDP arising in the service sector increased slightly in almost all countries, and total service-sector area increased slightly faster. The total area of buildings in the service sector, measured in m² of floor space per capita, grew to more than 24 m² in the U.S. vs. 13–18 m² in Europe or Japan. As with the residential and transportation sectors, then, the U.S. activity level implied by this relationship is significantly higher than that of any other country. However, the gap between the U.S. and Europe narrowed in the 1970s and 1980s.

The balance between electricity and fossil fuels or district heat in the U.S. is different from that of central Europe, and close to that of Sweden (over 35% of delivered energy as electricity). The high share of electricity in the U.S. is a result of high electric heating and cooling penetration, as well as high electricity use for lighting and other building services. Only Norway and Canada, with higher penetration of electric heating, have a higher share of electricity in delivered energy. After 1973, heating by oil lost its share of the market to gas, district heat, or electricity in every country, and the share of oil in overall delivered services-sector energy fell everywhere.

Schipper et al. (1986) found that U.S. space-heating intensity was considerably higher than that in other countries, and concluded that the use of electricity is somewhat less efficient than in other countries, particularly because of high lighting levels. Fuel intensity, measured in GJ/m², fell in most countries in the 1970s and 1980s, but the U.S. intensity still appears to be somewhat higher than those of colder, northern European countries. Electricity intensity increased in most countries, even where the penetration of electric heating did not. One important factor raising electricity use has been computers and other information technology. In spite of rising electrification, however, electricity intensity in U.S. buildings, measured in kWh/m²/yr for all purposes, did not rise significantly after 1979.

The outlook for energy savings in the services sector is bright. This is because increasing use of electricity is putting pressure on local and national supply authorities to work with building owners or operators to reduce peak and even average loads. Improvements in lighting technologies, optical coatings on windows to reduce undesirable heating gains or losses, and more efficient motors and compressors all promise to reduce electricity intensity or at least reduce growth. Unlike residential buildings, commercial buildings represent an opportunity where a small intervention by the user and supplier together can bring about an enormous absolute saving in electricity (Geller 1988b; Krause et al. 1988). The same is true for reducing oil and gas use for space and water heating.

CONCLUSIONS: INTERNATIONAL IMPROVEMENTS IN ENERGY EFFICIENCY

This brief comparison reveals that during the 1970s and 1980s wealthy countries reduced energy intensities and thus saved energy. In the case of the manufacturing sector, these savings represented an acceleration of historical trends. In the residential and service sectors, the savings broke with past trends. In transportation the record was mixed; indeed, only in the air travel sector were there clear energy savings. In the personal vehicle sector outside of North America, the overall efficiency of producing passenger-kilometers only increased marginally. In the truck freight sector, the improvement was either small or negative. But on balance, OECD countries used abut 32 EJ of delivered energy less than they would have in 1985 if 1972/1973 energy intensities had not fallen. About 90% of this decline could be attributed to energy conservation, the rest to structural changes that also reduced energy demands.

The brief comparison presented above also shows that, relative to GDP, larger homes and appliances raise U.S. residential energy use relative to that in Europe significantly. Larger cars, more driving, more flying, and more freight increased U.S. energy intensity even more relative to that of Europe or Japan. A greater service sector also increased U.S. energy intensity. Output of energy-intensive raw materials in the U.S., by contrast, is somewhat lower than in Japan or western Europe. On balance, these structural differences between the U.S. and other countries account for more than 50% of the difference in the energy/GDP ratio. However, increases in travel and increases in built space for homes and services have kept significant pressure on electricity and oil use in almost every country.

Since 1973, differences in energy use among wealthy countries have become smaller, particularly as other nations have approached U.S. levels of activity; however, considerable gaps remain. Additionally, some of the change in the U.S. energy/GDP ratio was caused by structural change: manufacturing output, home floor area, passenger travel, and freight all grew less than or equal to the rate of growth in GDP in the U.S. These considerations illustrate why it is dangerous to use changes in the energy/GDP ratio as indicators of energy savings.

REFERENCES

Geller, H. 1988a. Update on electricity use in the residential sector. Report to the U.S. Office of Technology Assessment. Washington, D.C.: American Council for an Energy-Efficient Economy.

Geller, H. 1988b. Update on electricity use in the service sector. Report to the U.S.

Office of Technology Assessment. Washington, D.C.: American Council for an Energy-Efficient Economy.
Howarth, R., and L. Schipper. 1991. Manufacturing energy use in eight OECD countries: trends through 1988. *Energy J.* **12(4)**, in press.
IEA (International Energy Agency). 1989. Electricity Conservation. Paris: Organization for Economic Cooperation and Development.
Jochem, E. et al. 1989. Energy Conservation Indicators. Heidelberg: Springer Verlag.
Ketoff, A., and L. Schipper. 1990. Energy conservation in OECD households: looking beyond the aggregate picture. In: Proceedings of the Summer Workshop on Energy Efficient Building. Washington, D.C.: Am. Council for an Energy Efficient Economy.
Krause, F. et al. 1988. Analysis of Michigan's demand-side electricity resources in the residential sector. 3 vols. LBL-23025. Berkeley: Lawrence Berkeley Lab.
Lashof, D., and D. Tirpak. 1990. Policy Options to Stabilize the Climate. Washington, D.C.: U.S. Environmental Protection Agency.
Mellde, R., I. Maasing, and T. Johansson. 1989. Advanced automobile engines. *Ann. Rev. of Energy* **14**:425–444.
Meyers, S. 1987. Energy consumption and structure of the U.S. residential sector: changes between 1970 and 1985. *Ann. Rev. of Energy* **12**:81–97.
Ross, M. 1989. Energy in transportation. *Ann. Rev. of Energy* **14**:131–172.
Schipper, L. 1987. Energy saving policies of OECD countries: did they make a difference? *Energy Policy* **15**:538–548.
Schipper, L. 1988. Energy use in Germany in the long term. Berkeley: Lawrence Berkeley Laboratory.
Schipper, L., R. Steiner, P. Duerr, S. Stroem, and F. An. 1992. Energy use in passenger transport in OECD countries: Change between 1970 and 1987. *Transportation, an Intl. J.*, in press.
Schipper, L. and D. Hawk. 1991. More efficient household electricity use: an international perspective. *Energy Policy*, in press.
Schipper, L., R. Howarth, and H. Geller. 1990a. United States energy use from 1973 to 1987: the impacts of improved efficiency. *Ann. Rev. of Energy* **15**:455–504.
Schipper, L., R. Howarth, and D. Wilson. 1990b. A Long Term Perspective on Norwegian Energy Use. LBL 27295. Prepared for the Royal Norwegian Energy and Oil Ministry. Berkeley: Lawrence Berkeley Laboratory.
Schipper, L., and A. Ketoff. 1985. Household oil savings in the OECD: permanent or reversible? *Science* **230**:1118–1125.
Schipper, L., and A. Ketoff. 1989. Energy efficiency: the perils of a plateau. *Energy Policy* **17**:538–542.
Schipper, L., A. Ketoff, D. Hawk, and S. Meyers. 1987. Residential electricity consumption in industrialized countries: changes since 1973. *Energy* **12(12)**:1197–1208.
Schipper, L., A. Ketoff, and A. Kahane. 1985a. Explaining residential energy use by international, bottom-up comparisons. *Ann. Rev. of Energy* **10**:341–405.
Schipper, L., A. Ketoff, and S. Meyers. 1986. Energy use in the service sector: an international perspective. *Energy Policy* **14(3)**:201–218.
Schipper, L., and A. Lichtenberg. 1976. Efficient energy use and well being: the Swedish example. *Science* **197**:1013–1025.
Schipper, L., S. Meyers, and H. Kelly. 1985b. Coming in from the Cold: Energy-Wise Housing from Sweden. Washington: Seven Locks Press.
Schneider, S. 1989. The greenhouse effect: science and policy. *Science* **243**:771–781.

Sperling, D., and M.A. DeLuchi. 1989. Transportation energy fumes. *Ann. Rev. of Energy* **14**:375–424.

Wilson, D., L. Schipper, S. Tyler, and S. Bartlett. 1989. Policies and programs for promoting energy conservation in the residential sector: lessons from five OECD countries. LBL-27289. Berkeley: Lawrence Berkeley Laboratory.

Standing, left to right:
R.G. Richels, S. Büttner, L.L. Jenney, W.U. Chandler, W.B. Ashton, A.B. Lovins
Seated, left to right:
L. Schipper, I.A. Bashmakov, K. Yamaji

16

Group Report: What Are the Options Available for Reducing Energy Use per Unit of GDP?

W.U. CHANDLER, Rapporteur
W.B. ASHTON, I.A. BASHMAKOV, S. BÜTTNER,
L.L. JENNEY, E.K. JOCHEM, A.B. LOVINS,
R.G. RICHELS, L. SCHIPPER, K. YAMAJI

INTRODUCTION

Between 1973–1987, falling energy intensity[1] reduced overall U.S., Japanese, and F.R. German energy demand by 19, 21, and 18%, respectively. The ratio of energy to gross domestic product (GDP) in these countries fell by 26, 34, and 19%, respectively, suggesting that for Japan and the U.S., changes in the mix of goods and services also had a significant effect on the overall ratio of energy use to GDP (Schipper et al. 1992; Schipper et al. 1990; Morovic et al.1987).

The U.S. increased its GDP by more than one-third between 1973–1985 while holding energy demand virtually constant. The significance of this experience, which can be expressed as a rate of reduction in the energy to GDP ratio of 2–2.5%/yr, is evident in that the U.S. also held carbon dioxide emissions steady over the same period. To be sure, the oil price increases took a multibillion dollar toll on most of the world's economies, but improvements in energy efficiency reduced this damage considerably.

What were the driving forces behind this decoupling of energy and

---

[1] energy per unit of activity in 20 sectors.

Limiting the Greenhouse Effect: Options for Controlling Atmospheric CO$_2$ Accumulation
Edited by G.I. Pearman © 1992 John Wiley & Sons Ltd

economic output? What will be their impact in the future? Can they be accelerated and made more widespread and, if so, at what cost?

UNDERSTANDING THE POTENTIAL FOR ENERGY EFFICIENCY

The terms energy efficiency and energy conservation must be carefully defined in order to avoid misunderstanding. Energy conservation has been defined as embodying four concepts (Gibbons and Chandler 1981):

1. reduction in the use of energy services;
2. fuel switching to conserve scarce or otherwise valuable fuels;
3. shifts in the mix of goods and services to those that use less energy, and
4. reduction in the energy required to produce a given level of goods and services.

The fourth concept is synonymous with energy efficiency, and concepts 3 and 4 control strongly the potential for energy intensity reduction per unit of GDP.

The future potential for energy efficiency may be characterized by five aspects, or cases, of the system that demands and delivers energy services:

1. the reference case;
2. the reference case with market imperfections removed;
3. the reference case with market imperfections removed and with the external costs of energy use incorporated, thus stimulating further efficiency;
4. the reference case without any regard for the cost of saving energy (that is, the application of the best available technology regardless of its cost-effectiveness); and
5. the reference case with the application of all feasible technology, without regard to its commercial availability or even, in an extreme case, its existence (see Figure 16.1).

The reference case or "business-as-usual projection" is an essential first step in any estimate of future energy-savings potential. The reference case projection of energy demand, the first aspect of this characterization, indicates the expected level of energy efficiency for a given date projected into the future, assuming that market barriers are not overcome. This projection requires a variety of assumptions, including future energy price levels and expected consumer price response. The reference case incorporates prevailing obstacles and market imperfections that hinder the full application of the potential for energy efficiency. This case yields the least energy efficiency and the highest energy demand of those considered here.

The second aspect, the "no market imperfections case," represents the additional energy efficiency which may be added to the reference case by removing all of the various impediments to energy efficiency. The common examples of market imperfections include: average pricing of electricity (as opposed to marginal pricing); lack of price signals (no meters in Moscow flats, for example); and split incentives, exemplified by the classic tenant-landlord situation where the person who could invest in energy-saving measures has no incentive to do so because the renter pays the energy bills. Common examples of this potential include: refurbishing the existing building and equipment stock; increasing the application of cogeneration in industry; and increasing the use of district heating. In this aspect, cost-effective energy-efficient technology is assumed to penetrate the market to its fullest extent, and cost-effectiveness is defined by applying social, not the much higher private, discount rates (Goldemberg et al. 1987).

The third aspect, or case, builds on the second by adding to the cost of energy all of the external costs of using energy. Common examples of external costs are the military or national security costs of energy dependence from imported oil, or the health, environmental, and welfare effects from energy-related air pollution.

The fourth aspect takes case three and applies all available technology for saving energy, regardless of its cost.

The fifth aspect is the theoretical maximum potential case. This case refers to the energy which analysts think could be saved if ideas for energy-saving measures could be moved from their heads onto drawing boards, into laboratories, and into commercial application. Examples range from improved energy efficiencies in heat transformers, high temperature fuel cells, through innumerable reductions in demand for "useful" energy due to improved insulation, catalysts, materials, biotechnology techniques, to the substitution of low-energy for energy-intensive materials.

OVERCOMING BARRIERS TO MORE-EFFICIENT ENERGY USE

The current debate over the potential for energy efficiency improvements is centered around the cost and benefits of moving from case 1 to case 2. Barriers to energy efficiency include price distortion; lack of infrastructure; disparity between private and social discount rates; promotional practices among suppliers of energy and energy-using equipment; split incentives; lack of markets in saved energy; lack of technical knowledge of architects, craftsmen, planners, and engineers; and inadequate consumer information.

A vast literature has identified and described these difficulties but has been less successful in prescribing solutions (e.g., Carlsmith et al. 1990). Solutions have nevertheless been documented in successful regulatory, incentive, training, consulting, and information programs. New research in the social sciences could greatly improve our understanding of how to overcome these barriers and resistance to change by suppliers and consumers (see Table 16.1). This research work deserves the highest priority.

Energy analysts believe that changes in energy prices were responsible for at least two-thirds of the energy-efficiency improvements attained in the developed market economies in the period following the oil shocks of the seventies and early eighties. Though market pricing is an important mechanism for achieving efficiency, price distortion remains common throughout the world. Many countries, particularly planned and developing nations, price energy at a small fraction of its market value. In the U.S., electric power is often priced at less than replacement cost. Coal in Poland and China and natural gas in the Soviet Union have until recently been sold at less than half their value on open export markets. A related problem

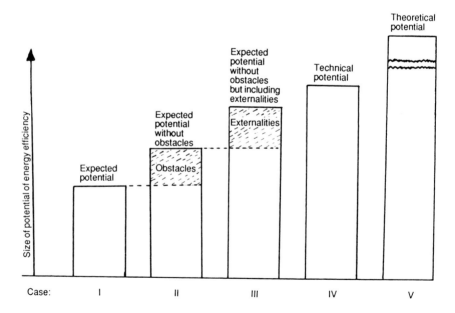

Figure 16.1

extends to state-owned enterprises in some countries which have received subsidies to offset rising energy prices in order to maintain output of critical goods for consumption or export. Price reform, and making prices matter, must provide the motivation for enterprise managers to renovate and replace equipment.

A problem arises in this context because the planned and post-planned economies must also develop or expand their capability to manufacture energy-efficient materials and equipment before efficiency opportunities can be captured. Lack of this infrastructure compounds pricing problems. The almost complete lack of measurement of domestic heat and hot water consumption at the household level in the Soviet Union and eastern Europe, for example, means that price reform could well have little positive effect in that sector. Individual consumers are unable to assess the consequences of their energy-efficiency activities and, in any case, unable to reduce their energy costs as a result of these activities. Lack of banking systems to finance investment, lack of communications equipment for gathering information and conducting business, and lack of high quality, reliable energy supplies, particularly for electricity, are also infrastructure problems which constrain progress in energy efficiency in much of the world. Given time and resources, these problems can be overcome (Makarov and Bashmakov 1990).

Problems with electric power reliability in developing countries complicates the use of advanced equipment and controls. Transferring energy-efficiency technology to developing and post-planned countries requires removing barriers necessary to facilitate development itself.

A common infrastructure problem in western developed countries is the lack of suitably sized lamp shade holders for use with highly efficient light bulbs. Such problems can easily be solved through minor changes in manufacturing, but only when the problem becomes a priority for manufacturers and retailers of lighting equipment. Indeed, smaller bulbs now available can eliminate this difficulty.

Capital replacement is a problem in all nations, but particularly in recently planned economies. For example, the current rate of capital replacement in the Soviet Union is only 2–3%/yr, while a rate of 7–8%/yr could be justified. Because capital replacement in many cases means the replacement of outmoded, energy-inefficient technology with much more efficient products and processes, stalled economies impede improvement in the energy to GDP ratio in many countries.

Tax incentives have sometimes been proposed as a means to overcome investment barriers, but there is no evidence that tax credits for home or industrial energy-efficiency investment in the U.S. have been worth their cost. However, some U.S. utilities have sponsored incentive plans that have worked quickly. In 1983–1984, when Southern California Edison Co.

experienced a 14 GW peak load, it reduced the 10 year forward peak by 1.2 GW/yr. About 45% of this savings came from the utility's own programs, which saved energy at a cost of 0.3 cents/kWhr, or $32/kW. Fifty-five percent came from state programs, at roughly no cost. The utility's total average cost of savings was 0.1 to 0.2 cents/kWh, while the total cost to society was only 0.5–0.8 cents/kWh. Other utilities have recently captured up to 90% of selected demand-side markets in just 1–2 years or even in a few months. Recent regulatory reforms, chiefly in the U.S., seek to conserve and reward efficient utility behavior, for example by decoupling profits from sales and by letting utilities keep as extra profits part of what they save their customers (Fickett et al. 1990).

Regulatory standards can effectively promote energy efficiency where there are clear market failures or where discount rates are unusually high. The level of fuel economy in private automobiles illustrates this effectiveness. Research and experience suggest that car-buyers will be indifferent to fuel economy over a broad range. While levels even doubling the current new car average would yield net savings to consumers, the savings would be quite small and can be ignored from the buyer's point of view. Evidence of the constant increase in average U.S. fuel economy over the last 15 years despite wildly fluctuating gasoline prices presents a strong case for the effectiveness of standards.

Regulatory policy, however, can fail unless continually evaluated and revised. For example, the fleet of light trucks in the U.S. has grown rapidly in part because trucks were treated differently from cars in the corporate average fuel-economy standards. These vehicles today are generally used not as trucks per se, but as automobiles. They generally are more powerful than cars, use far more fuel per unit distance, and remain in use longer. This fact suggests that a combination of policies might be required to accomplish a given objective. For example, U.S. appliance policy was crafted from a combination of efficiency labeling, regulatory standards, utility incentive programs for the purchase of efficient models, and rising energy prices. A hypothetical example of such a policy in the transport sector combines fuel-economy regulation with meaningful gasoline taxes, information programs designed to appeal to consumer self-interest, and perhaps even revenue-neutral fees and rebates which reward the purchase of efficient vehicles while correspondingly penalizing inefficient ones.

We do not know the relative contribution of each of the above barriers to reducing energy efficiency. We do know that small and large businesses alike often achieve savings far below the economic potential. This raises the question of how policies can be combined to overcome the barriers to energy efficiency effectively. Many regions will need cooperation in creating a system of energy-efficiency consulting, motivation, training, and regulation.

MODELING THE POTENTIAL

These complex variables are usually understood best in the context of models that interrelate energy-using behavior and technology. Valid models must be reproducible, transparent, and based on valid principles and assumptions. Recently, strong debate has revolved around the validity of the modeling tools themselves as well as the assumptions used in them. Strikingly different projections of future energy demand, and carbon dioxide emissions, can result from the modeler's choice of both models and inputs.

Confusion can arise from differences in modeling tools, particularly between the "top-down" and the "bottom-up" approaches. The former type of model is driven primarily by economic growth, modified by econometrically estimated income- and price-elasticities of energy demand. These models may also include parameters that modify energy demand projections based on estimates of technological change and policy instruments, particularly pricing and financial incentives, but they typically have very little detail on end-use activities or technologies.

End-use of "bottom-up" modeling efforts also are driven by economic growth, but they permit more detailed investigation of non-price-induced technical and policy changes. For example, automobile fuel-economy or appliance-efficiency standards can be assumed in an end-use or bottom-up model, while such a change can only be crudely approximated in a top-down, principally economic model. End-use models can thus reveal additional detail for understanding past and future energy demand.

If the elements affecting demand are studied over time, the relative impact of changes in total activity, the structure of activity, and the intensity of each activity on total energy use can be seen. The forces driving changes in each of these components, including changes in energy and other prices, income, technologies, energy and related policies, and demographic changes, can then be linked to changes in energy use. From such analysis, we can identify the most important behavioral and technological components and causes of energy saving in the past, and thus estimate how they can be amplified in the future.

End-use or bottom-up models are handicapped, however, by their inability to provide equilibrium solutions. That is, economic theory suggests that reducing energy demand through regulatory policy should also lower energy prices somewhat. Depending on the magnitude of the price reduction, energy demand could be stimulated and thus offset some of the savings. Similarly, energy savings could have the effect of increasing the net income of consumers, and thus increase energy demand by additional consumption of energy-intensive products and services, such as air travel. A macroeconomic model would in principle capture this effect, but the energy end-use models in use today often do not. On the other hand, the end-use models may better represent saturation of energy services.

In principle, there is no reason why the strengths of the two approaches, top-down and bottom-up, cannot be merged. Efforts to develop and validate such an approach could have a significant payoff. Relatively small projects are underway in several institutes to create models based on such an approach. This work plus related research efforts needed to improve the basic art of energy-economic modeling should be very high priorities (see Table 16.1).

There is wide disagreement in estimates of energy-efficiency potential, even applying the same types of models (that is, among analysts using bottom-up modeling characterized as the second case, above). This disparity can be attributed to several analytical differences, for example:

1. Assumed technological vintage or "modernity": Most of the best electricity-saving technologies have been available for less than 1 year. Lower estimates of efficiency potential use older data.
2. Extent of technological characterization: Analysts, for example, may consider only two of 10 measured savings mechanisms for electronic ballasts, or they omit savings from correcting oversizing of motors.
3. Extent of disaggregation of energy-using devices and activities: The sum of many small savings can be large.
4. Consideration of system effects: Multiple individual improvements, when combined in a system, are not generally additive and, conversely, there may be multiple benefits from single improvements.
5. Differences in methodology: Whether reduced maintenance costs are included sometimes makes large differences in estimated cost of savings.
6. Differences in cost valuation, particularly choice of discount rate.

For each of these reasons, it is not surprising that well-informed analysts can reach quite different conclusions. However, the differences in estimates may be unimportant: the savings potential may be so large that society may not need it all to address the climate problem. The real question is whether society can effectively capture the available savings through the policy process.

TRANSFERRING TECHNOLOGY FOR ENERGY EFFICIENCY

Technology transfer is broadly accepted as an important means of improving energy efficiency, particularly in developing and post-planned economies. Although the term is ill-defined, the process must transfer not just machines, products, or processes, but the systems of knowledge and services that go with them, and must be mindful of the inevitable concomitant transfer of cultural assumptions and attitudes.

Financial, institutional, cultural, and technical problems must be overcome to make available the most efficient energy technology for all nations. Debt is a major constraint for technology transfer and is an obstacle that should be considered in addressing both modeling and policy. There is a particular institutional constraint in nations such as the Soviet Union, where people have not owned or controlled their flats. Occupants cannot make changes even if they desire them. There are cultural constraints for preferences for types of energy services, e.g., level of heating, acceleration of cars, and willingness to accept given policies and practices.

A key question is how to avoid the transfer of outmoded technology. One example of this problem is in China, where one western company recently transferred the most modern technology to make a product for export, and an outmoded facility to make the same product for domestic consumption. Both developed and developing countries export very inefficient energy-using equipment, chiefly to developing countries, and that equipment's huge indirect cost in energy supply infrastructure often hobbles development. Commerce in such apparently cheap but actually costly devices cannot be restrained without better inventories, labeling, and standards. The extent of this problem is not known but appears to be very large and warrants urgent investigation and corrective action.

Valuable research could be organized as case studies (success stories) for specific transfer activities in industrial as well as developing countries. That is, we need to identify "what works" in transferring technology. Case studies of failures in technology transfer can also be constructive.

Several proposed solutions have been offered to ameliorate the institutional difficulties in technology transfer. One proposal is to create institutions for transferring funds from large financial organizations such as the World Bank to small investment projects. "Energy conservation banks" in the target countries could receive funds in large blocks. These institutions could thus overcome the inability of the World Bank to make loans of the relatively small magnitude appropriate for energy efficiency.

Another proposal would provide resources for developing and transition economies to organize their own skills to solve technical and policy problems, rather than the common practice of sending western consultants to do that work in the targeted country, which is an expensive approach usually resulting in little technology transfer. The energy-efficiency center concept was proposed as an alternative. Such centers can provide resources for local development of policy and technology expertise, the development of joint ventures with foreign firms possessing needed technology, development of energy-efficiency measurement protocols essential for the development of efficiency labels and standards, and basic data as well as training and public education campaigns. Several observers also urge the creation of mechanisms for transferring technology from developing to developed countries.

COMMUNICATING OPPORTUNITIES FOR ENERGY EFFICIENCY

Despite the overwhelming benefits of energy efficiency that have been demonstrated, there remains an enormous amount of energy efficiency to be achieved. The understanding of this potential must be communicated at individual as well as national levels.

The extraordinary range of cultural, economic, and political differences around the world presents an obstacle to effective communication. Using efficiency to curb emissions would be an acceptable argument in Germany, but not in other countries. In developing countries, support for development would be an acceptable argument. Another problem in effective communication is finding the right people with whom to communicate. In many countries, we could work with the currently responsible people, but they may soon be out of power.

Five broad categories of policies and activities can be used to communicate energy-efficiency advantages and opportunities. The first is market mechanisms. Prices may be the most effective way of communicating to consumers the need to conserve, as evidenced by the post-oil shock experience. Incentives such as fees on inefficient equipment and rebates on purchases which beat the minimum performance standards by a specific amount provide strong signals for changing consumer choices.

Second, information feedback to consumers would also help modify behavior. Efficiency labeling of appliances or motor vehicles helps consumers make informed choices. Clearly, understandable monthly energy bills are needed, and real-time metering for behavioral feedback would be ideal for households, automobiles, and industry alike.

Third, new institutions such as energy-efficiency centers offer opportunities for communicating energy efficiency. They can provide guidance on the purchase of efficient equipment, educational tools, and media campaigns to reach the larger public as well as policymakers effectively. Fourth is demonstration projects. The actual demonstration of successful ideas is necessary to convince consumers and investors that new ideas will really work.

Fifth, new tools are needed for improving energy education and our ability to illustrate the concepts of energy efficiency. We do not have adequate educational facilities in this field, and we need to develop university departments, centers, and institutes to train experts (and trainers) in

Table 16.1 Research priorities

I. Behavior and Policy:
    1. What are the causes that are prompting changes in the modes of energy-using activities?
    2. What are the energy-efficiency improvements of modal shifts?
    3. What role should government play in effecting energy-efficiency improvements?
    4. What are the different discount rates that must be applied for decision-making? What factors determine the discount rates, and what factors could decrease them?
    5. What are the causes that initiate additional activity in the transportation sector?

II. Economic Tools: Understanding Demand and Barriers
    1. can we explain past energy demand elastics in terms of bottom-up models?
    2. How can we understand the elasticities implications of land use and materials policy?
    3. What kind of factors influence the autonomous energy efficiency improvement, for example expenditures on research and development funds? Energy security considerations might prompt such investments and yield efficiency improvements?
    4. How useful are historical data in energy use in trying to predict future consumption?
    5. What are the aspects of the historical problem that need to be used to understand fuel mix?
    6. What are the causal relationships in the efficiency and safety of light vehicles?

efficiency. At the same time, we need to develop tools for teaching. Simple comparisons (for example, a car produces its own weight in carbon each year) would help make the subject come alive. We also need mechanisms such as expert systems to help train people faster.

In the end, it is neither the physical nor economic potential for energy saving that limits its contribution to reducing carbon dioxide emissions. Reserves of energy efficiency are truly enormous. Instead, social and institutional barriers limit this contribution and create uncertainty. Uncertainty leads to an important strategic question: how much energy saving will be realized?

REFERENCES

Carlsmith, R.S., W.U. Chandler, J.E. McMahon, and D.J. Santini. 1990. Energy Efficiency: How Far Can We Go? Oak Ridge, TN: Oak Ridge Natl. Lab.

Fickett, A.P., C.W. Gellings, and A.B. Lovins. 1990. Efficient use of electricity. *Sci. Am.* April 1990.

German Bundestag. 1989. Protecting the Earth's Atmosphere: An International Challenge. Bonn: Deutscher Bundestag.

Gibbons, J.H., and W.U. Chandler. 1981. Energy: The Conservation Revolution. New York: Plenum.

Goldemberg, J., T.B. Johansson, A.K.N. Reddy, and R.H. Williams. 1987. Energy for a Sustainable World. Washington, D.C.: World Resources Inst.

Makarov, A.A., and I.A. Bashmakov. 1990. The Soviet Union: A Strategy of Energy Development with Minimum Emission of Greenhouse Gases. Richland, WA: Pacific Northwest Laboratory.

Morovic, T., F.-J. Grundig, F. Jager, E. Jochem, W. Mannsbart, H. Poppke, M. Schon, and I. Totsch. 1987. Energy Conservation Indicators. Berlin: Springer.

Schipper, L., R. Howarth, and H. Geller. 1990. United States Energy Use from 1973 to 1987: the impacts of improved efficiency. *Ann. Rev. Energy* **16**.

Schipper, L., and S. Meyers. 1992. Energy Efficiency and Human Activity: Global Trends, Future Prospects. Cambridge: Cambridge Univ. Press, in press.

17
Regimes for Reducing Greenhouse-Gas Emissions

K. VON MOLTKE
Adjunct Professor, Dartmouth College, Hanover, NH, U.S.A. Senior Fellow, World Wildlife Fund and The Conservation Foundation

ABSTRACT

After reviewing the status of negotiations towards a climate change convention, this chapter identifies a number of precedents. It emphasizes that the central challenge of these negotiations is to conclude an international convention which will impact the everyday lives of virtually every person on the planet. This will require action at all levels of government and a framework to link them together. It focuses in particular on the need to regionalize a number of the issues so as to render them more manageable. This may prove true in particular for research, research assessment, technology transfer, and policy targets. A great deal of negotiation and reflection will still be needed before an appropriate structure emerges, the current move towards conclusion of a framework convention notwithstanding.

INTRODUCTION

The race is on to define the regimes for managing greenhouse-gas emissions internationally. This is the most important international decision of the decade. Even most issues of war, peace, and economic policy pale in comparison to the potential impact of this decision, which can shape the international economic and social order far into the 21st century. The challenge is to identify a practical route to follow, charting a path between the extremes of inaction (or its analogue, the timid adaptation of existing models to the new demands) and attempts to redefine all relationships between states and within countries with one masterstroke of the pen.

It is clearly impossible to sit down at a table and outline solutions to issues as important as those implied by the international management of greenhouse gases. Solutions can only grow out of a process of debate

Limiting the Greenhouse Effect: Options for Controlling Atmospheric CO_2 Accumulation
Edited by G.I. Pearman © 1992 John Wiley & Sons Ltd

followed by hard negotiation to ensure that major decisions are truly necessary and that sufficient foresight is being exercised. Presumably the ultimate result will be a number of interlocking regimes rather than a single structure operating according to unitary principles.

It is essential that this process be open, contrary to the traditions of international negotiations. Such an open process, which is bound to be painful and controversial, is needed because it is the first precondition of successful implementation. The regime for reducing greenhouse-gas emissions represents the first international decision which has the potential to reach into the life of every individual on this planet, almost without exception. If there is to be any prospect of successful implementation, it is essential that as many people as possible understand the implications of decisions which are being taken as they are taken.

Long experience shows that policy decisions adopted by default or by subterfuge are more likely to be subverted than those which are openly arrived at. This maxim of democratic policy-making applies to international decisions in the same manner as to those at any other level. In the absence of this element of consent, the only alternative is coercion, and that will not succeed in the area of greenhouse gases. Thus, modesty is the first principle of those who would venture into this arena.

CURRENT STATUS

With the adoption of the report of the Intergovernmental Panel on Climate Change (IPCC), the way is now free for negotiations to begin (IPCC 1990). The approach to this task is modeled in large measure on the experience of the Vienna Convention–Montreal Protocol–London Protocol process for the reduction of emissions of ozone-depleting substances. There is a strong high-level commitment for the conclusion of an initial climate change convention by the time of the United Nations Conference on Environment and Development, scheduled for Rio de Janeiro in June 1992. This convention is likely to be modeled on the Vienna Convention, which provided the basis for successful negotiation of measures to protect the stratospheric ozone layer.

The success of the Vienna process should not obscure the fact that the Vienna Convention was and is an exceedingly weak instrument. It recognizes the importance of protecting the stratospheric ozone layer, provides for research and systematic observation, cooperation between parties, exchange of information, establishes a conference of parties and a secretariat, and provides for the adoption of protocols. However, there is not a single binding obligation for any of the signatories in the Vienna Convention. A climate convention modeled on the Vienna Convention would not change the current status on this issue beyond providing a more permanent legal basis for some of the ongoing activities of research and assessment.

A crucial aspect of the Vienna Convention was the side agreement on a negotiating calendar. While it was not possible to include this agreement in the legally binding text, it initiated the process of assessment and negotiation which ultimately led to the Montreal and London Protocols. The agreement on a deadline for agreement, in lieu of substantive agreement, represents a classic negotiating technique; however, it is not a recipe for success unless the deadline is firm and forces all parties to confront the issues.

While there is not much debate about the need for a framework convention for climate change, and widespread expectation that this will be ready during 1992, there is much less agreement on the protocols which may be needed, their interrelationship, and a calendar for further negotiations (Kaiser 1990). The list of possible protocols is long. Among those most likely to require early attention are the protocols on carbon dioxide (or alternatively two protocols on energy policy and forestry) and on financial issues. Protocols on other greenhouse gases are liable to be delayed, with the possible exception of a protocol on chlorofluorocarbons not covered by the Vienna Convention process which are, however, greenhouse gases.

PRECEDENTS

One outcome of the increasing attention being paid to global warming is that past experience with environmental management at the international level is being viewed in a new light (Sand 1990). Clearly, there is a substantial body of international law and practice in this area (Caldwell 1984; Young 1989). Nevertheless, the lessons to be gained for managing greenhouse gases will probably prove limited, as in most instances contracting parties have been as much intent on limiting commitments as on establishing an efficient international regime. It needs to be recognized that very few existing agreements for international environmental management (e.g., conservation conventions, the Regional Seas Program, or the regime governing nuclear power plants) have been particularly successful in achieving real improvement in the area they have focused on. None of them have actually arrested, let alone reversed, the underlying trends they were to have combatted.

The most important exceptions to this general observation concern the Vienna–Montreal–London process on stratospheric ozone depletion, to a certain extent the process related to the Geneva Convention on Long-Range Transboundary Air Pollution and the Antarctic Treaty System. Almost always forgotten is the highly significant experience of the European Community in the area of environmental policy (Haigh 1991). One salient characteristic of each of these regimes is that they represent a continuing negotiating process rather than single substantive determinations on goals and instruments. In several instances, the beginnings have been highly inauspicious; however, the existence of a forum and a process for step-by-step development of needed measures and the accumulating evidence of continuing deterioration of environmental conditions have combined to drive parties to the regime to propose and even agree to measures which go far beyond what most of them envisaged at the outset. Thus if there is one lesson to be gained from past experience, it is the need to establish a dynamic regime which is capable of responding reasonably quickly to new information about environmental conditions in the area of concern.

Given the limited nature of past experience with environmental management, precedents have been sought beyond the environmental arena. One case which comes to mind is the law of the sea convention, and some have called for a "law of the atmosphere." The case against such an approach has been made cogently (Tolba 1989). The experience with the Third United Nations Conference on the Law of the Sea (UNCLOS III) is hardly an encouraging precedent. Protection and use of atmospheric resources require a very different approach than marine resources given the absence of a long history of international law of the atmosphere. Further, the law of the sea convention is lacking in crucial areas, such as the ability to adjust to changing science (regime dynamics) and interaction with national and regional jurisdictions.

Precedents could also be sought in other areas of international management, notably security and economic relations. This represents a fertile area of research since few experts in environmental affairs are knowledgeable about security and economic relations, and vice versa. Most recent work on the relationship of environment and security or environment and international economic relations, trade and finance in particular, has focused on the relevance of environmental concerns for these areas (World Commission 1987); not much work has considered the manner in which international security and economic practices may prove useful for environmental management and the control of greenhouse gases, in particular. Examples could be the development of verification procedures based on security regimes, the use of debt management for environmental purposes, the use of trade rules as incentives or disincentives, or the development of international "commodity" markets for emission rights. While several of these areas have been explored in a preliminary manner, and debt conversions have indeed been used to fund conservation, a substantial amount of work still needs to be undertaken.

Recourse to precedent is the normal impulse of international negotiators and lawyers, and it is fundamentally a sound impulse. Nevertheless, recourse to precedent is unlikely to yield the necessary solutions in the case of greenhouse gases. The task in this area is to develop, step-by-step, an international regime which encompasses numerous levels of action and a wide range of activities.

THE CHALLENGE: AN INTERNATIONAL REGIME FOR EVERYDAY LIFE

The paradox of greenhouse-gas control policies is that they must be agreed upon internationally but implemented at all levels, down to individual households. While there are numerous precedents for many of the structural elements of the regime, there is no precedent for such an international process. Control of greenhouse-gas emissions must occur at all levels: international, international regional, national, subnational, national regional, intermediate, local, and in individual social and economic units.

The basic rule of good government applies: no action should be taken at a higher level of centralization than necessary. This criterion will need to be stringently applied because there exists a theoretical argument that all decisions must be coordinated at the global level because the phenomena which are to be controlled are indubitably global in nature. Balancing the need for global coordination with the need for rigorous decentralization of decisions will prove one of the most important conundrums in designing regimes for reducing greenhouse-gas emissions.

It is possible to visualize the range of decisions as a classical hierarchy from the international level down, or as a series of interlocking circles. Clearly, the latter image offers more chances of success if we accept the principle of consent as fundamental to successful implementation. It is worth noting that past international practice corresponds to neither of these models, but was based on the sovereign state as a hinge mechanism between the international level and all others. This simplified image of international relations has been significantly undermined, primarily through the operation of nongovernmental interests at the international level which "short-circuit" the static and unproductive structure of intergovernmental relations. Nevertheless, ultimate decision-making authority has continued to rest with sovereign states, to the virtual exclusion of any other actors, including subnational governmental agencies. This approach holds little prospect of success in relation to controlling greenhouse-gas emissions.

The future role of nongovernmental actors at the international level represents a crucial issue for the development of global warming regimes. Throughout the steps of negotiation, implementation, and verification, the need for openness will tend to be in conflict with traditional forms of intergovernmental relations. Nongovernmental actors offer the prospect of maintaining necessary openness without requiring major modifications of established diplomatic practices. In effect, governments will find themselves communicating with each other through public fora and with the assistance of nongovernmental organizations which can often say explicitly and in public what governments may only be able to say implicitly and behind closed doors. Moreover, governments will find an increasing need to communicate with private actors in other countries but are constrained from doing so by diplomatic practice. Precedents for this kind of government/nongovernment structure exist in the European Community and between Canada and the U.S. (on acid rain) as well as between Israel and the U.S. (on security policy). Clearly the increased role for nongovernmental actors will also require adjustments on their part if they wish to remain effective.

REGIONALIZATION

Approaching international regime formation in the fashion outlined above leads to a modified concept of regional organization: the establishment of several interacting regional structures, all of which contribute to managing greenhouse gases but none of which is capable of achieving satisfactory results on its own. For example, in Europe it will be necessary and appropriate to act on greenhouse-gas emissions in the Economic Commission for Europe, in the European Community, in the Nordic Council, possibly in the Organization for Economic Cooperation and Development (OECD), and in the G7 process. In each of these fora there will be some overlap, but there will also be issues relevant to the specific forum in question. The overall frame of reference clearly needs to be established at a global level.

To the extent that this model is appropriate, there is a striking lack of regional organizations outside the western European area. North America is characterized by the virtual absence of working institutional arrangements between the U.S. and its neighbors. The recently negotiated Free Trade Agreement between Canada and the U.S. does not represent a viable framework, and it is unlikely that negotiations with Mexico will lead to a more satisfactory result. In Central and South America there are the beginnings of regional structure, including the Amazon Cooperation Treaty, the Contadora group of countries, and the Economic Commission for Latin America. The North Pacific and Africa are lacking in regional structures, other than the United Nations regional commissions. Neither the Organization for African Unity nor the Arab League holds much promise with regard to greenhouse gases.

In the absence of viable regional structures, the alternatives are national action or global action, both of which are subject to important vicissitudes. Moreover, it is unlikely that each country will develop its own institutions for verification, research, technology assessment, and transfer while generalized international organizations will not provide the degree of specificity and local understanding which policy-making in a national context requires. Similar phenomena can be observed in other environmental policy areas already. Typically, smaller countries are disproportionately active in the OECD because this represents a more cost-effective means of obtaining needed information and participating in an important debate. In the area of chemicals policy, all OECD countries have agreed to a pooling of testing resources because even the largest countries cannot undertake the testing needed to ensure risk reduction from existing chemicals.

Regionalizing Research

The importance of decentralizing research on greenhouse-gas issues is generally overlooked. There are several possible reasons for this oversight. The research community tends not to attach great importance to the place where research is conducted since it subscribes to an ethic of universality, that is, scientific facts should be the same everywhere. This apparently self-evident principle does not, however, invariably apply to environmental affairs because of the complexity and specificity of ecosystems. Research on greenhouse-gas emissions is, like most environmental research, only in part a purely scientific activity. Insofar as real conditions outside the laboratory are the principal subject of investigation, it is difficult to exert reliable control over experimental conditions and virtually impossible to actually duplicate results in different locations because of environmental variables.

Policymakers respond differently to research generated in their own country or region than to research findings derived from another country or region. The ongoing dispute concerning the rate of Amazonian deforestation is a good example. Similarly there is evidence that European policymakers failed to grasp the significance of U.S. findings on stratospheric ozone depletion for lack of active research in the field carried out by a sufficient number of European scientists to generate a balanced internal debate.

It is unlikely that policymakers in less-developed countries will be willing to take politically difficult decisions (e.g., on forestry, energy, or rice production) on the basis of research generated in industrialized countries alone. To reach such decisions it will be essential to ensure the availability of local research findings and assessments. Some precedents exist for such a geographically dispersed research strategy (e.g., in the field of agricultural research), but virtually no attention is being devoted to this issue with regard to climate change.

The need to regionalize extends to both principal branches of climate change research, on global mechanisms and on impacts. The need to regionalize research on mechanisms can be subject to discussion. The case is compelling with regard to research on impacts.

Regionalizing Research Assessments

It is necessary to distinguish between scientific research and research-based assessments for policy-making. The latter represent a hybrid of research and policy-making. While conclusions must be scientifically based, they do not represent scientific findings and are subject to the constraints of time and place typical of political decision-making.

Not much attention has been devoted to the assessment process, certainly not at an international level. Because of the close link between scientific research and environmental policy, many countries have institutional arrangements for this purpose. The paradox is that while such assessments are scientifically based, they represent consensus documents and have proven virtually impossible to transport across jurisdictional boundaries. That is true between countries. For example, reports of the U.S. National Academy of Sciences are read with attention in other countries but there is no evidence that they are viewed as authoritative for political decision-making. More surprising is the fact that it is also true between individual countries and regional bodies. British assessments have differed dramatically from those on which the Commission of the European Communities based proposals; the Scandinavian countries sought unsuccessfully for several years to obtain acceptance of its assessments on acid rain and finally recognized that the domestic assessment process needed to be duplicated by an international one; the U.S. sought vainly to get its assessment of the processes of stratospheric ozone depletion accepted by the European Community.

Some examples of international assessment exist (von Moltke 1991). In the area of stratospheric ozone depletion and climate change, ad hoc international processes have been used including a series of conferences leading up to the Montreal Protocol and the IPCC. Experience with the Montreal Protocol indicates that participation in this assessment process is essential if countries are to be convinced by the subsequent negotiations. With respect to greenhouse gases, it is to be feared that participation by several key less-developed countries was not adequate to carry conviction into international negotiations, leaving a continuing need to develop more broadly based assessments as well as update the work of the IPCC in light of new research results.

Regionalizing Technology Transfer

It is widely recognized that technology transfer will constitute an essential element of any ultimate agreement on controlling greenhouse-gas emissions.

Again the Vienna–Montreal–London process provides some precedents, which will presumably need to be substantially developed. Despite this recognition of the importance of technology transfer, little understanding exists of the processes and mechanisms which will need to be created.

Presumably technology transfer is taken to define a process which focuses and accelerates commercially based activities. In economic terms this represents a program of subsidies, irrespective of the form in which these are provided, as payments to technology developers, to users of technology, or by financing the institutional arrangements for such transfers. Such subsidies are widespread in trade policy, for example, in trade missions, insurance schemes, information dissemination or similar activities. Occasionally they are also already used for environmental purposes. For some time, government funds have been used directly and indirectly to finance technological research, although the thrust of these programs is subject to criticism from the perspective of controlling greenhouse-gas emissions. The net economic effect of all these activities is unlikely to be large enough to create problems from the trade policy perspective.

A limiting factor on technology transfer is likely to be the acceptability under trade regulation. A program which seeks to change the fundamental economics through subsidies is unlikely to be acceptable and equally unlikely to be effective. Internal tax codes will need to be used to ensure that distortions which favor greenhouse-gas-emitting technologies are eliminated; international technology transfer mechanisms are unable to achieve that goal.

The focus of technology transfer must therefore be on providing information about available technologies, funding pilot programs in carefully selected cases, and creating conditions which foster the development of further technologies. Joint development schemes may prove useful. Under no circumstances can this activity be conceived of solely as making technologies from industrialized countries available in less developed ones. It is important to develop a process which is broadly based and allows technology flows in all directions. Certainly there will need to be South–North and South–South technology transfer in addition to more conventional North–South flows. The institutional structure for such a process does not currently exist. Problems associated with international property rights will need to be addressed.

In any process of technology transfer, it is essential to ensure the appropriateness of new technologies to local conditions. Much environmental damage in tropical regions is the direct consequence of employing inappropriate technologies developed for more temperate climates, and in the absence of rigorous assessment such problems will continue to exist. Such assessment should again be organized on a regional basis so as to maximize limited resources and access to information. In general, it needs to be recognized that technology transfer programs are liable to fail in the absence of viable technology development and assessment capabilities in less-developed countries.

Regionalizing Policy Targets

A number of options exist to define targets for the control of greenhouse-gas emissions (Swart et al. 1989). As negotiations develop, it is likely that complex agreements concerning regional variations in the goals and extent of control will emerge. This is shown by experience with the Convention on Long-Range Transboundary Pollution (Sand 1990) and in particular by the European Community Directive on large combustion plants (Haigh 1989). Such regionalization will have consequences for the design of verification procedures.

It is possible to envisage a range of trade-offs within a regional context which are unlikely to prove acceptable at a global level. For example, emissions trading is favored by relatively few countries, but a solution may be found which allows such procedures on a regional basis, provided adequate verification is available.

CONCLUSION

This paper has sought to describe some of the conditions for the development of regimes for reducing greenhouse-gas emissions. Presumably several other matters not discussed here will emerge in the course of further negotiations. Two conclusions can readily be drawn from the current evidence:

1. Developing appropriate regimes will require a number of years and can only be undertaken step-by-step; too many uncharted issues remain to be resolved while practical first steps can presumably be taken.
2. Regional arrangements will play a central role in ensuring that the regimes which evolve are effective and dynamic.

REFERENCES

Caldwell, L. 1984. International Environmental Policy. Emergence and Dimensions. Durham, NC: Duke Univ. Press.

Haigh, N. 1989. New tools for European air pollution control. *Intl. Env. Affairs* **1**:26–37.

Haigh, N. 1991. EEC Environmental Policy and Britain, 3rd ed. London: Longman.

IPCC (Intergovernmental Panel on Climate Change). 1990. IPCC First Assessment Report. Overview. *Intl. Env. Affairs.* **3(1)**:64–84.

Kaiser, K.E. von Weizsäcker, R. Balischwitz, and S. Comes. 1990: Internationale Konvention zum Schutz der Erdatmosphäre. Bonn: Economica Verlag.

Sand, P. 1990. Lessons Learned in Global Environmental Governance. Washington D.C.: World Resources Inst.

Swart, R.J., H. de Boois, and J. Rotmans. 1989. Targeting climate change. *Intl. Env. Affairs* **1**:222–234.

Tolba, M. 1989. A step-by-step approach to protection of the atmosphere. *Intl. Env. Affairs* **1**:304–314.

World Commission on Environment and Development. 1987. Our Common Future. Oxford: Oxford Univ. Press.

Young, O. 1989. International Cooperation. Building Regimes for Natural Resources and the Environment. Ithaca, NY: Cornell Univ. Press.

18

Reduction of Greenhouse-Gas Emissions: Barriers and Opportunities in Developing Countries

M. MUNASINGHE
Environmental Policy Division, The World Bank, Washington, D.C., U.S.A.

ABSTRACT

Global warming has emerged as a key issue that complicates existing energy problems. Increased energy use is a vital prerequisite for economic development and less-developed countries (LDCs) are struggling to meet energy needs at acceptable costs. LDC decision-makers share the worldwide environmental concerns but also face other urgent issues like poverty. The industrialized countries can afford to substitute environmental protection for further material growth, but the LDCs will need concessional funding to participate in addressing global environmental problems. Global financing issues may be analyzed and resolved through trade-offs among several criteria, including affordability/additionality, fairness/equity, and economic efficiency. The short-term LDC response will be limited mainly to conventional technologies in efficiency improvements, conservation, and resource development. The industrialized nations should provide financial resources to LDCs and develop the technology to be used in the 21st century. Pilot international funds like the Global Environmental Facility and the Ozone Fund will help LDCs participate in the effort to solve global environmental issues.

INTRODUCTION

Energy became a major international issue in the 1970s, following the oil crisis. More recently, global climate change induced by excessive greenhouse-gas accumulation in the atmosphere has emerged as a potential problem

Limiting the Greenhouse Effect: Options for Controlling Atmospheric CO_2 Accumulation
Edited by G.I. Pearman © 1992 John Wiley & Sons Ltd

that further complicates energy issues. In this paper, we will focus on energy-related environmental issues in the developing countries. The development-related energy needs of the Third World and their financial implications, the potential for better energy management, barriers to reducing greenhouse-gas emissions, and financial mechanisms and policies that can improve the performance of developing countries in this respect will be explored below.

The experience of the industrialized countries emphasizes that a reliable supply of energy is a vital prerequisite for economic growth and development. For example, the observed trends relating to electricity demand in developing countries (which indicate annual growth rates in the region of 6 to 12%) are consistent with the development objectives that these countries all share (Munasinghe 1990a; World Bank 1990). Up to the present time, many developing countries have been struggling with the formidable difficulties of meeting these needs for energy services at acceptable costs. If such needs cannot be met, economic growth is likely to slacken and the quality of life will fall.

Given these already existing handicaps, the growing additional concerns about the environmental consequences of energy use considerably complicate the policy dilemma facing the developing countries. In the past, industrial countries that faced a trade-off between economic growth and environmental preservation invariably gave higher priority to the former. These richer countries have awakened only recently to the environmental consequences of their economic progress, and only after a broad spectrum of economic objectives has been reached. This model of economic and social development has been adopted by many Third World regions. Therefore, until both developed and developing countries find a less material-intensive and sustainable development path, environmental protection efforts will be hampered.

The developing or lesser-developed countries (LDCs) share the deep worldwide concerns about environmental degradation, and some have taken steps already to improve their own natural resource management as an essential prerequisite for sustained economic development. However, they also face other urgent issues like poverty, hunger, and disease as well as rapid population growth and high expectations. The paucity of resources available to address all these problems constrains the ability of LDCs to undertake costly measures to protect the global commons.

The crucial dilemma for LDCs is how to reconcile development goals and the elimination of poverty, which will require increased use of energy and raw materials, with responsible stewardship of the environment, and without overburdening already weak economies. The per capita gross national product (GNP) of low income economies (with half of the world population) averaged U.S. $290 in 1987, or under one-sixtieth of the U.S. value of $18,530 (World Bank 1989). In the two largest developing countries, India

Barriers and Opportunities in Developing Countries 489

and China, per capital GNP was $300 and $290 respectively. Correspondingly, the U.S. per capita energy consumption of 7265 kg (oil equivalent) in 1987 was 35 and 15 times greater than the same statistic in India and China respectively.

The disparity in both per capita income and energy use among different countries also raises additional issues in the context of current global environmental concerns, and the heavy burden placed on humankind's natural resource base by past economic growth. Fossil-fuel-related CO_2 accumulation in the atmosphere is a good example. The developed countries accounted for over 80% of such cumulative worldwide emissions during 1950–1986 (North America contributed over 40 Gt of carbon, western and eastern Europe emitted 25 and 32 Gt respectively) and the share of the developing countries was about 24 Gt. On a per capita basis the contrasts are even more stark, with North America emitting over 20 times more and the developed countries as a whole being responsible for over eleven times as much total cumulative CO_2 emissions as the LDCs. The LDC share would be even smaller if emissions prior to 1950 were included. Clearly, any reasonable growth scenario for developing nations that followed the same material-intensive path as the industrialized world would result in unacceptably high levels of future greenhouse-gas accumulation as well as more general depletion of natural resources.

Up to now, scientific analyses have provided only broad and rather uncertain predictions about the degree and timing of potential global warming. However, it would be prudent for humankind to buy an "insurance policy" in the form of mitigatory actions to reduce greenhouse-gas emissions. Ironically, both local and global environmental degradation might affect developing countries more severely, since they are more dependent on natural resources while lacking the economic strength to prevent or respond quickly to increases in the frequency, severity and persistence of flooding, drought, storms, and so on. Thus, from the LDC viewpoint, an attractive low-cost insurance premium would be a set of inexpensive measures that could address a range of national and global environmental issues, without hampering development efforts.

The recent report of the Brundtland Commission (WCED 1987), which has been widely circulated and accepted, has presented arguments along the theme of sustainable development, which consists of the interaction of two components: (a) *needs*, especially those of the poor segments of the world's population, and (b) *limitations* which are imposed by the ability of the environment to meet those needs. The development of the presently industrialized countries took place in a setting which emphasized needs and de-emphasized limitations. The development of these societies have effectively exhausted a disproportionately large share of global resources to include both the resources that are consumed in productive activity (such as oil, gas, and minerals), as well as environmental assets that absorb the waste products of economic activity and those that provide irreplaceable life support functions (like the high-altitude ozone layer). Indeed, some analysts argue that this development path has significantly indebted the developed countries to the larger global community.

The division of responsibility in this global effort is clear from the above arguments. The unbalanced use of common resources in the past should be one important basis on which the developed and developing countries can work together to share and preserve what remains. The developed countries have already attained most reasonable goals of development and can afford to substitute environmental protection for further growth of material output. On the other hand, the developing countries can be expected to participate in the global effort only to the extent that this participation is fully consistent with and complementary to their immediate economic and social development objectives.

In the context of the foregoing, this paper identifies critical energy–environmental issues, using examples from the highly capital-intensive and pivotal power sector to illustrate specific points. It also explores some policy implications, constraints, and opportunities at the national and global level for both the developing countries as well as the wider international community, and examines the role of emerging mechanisms such as the global environmental fund and ozone defense fund, in the allocation of resources for addressing transnational environmental problems.

ELECTRIC POWER NEEDS OF THE DEVELOPING COUNTRIES

Despite some anomalies, the link between energy demand and GDP is well established (Munasinghe 1990b). Electric power, in particular, has a vital role to play in the development process, with future prospects for economic growth being closely linked to the provision of adequate and reliable electricity supplies. Figure 18.1 indicates the relationship between electricity use and income for both developed and developing countries. A more systematic analysis of World Bank and UN data over the past two or three decades indicates that the ratio of percentage growth rates (or elasticity) of power system capacity to GDP is about 1.4 in the developing countries.

Assuming no drastic changes in past trends with respect to demand management and conservation, the World Bank's most recent projections indicate that the demand for electricity in LDCs will grow at an average annual rate of 6.6% during the period 1989–1999 (World Bank 1990). This compares with actual growth rates of 10% and 7% in the seventies and eighties, respectively. Such rates of growth indicate the need for total

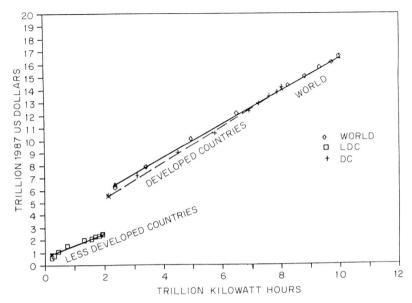

Figure 18.1 Gross product vs. kilowatt hours, 1960–1986 (from Starr and Searle 1988)

capacity additions of 384 GW during 1990–1999 (see Figure 18.2), and annual energy consumption of 3844 TWh by 1999. As indicated in Figure 18.3, the requirements of the Asia region dominate with almost two-thirds of the total anticipated new capacity. Coal and hydro are the main primary sources for meeting these demands, both of which have specific environmental problems associated with their use.

The investment needs corresponding to these indicative projections are also very large. Table 18.1 shows the projected breakdown of LDC power-sector capital expenditure in the 1990s. Of a total of $745 billion (constant 1989 U.S.$), Asia (which includes both India and China) again dominates, accounting for $455 billion or over $45 billion annually. In comparison with the total projected annual requirement for LDCs of $75 billion, the present annual rate of investment in developing countries is only around $50 billion (Munasinghe 1990a). Even this present rate is proving difficult to maintain. Developing country debt which averaged 23% of GNP in 1981, increased dramatically to 42% in 1987 and has not declined significantly since then. In low income Asian countries, outstanding debt doubled, from 8% in 1981 to 16% in 1987. Capital-intensive, power-sector investments have played a significant role in this observed increase.

If the developing countries follow this projected expansion path, the environmental consequences are also likely to increase in a corresponding

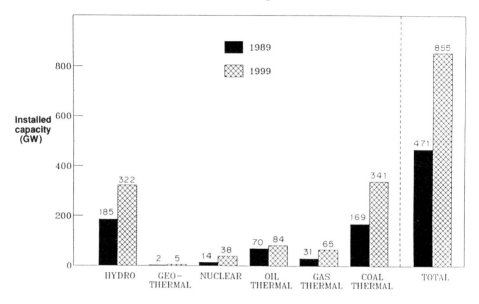

Figure 18.2 Comparison of 1989 and the required 1999 installed generating capacities in the lesser-developed countries (from World Bank 1990)

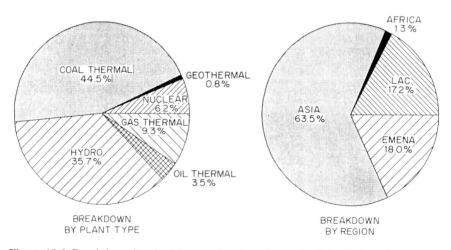

Figure 18.3 Breakdown by plant type and region of capacity (384 GW total) expected to be added in the LDCs in the 1990s (from World Bank 1990)

Table 18.1 Regional breakdown of LDC power capital expenditures in the 1990s (from World Bank 1990)

	Asia	EMENA[1]	LAC[2]	Africa	Total
Generation	277	82	83	6	448
Transmission	39	8	32	2	81
Distribution	100	23	27	2	152
General	39	11	13	1	64
TOTAL	455	124	155	11	745
Percent	61.1	16.6	20.8	1.5	100

[1]Europe, Middle East and North Africa (Mediterranean region)
[2]Latin America and the Caribbean

fashion. There is already a growing concern about the environmental consequences at the national level of energy use in developing countries. At a recent workshop on acid rain in Asia, participants reported on a wide range of environmental effects of the growing use of fossil fuels, especially coal, in the region (Foell 1989). For example, total 1985 sulfur dioxide emissions in Asia were estimated at around 22 million tons, and these levels, coupled with high local densities, have led to acid deposition in many parts of Asia.

The developing countries feel that any attempts to mitigate these environmental effects, however, cannot jeopardize the critical role played by electric power (and more generally, energy) in economic development. Similarly, the allocation of resources to environmental programs in developing countries cannot diminish the resources needed to fund projected expansion of supply. Energy and environmental policymakers in both developing countries and the global community are, indeed, confronted with a formidable dilemma.

THE ECONOMICS OF ENERGY–ENVIRONMENTAL ISSUES

The foregoing discussion has helped to establish a rational and equitable basis for addressing the problems of energy–environmental impact mitigation. In this section we present an integrating framework which ties together the issue of environmental protection with the existing energy-sector objectives of energy efficiency and economic growth.

It is convenient to recall here that traditionally, the specific prerequisites for economic efficiency have included both (Munasinghe 1990a): (a) efficient consumption of energy, by providing efficient price signals that ensure optimal energy use and resource allocation; and (b) efficient production of energy, by ensuring the least-cost supply mix through the optimization of investment planning and energy-system operation.

A new issue which has emerged in recent decades, as an area of particular concern, is the efficient and optimal use of our global natural resource base, including air, land, and water. Since there has also been much discussion about the key role energy efficiency and energy conservation might play in mitigating environmental costs, it is useful first to examine how these topics relate to economic efficiency. Specific issues dealing with the formulation and implementation of economically efficient energy policies are presented in the next section.

Major environmental issues vary widely, particularly in terms of scale or magnitude of impact, but most are linked to energy use. First, there are the truly global problems such as the potential worldwide warming due to increasing accumulation of greenhouse gases like carbon dioxide and methane in the atmosphere, high-altitude ozone depletion because of release of chlorofluorocarbons, pollution of the oceanic and marine environment by oil spills and other wastes, and excessive use of certain animal and mineral resources. Second in scale are the transnational issues like acid rain or radioactive fallout in one European country due to fossil-fuel or nuclear emissions in a neighboring nation, and excessive downstream siltation of river water in Bangladesh due to deforestation of watersheds and soil erosion in nearby Nepal. Third, one might identify national and regional effects, for example those involving the Amazon basin in Brazil or the Mahaweli basin in Sri Lanka. Finally, there are more localized and project specific problems such as the complex environmental and social impacts of a specific hydroelectric or multipurpose dam.

While environmental and natural resource problems of any kind are a matter for serious concern, those that fall within the national boundaries of a given country are inherently easier to deal with from the viewpoint of policy implementation. Such issues that fall within the energy sector must be addressed within the national policymaking framework. Meanwhile, driven by strong pressures arising from far-reaching potential consequences of global issues like atmospheric greenhouse-gas accumulation, significant efforts are being made in the areas of not only scientific analysis, but also international cooperation mechanisms to implement mitigatory measures.

Given this background, I discuss next some of the principal points concerning energy use and economic efficiency (Munasinghe 1990b). In many countries, especially those in the developing world, inappropriate policies have encouraged wasteful and unproductive uses of some forms of energy. In such cases, better energy management could lead to improvements in: (a) economic efficiency (higher value of net output produced); (b) energy efficiency (higher value of net output per unit of energy used); (c) energy conservation (reduced absolute amount of energy used); and (d) environment protection (reduced energy-related environmental costs). While such a result fortuitously satisfies all four goals, the latter is not always mutually consistent. For example, in some developing countries where the existing levels of per capita energy consumption are very low and certain types of energy use are uneconomically constrained, it may become necessary to promote more energy consumption in order to raise net output (thereby increasing economic efficiency). There are also instances where it may be possible to increase energy efficiency while decreasing energy conservation.

Despite the above complications, our basic conclusion remains valid, that the economic efficiency criterion which helps us maximize the value of net output from all available scarce resources in the economy (especially energy and the ecosystem in the present context) should effectively subsume purely energy-oriented objectives such as energy efficiency and energy conservation. Furthermore, the costs arising from energy-related adverse environmental impacts may be included (to the extent possible) in the analytical framework of energy economics to determine how much energy use and net output society should be willing to forego in order to abate or mitigate environmental damage. The existence of the many other national policy objectives, including social goals that are particularly relevant in the case of low income populations, will complicate the decision-making process even further.

The foregoing discussion may be reinforced by the use of a simplified static analysis of the trade-off between resource use and net output of an economic activity (see Figure 18.4).

Energy Efficiency

Curve Y in Figure 18.4 represents the usual measure of net output of productive economic activity in a country, as a function of some resource input (say energy), considering only the conventional internalized costs, i.e., not accounting for environmental impacts. Due to policy distortions (for example, subsidized prices), the point of operation in many developing countries appears to be at A, where the resource is being used wastefully. Therefore, without invoking any environmental considerations, but merely by increasing economic and resource use efficiency (i.e., energy efficiency), output as usually measured could be maximized by moving from A to B. A typical example might be improving energy end-use efficiency or reducing energy supply-system losses (see the section on Framework for Energy Environmental Policy Analysis, for practical examples).

Quantifiable National Environmental Costs

Now consider the curve EC_{NQ}, which represents the environmental costs associated with energy use which are nationally quantifiable. The latter

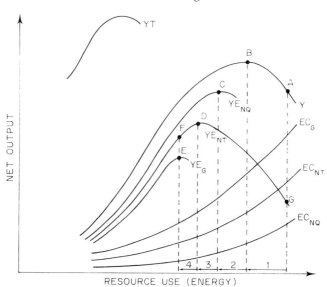

Figure 18.4 Net output, resource use, and environmental cost. $YE_{NQ} = Y - EC_{NQ}$; $YE_{NT} = Y - EC_{NT}$; $YE = Y - EC_C$

might include air pollution related health costs of coal power plant or the costs of environmental protection equipment (like scrubbers and electrostatic precipitators to reduce noxious gas and particulate emissions) installed at such a plant, or the costs of resettlement at a hydropower dam site. The corresponding corrected net output curve is: $YE_{NQ} = Y - EC_{NQ}$, which has a maximum at C that lies to the left of B, implying lower use of (more costly) energy.

Nonquantifiable National Environmental Costs

Next, consider the "real" national output YE_{NT}, which is net of total environmental costs, whether quantifiable or not. The additional costs to be considered include the unquantified yet very real human health and other unmonetized environmental costs. These total (quantifiable and nonquantifiable) costs are depicted as EC_{NT}; and once again, $YE_{NT} = Y - EC_{NT}$. As shown, the real maximum of net output lies at D, to the left of C.

Transnational Environmental Costs

Finally, EC_G represents the globally adjusted costs, where the transnational environmental costs (to other countries) of energy use within the given country have been added to EC_{NT}. In this case, $YE_G = Y - EC_G$ is the corresponding net output corrected for both national and global environmental costs, which implies an even lower level of optimal energy use.

For example, consider the costs imposed on other countries (such as transborder impacts of a major dam or global climate impacts of carbon dioxide emissions). If it is decided to reduce resource use within this country further in order to achieve the internationally adjusted optimum at E, then a purely national analysis will show this up as a drop in net output, i.e., from D to E. Since other countries benefit, this drop in net output may justify compensation in the form of a transfer of resources from the beneficiary countries. Note that the (reverse) transnational costs imposed by other countries on the nation in question will be a function of external resource use rather than the national resource use shown on the horizontal axis.

The additional curve YT shows net output for a technologically advanced future society that has achieved a much lower resource intensity of production.

Policy Measures for the LDCs

The foregoing analysis illustrates the crucial dilemma for developing countries. In Figure 18.4, all nations (including the poorest) would readily adopt measures that will lead to shift (1) which simultaneously and unambiguously provides both economic efficiency and environmental gains. Most developing countries are indicating increasing willingness to undertake shift (2). However, implementing shift (3) will definitely involve crossing a "pain threshold" for many Third World nations, as other pressing socioeconomic needs compete against the costs of mitigating nonquantifiable adverse environmental impacts.

We note that real economic output increases with each of the shifts (1), (2) and (3), as shown by the movement upward along the curve YE_{NT} from G to D. However, these shifts are mistakenly perceived often as being upward only from A to B (energy-efficiency improvements), followed by downward movements from B through C to D. It is therefore important to correct any misconceptions that environmental protection results in reduced net output. However, it is clear that shift (4) would hardly appeal to resource-constrained developing countries unless concessionary external financing was made available, since this movement would imply optimization of a global value function and costs that most often exceed in-country benefits. In the foregoing, we have neglected considerations involving reciprocal benefits to the given country due to energy-use reductions in other countries.

Therefore, we may conclude that, while the energy required for economic development will continue to grow in the developing countries, in the short to medium term there is generally considerable scope for most of them to practice better energy management, thereby increasing net output, using their energy resources more efficiently and contributing to the effort to reduce global warming. In the medium to long term, it will become possible for the developing countries to adopt newer and more advanced (energy efficient) technologies that are now emerging in the industrialized world, thus enabling them to transform their economies and produce even more output using less energy.

In other words, the developing countries could be expected to cooperate in global environmental programs only to the extent that such cooperation is consistent with their national growth objectives. The role of the developed countries, on the other hand, is to incur the risks inherent in developing innovative technological measures which are the prerequisites for the next level in environmental protection and the mitigation of adverse consequences. These risks include the possibility that the more extreme measures may turn out to be unnecessary or inapplicable after all, given the prevailing uncertainty about the future impact of current environmental developments.

FRAMEWORK FOR ENERGY–ENVIRONMENTAL POLICY ANALYSIS AND OPTIONS AT THE NATIONAL LEVEL

The previous section introduced a way of considering energy–environmental issues in terms of four shifts. In this section a rational framework for energy–environmental policy analysis within a country is presented in terms of these shifts. This coverage is expanded to the global level in the next section, focusing particularly on the interaction between developed and developing countries.

Developed countries generally differ from the developing countries in the extent to which the shifts have already been made. The LDCs are more likely to be characterized as being at point A in Figure 18.4, while the developed countries are somewhere between C and D. This means that for developing countries there is still considerable scope for environmental improvement by undertaking programs that are consistent with the national objective of increasing overall output. The challenge for national decision-makers and the international community is to find as many areas as possible where such consistency and complementarity exist, between growth and environmental protection goals.

Advantages of an Integrated Approach

Successful policy analysis requires an integrated approach, unified in an economic sense so that all feasible options can be balanced and traded off, if necessary, in the search for an overall optimal strategy. It is within such a comprehensive framework that barriers to and opportunities for making various choices will become apparent. For successful energy-policy analysis, a better understanding of economy-wide linkages is useful, whatever the prevailing political system. It will help decision-makers in formulating policies and providing market signals and information to economic agents that encourage more efficient energy production and use, as well as better protection of the environment. In Figure 18.5 a hierarchical framework for integrated national energy planning (INEP), policy analysis and supply-demand management to achieve these goals is summarized (Munasinghe 1990b).

Although the INEP framework is primarily country-focused, we begin by recognizing that many energy–environmental issues have global linkages. Thus individual countries are embedded in an international matrix, while economic and environmental conditions at this global level will impose a

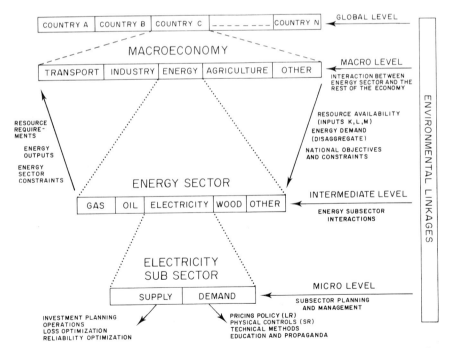

Figure 18.5 Integrated conceptual framework for energy and environmental policy analysis

set of exogenous inputs or constraints on decision-makers within countries. The next hierarchical level in Figure 18.5 treats the energy sector as a part of the whole economy. Therefore, energy planning requires analysis of the links between the energy sector and the rest of the economy. Such links include the energy needs of user sectors, input requirements of the energy sector, and impact on the economy of policies concerning energy prices and availability.

The next level of the integrated approach treats the energy sector as a separate entity composed of subsectors such as electricity, petroleum products, and so on. This permits detailed analysis, with special emphasis on interactions among the different energy subsectors, substitution possibilities, and the resolution of any resulting policy conflicts. The final and most disaggregate level pertains to analysis within each of the energy subsectors. It is at this lowest hierarchical level that most of the detailed energy resource evaluation, planning, and implementation of projects is carried out by line institutions (both public and private). In practice, however, the three levels of INEP merge and overlap considerably. Thus the interactions of electric power problems and linkages at all three levels need to be carefully examined.

Energy–environmental interactions (represented by the vertical bar) tend to cut across all levels and need to be incorporated into the analysis. Finally, spatial disaggregation may be required also, especially in larger countries. Such an integrated framework facilitates policy-making and does not imply rigid centralized planning. Thus, such a process should result in the development of a flexible and constantly updated energy strategy designed to meet the national goals mentioned earlier. This national energy strategy (of which the investment program and pricing policy are important elements) may be implemented through a set of energy-supply and demand-management policies and programs that make effective use of decentralized market forces and incentives.

Policy Tools and Constraints

To achieve the desired national goals, the policy instruments for optimal and energy management available to Third World governments must include:

- physical controls;
- technical methods;
- direct investments or investment-inducing policies;
- education and promotion; and
- pricing, taxes, subsidies and other financial incentives.

Since these tools are interrelated, their use should be closely coordinated for maximum effect.

The chief constraints that limit effective policy formulation and implementation are:

- poor institutional framework and inadequate incentives for efficient management;
- insufficient manpower and other resources;
- weak analytical tools;
- inadequate policy instruments; and
- other constraints such as low incomes and market distortions.

Technological Options

The INEP framework is particularly appropriate for considering shift (1) in Figure 18.4, which implies improvement in overall economic efficiency but without any explicit consideration of the external (environmental) costs. Such efficiency improvements require better energy supply and demand management. More specifically, the former category includes more accurate demand forecasting, improved least-cost investment planning, and the optimal operation of energy systems, which implies that plant performance, operating and maintenance procedures, loss levels, etc. are optimized. The latter comprises efficient electricity end-use, load management, and pricing (described in the next section). All of the above options constitute an attractive policy package for most power utilities in both the developing and developed countries.

There is a spectrum of technological options which the developing countries could potentially utilize in order to improve energy efficiency and thereby reduce environmental effects arising from energy-sector activity. These range from simple infrastructural retrofits to the use of advanced energy-supply technologies. Among the short-term technological options for the developing country power sector, reducing transmission and distribution losses and improving generation plant efficiencies appear to be the most attractive. Recent studies show that up to a certain point these measures for enhancing energy-supply efficiency yield net economic savings or benefits that are several times the corresponding costs incurred (Munasinghe 1990a). While estimates of such power-system losses vary, they all point to levels which are far in excess of accepted norms. Table 18.2 presents estimates for some Asian countries in comparison with industrialized countries. While acceptable loss levels may be about 6–8% in transmission and distribution as a percentage of gross generation, these losses in Third World power systems are estimated to average in the 16–18% range (of which about one-third could be theft).

The consequences of reducing these losses can be quite important. On

Table 18.2 Electrical transmission and distribution losses (% of gross generation)

Pakistan	28%
India	22%
Bangladesh	31%
Sri Lanka	18%
Thailand	18%
Philippines	18%
South Korea	12%
Japan	7%
US	8%

These loss estimates include nontechnical losses (i.e., due to deficient metering and theft). From the World Bank (internal documents) and AID (1988).

the basis of our previous estimates of capacity requirements, a one percent reduction in losses per year would reduce required capacity by about 5 GW annually in the developing countries. The estimated saving in capital investment would be around 10 billion dollars per year. Meanwhile, the Agency for International Development (AID 1988) has estimated that the average heat rate of LDC power plants is around 14,000 MJ/kWh, compared to 10,000–12,000 MJ/kWh if these plants were operated efficiently. The energy savings (and positive environmental consequences) implied in these figures are quite significant also.

Similar gains are possible by conservation on the demand side. Johansson et al. (1987) provide an insightful review of the developments that have been taking place in end-use technologies which can have a major impact on energy efficiency. These technologies (which developed in the industrialized countries as a response to the oil price escalation in the seventies) can be easily applied towards more efficient lighting, heating, refrigeration, and air conditioning around the developing world, as described below.

Substitution of primary-energy sources in power generation is another potential means of achieving dual benefits. In the developing world, natural gas is the most likely candidate for coal or oil substitution. The economic benefit of natural-gas substitution comes from either import substitution for petroleum products or releasing these products for export. On the environmental front, natural gas firing typically achieves reduction in carbon emissions of 30–50%. Many Asian countries are endowed with significant resources of natural gas, including Malaysia, Indonesia, and Thailand.

In the longer term, developing countries will need to rely on more advanced technological options which are currently being developed in the industrialized countries. As we have discussed above, power generation capacity in developing countries is expected to nearly double by the turn of the century and will increase further thereafter. This provides opportunities to add state-of-the-art technologies which have been designed with regard to both economic and environmental criteria. Clean coal technologies, cogeneration, gas turbine-combined cycles, steam-injected gas turbines, etc. are all part of this menu of technologies which have important potential in developing countries. Similar applications will become available for emission control technologies. However, as argued previously, developing countries will look to the industrialized nations to provide the leadership in refining and proving these technologies before they are implemented in the developing world.

One indicative example of the state-of-the-art in supply planning and demand-management possibilities are the results of a recent Swedish study (Johannson et al. 1989). The power sector in Sweden, currently supplied half by hydro and half by nuclear generation, faces the following severe restrictions: (a) hydro expansion limited by environmental constraints; (b) mandatory phasing out of all nuclear units by 2010; and (c) no increase permitted above present CO_2 emission levels. The demand for electricity-derived services is projected to increase by 50%, from 1987 to 2010. If end-use efficiency remained unchanged, then under this "frozen efficiency" scenario, the electrical load also would increase by 50%, from 129 TWh in 1987 to 195 TWh in 2010, at an average annual growth rate of 1.8%.

The same output of electrical services could be provided, but with steadily declining electricity input needs and load levels given the increasingly energy-efficient scenarios A, B, C, and D, shown in Figure 18.6. The corresponding loads in 2010 would be 140, 111, 96, and 88 TWh, respectively. Only options C and D permit the load to be met after all the nuclear plants are retired in 2010. Figure 18.7 indicates that the total costs of energy supply also fall progressively under the scenarios A through C (some of the costs are undefined for scenario D). In addition, there are three supply scenarios based on different selection rules for generation, including that supply costs could rise steadily as we move through the economic dispatch, natural gas/biomass, and environmental dispatch options. These costs exclude taxes and subsidies and are based on a 6% discount rate, 1987 world oil and coal prices, and coal equivalent gas prices for steam power generation.

Economic Incentives and Related Options

Providing the correct economic signals, or more specifically price rationalization, offers the most attractive demand-side option for improving energy-sector efficiency and corresponds to the shift (1) in Figure 18.4. While the economic principles of energy pricing are now well understood, pricing

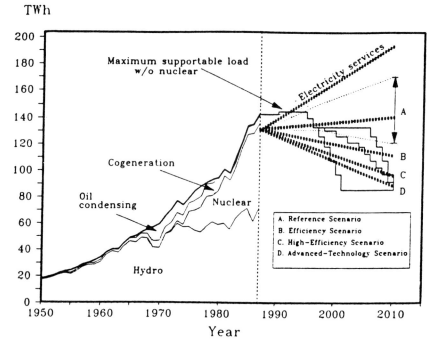

Figure 18.6 Generation capability and load under various scenarios (from Johansson et al. 1989)
Notes:
1. Electricity services = end-use efficiency frozen at 1987 level.
2. Scenario A = normal penetration of energy-efficient end-use technologies (vertical arrows show range of uncertainty).
3. Scenario B = high penetration of energy-efficient end-use technologies that are cost effective and commercialized.
4. Scenario C = same as scenario B but includes uncommercialized newly developed technologies.
5. Scenario D = same as scenario C, but includes advanced technologies still in the research and development stage.
6. Downward stepped curves indicate generation capability with different options for phasing out 12 nuclear plants.

policy in developing countries is guided by a trade-off between economic efficiency on the one hand and a series of financial and socioeconomic considerations on the other. It is widely accepted that energy is fundamental to productivity and economic growth. However, the strong perception among decision-makers, that access to energy is a basic need that improves living standards of the people, has driven a policy in which affordability competes with economic efficiency as the criterion for energy pricing. Furthermore, in practice, it has been difficult to separate social and economic criteria

Barriers and Opportunities in Developing Countries

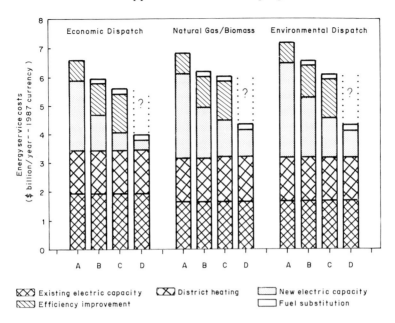

Figure 18.7 Total energy supply costs under various scenarios (from Johansson et al. 1989)

Notes:
1. Scenarios A to D are the same as in Figure 18.6, but efficiency improvement costs are unknown for scenario D.
2. Economic dispatch = traditional least-cost generation expansion and operation.
3. Natural gas/biomass = intensive use of gas and biomass, with coal use banned (to limit CO_2 emissions).
4. Environmental dispatch = generation expansion and operation in order of increasing CO_2 emissions per kWh produced.

within the same pricing structure, leading to poorly designed policies that may be both economically inefficient and socially regressive (or perverse).

A recent study of over 350 electric power utilities in developing countries (Munasinghe et al. 1988) indicates that electricity tariffs have not kept up with cost growth. The operating ratio (defined as the ratio of operating costs before debt service, depreciation and other financing charges, to operating revenue) for the extensive sample studied deteriorated from 0.68 in the 1966–1973 period to 0.80 between 1980 and 1985. At the same time, the financial rate of return on fixed assets has decreased steadily from over 10% in the mid-1960s to around 5% in the mid-1980s. In some countries these declines are significantly greater.

This study and other available evidence indicates that a significant shift towards economic criteria in electricity pricing would be possible without creating undue hardship to the poorer segments of the population. While extensive information (in the form of price elasticities) is not available for most developing countries, several recent studies provide a reasonable basis for projections. Assuming the price elasticity for electric power to be -0.3, a 20% real increase in electricity prices (which would restore the above operating ratio to its 1960/1970s level) would result in a 6% reduction in electricity demand.

Apart from price rationalization, there is also scope for applying a coordinated package of other measures aimed at improving the efficiency of energy use. These would include taxes and subsidies based on fuel type, technology, research and development, retrofits, conservation programs, etc. In most instances these programs are likely to achieve desirable effects on both energy use and environmental impacts. Fiscal instruments such as emission fees and carbon-based user fees can be used to control environmental impacts more directly. In many LDC applications, however, problems of implementing and monitoring such mechanisms are significant.

To conclude, generally the first priority is to improve the efficiency of energy supply and use, through technical means and pricing energy at marginal economic cost. Improvements in energy-intensive industries could yield significant efficiency gains. In transport, considerable opportunities exist to reduce fuel consumption by improving traffic management methods, using less energy-intensive travel modes, and phasing out inefficient vehicles. Substitution of coal and oil with less-polluting fuels (like natural gas, where available), and increasing the efficiency of fuelwood use could also be environmentally and economically sound. Agriculture presently contributes about 14% of all CO_2 emissions, mainly due to forest clearing and burning of wastes. Strengthening property rights, protecting forest lands, and improving land-use planning and management (e.g., through agro-ecological zoning), are some key actions in this sector.

Examples of more expensive options are nonconventional energy sources, large-scale reafforestation, advanced energy technologies, and substitution of chlorofluorocarbons. LDCs will need significant financial assistance to implement such measures, once the less costly options have been exhausted.

National Level Organizations/Institutional Options

The energy sector in developing countries is typically owned and controlled by the government and is characterized by large monolithic organizations. While there is some rationale for this centralization, it could be a critical barrier in the path of greater efficiency and improved flexibility. The desperate circumstances of the energy-supplying enterprises of many developing countries have generated pressures for new approaches to organizing the sector. In particular, there appears to be considerable interest in the scope for more decentralization and greater private participation. Power sector officials in developing countries have been very active in studying this option, and some countries have already prepared the necessary legislative and institutional groundwork for this transition. India plans to install as much as 5000 MW of private power capacity over the 1990–1995 period, and similar plans are underway in Indonesia, Malaysia, Thailand, Philippines, and Pakistan. In Sri Lanka, a private company has been distributing power since the early 1980s, and significant efficiency and service improvements have been observed during this period.

Despite these trends, enhanced private participation in the energy sector is likely to be more successful when it is one element in a broader economic package involving policy reforms in other parts of the economy. Market forces confined to the power sector in a highly distorted economy may not necessarily improve the power-sector situation, since private participants will try to maximize financial rather than economic costs. Thus, private-sector participants would make full use of cheaper generation inputs such as coal even when this is potentially detrimental (both economically and environmentally). Even in a reasonably market-oriented economy, the introduction of private participation in the energy sector is unlikely to lead to environmental benefits unless the costs of pollutants can be fully captured (i.e., internalized) in the financial cost to the participant. Thus, while private participation is likely to bring significant gains by the infusion of new capital and innovative management methods, it is likely to remain one of several methods aimed at restructuring the sector.

Environmental Costs

With regard to environmental issues, the national environment in which the utility functions is likely to play an equally important role. Actions of the utility need to be backed up by a set of consistent national policies and legislative support. The development of environmental standards and regulations is likely to (and should) take place outside the utility, and the public needs to understand the importance of a commitment to a program of environmental mitigation.

Our integrated framework also provides an appropriate starting point for consideration of shift (2) in Figure 18.4, that seeks to incorporate the quantifiable environmental costs. A number of techniques exist for valuing environmental impacts of energy projects (Munasinghe and Lutz 1990). Such evaluation approaches may be used to incorporate environmental costs into methodologies mentioned above for least-cost planning and estimating marginal costs of energy production. However, one should be aware of the uncertainties in such estimates of value and perform sensitivity tests where appropriate.

Going beyond the quantifiable environmental costs is of course problematical, but to the extent that these costs are significant, an attempt must be made rather than implicitly assuming that these costs are negligible. Nonquantifiable environmental costs can be incorporated in various ways, such as adding new constraints on the optimization that reflect social concerns or absolute environmental standards, or even by using an entirely different methodology than least-cost planning, e.g., a type of multi-attribute assessment. This is still consistent with INEP, though the various trade-offs would be made explicitly on social-environmental criteria rather than implictly in economic terms.

GLOBAL ENVIRONMENTAL ISSUES IN THE LDC CONTEXT

The developed and the developing countries are at different points in resolving domestic energy–environmental interactions, and this is an important difference which must be taken into account when devising forms of cooperation for solving transnational and global energy–environment problems. Developing countries still have considerable scope for environment-improving activities that are economically attractive for them, e.g., energy conservation and ameliorating the domestic environmental consequences of energy use. These actions will, of course, have positive global environmental benefits also. While no country can be said to have exhausted the potential for shifts (1), (2), and (3), the developed countries are generally closer to D in Fig. 18.4, at which point they explictly need to consider trade-offs in domestic policy options to improve the global commons.

A second aspect is that developing countries are less able to afford actions to protect the global environment. This is also an equity issue, since much of the responsibility for cumulative damage to the global environment lies with the developed world.

A third aspect is the extent to which developing countries contribute to and are affected by global environmental problems. This is not a clear-cut issue since some countries may feel that they are highly vulnerable while contributing little to global climate changes (e.g., Bangladesh or Maldives who will suffer from sea-level rises); still others will contribute far more and experience varying degrees of impact (e.g., China and India).

The foregoing sets the context within which the developing countries are capable of participating in environmental mitigation efforts at the global level. It is quite obvious that LDCs do not have the ability to contribute financially for global environmental cleanup efforts where the measurable benefits to the national economy are too low to trigger investment. Indeed, this paper has argued that many LDC projects which do have positive measurable benefits at the national level are being bypassed on account of capital constraints.

The principle of assistance to developing countries for environmental mitigation efforts, in terms of technology transfer, financial support and other means, is already well established. The Montreal Protocol, which was adopted in 1987 as a framework within which reduction in the consumption and production of certain types of chlorofluorocarbons is to be achieved, recognized the need for global cooperation and assistance to the developing countries. Subsequent ministerial conferences on various aspects of global environmental issues have reinforced the idea of protecting the global commons.

Currently, discussions are underway among world bodies and governments to define effective criteria and mechanisms for both generating and disbursing funds from a global environmental fund. While a broad workable agreement will not be easy to reach, global financing issues might be analyzed and resolved through a trade-off involving several criteria: *affordability/additionality*, *fairness/equity*, and *economic efficiency*.

First, since LDCs cannot afford to finance even their present energy-supply development, to address global environmental concerns they will need financial assistance on concessionary terms that is additional to existing conventional aid. Second, as noted in the recent Brundtland Commission report (WCED 1987), past growth in the industrialized countries has exhausted a disproportionately high share of global resources, suggesting that the developed countries owe an "environmental debt" to the larger global community. This approach could help to determine how the remaining finite global resources may be shared more fairly and used sustainably. Finally, the economic efficiency criterion indicates that the "polluter pays" principle may be applied to generate revenues, to the extent which global environmental costs of human activity can be quantified. If total emission limits are established (e.g., for CO_2), then trading in emission permits among nations and other market mechanisms could be harnessed to increase efficiency.

One specific mechanism that has recently been implemented includes a core multilateral fund of about U.S.$ 1.5 billion (the Global Environment Facility) to be implemented as a pilot over the next three years. This fund would finance investment, technical assistance, and institutional development activities in four areas: global climate change, ozone depletion, protection of biodiversity, and water resource degradation. A more narrowly focused Ozone Defense Fund of about U.S.$160 to 240 million has been set up also to help implement measures to reduce chlorofluorocarbon emissions under the Montreal Protocol. Both funds are being managed under a collaborative arrangement between the United Nations Development Program, United Nations Environment Program, and the World Bank. In particular, they

would fund those investment activities that would provide cost-effective benefits to the global environment, but would not have been undertaken by individual countries without concessions. Thus, these funds are being specifically designed to fill the void which is created by the lack of individual national incentives for those activities which would, nonetheless, benefit us all.

CONCLUSIONS

International pressures to implement environmentally mitigatory measures place a severe burden on developing countries. The crucial dilemma this poses to LDCs is how to reconcile development goals and the elimination of poverty when the latter will require increased use of energy and raw materials yet at the same time exhibit responsible stewardship of the environment, and avoid overburdening economies that are already weak. This paper has argued that in view of the severe financial constraints already facing developing countries, the response of these countries in relation to environmental preservation cannot extend beyond the realm of measures that are consistent with near-term economic development goals. More specifically, the environmental policy response of LDCs in the coming decade will be limited to conventional technologies in efficiency improvement, conservation, and resource development.

The developed countries are ready to substitute environmental preservation for further economic expansion and should, therefore, be ready to cross the threshold, providing the financial resources that the LDCs need today and developing the technological innovations and knowledge-base to be used in the 21st century by all nations. The Global Environmental Fund and Ozone Fund, presently being established, will facilitate the participation of LDCs in addressing issues at the global level.

ACKNOWLEDGEMENTS

The opinions expressed in this paper are those of the author and do not necessarily represent the views of any institution or government. Assistance provided by Chitru Fernando and Ken King is gratefully acknowledged.

REFERENCES

AID. 1988. Power Shortages in Developing Countries. Washington, D.C.: U.S. Agency for International Development.

Asian Development Bank. 1988. Private sector participation in power development. Manila: Asian Development Bank.

ESCAP (Economic and Social Commission for Asia and the Pacific). 1987. Structural change and energy policy. New York: United Nations.

Flavin, C. 1989. Slowing global warming: a worldwide strategy. Paper No. 91. Washington, D.C.: World Watch Institute.

Foell, W. 1989. Report on the workshop on acid rain in Asia. Bangkok: Asian Institute of Technology.

Johansson, T., B. Bodlund, and R. H. Williams. 1989. Electricity. Lund, Sweden: Lund Univ. Press.

Krause, F., W. Bach, and J. Koomey. 1990. Energy Policy in the Greenhouse, vol. 1. El Cerrito, CA: International Project for Sustainable Energy Paths.

Munasinghe M. 1990a. Electric Power Economics. London: Butterworths.

Munasinghe M. 1990b. Energy Analysis and Policy. London: Butterworths.

Munasinghe, M., J. Gilling, and M. Mason. 1988. A review of World Bank lending for electric power. Industry and Energy Dept., Energy Series Paper No. 2. Washington, D.C.: World Bank.

Munasinghe, M., and E. Lutz. 1990. Economic evaluation of the environmental impact of investment projects and policies. ENV WP4Z. Washington, D.C.: World Bank.

Starr, C., and M. Searl. 1988. Global projections of energy and electricity. Paper presented at American Power Conference Annual Mtg. Chicago.

WCED (World Commission on Environment and Development). 1987. Our Common Future. London: Oxford Univ. Press.

Wilbanks, T., and D. Butcher. 1990. Implementing environmentally sound power sector strategies in developing countries. *Ann. Rev. of Energy*.

World Bank. 1989. World Development Report 1989. London: Oxford Univ. Press.

World Bank. 1990. Capital expenditures for electric power in the developing countries in the 1990s. Industry and Energy Working Paper No. 21. Washington, D.C.: World Bank.

Standing, left to right:
J. Ausubel, O. Davidson, J.-H. Kim, E. Arrhenius
Seated, left to right:
S. Rayner, A.K. Jain, B. Schlomann, K. von Moltke

19

Group Report: Social and Institutional Barriers to Reducing CO_2 Emissions

J. AUSUBEL, Rapporteur
E. ARRHENIUS, R.E. BENEDICK, O. DAVIDSON,
A.K. JAIN, J.-H. KIM, S. RAYNER, B. SCHLOMANN,
R. UEBERHORST, K. VON MOLTKE

SUMMARY

Seven subjects warrant special attention to increase understanding of ways to reduce social and institutional barriers to reduction of CO_2 emissions. These are:

1. the diverse or plural rationalities for decision-making;
2. processes for consensus formation;
3. time horizons for social decision-making and action;
4. economic distortions of environmental and energy services;
5. design of organizations for research, assessment, and evaluation;
6. diffusion of environmentally relevant technology in developing countries and technological leapfrogging;
7. lifestyle trends and changes related to climate and energy.

INTRODUCTION

Discussion in our group moved from general issues of world views to how these views affect consensus formation, to the constraints from temporal and economic factors in decision-making, to the issues of the relationship of science to policy, decision-making for development in the energy field, and consumer choices that maintain characteristic lifestyles (Figure 19.1). The point of departure for each subject was a question, as presented in this report. The common thread throughout was the emphasis on decision-making, including inputs, processes, and outcomes.

For most of the questions, it was found useful to explore the following dimensions:

1. *system levels*, i.e., international, national, subnational;
2. *organizational types*, i.e., intergovernmental, governmental, nongovernmental (including corporate);
3. *time horizons*, i.e., less than 15 years, 15–50 years, and beyond 50 years;

WHAT ARE THE LIMITATIONS OF DIFFERENT VIEWS OF NATURE, APPROACHES TO RATIONALITY, RISK MANAGEMENT, AND FRAMING OF INFORMATION, AND HOW CAN THESE BE TURNED TO ADVANTAGE?

The climate change issue provides ample evidence that there are abiding and sometimes contradictory views of nature and philosophies of risk

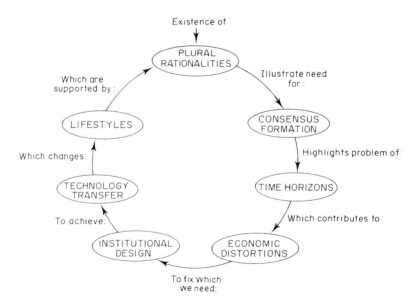

Figure 19.1 Schematic overview of social and institutional barriers to CO_2 reduction

management, in short *plural rationalities*. For some hazards, risk can be adequately defined by multiplying the probability of an event times its magnitude. For problems in which there is precision in measurement of the risk and for which the stakes are largely local, few difficulties arise. When risks become more difficult to quantify, when assessment relies on less-developed methods, and when the spatial extent of risk enlarges, understanding of underlying views of nature assumes greater significance.

One perspective on views of nature is provided by "cultural theory," as developed in anthropology (Douglas 1978; Douglas and Wildavsky 1982; Gross and Rayner 1985; James et al. 1986). Cultural theory suggests that there are primary "nature myths" (Holling 1986; Schwarz and Thompson 1990). These may be that nature is either fragile, robust, resilient, or capricious. As shown in Table 19.1, where nature is represented as a ball on a differently shaped surface, myths leads to a particular moral imperative, preference in response strategy, and type of social organization. This typology is a heuristic device, defining pure or ideal types that are rarely

Table 19.1 Nature myths. For a thorough discussion of such myths and a somewhat different categorization, see Schwarz and Thompson (1990)

FRAGILE	ROBUST	RESILIENT	CAPRICIOUS
DON'T MESS WITH NATURE	DON'T CURB GROWTH	PRESERVE CHOICE	DON'T TREAD ON ME
PREVENTION	ADAPTATION	SUSTAINABLE DEVELOPMENT	FATALISM/ DENIAL
"GREEN" ENVIRONMENTAL GROUPS	MARKETS	BUREAUCRACIES	NIMBYS* *Not In My Backyard

found in practice. Individuals and organizations are in fact likely to be hybrids of these views and characteristics.

The approach, nevertheless, does suggest the extent to which notions of trust, liability, and consent are integral to the definition of risk. As suggested in Table 19.2, each form of social organization is likely to have preferences in regard to how trust is established, how liability is characterized, and how consent is obtained. The existence of plural rationalities creates structural obstacles to social learning, because it is hard for individuals and groups, dwelling primarily in one or another paradigm, to interact.

The diversity of views is also a resource. It provides society at large with sources of warning and with explorers, as well as with more stable organizations that can take a longer view. It also creates some ambiguity about the role of expertise. From the viewpoint of cultural theory, every social group has its own experts and its own characteristic demands for information. Because experts may hold fundamentally different views of nature, their views may remain in conflict, both about what the facts are and whether the facts make an argument. There are also structural reasons for the subversion of "rationality" (Elster 1983).

Understanding more about plural rationalities could have several benefits with regard to CO_2 reduction. It could help in design of more viable

Table 19.2 Characterization of trust, liability, and consent by various social organizations

	"GREEN" ENVIRON-MENTAL GROUPS	MARKETS	BUREAUCRACIES	NIMBYS
TRUST	PARTICI-PATION	SUCCESSFUL INDIVIDUALS	INSTITUTIONS	DUMB LUCK
LIABILITY	STRICT	INSURANCE	DEEP POCKETS	AVOIDANCE
CONSENT	EXPLICIT	REVEALED PREFERENCES	HYPOTHETICAL (REPRESENT-ATIVE)	VETO

and customized implementation strategies. It could help to facilitate communication between groups and to recognize where communication or synthesis will never be achieved. It argues for the preservation of cultural diversity as a resource from which behavioral and technological solutions may arise. It helps define balanced institutional development and explain when and where it may be useful to catalyze the formation of new organizations (Carroll 1988). For example, some societies may be particularly deficient in market organizations, others in small, critical nongovernmental organizations, and others in stable and effective bureaucracies. Finally, study of human views of nature and culture reminds us of the limits of particular notions of reason, including efficiency, and to take into account fairness and justice, administrative feasibility, and beauty.

Among the salient research questions are:

1. Can we describe better and more fully world views and their configuration?
2. How do diverse world views constrain action at the global level?
3. At the national level, how do diverse world views influence choice of policy instruments?
4. At the subnational level, are there ways to invert the tragedy of the commons so that local or community goals favor the global good?
5. Where is there a need to stimulate the growth of "missing" institutions?
6. What organizations are best adapted for the range of functions required to respond to global change, for example, functions of monitoring and verification of international environmental agreements?
7. How do diverse world views influence the sense of urgency that different groups hold about global environmental change?
8. How does culture structure the use and perception of time?
9. What environmental goals are best pursued through *explicit* consensus?
10. What environmental goals are best pursued through *informal* agreement?
11. To what extent can environmental and energy technologies be designed around world views?

WHAT ARE THE BARRIERS TO COLLECTIVE ACTION ARISING FROM KNOWLEDGE AND IGNORANCE, TRUST, SELF-INTEREST, POLITICAL MOTIVATION AND CONSENT, AT THE LEVELS FROM LOCAL TO GLOBAL DECISION-MAKING?

The existence of competing and differing world views rises to great importance in the *process of consensus formation*. In facing challenges such as reduction of carbon emissions, it becomes desirable not only for the views to interact but also in key situations to arrive at a common view, or at least a view that is sufficiently broad for widespread social action to take place. In this context, it also becomes necessary to ask which world views are politically feasible or best adapted to action.

Barriers to collective action may arise because of deficits in knowledge with regard to underlying scientific questions, or with regard to motivations, attitudes, and willingness to act in the larger society of which the scientific community forms one element. The traditional model of "science advising" addresses the production of knowledge by the scientific community for an external "receptor" (Plowden 1987). It is increasingly necessary to consider new roles of the scientific community as all fields of human action become pervaded with technical questions and the more embracing question of the interaction of the scientific community with other groups (Burns and Ueberhorst 1988).

In many instances, work from the scientific community has been judged too narrow to be of use in politics. In politics it is often a mistake to isolate certain aspects of a question (for example, to consider only alternatives to gasoline without considering other changes in transport systems). Scenarios developed internally to science often have heuristic power but do not represent sufficient or socially urgent viewpoints. A key in politics is provision of alternatives, particularly in terms of what can be accepted by the public. There is a need to develop institutions that can engage in cooperative conceptualization of complex processes in such a way that social learning by all groups participating is enhanced. As long as there is a lack of a "common problematique" between the political system and the scientific community, proposals worked out in the scientific community will be of little use.

The essence of the stiuation is that many contemporary problems can only be defined with the help of scientists but cannot be solved by them. Solutions rest with intercommunitarian processes and, in some cases, with normative consensus formation. This requires a willingness to participate in political processes by scientists, changing roles for experts, and recognition that qualities of work by scientists can themselves be barriers to action. It also implies that sometimes consensus is more important than selection of a particular strategy, that the greatest need can be to create an action coalition to implement at least one of several reasonable strategies.

The situation is further complicated by the differing orientations of science and politics. Politics remains oriented primarily to decisions and concerns within the nation-state, although notions and limits of national sovereignty and security are changing, largely because of technology (Shultz 1990) and environment (Goldemberg and Durham 1989). The values of science tend to be "universalistic." Many nongovernmental environmental organizations share this universalist viewpoint, sometimes irking governments, as in Brazil.

The challenge in the situation is to preserve the rights of science and the rules of democratic decision-making while recognizing that traditional argumentation rules used by society are not sufficient. It is necessary to develop processes that encourage the ability to work out views in detail, especially minority views, for those who may lack certain argumentative skills. It is desirable to work out alternative policies in a high quality way that does not rely on bringing every issue to a vote or a court trial. It is important to have a process that represents fairly both the middle and the tails of the distribution of opinion, or whatever shape the distribution may have.

One interesting example of a mechanism of this type has been the Enquête-Kommission of the German Parliament on protection of the atmosphere (Enquête-Kommission 1989). Parliaments can serve as a mediating link between scientists and the general public. Through public hearings and published results, parliaments can provide a means for translating scientific findings into policy-relevant concepts and policy options.

The Enquête-Kommission broke new ground in the scope of its hearings and the balance of membership involving both scientists and politicians as full participants. Although primarily aimed at an audience of German decision-makers, the Enquête-Kommission made considerable efforts to broaden its influence. It invited witnesses on an international scale and designed a comprehensive report not limited to the German perspective but adopted a global viewpoint. The international significance of the Enquête-Kommission's process and work was reflected in the decision to translate the report into foreign languages and its use and citation by German political leaders, not only as the basis for German national policy but explicitly to serve as an example to other industrialized nations.

A paradoxical question is that bargains may be more likely to occur in an information-poor setting. While we live behind a veil of ignorance about winners and losers (Rawls 1971), the dominant strategy may be cooperative and collectively oriented. Since no participant can guarantee that he or she will not receive a devastating blow, risk averse players will want to minimize such possibilities. Thus, ironically, ignorance as much as information, may motivate environmentally (and globally) desirable outcomes.

Specific research questions:

1. How can processes for consensus formation in the field of environment and technology field be enhanced at all levels?
2. What changes are required in the processes internal to the scientific

community to make its work in its roles with a normative orientation more useful?
3. Should there be more discussion of guidelines or norms of advocacy for scientists to facilitate communication with the political system?
4. How is communication between science and politics affected by the universalist orientation of science in contrast to orientation of politicians toward the sovereign state?
4. To what extent is it the case in environmental negotiation that stakeholders facing decisions with highly uncertain outcomes will emphasize a fair process of decision-making rather than gambling for advantage?

WHAT OPPORTUNITIES AND CONSTRAINTS ON COLLECTIVE ACTION ARISE FROM DIFFERING PLANNING AND DECISION CYCLES AMONG INSTITUTIONS, INCLUDING PARTS OF GOVERNMENT, PRIVATE FIRMS, NONGOVERNMENTAL ORGANIZATIONS, AND THE MEDIA?

The relevance of the issue of *time horizons* originates in the fact that many of the costs potentially associated with reducing carbon emissions may appear in the short term, while benefits emerge over the longer run. The high level of uncertainty about both costs and benefits further creates a preference in many organizations and groups for a myopic strategy and for a search for "no regrets" policies.

Various time horizons characterize parts of government, consumers and voters, and industry, as well as science itself. Some parts of government, especially elected officials, tend to have relatively short-term perspectives. Many politicians will avoid taking decisions on sensitive or volatile issues during their term of office. Factors that influence the time horizons of elected officials include the power of narrow special interests and the costs of campaigns. However, there are other quite different factors that also lead toward short-term horizons. One is the failure of effective central planning, which often seeks to have a longer-range character through 5-year plans and other programs. Another is competition among priorities. Many governments, especially in developing countries, face immediate issues of survival of such large dimensions that there is little or no room on the political agenda for the long term.

It is important to note that there are instances of far-sighted decisions of governments, for example the establishment of national systems of agricultural research and the building of an infrastructure for water supply, wastewater treatment, and transportation. Over the past 100 years almost all nations of the world have chosen to set aside areas as nature preserves and parks, and many of these decisions have not been based on economic assessments. There have also been long-range studies such as the "Global 2000" project of the U.S. government in the late 1970s (Barney 1980), which sought to encourage all departments of government to look forward to the issues that they would need to address more than 20 years in the future.

Consumers and voters also often opt for the near term. The tremendous expansion of consumer debt in all countries is one indicator of the desire for short-term gratification. The experience of the public with changes in scientific knowledge also leads to a certain skepticism about long-term commitments.

In industry, time horizons are determined in large part by the depreciation structure of capital stock, tax rules, and features of financial markets that favor optimizing for periods that are often less than two years and rarely extend for more than 7–10 years. Businesses face the very real risk of bankruptcy and thus must take decisions in the interest of survival.

At the same time, some sectors of industry demonstrate long-term horizons. Decisions to develop entirely new products in such sectors as pharmaceuticals, to build automobile factories and power stations, to develop a mine, or to plant timber resources imply horizons of a decade or more. At the highest levels of industry, there may often be more long-term vision about what is best for an enterprise, which involves not only short-term financial results but a long-term flow of products and a positive public image.

Nongovernmental organizations appear to span a range of time horizons. Some follow fads and fashions and can sustain themselves only by fund-raising strategies that require abandoning an issue if it will not attract contributions, dues, and membership. Others are explicitly oriented toward long-term considerations and may be insulated by endowments or stable memberships. There are also organizations like churches and universities that have displayed longevity measured in many centuries despite taking many short-run decisions that appear uneconomic.

At the international level, the United Nations system has sought to provide long-range perspectives and has designed many decadal programs. However, in practice, unreliable UN budgets have meant that many more programs are announced than carried through. Moreover, there is just as much mistrust of centralized management and planning at the global level as there is on the national level.

The scientific community is unusual in its comfort with long-term considerations. A period of one or two hundred years is short compared to the time horizons of disciplines such as cosmology, geology, or ecology, and scientific agendas are routinely pursued over decades. However, science is always uncertain about what constitutes usable knowledge for today.

Consideration of time horizons suggests that we may need not only environmental laboratories and environmental ministries but also environmental churches. If "Green" is the new religion, it may be because "peace with nature" (Meyer-Abich 1986) can be maintained only by institutions and taboos with extraordinary durability and longevity.

Specific research questions:

1. What are the time horizons characteristic of the organizations most important to global environmental change and why?
2. What are the institutional factors and cultural beliefs that shape the spectrum of time horizons and enable them to change?
3. Why are some political systems more open to change and long-term perspectives?
4. What is the time required to reach various kinds of international agreements? Does it differ systematically for broad and specific agreements?
5. How can methods be improved for the conduct of studies that extend decades and generations ahead?
6. What establishes the calendar of science and are there ways to accelerate the production of usable knowledge from science important for global environment?

WHAT CHANGES IN PRICING RESOURCES AND DECISION-MAKING FOR RESEARCH AND DEVELOPMENT SHOULD BE GIVEN PRIORITY?

Important aspects of environmental protection and natural resource management are hindered by reliance on an antiquated and flawed economics (Baumol and Oates 1988; Pearce 1987; Solow 1973). At the same time, economic instruments can be powerful tools for environmental protection and for harmonizing goals for energy, environment, and growth (IEA 1989). Reducing *economic distortions* and shifting the economic system to reflect more accurately current, shared values about environment would be an important instrument for substantially reducing carbon emissions.

The problem for environment of the evaluation of time within the discipline of economics has been widely discussed for 20 years (Ausubel 1980). The practice of discounting, which correctly reflects that for many economic decisions a dollar today is worth more than the same dollar held at a time in the future, systematically diminishes the value of environmental assets. To take the extreme case, a profitable activity today that would destroy the environment 100 years from now would still be assessed favorably in narrow economic terms, as any positive discount rate applied to an asset 100 years in the future would render it trivial.

Values of stocks of natural resources (as well as important social functions such as housework) are largely omitted from the systems of accounting generally used in national economic planning and in business decisions (Nordhaus and Tobin 1972). Partly the reason is that the systems were established for purposes quite distant from environmental protection. Partly the reason is that some environmental goods, which can be assessed in monetary terms with reasonable accuracy, have not been internalized into economic analyses. Partly it is because some environmental goods, such as genetic diversity or the assimilative capacity of the environment for wastes, are difficult to monetize at all.

The result is a set of energy prices that are particularly distorted from an environmental point of view. Prices have not reflected true *internal* costs, whether in central and eastern Europe, the Soviet Union, North America, western Europe, or less-developed countries. They have also not included *external* environmental costs and benefits. This is true not only for climate change but also for effects of energy sources on human health, materials and the built environment, and ecology. The unrealistically low monetary price of energy is associated with high levels of energy consumption.

The distortions affect not only overall consumption of energy but also the allocation within energy sources. Subsidies have tended to favor coal, oil, and nuclear energy over natural gas, solar energy, and energy efficiency, even though the latter are all more favorable from an environmental perspective. Pricing structures in energy are also strongly influenced by a tendency to use energy pricing for other goals of social welfare. The difficulty of separating social and economic criteria in practice in a single pricing structure has led to a dominance of social goals. It is important to note that subsidies to maintain employment in the coal industry or to insure against hazards of nuclear accidents were social choices of the same kind that are now needed to favor protection of the environment. It is also important to recognize that the structure of subsidies and incentives pervades not only energy use but also energy research and exploration. Thus, we may tend to develop the wrong fuels for the future, as well as use the wrong ones today.

The difficulty is that while there is agreement on the distorted pricing of the present system, there is much less agreement on how to improve it. There is general agreement on the need to internalize more environmental costs, and there is general agreement that changing prices can beneficially effect both the sources of supply and the level of demand. However, it will be useful to provide much more insight into the potential for environmental management of various economic instruments and incentives.

The economic system must also be adapted to concerns for the resilience of ecosystems. The risk of sudden irreversible events in ecological systems places an increased value on early action. The absence of efficient economic evaluation systems considering abrupt human-induced changes and associated economic tools giving a bonus for early action, is a constraint in greenhouse-gas management. In adapting to greater instability and increased change in ecosystems under climate change, there is also a need for flexibility in future infrastructure establishment. Efficient economic tools for stimulating such flexibility must be developed.

Economics has been under pressure in recent years because it has neglected to look inside the "black box" of technology that is in fact responsible for much economic growth over the long run (Dosi et al. 1988). Now it is clear that the movement to achieve an economics that is more holistic, systemic, and evolutionary also requires that it operate intelligently inside "the green box."

Specific research needs:

1. To what extent will provision of better information about environmental costs of energy use change behavior?
2. How broad a definition of externalities can be functionally applied to current pricing structures?
3. What are long-run elasticities of energy demand, how high will taxes or charges need to be to exert a sustained influence on behavior, and are these best applied in gradual or abrupt price changes?
4. What are the strengths and weaknesses, by criteria including fairness and efficiency, of various economic regimes for limitation of carbon emissions (fixed reductions, per capita targets, carbon taxes, tradable permits, etc.)?
5. What are the relative benefits of approaches to the energy system as a whole versus approaches focusing only on carbon dioxide?
6. Why are more costly instruments for economic control often selected by society than the instruments judged superior by economists?
7. What is the shape of the "supply curve" for carbon reductions for different nations, regions, and the world as a whole?
8. To what extent will formal action at both the international and national level be needed in order to bring about changes in energy pricing sufficient to achieve major carbon emission reductions?

WHAT IS THE RELATION OF SCIENCE AND POLICY IN THE DESIGN AND EVALUATION OF INSTRUMENTS FOR REDUCING EMISSIONS?

The effectiveness of the relation between science and policy depends critically on *organizational design* for research, assessment, and evaluation regarding climatic change, its causes, and efforts for prevention and adaptation. The importance of design of organizations and decision-making processes has been highlighted by the creation of the Intergovernmental Panel on Climate Change (IPCC), whose results have been widely accepted as authoritative. Of course, structural aspects of decision-making processes cannot be separated from substantive aspects and performance. The relation between science and policy is determined not only by the vehicles for interaction but also by the quality, relevance, and timeliness of results, which along with the process employed contribute to credibility and legitimacy.

There are many functions requiring scientific or analytic skills that need to be fulfilled with regard to climatic change and carbon emission reduction. These include basic research, monitoring, assessment, policy design and implementation, verification and compliance, and policy evaluation. We highlight three gaps in an organizational landscape that merits careful study in its entirety (Tolba 1990). The three gaps are in the joint international conduct of basic environmental research, the joint international assessment of environment issues, and the evaluation of the effectiveness of programs intended to address environmental problems.

The probability of international agreement and action on climate change will likely be increased if scientists from many nations have the opportunity to participate in basic research related to global environment. However, most nations lack sufficient financial, technical, and human resources at the national level to develop autonomous research programs at the frontiers of environmental science. Equally important, it is necessary for the scientific community of nations and regions to be able to understand local and regional implications of global analyses. Ultimately, global issues are local problems, such as drought.

Already there are several useful programs, such as the World Climate Program, that coordinate national research efforts to achieve larger goals. A powerful means to achieve scientific advance and greater participation may be the establishment of a network of international environmental research centers. These centers would be governed internationally and have scientific staff members from many countries. In some ways, the network would be a "Green" version of the Consultative Group on International Agricultural Research. The centers would be responsible for both research and advanced training and would be located in both developing and developed countries. They should seek to strengthen national systems of environmental research, as well as to perform regional and global analyses in order to fill gaps likely to remain from national systems. Recommendations for centers of this kind have been made as part of the International Geosphere-Biosphere Program (IGBP 1991) and by the Second World Climate Conference. However, as yet, little careful thought has been given to how the network might most usefully be designed, taking into account not only the goals of the scientific community but other communities as well.

While a network of centers might generate and diffuse new knowledge and strengthen the human resource base in environmental sciences for many regions, the question remains how to synthesize what is known at the international level. National efforts such as the Enquête-Kommission are unlikely to touch the full spectrum of issues and people concerned. Although the IPCC was a remarkable step forward in this regard, the IPCC reports leave many questions unanswered, especially with regard to impacts of climate and mitigation strategies for emissions. The IPCC analyses also say little at the regional level. The IPCC is likely to continue in some form. Nevertheless, to establish a more consistent and comprehensive capability, it might also be useful for several international scientific organizations to explore and develop their potential to perform similar assessments. Joint international assessments are integral to the process of consensus formation discussed earlier (see section on WHAT ARE THE BARRIERS. . .).

Among the organizations that might play a stronger role in international scientific assessments are the International Council of Scientific Unions, which embraces more than 40 national academies of sciences, the Council of Academies of Engineering and Technological Sciences, whose membership includes most national academies of technology; the Third World Academy of Sciences; and the African Academy of Sciences. Such organizations need to clarify the processes that they would use to assure high quality, credible, and independent results, as well as their relationships to governments and intergovernmental organizations in their role as conveners of experts to carry out assessments.

The third functional gap, evaluation, is one that is often neglected. Most organizations and sponsors prefer planning and making promises to evaluation. The need for evaluation is great for the larger society to accelerate social learning.

Historically there has been a rather weak connection between cause and effect in broadly defined formal social policy. This has been evident in areas such as urban policy, migration, and energy itself (Landsberg 1980). There have been many perverse and unexpected outcomes of policy interventions. It is important to have realistic expectations about our ability to create alternatives for human societies and move deliberately toward them. There has probably been a gradual increase in the ability to do so and there may be a need for a great increase in this ability, not only because of climate change, but because of needs in development, population, health, and other areas. As we consider substantial escalation of policy interventions to achieve policy goals in environment, it is necessary concurrently to put in place mechanisms at the national and international levels to assess the efficacy of programs and policies and how it might be improved.

Specific research questions include:

1. How can progress in basic environmental research be accelerated to the benefit of many nations? Would a network of international environmental research centers be useful, and if so, how should it be designed?
2. How can the joint international conduct of scientific assessments be improved? Are there new roles for international nongovernmental scientific organizations in this regard?
3. How can evaluation of programs and policies designed to achieve reduction of carbon emissions be reliably assured? What combination of existing and new independent organizations might best carry out this function? How can institutional learning be accelerated?

HOW DO DIFFERENT LEVELS OF DEVELOPMENT, OUR UNDERSTANDING OF DEVELOPMENT PROCESSES, AND TRANSFER OF INFORMATION AND TECHNOLOGY CONSTRAIN CO_2 REDUCTION?

The ability of 80% of the global population to develop in a way that generates substantial income and employment, yet low levels of carbon emissions, will depend in large part on *diffusion of technology* and the possibility of *technological leapfrogging*. It is obvious that if the bulk of the developing world repeats the 20th century pattern of the advanced industrialized nations with their reliance on fossil fuels, and motorization and electrification based on these fuels, the atmospheric burden of CO_2 will grow substantially.

In general, countries of the world can be analyzed in three groups: (a) developing countries, those countries using energy for survival; (b) newly industrialized countries, those using energy for development and industrialization; and (c) those using energy to sustain an industrialized lifestyle (Figure 19.2). Each type of country needs a new growth strategy in light of concern about carbon emissions and is likely to have a different response to the challenge to reduce emissions.

Countries differ in the extent to which their economies and administrations are oriented toward local matters, national concerns, and the international

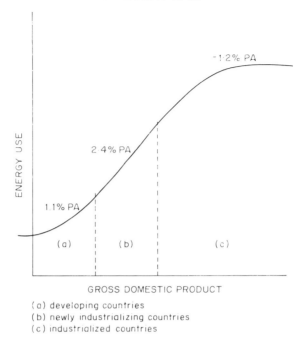

Figure 19.2 Generic energy/economy relationship

system. For many developing countries, subnational units are very important; for newly industrializing countries, there is often a very strong emphasis on national interest; the industrialized nations are most integrated into a global economy and participate most fully in regional blocs and international governance. The strength and abundance of organizational types also tend to vary with development. Government is often very weak in developing countries, and in many newly industrializing countries there are still few independent nongovernmental organizations. The time horizons of the poorest nations tend to be short, while the industrialized nations can usually afford to look further into the future. Moreover, informal links and decisions are often dominant in developing countries, while the industrialized nations tend to make the most explicit and formal decisions.

For countries of type (c), the challenge is to harvest the many technological opportunities that appear to exist. In many cases, the potential for efficiency has begun to be better utilized in the past 10–15 years. Countries of type (b) have tended to equate growth in energy use with economic growth. For these countries, it may be hard to change paths because of the recently installed capital infrastructure. Their carbon emission growth is likely to be steep but incremental.

In many ways countries of type (a) present the greatest challenge. They need new energy sources the most and also need to make the clearest choices about paths to follow and organizational structures to foster them (Wilbanks 1990). There is often little political means to do much. The processes that can bring about more rapid growth are not well understood. In addition, there are many barriers to diffusion of technologies that might be most helpful to a more carbon-friendly pattern of development. Conversely, there is much transfer of inefficient and obsolete technology that looks inexpensive but can bring crippling infrastructure costs.

A significant barrier rests in the current status of intellectual property rights. Such rights are necessary to create and maintain incentives for innovation. However, the current system, or absence of system, may simultaneously harm industrialized countries and developing countries. The lack of rules in some less-developed countries discourages technology transfer and development of markets. However, simply expanding the present system of the developed nations may place excessive hurdles on less-developed countries and also undervalue some of their assets, for example, in biological resources and traditional knowledge. The London Ozone Convention, which includes a mechanism for financing replacement of chlorofluorocarbons in less-developed countries, is an illustration of an innovative approach to technology transfer in the environment field. Consideration needs to be given to appropriate mechanisms for joint ventures between countries of North and South in energy and environment. The role of international development organizations, which have only recently become concerned with global environment, also needs to be considered in this regard.

A further important barrier is the difficulty of interaction between science and government in many less-developed countries. The traditions and mechanisms of constructive relations between the communities are at an early stage. Establishment of national research councils, strengthening of the independence of universities, and strengthening of regional organizations such as the African Academy of Sciences can be valuable ways to lessen this barrier.

Specific research questions include:

1. Can we understand better the relative importance of various factors affecting response to carbon emission reduction at various levels of development?
2. Recognizing that in many countries there will be governments that lack leverage over the national economy, how can obstacles to action by developing countries to reduce carbon emission be overcome most effectively?
3. What are the possibilities for informal action to reduce carbon emissions in developing countries?
4. What is the possibility for developing countries to leapfrog in energy

and environmental technologies in order to avoid the pattern of development that has traditionally characterized industrialization?
5. How can international arrangements for transfer of technology and intellectual property rights be modified in a way that maintains incentives conducive to vigorous innovation globally and also equitably recognizes the assets and needs of less-developed countries?
6. What can be done to improve the weak performance of international organizations with regard to environmental protection in developing countries?
7. What can be done to strengthen indigenous environmental research capability in developing countries and to strengthen participation of scientists from developing countries in regional and global evaluations of environmental issues?

WHAT DO WE UNDERSTAND ABOUT THE POTENTIAL FOR CHANGES IN CONSUMPTION PATTERNS AND SOCIAL BEHAVIOR OVER DIFFERENT TIME HORIZONS?

Quantitative analysis (see Schipper, chapter 15, this volume) shows that differences in *lifestyle* account for differences in final energy consumption and patterns of energy use at least as large as those caused by the technologies employed. Because habits define energy demand to such a great extent, it is necessary to consider their flexibility. It is also not unreasonable for individuals to be concerned that efforts to reduce carbon emissions will have impacts on preferred behavior in such fundamental and sensitive areas as diet, movement, living area, and reproduction, and family size.

Although economic factors may heavily influence energy consumption, it is also important to recognize the limits of these factors. For most individuals, firms, and countries, expenditures on energy require less than 10% of income. Thus, in many circumstances, even large increases in energy costs can be absorbed without major disruption of overall consumption. For a few activities, such as production of aluminum, energy costs constitute such a large proportion of total budget that the conduct of the activity is highly sensitive to economic factors. There is evidence that over the past two decades lifestyle preference has overridden several major price shocks to the energy system.

Most lifestyle changes appear to be moving in the wrong direction from the point of view of carbon emissions. For example, people in many societies have been spending somewhat more time traveling, and often in ways that demand more energy. On average, time spent in travel consumes about ten times as much energy as time spent in a stable location, whether in work or leisure. The increase in mobility during the past decades has roughly canceled gains from greater efficiency of vehicles. Moreover, trends are moving toward fewer passengers per vehicle, larger vehicles, and kinds of vehicles that consume more energy, such as aircraft and cars.

Changes in population profile are also tending to increase energy demand. Along with population increase itself, the shrinking of households and the aging of the population tend to raise energy demand. Two households with three persons each will consume considerably more energy than one household of six. In industrialized societies, a significant and growing fraction of the population may now live on pension for a period of 20 years or more. This older segment of the population has a historically unprecedented amount of time and income to travel and maintain residences. Moreover, the population that is aging now is the first population in which possession of driver's licenses is prevalent.

A central question is the extent to which leisure activities will prove to be energy-intensive. With people in the industrialized societies living longer and steadily spending fewer lifehours at work, the question of how non-work time is used throughout life may become a main determinant of trends in energy consumption.

For developing countries, the question must be asked whether the consumption pattern of the industrialized nations will be repeated. So far, the pattern of urbanization, motorization, and unbundling of families appears similar.

For all societies, transportation and communication have advanced in lockstep as complementary goods. Increases in communication increase demand for transport, and increases in transport increase demand for communication. If one wishes to travel less, a good strategy is to give up the telephone. There is no evidence yet that communication substitutes for travel (Grübler 1990).

Ultimately, the question is to what extent it is possible to have patterns of development genuinely alternative to those with which we are familiar in the industrialized countries. In the 1920s, Trotsky and Stalin parted ways over the question of whether there could be socialism in one country or whether a world revolution was needed for the alternative to flourish. The year 1989 seemed to suggest that there is only one economic system for the whole planet and that it is not possible to maintain a separate development. A traveler to Bangkok, Sydney, Honolulu, Lagos, Bombay, Warsaw, and Berlin might well agree that amidst the cultural diversity there is still only one system of large buildings, cars, and urbanization. From an environmental point of view, global economic integration is both risk and opportunity. If the system becomes more homogeneous, if everyone must use the same standards, and if lifestyles all tend in the same direction, then the task of change is immense though also clear and well-defined.

Specific research questions include:

1. Where and how will we live? What will be the size and kind of homes and households?
2. Where and how will we work? Can the relationship between transportation and communication be changed?
3. Where and how will we play? Will leisure be energy-intensive or not?
4. What have been the most significant behavioral changes in recent decades that have been favorable for energy efficiency?
5. How can lifestyles and behavior be changed while respecting individual rights? What is the potential of education in this regard? How is it best to project notions of needs for lifestyle change so that acceptance may be encouraged?
6. Can market research and consumer psychology be employed more constructively from the perspective of global environment?
7. What are the implications for energy demand of a continuation of recent trends in lifestyle?
8. What are the implications for lifestyle of various goals for carbon emission reduction?
9. How is it possible to foster the differentiation of societies to explore different evolutionary paths that may be more benign with regard to environment sufficient to establish the viability of the paths?

CONCLUSION

We have identified seven major barriers to reduction of carbon emissions: plural rationalities, consensus formation, time horizons, economic distortions, organizational design, technology diffusion, and lifestyles. In each area it is evident that proposals for action and change could be made now. At the same time, it is evident that much more remains to be learned about the barriers to action, and that contributions can come from anthropology, philosophy, religion, sociology, political science, psychology, organizational behavior, market research, economics, history, statistics, demography, geography, and development studies, and from the integration of all these disciplines (Meyer-Abich 1988; NRC 1990). The task of addressing the social and institutional barriers to carbon emissions will be with us for decades and perhaps centuries. It is necessary to deepen our partial views of the barriers, to study the actions to reduce them for efficiency, flexibility, fairness, affordability, administrative burden, and sustainability, and to try to come to collective views of how action should proceed.

REFERENCES

Ausubel, J.H. 1980. Economics in the air: an introduction to economic issues of the atmosphere and climate. In: Climatic Constraints and Human Activities, ed. J.H. Ausubel and K. Biswas, pp. 13–59. Oxford: Pergamon.
Barney, G.D., ed. 1980. The Global 2000 Report to the President of the U.S. New York: Pergamon.
Baumol, W.J., and W.E. Oates, 1988. The Theory of Environmental Policy, 2nd ed. New York: Cambridge Univ. Press.
Burns, T.R., and R. Ueberhorst. 1988. Creative Democracy: Systematic Conflict Resolution and Policymaking in a World of High Science and Technology. New York: Praeger.
Carroll, G.R., ed. 1988. Ecological Models of Organizations. Cambridge: Praeger.
Dosi, G., C. Freeman, R. Nelson, G. Silberberg, and L. Soete. 1988. Technical Change and Economic Theory. London, New York: Pinter.
Douglas, M. 1978. Cultural Bias. Occasional paper 35, Royal Anthropological Institute. Reprinted in: In the Active Voice. London: Routledge and Kegan Paul.
Douglas, M., and A. Wildavsky. 1982. Risk and Culture: An Essay on the Selection of Technological and Environmental Dangers. Berkeley: Univ. of California Press.
Elster, J. 1983. Sour Grapes: Studies in the Subversion of Rationality: Cambridge: Cambridge Univ. Press.
Enquête-Kommission. 1989. Protecting the Earth's Atmosphere: An International Challenge. Interim report of the study commissions of the 11th German Bundestag. Bonn: Economica Verlag, Verlag C.F. Müller.
Goldemberg, J., and E.R. Durham. 1989. The Amazonia and national sovereignty. Workshop on: Our Changing Atmosphere: Sources of Stress and Challenges to Cooperation. Cambridge, MA: Am. Acad. Arts and Sciences.
Gross, J.L., and S. Rayner. 1985: Measuring Culture: A Paradigm for the Analysis of Social Organization. New York: Columbia.
Grübler, A. 1990. The Rise and Fall of Infrastructures. Heidelberg: Physica.
Holling, C.S. 1986. The resilience of terrestrial ecosystems: local surprise and global change. In: Sustainable Development of the Biosphere, ed. W.C. Clark and R. E. Munn, pp. 292–316. Cambridge: Cambridge Univ. Press.
IEA (International Energy Agency). 1989. Energy and the Environment: Policy Overview. Paris: Organization for Economic Cooperation and Development.
IGBP (International Geosphere-Biosphere Program). 1991. Global change system for analysis, research and training (START). Report of a meeting at Bellagio, Dec. 3–7, 1990. Report No. 15. Stockholm: IGBP.
James, P., P. Tayler, and M. Thompson. 1986. Plural Rationalities. Coventry: Institute for Management Research and Development, Univ. of Warwick.
Landsberg, H.H. 1980. Let's All Play Energy Policy! Daedalus **109(3)**:71–84.
Meyer-Abich, K.M. 1986. Peace with nature, or plants as indicators to the loss of humanity. Experientia **42**:115–120.
Meyer-Abich, K.M. 1988. Holism in science, technology, and society. In: Complexities of the Human Environment: A Cultural and Technological Perspective, ed. Karl Vak, pp. 41–48. Vienna: Europa Verlag.

Nordhaus, W., and J. Tobin. 1972. Is growth obsolete? Economic Growth Fiftieth Anniversary Colloquium V. New York: Natl. Bureau of Economic Research.
NRC (National Research Council). 1990. Research strategies for the U.S. global change research program. In: Human Sources of Global Change, pp. 108-130. Washington, D.C.: Natl. Academy Press.
Pearce, D.W. 1987. Foundations of an ecological economics. *Ecological Modelling* **38**:9–18.
Plowden, W., ed. 1987. Advising the Rulers. Oxford: Basil Blackwell.
Rawls, J. 1971. A Theory of Justice. Cambridge, MA: Harvard Univ. Press.
Rayner, S. 1987. Risk and relativism in science for policy. In: The Social and Cultural Construction of Risk, ed. B.B. Johnson and V.T. Covello. Boston: Riedel.
Schwarz, M., and M. Thompson. 1990. Divided We Stand. Philadelphia: University of Pennsylvania.
Shultz, G. 1990: On sovereignty. In: Engineering and Human Welfare, ed. H.E. Sladovich; pp. 31–35. Washington, D.C.: Natl. Acad. of Engineering.
Solow, R.M. 1973. The economics of resources or the resources of economics. *Am. Econ. Rev.* **64**:1–14.
Tolba, M.K. 1990: Building an environmental institutional framework for the future. *Env. Conservation* **17**:105–110.
Wilbanks, T.J. 1990. Implanting environmentally sound power sector strategies in developing countries. *Ann. Rev. Energy* **15**:255–276.

20
Implications for Greenhouse-Gas Emissions of Strategies Designed to Ameliorate Other Social and Environmental Problems

D.A. TIRPAK[1] and D.R. AHUJA[2]
[1] Office of Policy Analysis, PM-221 Environmental Protection Agency, Washington, D.C. 20460, U.S.A.
[2] The Bruce Company, 1100 6th Street, S.W., 215 Washington, D.C. 20024, U.S.A.

ABSTRACT

This chapter deals with repercussions of strategies designed to tackle primarily other social and environmental problems on greenhouse warming, which is caused by the emissions of several gases, each of which has other effects besides those on Earth's radiative balance. Because of the uncertainties on the timing, rate, and magnitude of anticipated climate change, the strategies being considered now by many countries are those that are justified because of their benefits for other social and environmental problems. Situations where benefits complement one another are, of course, the most attractive. Where instead of multiple benefits there are multiple trade-offs, there is little guidance for action. There are few strategies that will not involve some form of a trade-off, either with other social and environmental problems, between greenhouse gases themselves, or requiring some modifications in lifestyles. The U.S. has under considerations several commitments to environmental protection for reasons unrelated to climate change which could hold the emissions of greenhouse gases in 2000 to their levels in 1987, if chlorofluorocarbons are included in the total budget. Finally we point out that there are social and economic costs of both action and inaction.

Limiting the Greenhouse Effect: Options for Controlling Atmospheric CO_2 Accumulation
Edited by G.I. Pearman published 1992 John Wiley & Sons Ltd

INTRODUCTION

There is an increasing appreciation today among analysts, policy makers, and the informed public that the causes of most societal problems are multiple and hence their solutions are rarely clear-cut. This is true for most pressing problems including those that threaten the global environment. We live in times of interdependence and muddled causality, each action results in multiple consequences and each consequence is a result of multiple causative agents. Thus greenhouse warming is caused by the emissions of several gases, each of which has other effects besides those on the Earth's radiative balance. This is one reason, therefore, for advocating a "comprehensive" approach (Stewart and Wiener 1990) for dealing with strategies to limit climate change.

The direct consequence of an action, in turn, triggers a series of other changes, some of which will either amplify or attenuate the original direct consequence. These are the positive and negative feedback effects. There also will be positive or negative repercussions on other societal problems as well. The greater the number of intervening steps between the original "cause" and the effect, the higher the order of the effect is while the magnitude of the modification caused is perhaps lower. This paper deals with first-order repercussions that strategies, designed to solve primarily other social and environmental problems, have on the global warming issue. Because of the uncertainties (in gains and in time constants) involved with feedbacks, the only strategies most countries are considering now are those that are justified because of their benefits for other social and environmental problems. These actions, the elements of "no regrets" policies, also happen to have simultaneous benefits for the global warming problem. We shall also point to the negative implications some strategies have for global warming.

In open societies, the burden of proof, with very few exceptions, lies on those attempting to change the status quo. Because existing activities have several beneficiaries, every change is resisted by those who may have to bear the costs or forego existing benefits. On the one hand, taking actions to mitigate the risk of global warming imposes costs that are *local, certain, and immediate* to avoid costs that are *global, uncertain, and in the future*. Doing nothing, on the other hand, has global, uncertain risks in the future that must be traded against local, certain benefits in the present. Though these local benefits (or avoided costs) may seem so only because of a flawed calculation that assumes infinite environmental resources, this asymmetry forces proponents of action to push for measures that ought to be done for other reasons to derive greenhouse benefits. The opponents of even these actions will point to the negative implications or trade-offs for other sectors to buttress their do-nothing-as-yet stand. Thus the role of analysis is to confine the problem (which effects to include or exclude) after assessing the relative importance of each effect.

The appropriate system boundaries for a satisfactory analysis of the consequences of a societal action shift with time, increasing knowledge and experience. Some concerns are incorporated when found to be relevant while others are dropped when proved to be inconsequential. Actions which seemed "unalloyed goods" turn out to be mixed blessings upon deeper analysis (Daly and Cobb 1990). A belching smokestack, once universally taken as a sign of a healthy local economy, is now almost as universally deemed unacceptable. Also, air pollution control strategies often resulted in significant solid waste disposal problems. The earlier we are on the learning curve, the more frequently the boundaries are apt to shift. This is especially true in the environmental arena and will be so with global environmental problems. So while two decades ago, comparisons of pollution caused by coal-fired boilers and nuclear reactors were common, now the appropriate comparison for impacts is the entire cycle from fuel extraction to waste disposal. Still, even today, effective ways of analyzing and quantifying the full environmental costs of full fuel cycles are not available (EIS 1990).

The credibility of an analysis is largely dependent upon how deftly the boundaries of analysis are drawn to include simultaneous and other higher-order effects. Besides deciding what to include and exclude, the analysis must attempt to evaluate the relative magnitudes or contributions of different effects to the problem under study. In this chapter, we identify the more obvious implications for emissions of greenhouse gases of strategies suggested to solve other social and environmental problems.

ELEMENTS OF A COMPREHENSIVE ASSESSMENT

Seldom are the consequences of an action circumscribed in the sector for which the action was intended. Thus when strategies designed to control greenhouse-gas emissions are implemented, they are likely to have effects upon other social and environmental problems. Similarly actions taken to solve other social and environmental problems may have a bearing on the climate change problem. These interactions could be synergistic, neutral, or antagonistic. The analysis of the implications of strategies to control social and environmental problems must consider at least three classes of questions:

- To what extent are other social or environmental problems likely to be ameliorated or exacerbated by the proposed action?
- Will the proposed action increase or reduce the emissions of greenhouse gases? If yes, are there trade-offs between two greenhouse gases? If yes,

are there trade-offs between immediate effects and those in the long term?
- Is the action a technological fix or does it involve significant "life-style" changes?

These are a part of comprehensive impact assessment akin to that now routinely undertaken for projects (Montgomery et al. 1990). Additionally, there will be other economic and national considerations that we do not address in this paper.

In answering the first of these questions, one must attempt to quantify the magnitude of the effects. As already stated, the action designed primarily to solve one set of problems could either help or hinder another set of problems, including climate change. Figure 20.1 enumerates several such options, selected for their positive or negative implications for greenhouse warming.

The climatic consequences of a given amount of a greenhouse gas will depend upon the amount and the rate of warming caused. The warming in turn depends upon the radiative absorptance, atmospheric lifetimes, indirect effects on the concentrations of other radiatively active trace gases, and on the past, present, and future emissions of other greenhouse gases and their resulting concentrations. Because of these differences among greenhouse-related gases, the full potential for warming from such a gas as methane (CH_4) is realized in the first few decades after emission, while that from other long-lived gases, such as carbon dioxide (CO_2) or nitrous oxide (N_2O), is realized over centuries. Thus because of the differing relative effects, actions to reduce greenhouse gases may need to serve two different, important and complementary goals—that of *slowing* the rate of warming and that of *limiting* the amount of ultimate warming caused. The early analyses of response strategies considered CO_2 effects only; analyses now must evaluate the net greenhouse impact.

Remedial action almost invariably places restrictions on some group in the near term so that a larger group may benefit. These benefits could either accrue in the present or in the future. Thus, as should be clear, mandating that a utility use only gas reduces the utility's ability to switch fuels. Besides having economic consequences, this involves some behavioral changes. It does not require, however, the same sort of behavioral change as asking some Americans to drive smaller cars or bushmen to stop burning savannah.

Some mixture of multiple trade-offs and benefits is involved in most societal actions. The next section below illustrates this. Situations where benefits complement one another are, of course, the most attractive, especially where action is contemplated in the face of scientific uncertainty.

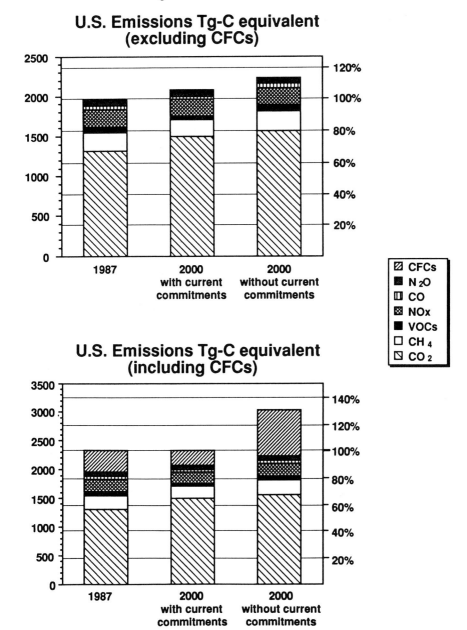

Figure 20.1 U.S. emission of greenhouse gases in 1987 and projected emissions for 2000, with and without current commitments to other environmental programs, including and excluding CFCs. The units are Tg (million tonnes) of carbon-equivalent (from Cristafaro and Scheraga 1991)

Following this section we discuss some of the principles that may help to weigh competing claims in situations with inherent tradeoffs.

We must first emphasize that this chapter deals only with mitigation options and not adaptation options. In general, a society will spread its expenditures between those activities that reduce vulnerability, such as coastal defenses and increasing farming resilience, and those that reduce emissions. Second, we do not discuss economic consequences that sometimes could be important.We assume that the examples or strategies identified are cost-effective or have small adverse effects, if at all, on employment or economic growth. This is something that, of course, will need to be shown. A third consideration ignored is the choice in cost-effective investments in mitigation overseas versus less cost-effective strategies on-shore. Again this is a symptom of improper "social accounting" and not different from other forms of economic protectionism (Smith et al. 1991).

SELECTED EXAMPLES

In this section we outline a few examples of the interactions between strategies to control greenhouse-gas emissions and other social or environmental problems. The list is not exhaustive, and the examples are chosen to illustrate the ideas of the previous section. The focus is on anthropogenic sources and not on modifications of natural sources and sinks. A more detailed description of most of these options can be found in chapter 5 of the Environmental Protection Agency's (EPA) Report to U.S. Congress on Policy Options for Stabilizing Global Climate (Lashof and Tirpak 1990). Table 20.1 summarizes the discussion relating to a few selected stragegies in the energy and industry sectors; Table 20.2 does the same for the agriculture and forestry sectors.

Utilities Sector

Fuel Switching from Coal to Gas

Using natural gas rather than coal to fuel boilers in utilities has several advantages for local air quality (German 1989). In addition, acid deposition is reduced. In switching from coal to gas, there is also an efficiency improvement and so CO_2 emissions will be reduced. Leakages of natural gas (CH_4) during production and transport could off-set some of the CO_2 reductions made possible by the switch. Methane emissions from the coal that would now not have to be mined must also be considered in determining net impact. A few studies have calculated the break-even leakage rate beyond which the CH_4 emissions will outweigh the carbon savings. Rodhe

Ameliorating Social and Environmental Problems

(1990) has estimated this rate to be 5–6%; Isaksen and Roland (1990) claim leakage rates as high as 14% justify switching from coal to gas. The number is strongly dependent on the global warming potential index chosen for CH_4 (Lashof and Ahuja 1990). Since typical leakage rates are generally in the range 1–5%, usually such a switch would be justified.

Fuel choices in a country, however, are determined by several other factors. Often, a premium is placed on natural gas as a feedstock for the manufacture of chemicals and fertilizers rather than its use as a power source for which generally cheaper alternatives are available. If downstream fertilizer and chemical plants are available, that use might be preferred. If they are not, as in India, then use as a fuel is preferable to flaring. Assured availability of the fuel for the lifetime of the plant, comparative prices, and lead construction times are other factors that will influence fuel choices.

Fuel Switching from Coal to Oil

The considerations in switching from coal-fired to oil-fired boilers are similar to those just described in switching to gas-fired boilers. Local air quality improves in terms of particulate emissions. Whether the acid deposition problem is made better or worse will depend upon the respective sulfur contents of the coal and the oil that replaces it. There are some carbon gains, though less than those obtained from shifting to gas, from the improvements possible in efficiency. Again, these must be compared with the sum of releases of associated gas at oil wells and CH_4 emissions avoided in coal mining.

Conservation

Conservation of electricity and other demand-side reductions provide multiple benefits for the environment. By avoiding the need for generation or the commissioning of new capacity, local environmental quality (reduced acid deposition) is improved. The emissions of greenhouse gases associated with the generation are also reduced. However, depending on the technology employed, some life-style changes are required. New technologies, such as smart sensors, do overcome some of the earlier objections based on concerns for safety, for example, in an inadequately lit house.

Building New Capacity for Efficient Lighting Devices

Cutting electricity consumption in lighting devices offers many cost-effective opportunities for saving energy both in the industrialized and developing countries. Compact fluorescent lamps (CFL) are 60–70% more efficient than the incandescent lamps they replace. Gadgil and Rosenfeld (1990) have

Table 20.1 Selected examples of implications for greenhouse-gas emissions of strategies primarily designed to solve other social or environmental problems (energy and industries sectors)

	Problems		Greenhouse-related gases that might:		Significant lifestyle changes?
	Ameliorated	Exacerbated	Decrease	Increase	
UTILITIES					
1) Switch from coal to gas	Local environmental quality, regional air quality, acid deposition	Premium feed stock	CO_2	CH_4	No
2) Switch from coal to oil	Local environmental quality, acid deposition(?)	Premium fuel, acid deposition(?)	CO_2	?	No
3) Conservation	Energy security; local, regional environmental quality	—	CO_2, CO, CH_4, NO_x	—	Yes
4) Build CFL plants	Reduce capital expenditures; local, regional environmental quality	—	CO_2, CO, CH_4, NO_x	—	No
5) Renewables	Energy security, local, regional, environmental quality	Minimal impact (depending on source)	CO_2, CO, CH_4, NO_x	—	No
6) Scrubbers	Acid deposition, local air quality	Sludge disposal, reduction in cooling	SO_2	CO_2	No

543

AUTOMOBILES					
(1) Higher efficiency through					
– drivetrain/styling improvements	Urban air quality, energy security	—	—	—	No
– weight reductions	"	Safety	CO_2, CO, NO_x, VOCs	—	Yes
(2) Strengthen vehicle emissions controls	Urban air quality	Modest decline in fuel economy	CO, HC_x, NO_x	CO_2 (slightly)	No
(3) Alternate Fuels					
– Diesel	Energy security, safety	Urban air quality, cancer risk	CO_2	VOCs	Yes
– CNG	Urban air quality, energy security	—	CO_2, CO, VOCs	CH_4	No
– Ethanol from crops	"	Competition for food, land	?	?	No
– Ethanol from woody biomass	"	Competition for land	?	?	No
Methanol from coal	"	—	—	CO_2	No
– from natural gas	"	—	—	—	No
– from biomass	"	Competition for land	CO_2	—	No
4) Effective driving restrictions	Urban air quality, energy security	—	CO_2, CO, VOCs, NO_x	—	Yes
CFCs					
1) Replace CFC-12 by FC-134a	Ozone depletion	—	CFC-12	CO_2	No

Table 20.2 Selected examples of implication for greenhouse gas emissions of strategies primarily designed to solve other social or environmental problems (agriculture & forestry sectors)

	Problems		Greenhouse-related Gases That Might:		Significant Lifestyle Changes?
	Ameliorated	Exacerbated	Decrease	Increase	
LANDFILLS					
1) Waste minimization	Local environmental quality, land requirements	—	CO_2, CH_4, VOCs	—	Yes
2) Incinerators	Land requirements	Local air quality, dioxin emissions, ash disposal	CH_4, VOCs	CO_2	No
3) Flaring	Explosion hazard	—	CH_4	CO_2	No
4) Energy recovery	Energy security, explosion hazard	—	CH_4, VOCs, CO, NO_x	—	No
LIVESTOCK					
1) Feed additives	Increased milk and meat production	Food safety	CH_4	—	No
2) Feedlot waste management with energy recovery	Local environmental quality, energy security	—	CH_4, VOCs, CO, NO_x	CO_2	No
3) Biogas plants	Increased clean energy supply, local environmental quality	—	CO_2	CH_4	No

Option	Positive effects	Negative effects			Win-win?
4) Vegetarianism	Deforestation, ill health due to malnutrition(?)	Protein malnutrition(?)	CO_2, CH_4	–	Yes
FERTILIZERS					
1) Use of nitrogenous fertilizers	Increased food production	Local and regional water pollution	–	N_2O, CO_2	No?
FORESTRY					
1) Reduce deforestation	Biodiversity, soil erosion, desertification	Pressure for grazing and cultivable lands	CO_2, CH_4, CO, NO_x	–	?
2) Grow trees in urban areas	Reduced air conditioning in summer, aesthetics, local air quality	Increased heating loads in winter, safety(?)	–	CO_2	No
3) Reduce burning of agricultural residues	Local air quality, soil conditioning	Disposal problem	?	?	Yes
4) High-efficiency cookstoves	Health (if also low exposure)	Local air quality (if high emissions)	CO_2	CO, CH_4 VOCs	?
WETLANDS					
1) Drain wetlands	Increased availability of agricultural land, reduced vector-borne diseases	Loss of biodiversity	CH_4	–	–
2) Irrigated rice cultivation	Increased rice production	Vector-borne diseases(?)	N_2O	CH_4	–

shown that a new CFL plant costing $8 million producing 5000 lamps per day could replace a 200 MW plant costing over $300 million. In addition, simply by carrying out the available cost-effective lighting efficiency improvements, 40 large U.S. power plants could be retired (Rosenfeld and Hafemeister 1985) with all the consequent benefits for local, regional, and global environmental quality. Earlier objections about differences between warmth, color, and other light characteristics have been removed by the new lamps.

Renewable Sources of Energy

The use of renewable sources of energy, such as solar, wind, mini-hydro, geothermal, etc., to produce electricity often makes sense on economic grounds, with the high capital costs offset by the low operating costs. For remote locations unconnected to an electric grid, renewable sources of energy may provide a cheaper alternative to grid extension. The localized environmental effects of renewable sources will depend on the source and the location but are generally not as serious as those from coal- or oil-based generation. To the extent renewables displace the need for fossil-based generation, emissions of all associated greenhouse-gas emissions will be reduced.

Flue-gas Desulfurization Systems

The examples thus far have emphasized strategies that result in the reduction of greenhouse gases. Examples of trade-offs also exist. Flue-gas desulfurization systems, commonly called scrubbers, remove oxides of sulfur from exhausts of coal- and oil-fired plants. However, the energy required to operate the scrubbers imposes a penalty of up to 10% (Martin 1989). Thus controlling acid deposition makes the greenhouse problem slightly worse. Not only are the CO_2 emissions increased, it might be possible that sulfur emissions may have had a negative influence on radiative forcing through their effect on the cloud condensation nuclei (Houghton et al. 1990). In addition, scrubbers generate sludge that needs to be disposed of on land. Some countries have decided that the advantages of controlling acid deposition over large areas outweigh the costs of scrubbers, the slight costs imposed by the land disposal problem, or the additional carbon emissions that exacerbate the global warming problem. However, this could be because the adverse effects of acidification have manifested earlier than those expected from anthropogenic greenhouse warming. New systems that remove NO_x and SO_2 will have the additional effect of reducing warming through the reduction of tropospheric ozone formation.

Automobiles

Higher-efficiency Standards

Automobile fuel economy standards improve both local air quality and energy security as well as reduce gasoline consumption and trade imbalance in oil-importing countries. Yet, consumers value other factors such as cost, safety, aesthetics, reliability, and performance in their decisions to purchase cars. Depending upon which of the technologies are used to improve fuel efficiency, different trade-offs are involved. Much of the early improvements in automobile efficiency came from weight reductions and the use of alternative materials. There is some evidence that there is a correlation between reductions in size and weight, and increases in injuries and fatalities (OTA 1982). It is possible that alternative materials could reduce maintenance costs. But clearly, evaluations of technical alternatives for improving fuel efficiency must assess effects on safety. There is also some evidence, contrary to popular belief, that use of plastic composites sometimes increases safety (Bleviss 1988).

Driving a smaller or less-responsive automobile requires at least a modicum of change in consumer behavior. Increases in fuel economy due to improvements in engines or transmissions or aerodynamic styling are less likely to elicit consumer resistance. Predicting savings in emissions due to stricter fuel efficiency standards, however, is complicated by other higher-order changes in consumer responses, such as a shift to unregulated light trucks, slower rate of stock turnover, etc. Yet this is part of a dynamic learning process: trucks also can be regulated, and stock eventually turns over. Alternative policies can be devised to respond to such shifts.

Strengthen Emission Controls

The U.S. currently regulates the emissions of hydrocarbons, CO, NO_x, and particulate matter from light-duty gasoline vehicles on a gram per kilometer basis. This is achieved through precise control of air–fuel ratio, exhaust gas recirculation, and catalytic convertors. Emission standards in most European OECD countries are less stringent than those in the U.S. Developing countries have even laxer standards, if at all. Significant reductions in global emissions of NO_x and CO could result from the extension of U.S. standards to the rest of the OECD and ultimately, to the rest of the world (Lashof and Tirpak 1990). The U.S. emissions could be reduced further if standards were based on a grams per liter basis, encouraging the development of technologies that simultaneously would improve fuel efficiency and reduce emissions of air pollutants (Bleviss 1988; Khazzoom et al. 1990).

Alternative Fuels

Gasoline-powered vehicles have been regulated in OECD countries mainly because of the deterioration of urban air quality and because of the volatility of oil supplies from the Middle East. Light duty vehicles emit CO, NO_x, and volatile organic carbons (VOCs) all of which directly and indirectly affect both urban air quality and climate change by their effects on ozone concentrations. The trade-offs or benefits are dependent upon which fuel is recommended as a replacement for gasoline. It is critical to compare the entire fuel cycle before recommending a shift. For electrical vehicles, which will improve urban air quality, for example, the net greenhouse benefit is dependent of the source used to generate the electricity. We discuss briefly the implications of a few of the suggested alternatives.

Diesel—Diesel engines have a higher compression ratio and are more fuel efficient than gasoline-powered automobiles, but diesel engines tend to produce greater emissions per mile traveled of particulates, some of which are carcinogenic. An early study (NRC 1982) showed that the slight increase in cancer risk is offset by a reduction in risk of serious injury as diesel engines tend to be heavier. However, these are not the only factors that will decide the mix between diesel and gasoline: diesel engines also are sluggish and tend to require more frequent maintenance.

Compressed Natural Gas (CNG)—CNG, currently used in New Zealand and Canada, is among the cleanest transport fuels available considering the emissions of CO and VOCs. It also produces lesser CO_2 emissions per unit of energy released than any other fossil fuel. Yet, leaks of natural gas, primarily CH_4, from production, distribution and refueling processes could offset the advantages of lower carbon emissions (Victor 1990).

Ethanol—Currently, some governments (e.g., the U.S. and Brazil) subsidize ethanol production from biomass sources. These technologies rely on feedstocks such as corn and sugarcane, both food crops. This competition with food production will vary among countries but raises doubts about the long-term viability of this approach. As the technology of producing ethanol from woody biomass matures, this concern would be lessened. To the extent fossil fuels are used in the production process (fertilizers, farming operations, distillation, etc.) the greenhouse-gas benefits of biomass as a feedstock are reduced.

Methanol—Methanol can be produced from biomass, coal, or natural gas. The greenhouse implications are again dependent upon which is the feedstock. The estimated net CO_2 contributions to greenhouse gases from cars powered with methanol derived from natural gas as a feedstock is roughly equivalent to that from gasoline-driven automobiles. From coal, however, emissions over the entire fuel cycle are twice those from crude oil-derived gasoline (DeLuchi et al. 1987). These estimates were made before the concept of global warming potentials was developed. This concept is still being refined, and these comparisons would need to be redone to examine net greenhouse-gas impacts.

Driving Restrictions—Some improvement in urban air quality and reductions in greenhouse-gas emissions can be obtained by restricting driving in private automobiles. Different regions have attempted to do this in various ways, but to a different extent, all are disruptive of people's life-styles. Mexico City, for example, prohibits some cars from roads on some days of the week; Washington reserves some lanes on highways for cars carrying three or more passengers; and many communities encourage car and van pools, mass transit, park and ride lots, etc. Whether a particular scheme will succeed in a particular place cannot be predicted easily in advance. Where most families have more than one car, day of the week restrictions may not result in significant decrease in total miles driven.

Chlorofluorocarbons

Chlorofluorocarbons (CFCs) are used widely in industry for many applications such as refrigerants, in fire extinguishers, as solvents, and in the manufacture of foam products. They are often highly suited to the purpose for which they are being used, e.g., CFC-12 as a refrigerant. CFCs exacerbate both the greenhouse problem and ozone-depletion problem. For air-conditioning and refrigeration applications, substitutes impose an energy penalty and thus like scrubbers will exacerbate greenhouse warming. The substitutes sometimes have higher radiative forcing than the compounds they replace but are more short-lived, thus their global warming potential is less than the compounds currently being used when considered over a period of a number of decades or more.

Substituting CFC-12 by a compound such as FC-134a offers an opportunity for redesigning refrigerators and mobile air conditioners, which will then avoid the energy penalty. It can be argued, though, that a penalty would exist if the comparison is made to a redesign with the original substance. It could be countered that perhaps the redesign would never have taken place if the pressure to find substitutes were not there. Yet the energy penalties are slight, a few percent, and should not deter action that clearly responds to the greater danger of stratospheric ozone depletion.

Landfills

Waste Minimization

Landfills and the currently available alternatives (CH_4 recovery, incineration) involve some potential trade-offs. Minimizing waste going to landfills by reducing consumption or recycling efforts perhaps avoids adverse effects but requires (modest) behavioral changes. The separated food and garden wastes can be aerobically composed into fertilizer without the concomitant emissions of CH_4. Anaerobic decomposition at landfills results in the emission of CH_4 to the atmosphere over a prolonged duration (decades). Eventually, this CH_4 gets converted to CO_2 in the atmosphere.

Incinerators

Some cities have resorted to burning municipal wastes because of the shortage of suitable areas for landfills. These municipal waste combustors or incinerators convert the carbon directly to CO_2 but also require some supplemental fossil fuel to operate. However, for Sweden, Levander (1989) calculates that a significant reduction in net greenhouse warming potential is obtained even after considering only that part of the refuse that is of fossil origin and accounting for the CH_4 that escapes from the incinerator unburnt. If the incinerators are coupled with energy recovery units, this could result in a greater reduction of net greenhouse-gas emissions to the atmosphere. The emissions of dioxin, however, may be grounds for concern at some sites.

Flaring

Capturing CH_4 at landfills and flaring it (which often is mandated to reduce the risk of explosion and the emissions of trace amounts of volatile organic compounds) would reduce the amount of CH_4 emitted into the atmosphere. A better option, of course, would be the recovery of the gas and use as an energy source, if an economic use can be found for it.

Energy Recovery

Recovering and using the CH_4 captured at the landfill for energy to displace other fossil energy would perhaps be the best option after waste minimization and resource recovery and recycling. However, because of several constraints, less than 2% of sites in the U.S. recover CH_4 for energy use. The economic viability of CH_4 recovery depends upon the size of the landfill, location, proximity to potential users or existing pipelines, current competing energy costs, and government regulations. New regulations proposed by the U.S. EPA have the potential to reduce significantly the emissions of CH_4 from landfills.

Livestock Management

Feed Additives

Methane emissions from animals can be reduced significantly by modifying their diets with alternative food stocks or by adding hormones or other supplements to their diet. In developing countries, significant emission reductions may be achieved through increased animal productivity (of milk, for example) and reproductive efficiency (reduced inter-calving interval) through the addition of supplements, such as urea and molasses, to correct nutritional deficiencies. In industrialized countries, hormone implants are currently available for nonlactating cattle (Gibbs et al. 1989). Emissions reduction is achieved through faster weight gain and increased feed efficiency. Bovine growth hormones, currently being developed, may increase milk production significantly, thereby helping to reduce the size of animal populations. In all such strategies, however, food safety (for humans) is a primary prerequisite.

Feedlot Waste Management

In feedlots where wastes are disposed of in lagoons or other anaerobic environments, capturing CH_4 with plastic covers or digestors significantly reduces CH_4 emissions and improves local air quality. The CH_4 can be burnt to produce electricity (which displaces other fossil fuels) and the CO_2 produced can be used to fertilize plant growth in greenhouses or algae ponds.

Biogas Plants

In developing countries, cattle dung is used as a solid fuel, as a fertilizer and soil conditioner, and as a building material. One alternative is to use it in anaerobic digestors to produce biogas (mainly CH_4) that can be used for cooking and lighting. In normal circumstances when dung decomposes aerobically there would be little emission of CH_4. Khalil et al. (1990) found leakage rates from biogas plants around Chengdu in Sichuan province of China to be low, probably around ~1%, and not more than 5%. Improvements in air quality, both indoor and outdoor, and increases in convenience over using dung directly as a fuel are substantial. Smoke from dung burning has insect repellent and thatch-roof preservation properties, which are absent in biogas. Periodic planned fumigation could reduce this objection to the use of biogas.

Vegetarianism

It is possible that with increasing popularity of animal rights movements, anti-vivisectionists, and some eastern religions, there would be a shift in diets toward species lower in the food chain. Such a shift may be promoted also on strictly health grounds. To the extent that this happens, standing livestock populations, and CH_4 emissions from it, would be reduced also. Diets are habitually entrenched and notoriously difficult to change.

Use of Nitrogenous Fertilizers

The global agricultural system also provides some examples of trade-offs. The use of nitrogenous fertilizers results in emissions of N_2O which are dependent on the amount, type, and method of application of the fertilizers. Their use may also cause localized ground and surface water pollution. Despite this, countries subsidize fertilizers, and therefore the true cost to society of their use is not apparent and encourages their overuse. Their role in helping increase food production, on the other hand, seems undeniable.

Forestry

Reduce Deforestation

Both the causes and consequences of deforestation are well known. The causes vary among regions; the consequences perhaps are similar: soil erosion, desertification, loss of biodiversity, and carbon storage. The causes include population pressure for agricultural land, the demand for industrial timber, and inappropriate government policies regarding land tenure, economic incentives, forest settlement, and other population issues (Bellagio 1987). If deforestation can be reversed, tropical forests could serve as a vast carbon sink, reducing global CO_2 levels. The first critical steps in accomplishing this would be to support local people in introducing sustainable forest management and reforestation techniques that would provide for food, fiber, fodder, and fuelwood needs without mining the primary forest (Lashof and Tirpak 1990).

Grow Trees in Urban Areas

Trees have many benefits besides being stores of carbon. They modulate the flow of groundwater, prevent soil erosion, and provide habitats for birds, shade, aesthetic benefits, fuel, and fodder besides other commercial benefits such as maple syrup, gum, and medicines. Appropriately placed, because of shading, they can reduce air conditioning loads in the summer. Similarly they can increase heating loads in the winter. Thus a recommendation of tree planting as an option to reduce energy usage would have to consider local weather patterns, that is, to be promoted where cooling demands dominate (such as in the southwestern U.S.) and evaluated for net benefits elsewhere.

Reduce Burning of Agricultural Residues

There are several paths for the disposal of agricultural residues. They may be burnt in cookstoves or in the fields, they may be used for fodder, building materials, composted, ploughed under or used in energy extraction devices such as boilers and gasifiers. The set of greenhouse gases emitted is different for each of these paths. The recommendation to reduce burning of agricultural residues can be made only after a comparison with the emissions in the alternative disposal path.

Biomass-burning Cookstoves

Traditional cookstoves in developing countries contribute to the emissions of non-CO_2 greenhouse gases and perhaps also to the flux of CO_2 through deforestation or reduction of biomass densities. Replacing traditional stoves by more fuel-efficient stoves could save carbon and reduce the flux of CO_2 to the atmosphere but also could increase simultaneously the flux of CO, CH_4, and other VOCs. This can happen if the increase in fuel efficiency comes from an increase in thermal efficiency, which in turn comes from a reduction in burning rate and a consequent decrease in combustion efficiency.

Wetlands

Drain Wetlands

Thus far we have dealt only with strategies that seek to control emissions caused by anthropogenic sources. In principle, opportunities exist for altering natural sources and sinks. Some, such as the possibility of fertilizing the oceans, appear to be impractical. Others appear to be possible. Draining wetlands, for example, is often undertaken to reclaim "unproductive" land for human use. But this will also reduce the emissions of CH_4. Wetlands serve so many other ecological functions that draining them may do more harm than the harm prevented by the reduction in CH_4 emissions. Clearly, all suggestions to manipulate natural systems on a large scale raise large philosophical questions.

Promote Irrigated Rice Cultivation

Shifting from natural intermittent flooding for rice cultivation to irrigated fields, which are flooded for prolonged periods, will increase rice production and reduce the emissions of N_2O (Eriksen et al. 1985) but increase the emissions of CH_4. If accompanied by a shift to high-yielding, fertilizer-responsive modern varieties, the total emissions per kilogram of rice produced could be reduced, although CH_4 emissions per hectare would increase.

PRINCIPLES FOR EVALUATING RESPONSE STRATEGIES

The Response Strategies Working Group (IPCC 1991) of the Intergovernmental Panel on Climate Change has concluded that given the uncertainties on the timing, rate and magnitude of climate change, the costs and effectiveness of response options in averting climate change, the most effective immediate response strategies will be those that are:

- beneficial and justifiable in their right for reasons other than climate change; this means that the strategies must serve multiple social, economic, and environmental goals;
- economically efficient, cost-effective, and compatible with sustainable economic growth and development; and
- flexible and administratively practical, observable and enforceable.

These are all unexceptionable and laudable principles. In situations where instead of multiple benefits there are multiple trade-offs, they do not provide any guidance for action. As Tables 21.1 and 21.2 have shown, there are going to be few strategies that will not involve some form of a trade-off, either with other social and environmental problems, between greenhouse gases themselves, or through modifications in life-styles.

When faced with a trade-off between two effects, it would be a somewhat trivial matter to evaluate whether the proposed action does more harm than good if the total damage functions for both the effects were available in a common metric (Ott 1978). However, situations where even partial damage functions will be available will be exceptional. Here the policymaker can decide on the appropriate weight to ascribe to each effect and could do so by asking the following questions about the effects.

Ameliorating Social and Environmental Problems 555

- Is the effect associated with an increase in entropy? (Pollutants confined are preferable to pollutants dispersed if the concentrations exceed the assimilative or cleansing capacity of the environment.)
- Is the effect irreversible?
- Is there a finite probability that the effect may be potentially catastrophic? (Most open societies tend to be risk averse to large adverse events.)
- Will the effect threaten systems that have unique heritage or pristine natural values?

The effect associated with more positive answers to these questions will tend to receive a greater weight while deciding upon remedial measures.

INITIATIVES IN THE UNITED STATES

As an example of the inter-relatedness of strategies to address environmental problems, we present here a few of commitments to environmental protection being considered in the U.S. for reasons unrelated to climate change. To study the implications of these commitments for potential climate change, Cristafaro and Scheraga (1991) have compared recently, on a carbon-equivalent basis, what the net U.S. emissions of greenhouse gases would be with and without these commitments. The methodology uses the 100-year global warming potentials developed by the IPCC Working Group 1 (Houghton et al. 1990). The proposed initiatives are a tree planting program, amendments to the Clean Air Act, CFC phase-out, U.S. EPA landfill regulations and Department of Energy appliance standards and the energy efficiency and renewables programs. Each of these is described briefly below; Cristafaro and Scheraga (1991) provide further details.

Appliance Standards, Energy Efficiency, and Renewables Initiatives

Revised and more stringent standards for refrigerators, freezers, dryers, and dishwashers are expected to save approximately 4 Tg C-equivalent per year by 2000. Other efficiency initiatives include the use of more efficient lighting, promotion of least-cost utility planning, energy analysis and diagnostic centers and standards for buildings. Hydroelectric power and photovoltaic technologies are also anticipated to increase in use. Not only will the emissions of greenhouse gases be reduced with these programs, the emissions of other conventional pollutants, such as SO_2 will be reduced as well.

Clean Air Act

Implementation of the new Clean Air Act is expected to result in reductions of CO_2, NO_x, CO, VOCs, and other regulated air pollutants. NO_x reductions are projected to come from controls on motor vehicles, utilities, nonutility point sources, and area sources. VOC reductions are expected to come from tighter control on motor vehicles. Reductions in CO are expected from automobile fleet turnover, an oxygenated fuels program, and others.

CFC Phase-out

After the London Amendments to the Montreal Protocol on Substances that Deplete the Ozone Layer, a phase-out of the controlled CFCs and halons has been agreed to by the U.S. and several other countries. This is expected to eliminate by 2000, 95% of the production of currrent CFCs.

Landfills

There are about 6000 landfills in the U.S. of various sizes generating about 255 Gg/yr of non-CH_4 organic compounds and 10.5 Tg/yr of CH_4. Regulations designed to control VOCs by collection and combustion will reduce also methane emissions by approximately 9 Tg/yr by 2000.

Tree Initiative

President Bush's reforestation initiative hopes to plant a billion trees per year on 0.6 million ha, for 10 to 20 years beginning in 1991, and to improve forest management practices on an additional 40,000 ha per year. The program, under consideration in the U.S. Congress, could sequester approximately 50 Mt/yr of carbon by 2010, amounting to 4–5% of current net U.S. emissions of CO_2.

Table 20.3 summarizes the expected reductions from each of these programs by the year 2000.

Cristafaro and Scheraga conclude from their analysis that total emissions of greenhouse gases in the year 2000 could be held to their level in 1987, *if CFCs are included in the total budget*. If CFCs are excluded, U.S. emissions increase by 5% on a carbon-equivalent basis. Emissions of direct CO_2 on the other hand are expected to increase by 15%. Figure 20.1 summarizes these calculations.

Table 20.3 Expected reductions by the year 2000

Current Commitments	Reduction (Tg C -equivalent)
Appliance Standards	4
Energy Efficiency	28
Renewable Energy	4
Clean Air Act	17
CFC Phase-out	551
Landfill Controls	44
Tree Initiative	9

CONCLUDING REMARKS

There is a tendency among those who would like to see action for climate reasons postponed to inject into the debate something that has been ignored hitherto, e.g., an overlooked trade-off or a negative feedback that will take care of the problem. While some trade-offs will be minor and others simply red herrings, the proponents of action have a responsibility to incorporate and evaluate those and as many other relevant factors into analysis as is possible. As the Energy and Industries subgroup of the Response Strategy Working Group of IPCC concluded, there are social and economic costs of both action and inaction. The no–action approach is a "decision" to "freeze" the world on a path as it exists today. It is a call for a dynamic status quo—an oxymoron world. It commits us indefinitely to a future with unforeseeable risks while awaiting the removal of all uncertainties. The risks associated with inaction will increase as greenhouse-gas emissions and concentrations will continue to rise causing disruptive climate change with ecological and societal consequences. Also, the costs of effective remedial action may rise with increasing delay if societies get committed to old, inefficient, and polluting technologies. Obviously, new technologies could reduce costs also, but only if actions are taken to develop and promote them.

We endorse a comprehensive approach that takes a long-term view and favors incremental decision-making following the assessment of identifiable (and significant) costs and benefits, yet an approach that recognizes a need for action and abjures its postponement under the pretext of evermore analysis, even in the face of clear risks and multiple benefits.

ACKNOWLEDGEMENTS

We would like to thank Paul Schwengels for bringing to our attention some studies mentioned in Section III of this paper. Mr. Schwengels is the primary author of Chapter 5 on Technical Options for Reducing Greenhouse Gas Emissions of EPA's Draft Report to Congress on *Policy Options for Stabilizing Global Climate* (Lashof and Tirpak, 1990). The views expressed are those of the authors and do not represent the official position of the Environmental Protection Agency.

REFERENCES

Bellagio. 1987. Statement of the Bellagio Strategy Meeting on Tropical Forests, July 1–2, Bellagio, Italy.

Bleviss, D. 1988. The New Oil Crisis and Fuel Economy Technologies: Preparing the Light Transportation Industry for the 1990's. New York: Quorum Press.

Cristafaro, A., and J. Scheraga. 1991. Policy implications of a comprehensive greenhouse gas budget. *Forum for Applied Research and Public Policy*, in press.

Daly, H.E., and J.B. Cobb, Jr. 1989. For the Common Good: Redirecting the Economy Toward Community, the Environment, and a Sustainable Future. Boston: Beacon Press.

DeLuchi, M.A., R.A., Johnston, and D. Sperling. 1987. Transportation Fuels and the Greenhouse Effect. Berkeley: Univ. Wide Energy Research Group, Univ. of CA.

EIS. (Energy and Industries Subgroup) 1990. Report of the EIS of the Response Strategies Working Group of the Intergovernmental Panel on Climate Change (IPCC). May 31.

Eriksen, A., M. Kjeldby, and S. Nilsen. 1985. The effect of intermittent flooding on the growth and yield of wetland rice and nitrogen-loss mechanism with surface applied and deep placed urea. *Plant and Soil* **84**:387–401.

Gadgil, A., and Rosenfeld, A. 1990: Conserving energy with compact fluorescent lamps. Berkeley: Lawrence Berkeley Lab.

German, M.I. 1989. The environmental attributes of natural gas. In: Global Climate Change Linkages: Acid Rain, Air Quality and Stratospheric Ozone, ed. J.C. White, pp. 231–254. New York: Elsevier.

Gibbs, M.J., L. Lewis, and J.S. Hoffman. 1989. Reducing methane emissions from livestock: opportunities and issues. Washington D.C.: U.S. EPA, Office of Air and Radiation.

Houghton, J.T., G.J. Jenkins, and J.J. Ephraums, ed. 1990. Climate Change. The IPCC Scientific Assessment, Cambridge: Cambridge Univ. Press.

IPCC. 1991. Response strategies, working group III. In: Climate Change. Washington, D.C.: Island Press.

Isaksen, I.S.A., and K. Roland. 1990. Metanutslipp Fra Naturgasskjeden: Om Metan Som Klimagass. Copenhagen: Nordisk Ministerrad.

Khalil, M.A.K., R.A. Rasmussen, M.-X. Wang, and L. Ren. 1990. Emissions of trace gases from Chinese rice fields and biogas generators: CH_4, N_2O, CO, CO_2, chlorocarbons and hydrocarbons. *Chemosphere* **20**:207–226.

Khazzoom, J.D., M. Shelby, and R. Wolcott. 1990. The conflict between energy conservation and environmental policy in the U.S. transportation sector. *Energy Policy* **18**:456–458.

Lashof, D.A., and D.R. Ahuja. 1990. Relative contributions of greenhouse gas emissions to global warming. *Nature* **344**:529–531.

Lashof, D.A., and D.A. Tirpak. 1990. Policy options for stabilizing global climate. Draft report to U.S. Congress. Washington, D.C.: U.S. EPA.

Levander, T. 1989. The importance of greenhouse gases other than carbon dioxide and other possible differences between various fuels. *Statens energiverk* memorandum dated 14 September sent to Delegates of the OECD Group on Energy and Environment by B. Assarsson. Stockholm: Nat. Energy Admin., Dept. of Policy Analysis.

Martin, H.C. 1989. The linkages between climate change and acid rain. In: Global Climate Change Linkages: Acid Rain, Air Quality and Stratospheric Ozone, ed. J.C. White, pp. 59–66. New York: Elsevier.

Montgomery, J., B. Solomon, J. Smith, and C. Berish. 1991. Addressing Global Climate Change in NEPA Reviews. *Env. Prof.*, in press.

NRC (National Research Council). 1982. *Diesel cars: benefits, risks, and public policy*. Final report of the Diesel Impacts Study Comm. Washington, D.C.: Nat. Acad. Press.

OTA (Office of Technology Assessment). 1982. Increased automobile fuel efficiency. Washington, D.C.: OTA.

Ott, W.R. 1978. Environmental Indices: Theory and Practice. pp. 329–348. Ann Arbor: Ann Arbor Science Pub.

Rodhe, H. 1990. A comparison of the contribution of various gases to the greenhouse effect. *Science* **248**:1217–1219.

Rosenfeld, A., and D. Hafemeister. 1985. Energy conservation in large buildings. In: Energy Sources: Conservation and Renewables, ed. D. Hafemeister, H. Kelly, and B. Levi. Conference Proceedings No. 135. New York: American Inst. of Physics.

Smith, K.R., J. Swisher, R. Kanter, and D.R. Ahuja. 1991. Indices for a greenhouse gas control regime: incorporating both efficiency and equity goals. Report prepared for the Env. Division, The World Bank. Honolulu: East West Center.

Stewart, R.B. and J.B. Wiener. 1990. A comprehensive approach to climate change. *Am. Enterprise*, Nov.

Victor, D. 1990. Leaking methane from natural gas vehicles: implications for transporattion policy in the greenhouse era. *Climatic Change*, in press.

21
How Will Climatic Changes and Strategies for the Control of Greenhouse-Gas Emissions Influence International Peace and Global Security?

P.H. GLEICK
Pacific Institute for Studies in Development, Environment, and Security
Berkeley, CA 94709, U.S.A.

ABSTRACT

Just as the prospects are looking up for peace and military security between the superpowers, a new set of challenges faces us: widespread underdevelopment and poverty, and large-scale environmental problems that threaten human health, economic equality, and international security. These threats have global implications. As a result, finding satisfactory solutions will be especially difficult.

A strong argument can be made linking certain resource and environmental problems with the prospects for war or peace. Perhaps the most important of these is the likelihood of major climatic changes resulting from growing atmospheric concentrations of carbon dioxide and other trace gases. Given the extent and severity of the climatic changes now anticipated, we must ask how such changes will affect international relationships, economics, behavior, and security. This chapter assesses the most likely links between the societal impacts of climatic changes, strategies to reduce the emissions of greenhouse gases, and international peace and security.

INTRODUCTION: ENVIRONMENT AND SECURITY

Unprecedented political changes have occurred recently, reducing tensions and long-standing conflicts between the East and the West. But just as we

Limiting the Greenhouse Effect: Options for Controlling Atmospheric CO_2 Accumulation
Edited by G.I. Pearman © 1992 John Wiley & Sons Ltd

make progress in the fields of arms control and traditional military security, a new set of challenges now complicates the international scene: growing underdevelopment and poverty, and large-scale environmental problems that threaten human health, economic equality, and international security. Unlike earlier worries about air and water pollution, these new threats have global implications; they know no political distinctions and they obey no international boundaries. As a result, finding satisfactory solutions will require new thinking and international cooperation.

While many of the past, present, and future causes for conflict and war have little or no direct connection with the environment or with resources, a strong argument can be made linking certain resource and environmental problems with the prospects for war or peace. There is a long history suggesting that access to resources is a proximate cause of war, that resources have been both tools and targets of war, and that environmental degradation and the disparity in the distribution of resources can be roots of conflict (Gleick 1990b).

Recently, international attention has focused on the possibility that major climatic changes will result from growing atmospheric concentrations of carbon dioxide (CO_2) and other trace gases (Houghton et al. 1990). Given the extent and severity of the climatic changes now anticipated, we must ask how such changes will affect international relationships, economics, behavior, and security. This paper assesses the most likely links between the societal impacts of climatic changes, strategies to reduce the emissions of greenhouse gases, and international peace and security.

I do not mean to suggest that the political and ideological questions that now dominate international discourse will become less important in the future. Rather a new set of issues that were of only local or regional importance in the past, such as population growth, transborder pollution, poverty, and inequitable social systems, are now taking on international importance. National energy policies will come to depend not only on the price and supply of fossil fuels, but also on the global environmental consequences of their combustion. The stability of regional peace and security may be undermined by refugee populations migrating in search of more benevolent environmental and social conditions. International tensions over shared freshwater resources may be worsened by rapidly growing populations, growing demands for irrigation water, and future climatic changes. In short, regional and global environmental deficiencies will increasingly produce conditions that make conflict more likely.

In extreme cases, actual interstate violence, caused or worsened by environmental factors, may occur. Understanding this problem does not require that "international security" be redefined in the academic sense; rather it requires a better understanding of the nature of certain new threats to security, specifically the links between environmental and resource problems and international politics. If we then accept that these problems are a legitimate cause for concern, we must ask whether traditional means and institutions for resolving international political problems and conflicts are adequate for addressing them. For more detailed discussions, see Brown (1977), Ullman (1983), Ehrlich and Ehrlich (1989), Mathews (1989), and Gleick (1990b).

CLIMATIC CHANGES AND SECURITY

Among all the major environmental threats, global climatic change appears most likely to affect international politics because of its wide scope and magnitude. Every nation and region on Earth is responsible for the production of greenhouse gases; every spot on the planet will be affected by altered climatic conditions; every individual may play a role in reducing greenhouse-gas emissions.

Any analogies linking climate change with international political effects are necessarily imperfect. One problem is that the links between political behavior and climatic conditions are often tenuous, given the other relevant factors that affect such behavior. A further complication is that the magnitude and severity of future climatic changes will most likely considerably exceed what we have experienced before, making it difficult to use lessons from the past to set future policy. Nevertheless, recent major developments indicate that the problem of climatic change has reached the international political agenda (Table 21.1).

Finally, comments from many of the world's leaders, including Gorbachev, Reagan, Bush, Thatcher, Mulroney, Brundtland, von Weizacker, and others, indicate that the issue of climatic change is being discussed at the highest political levels. A series of high-level international conferences and negotiations on the subject began in the late 1980s and are continuing today. In February 1991, the United States hosted the first negotiating session on a Framework Convention on Climate Change, which is expected to be signed in 1992 in Brazil at the United Nations Conference on Environment and Development.

Climatic changes will affect more than just traditional East–West concerns. Indeed, several fundamental factors suggest strongly that North–South tensions over climatic changes will be the most problematic (Gleick 1989a). First, the industrialized countries are responsible for nearly 80% of the increase in greenhouse-gas concentrations, while having only 25% of the world's population. Although rapid population growth in developing countries will lead to large increases in energy use and greenhouse-gas emissions, this disparity in per capita production of greenhouse gases will persist for a long time. Second, the consequences of climatic changes, and many other large-

Table 21.1 Global climate change policy developments: a partial history

1972	• United Nations Conference on the Human Environment, Stockholm, Sweden UN General Assembly establishes United Nations Environment Program
1979	• First World Climate Conference, Geneva
1985	• Vienna Convention for the Protection of the Ozone Layer • 1985 "Villach" Conference: International Conference on the Assessment of the Role of Carbon Dioxide and of Other Greenhouse Gases in Climate Variations and Associated Impacts, Villach, Austria • Creation of Advisory Group on Greenhouse Gases (AGGG) by UNEP and WMO.
1987	• Montreal Protocol on Substances that Deplete the Ozone Layer • 1987 Villach/Bellagio Conference Process
1988	• Conference on the Changing Atmosphere, Toronto, Canada • Creation of Intergovernmental Panel on Climate Change by WMO and UNEP. • UN General Assembly Special Resolution No. 43/196 calling for adoption of Framework Convention on Climate Change • UN General Assembly Resolution No 43/53 calling climate change a common concern of humankind.
1989	• Hague Ministerial meeting, the Netherlands • Nairobi Conference, Kenya • G-7 Meeting, Paris, France • Tokyo Conference on the Global Environment, Japan • Cairo Climate Conference, Egypt • Noordwijk Declaration, the Netherlands (Supports a 20% goal for CO_2 reductions; US/USSR/Japan dissent.) • UN General Assembly Resolution 44/207 calling for a Framework Convention as best solution

1990	• IPCC final activities all year
	• Final IPCC documents approved Sundsvall, Sweden (August)
	• Final AGGG documents approved Stockholm, Sweden (September)
	• European Community (EC) Declaration on targets for emissions, Luxembourg (Oct 29-30)
	• Second World Climate Conference, Geneva, Switzerland
	      Scientific Meeting (Oct 29-Nov 3)
	      Ministerial Meeting (Nov 6-7)

1991	• First meeting on negotiating a Framework convention (Feb) • Conference on Climate Impacts and Society, Japan

scale environmental disruptions, will be distributed among both rich and poor countries, but there is growing evidence that climatic impacts will be felt more severely by poor countries (Tegart et al. 1990). And third, developing countries have far fewer technical and economic resources at their disposal for adapting to or mitigating the impacts of climatic changes than do the industrialized nations. These discrepancies have the potential to worsen existing tensions and to raise new disputes between industrialized and developing nations and between poorer nations themselves (Wilson 1983; Tickell 1986; Gleick 1989a, b, 1990b).

Impacts of Climatic Changes

The distribution of impacts from greenhouse warming will be global in extent and delayed in time. Among the victims of climatic change will be those nations and generations least responsible for the production of greenhouse gases, least able to either adapt to or mitigate the changes, and with little international economic or political clout. While there are many differences among countries, developing countries are particularly vulnerable to disruptions in the availability and quality of freshwater resources and mineral resources, alterations in agricultural productivity and trade, and increases in sea-level rise. These impacts may lead to severe disruptions of economic and political activities, and to disputes between neighboring nations and regions.

Agricultural productivity varies with temperature, precipitation, and water availability; water resources are sensitive to both floods and droughts and are already limited in many regions due to natural variability or high societal demand; mineral resources, including oil and gas, are found in significant amounts in regions constrained by climatic conditions; sea-level rise will affect coastal developments and the numbers of refugees crossing borders. Each of these problems is discussed briefly below.

Food

Threats to the food supplies of a country are cause for frictions and tensions between nations and have been used for political and military purposes (Schneider 1983; Wallensteen 1986). The most recent example is the embargo of Iraq in response to its invasion of Kuwait. Mechanisms for such threats include trade embargoes (such as was imposed by the U.S. on the Soviet Union in response to their invasion of Afghanistan) or other forms of political manipulation of access to food, environmental degradation such as loss of soil fertility, or competition among conflicting land uses. Because regional scarcity and dependence on imports are fundamental conditions for food to become a political tool, the extensive reliance on imported agricultural goods in the developing world strongly suggests their vulnerability to embargoes and food-related threats. Even today, natural climatic variability can cripple food production in developing countries dependent on the vagaries of rainfall. Under conditions of changing climate and growing population, this situation may grow more precarious.

Analysis of the net effect (both regionally and globally) of climatic changes on food production is complicated by the difficulties of estimating the effect of changes in yields on world agricultural markets. Indeed, short-term reductions in yields alone are not necessarily bad for overall long-term productivity and food availability. Moreover, some producers may benefit economically from short-term yield reductions, as the market prices adjust upward to account for shortages. Evaluating winners and losers requires, therefore, defining the different possible comparison criteria, such as economics, human health, other measures of well-being, etc. Other confounding factors include the size of stocks, subsequent investments in other regions, planting patterns, international prices, and the character of trading agreements.

Freshwater Resources

History provides numerous examples of conflicts over the quality or quantity of freshwater resources (Naff and Matson 1984; Falkenmark 1986; Gleick 1988, 1990a). Even in the absence of climatic changes, pressures on existing water resources are growing due to increases in population, industrial water demand, and development in semi-arid and arid regions. Where water resources are shared, as in international river basins or bodies of water bordering more than one country, the possibility of friction and conflicting demands exists. The nature of such frictions varies from region to region: from disputes over water quality in humid regions to competition for scarce resources in arid and semi-arid regions.

Regions with a history of international tensions or competition over water resources include the Jordan and Euphrates Rivers in the Middle East; the Nile, Zambezi, and Niger Rivers in Africa; the Ganges in Asia; and the Colorado and Rio Grande Rivers in North America. During 1990 alone, political disputes have flared up over the Euphrates River between Turkey, Syria, and Iraq; the Han between North and South Korea; the Brahmaputra between India and Bangladesh; and the Nile. As water demands increase, the probability of conflict over remaining water resources will also increase. Another sign that water resources may increasingly be used for military purposes was the proposal to cut off the flow of the Euphrates River to Iraq as a way to tighten the pressure on them to withdraw from Kuwait.

Future climatic changes can reduce or exacerbate these water-related tensions. Among the critical concerns are changes in (1) water availability from altered precipitation patterns or higher evaporative losses due to higher temperatures, (2) the seasonality of precipitation and runoff, (3) flooding or drought frequencies, and (4) the demand for and the supply of irrigation water for agriculture. International water law is not sufficiently developed or accepted to successfully mitigate these problems.

Northern Mineral Resources

Access to certain strategic minerals is already constrained in some regions by climatic conditions. In particular, the ability to extract oil and natural gas in Arctic continental and offshore regions depends on expensive and vulnerable methods and materials. Yet significant resources underlie these regions and they are vital elements in national economies and world trade markets. Any change in climate that affects the ease of extracting these resources could play a role in the response of international actors to initiatives to control climatic change (Gleick 1989b). Indeed, we have already seen that the presence of mineral resources, and perceptions about their value and ease of extraction, has played a major role in the positions of nations negotiating over the fate of Antarctica.

Sea-level Rise

Among the impacts of the greenhouse effect will be an increase in the level of the oceans, with subsequent increases in coastal flooding and damaging storms. Some regions, such as estuarine habitats and river deltas are particularly vulnerable, and risks are especially great where large populations reside, such as the deltas of the Nile and the Ganges/Brahmaputra. In addition to the intrusion of salt water into freshwater habitats, there is growing worry that large numbers of "environmental" refugees will be created, putting economic and political pressures on neighboring countries and worsening regional tensions.

WINNERS AND LOSERS: THE COMPLICATION OF PERCEPTIONS

There has been considerable talk in the last several years about "winners" and "losers" from climate change, although much of this has been polemic posturing rather than informed discussion. At one extreme, some individuals and organizations reject any talk about possible benefits of climate change saying that there will only be losers, while a smaller number suggest that there will be enormous, and perhaps universal, benefits (see UNEP 1990, for a discussion). In fact, determining whether there will be unambiguous winners or losers is an almost impossible proposition. How are such categories determined? Over what period of time? Over what range of environmental effects? Using what kinds of indicators and measures?

Ultimately, the problem of winners and losers, and its relevance for international security and conflict, is a problem of perceptions rather than reality, since perceptions often drive policy, and "reality" in this case will likely always be elusive.

Some important differences in the perceptions of developed and developing countries are worth noting. In the absence of definitive advance evidence of prompt and severe climatic impacts, industrialized nations are likely to prefer the "wait and see" strategy of adaptation, which produces the fewest near-term costs. The present international environmental policy of the U.S. is perhaps the best example of this problem. By ignoring or discounting the evidence and recommendations of the international scientific community and focusing on the short-term conventional economic costs of emissions reductions, the U.S. has chosen a position that slows progress on international negotiations to reduce the risks of climatic changes. This position also ignores the more poorly quantified, but potentially far greater longer-term costs of these changes.

Conversely, developing countries have the most to gain by convincing greenhouse-gas producers to cut back on emissions and the most to lose by a failure to achieve emissions reductions. Poorer nations have limited resources for responding to major climatic impacts and have fewer responsibilities for greenhouse-gas emissions. Recently, the developing nations have begun to argue forcefully that the burden for prevention should rest with those industrialized countries that bear the greatest responsibility for the production of greenhouse gases (see Cheng 1990; MaGraw 1990).

A further complication may be a desire of certain actors to capitalize on perceived regional advantages of climatic changes. If some international actors believe that they will benefit from climatic changes while others suffer, such perceptions, correct or not, will drive policy actions and decisions. The views of those with the financial and technological means at their disposal to affect the outcome or mitigate the impacts of climatic changes are especially important. Arguments for international action are complicated by individual nations taking positions dependent not on the global good, but on the perceived advantage or disadvantage to them of the likely change and impact. If these actors are among those most responsible for greenhouse-gas production, and hence most able to affect the outcome, others may be forced into an adaptation strategy.

There are two serious problems with setting policy today using forecasts of possible future benefits. First, the net societal impact depends on not just regions that benefit but on those that suffer as well. Second, the uncertainties about actual regional effects are so great that policies made on the basis of regional predictions will be both highly unreliable and difficult to justify. For example, poor conditions in one area are sometimes compensated for by more favorable conditions in another. Similarly, there is a major distinction between short-term and long-term changes. What may appear to be a short-term gain could subsequently become a long-term loss as climatic changes become more severe. There is a parallel risk to setting policy using forecasts of future costs: if the actual costs of climatic changes are less than the costs of taking actions to prevent them, society may be worse off. By late 1990, however, it was becoming increasing apparent that extensive actions could be taken to reduce greenhouse-gas emissions at either little cost or at an actual economic gain. This strongly suggests that the assumption that society as a whole will gain from certain actions to restrain climatic changes (though some individuals or regions might lose in the short or long term) is a conservative and sound assumption. Convincing those most responsible for greenhouse-gas emissions that this assumption is correct thus becomes a political rather than a scientific issue.

STRATEGIES TO REDUCE GREENHOUSE-GAS EMISSIONS

There are many strategies for reducing greenhouse-gas emissions, including technical, economic, and political options. Not every strategy to reduce greenhouse-gas emissions will affect international peace and security. Far more analysis is needed to separate out the direct and indirect impacts of these options.

Positive Outcomes

The inequitable use of resources and the inequitable responsibility for global environmental degradation threatens international relationships and, potentially, peace and security. By addressing these global problems, specifically global climatic change, we have an opportunity to reduce directly these inequities and to reduce the international disputes they engender. Two approaches have received particular attention: increased efficiency of energy use primarily in industrialized nations, and increased development and deployment of alternative energy resources everywhere. Both actions would have positive outcomes politically, by reducing the disparity in per capita energy use between rich and poor countries, and by providing energy with more environmentally sensitive technologies. Several other chapters in this volume deal with these issues more comprehensively.

Several secondary benefits would be realized from these actions, including enhancement of mechanisms for the transfer of technology and information from industrialized to developing nations, and decreases in other environmental problems associated with the way we presently find and use energy.

An additional positive outcome of efforts to reduce the risks of greenhouse-gas emissions would be the "internationalization" of the process of environmental negotiations and a broader acceptance of the rule of international law in dealing with these disputes. The recent role of the United Nations in regional dispute resolution suggests a new willingness to accept more internationalism in affairs that transcend traditional borders. Environmental problems are particularly ripe for such an approach.

Negative Outcomes

Unfortunately, international negotiations over the issue of climatic change can also lead to a worsening of tensions over these issues. Disputes can arise over the responsibility for the problem, over the costs of addressing it, and over the timing of the response. As discussed above, we are already witnessing a split in the international community over the most appropriate role for nations to take.

Additional complications can arise. Considerable details still remain to be worked out over the "tradeability" of emissions, limitations on greenhouse gases other than CO_2, who pays for technology transfer and climatic impacts, and a range of other issues. Unless care is taken, solutions may be chosen that worsen inequitable resource use and the problems of technology transfer, and increase the risk of tensions.

DISCUSSION: REDUCING THE RISK OF CONFLICT OVER CLIMATIC CHANGE

Of all the pressing large-scale environmental problems facing society, global climatic changes appear to have the greatest potential for provoking disputes, worsening tensions, and altering international relations between developed and developing countries. Several routes to conflict have been identified. Climatic changes associated with global warming will affect freshwater availability, food productivity, and access to other resources, goods, and services. The societal impacts of these climatic changes will be widely distributed, but they are likely to be felt far more severely by poorer nations with fewer resources for adaptation and prevention (see also Meyer-Abich, Chapter 22, this volume). Methods are available to reduce the severity of these impacts. For world food markets, for example, many methods are already used to reduce the severity of climatic variability. These include food-storage programs, distribution systems, and famine identification and assistance plans. Each of these mechanisms could be considerably strengthened and expanded. Long-term options include the development of climatically resilient strains of crops, better long-term forecasting abilities, and strategies for reducing the vulnerability of mono-cropping to specific climatic events.

Where existing tensions may be exacerbated by climatic change, such as in disputes over water resources, advances are needed in both conflict resolution among states and in the development of international resource law. The International Law Commission, for example, has been debating questions of international water law for twenty years, without reaching any final formulation. Efforts to develop general principles concerning resource use and competition could be considerably strengthened and accelerated. Such advances would be useful not only for resolving international resource controversies, but for addressing the very issue of future climatic change.

The inequitable responsibility for the production of greenhouse gases has begun to raise important and still unresolved questions about equity, fairness, and international environmental ethics, and to provoke calls from developing countries for prompt actions to reduce these emissions. With few exceptions, developing countries are least able to mobilize the resources for adapting to severe impacts, while developed countries are more likely to prefer to adapt to climatic changes than to alter their energy and industrial infrastructures drastically.

Finally, a wide range of policy options is available to reduce the severity of climatic changes, but these options themselves can have both positive and negative international ramifications. This fundamental dichotomy is likely to lead to international frictions and disputes between developed and developing countries, unless mechanisms can be worked out to more equitably share the costs of limiting climatic change.

REFERENCES

Brown, L.R. 1977. Redefining national security. Paper No. 14, 46 pp. Washington, D.C.: Worldwatch Inst.

Cheng, Z.-K. 1990. Equity, special considerations, and the third world. *Colorado J. Intl. Env. Law and Policy* **1(1)**:57–68.

Ehrlich, P., and A. Ehrlich. 1989. The environmental dimensions of national security. In: Global Problems and Common Security, Annals of Pugwash, 1988, ed. J. Rotblat and V.I. Goldanskii, pp. 180–190. Berlin: Springer.

Falkenmark, M. 1986. Fresh waters as a factor in strategic policy and action. In: Global Resources and International Conflict: Environmental Factors in Strategic Policy and Action, ed. A.H. Westing, pp. 85–113. New York: Oxford Univ. Press.

Gleick, P.H. 1988. The effects of future climatic changes on international water resources: the Colorado River, the United States, and Mexico. *Policy Sci.* **21**: 23–39.

Gleick, P.H. 1989a. Climate change and international politics: problems facing developing countries. *AMBIO* **18(6)**:333–339.

Gleick, P.H. 1989b. The implications of global climatic changes for international security. *Climatic Change* **15**:309–325.

Gleick, P.H. 1990a. Climate changes, international rivers, and international security: the Nile and the Colorado. In: Greenhouse Glasnost, ed. T.J. Minger, pp. 147–166. New York: Ecco Press.

Gleick, P.H. 1990b. Environment, resources, and international security and politics. In: Science and International Security: Responding to a Changing World, ed. E. Arnett, pp. 501–523. Washington, D.C.: Am. Assoc. for the Advancement of Science (AAAS).

Houghton, J.T., G.J. Jenkins, and J.J. Ephraums, eds. 1990. Climate Change: The IPCC Scientific Assessment. Prepared for the Intergovernmental Panel on Climate Change by Working Group I. WMO/UNEP. 365pp. Cambridge: Cambridge Univ. Press.

MaGraw, D.B. 1990. Legal treatment of developing countries: differential, contextual, and absolute norms. *Colorado, J. Intl. Env. Law and Policy* **1(1)**:69–99.

Mathews, J.T. 1989. Redefining security. *Foreign Affairs* **68(2)**.

Naff, T., and R.C. Matson, ed. 1984. Water in the Middle East: Conflict or Cooperation. Boulder, CO: Westview Press.

Schneider, S.H. 1983. Food and climate: basic issues and some policy implications. In: World Climate Change: The Role of International Law and Institutions, ed. V.P. Nanda, pp. 46–63. Boulder, CO: Westview Press.

Tegart, W.J., G.W. Sheldon, and D.C. Griffiths, eds. 1990. Climate Change: The IPCC Impacts Assessment, (Working Group II), WMO/UNEP, 230pp. Canberra: Australian Government Pub. Service.

Tickell, C. 1986. Climatic Change and World Affairs. (rev. ed.) Center for Intl. Affairs, Harvard Univ. Lanham, MD: Univ. Press of America.

Ullman, R.H. 1983. Redefining security. *Intl. Security* **8(1)**:129–153.

UNEP (United Nations Environment Programme). 1990. On assessing winners and losers in the context of global warming, ed. M.H. Glantz, M.F. Price, and M.E. Krentz. Report of the workshop, St. Julians, Malta, 18–21, June. Boulder, CO: Environmental and Societal Impacts Group, Natl. Center for Atmospheric Research.

Wallensteen, P. 1986. Food crops as a factor in strategic policy and action. In: Global Resources and International Conflict: Environmental Factors in Strategic Policy and Action, ed. A.H. Westing, pp. 143–158. Stockholm Intl. Peace Research Inst. U.N. Env. Prog. Oxford: Oxford Univ. Press.

Wilson, T.W., Jr. 1983. Global climate, world politics and national security. In: World Climate Change: The Role of International Law and Institutions, ed. V.P. Nanda, pp. 71–78. Boulder, CO: Westview Press.

22

Winners and Losers in Climate Change: How Will Greenhouse-Gas Emissions and Control Strategies Influence International and Intergenerational Equity?

K.M. MEYER-ABICH
Kulturwissenschaftliches Institut im Wissenschaftszentrum NRW, 4300 Essen, F.R. Germany

ABSTRACT

Equity is not a term which suggests a description of the present international relations or of the intergenerational terms of community. It is rather a goal, an urgent one in view of the current inequities. The question that arises, therefore, is how are our chances of meeting this goal in the future, with respect to future generations, going to be affected by the expected climate change as well as by responses to climate change, or to corresponding forecasts. This chapter takes up this question and concludes that international as well as intergenerational inequities must be expected to increase with climate change as well as with probable responses to climate change. This implies that humankind's political organization for the time being is inadequate to meet the challenge of anthropogenic climate change.

PRESENT INTERNATIONAL EQUITIES AND INEQUITIES

Industrialization has brought hitherto unknown wealth to some parts of the world. Using the gross national product (GNP) as a first approximation of wealth, in 1987 the 19 high income OECD industrial market economies (15% of the world's population) had an average income of U.S.$ 14,670 per capita, while the 42 poorest countries (56% of humankind) received only U.S.$ 290 (Weltentwicklungsbericht 1989, p. 194f). This is a relation of 50:1. Moreover, income distribution in western-type democracies is more even than in most countries of the Third World. Assuming a factor of three between the average income of the wealthier people and the general average in western Europe's welfare states, for the poorest countries the corresponding figure between the average poor and the national average may be about six. Thus incomes differ globally by a factor of about 1,000. Even if the GNP is a very doubtful indicator of wealth and equity is not taken to mean equality, considering humankind as a community, this is certainly not a situation of equity. Two centuries ago, Adam Smith (1759) still thought: "Take the whole earth at an average, for one man who suffers pain or misery, you will find twenty in prosperity and joy, or at least in tolerable circumstances" (Smith 1969, p. 238 [part III, chapter 3]). Even without the quantitative statistical figures nobody can feel like that in our times.

After three decades of "development," this result is not particularly encouraging. Even for the future there are no signs of change. In fact, contrary to current notions that the rich should help the poor through "development aid," the transfer of wealth between the industrialized and the "developing" countries since 1983 has reversed to a situation reminiscent of colonial times. Official figures are given in Figure 22.1 (WDR 1988, p. 30). Others estimate that in 1984–1988 the net transfer of resources from the Third World to the industrialized countries amounted to 140 billion U.S.$ (Deutscher Bundestag 1990a, p. 243).

Colonialism also still prevails, in that mainly raw materials are transferred from Third World to industrialized countries. Since, apart from oil, prices of raw materials have generally fallen since 1980 (Weltentwicklungsbericht 1989, p. 14) while interest rates rose, the share of Third World's exports devoted to interest and amortization payments rose from 8% in 1975 to an unprecedented level of 32.5% reached in 1986 (Deutscher Bundestag 1990a, p. 208). By now many countries of the Third World are exploiting or at least overstraining their natural resources to make the rich even richer. This is done at the expense of (a) the present population, (b) future generations, and (c) the connatural world.[1]

Rising international inequities between the rich and poor parts of the world are frequently said to result from population growth or the indebtedness of Third World countries. Both reasons can, however, be questioned.

---

[1] I introduced the term connatural world (*natürliche Mitwelt* in German, (Meyer-Abich 1984) to denote that we are part of nature and that the other parts belong *with* us in the community of nature. They are not *around* us to serve our purposes, as suggested in calling them our "environment."

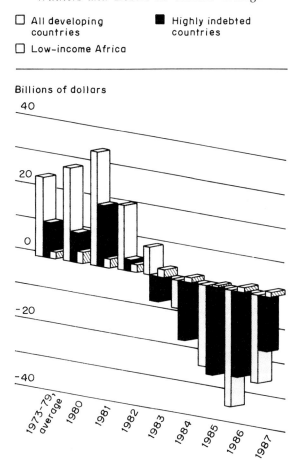

Figure 22.1 Net resource transfers between developing and industrialized countries, 1973–87. Note: Net resource transfers are defined as disbursements of medium- and long-term external loans minus interest and amortization payments on medium- and long-term external debt

Population Growth

It is true that in the industrialized countries, and perhaps also in China, population growth is 0–1%/yr, while Third World figures are between 2–3%/yr (Southeast Asia 2.2%, Latin America 2.3%, India 2.4%, Middle East 2.8%, Africa 2.8%; in 1986 [Deutscher Bundestag 1990a, p. 219, tab. 7]). Equity, however, is not affected by the fertility of the poor but by the consumption of the rich, and this amounts to about 80% of the world's steel consumption (even more for other metals) or to about three-quarters of humankind's energy consumption for less than one-quarter of today's world population. If we ask the question whether there are too many people, the most obvious conclusion is that there are too many rich people, since, if their standard of living were generalized to all of humankind, nature definitely could not endure us any more. As will be shown below, about 75% of the problems of climate change are caused by the 25% of humankind without population growth. The remaining 75% of humankind are responsible for only about 25% of the problem but will have to suffer the implications.

Indebtedness

This also is not the real issue. If a credit is used for an investment that brings enough profit to pay interests and amortization, there is no debt problem. The present debt crisis has emerged because:

- Interest rates in the 1980s rose above 10 or 15%/yr, caused mainly by the U.S. fiscal policy. This refers to market conditions, but in 1983 these applied already to about 80% of all Third World credits. Also a credit is already termed "development aid" when its cost is 25% below market conditions and this reduction is often paid for by restrictions in the availability of the money;
- Oil prices rose and solar alternatives were not developed soon enough;
- Unproductive investments were made, particularly for military armament and to some extent also for luxuries. However, regarding the "flow back" of capital to the industrialized world through armament expenses, it is not only the developing countries which must be held responsible for them. A shared responsibility is even more true for many industrialization projects which "secured jobs" in some industrialized countries and brought advantages to some people in a developing country without contributing to the general welfare. Specifically, I am thinking of projects like the Balbina dam in Brazil, which seems to be the world's most inefficient hydropower plant (0.06 W/m^2, several hundred times less efficient than a photovoltaic cell), or of iron ore mining in Carajas, Brazil. These two examples are in Brazil; however, similar cases can be found anywhere that has been hit by "development." In Carajas, 35 Mt of iron ore were to be produced in 1989. In the preceding year, 10 Mt were exported to the European Community, with a further 10 Mt to Japan. Part of the production is processed in Carajas itself in blast furnaces of German origin, heated with charcoal from the Amazon tropical forest (Deutscher Bundestag 1990a, p. 186f.) Part of the production is supposed to be exported as well. It is, therefore, easily seen how expensive "cheap steel" can be. The program Grande Carajas covers 900,000 km^2, of which 70% is currently tropical rain forest.

Without these factors, debts might not have led to a debt crisis.

My general conclusion is to suggest that we do not point primarily at population growth, or at the debt crisis, but at the underlying notion of development. The idea that the whole world should be subjected to one type of development has only replaced the colonial attitude of the Third World countries being "below" the European countries by the idea that they are economically "behind." This idea preserves the former hegemony (Sachs 1990) in a form which fitted into the "century of economics" (von Weizsäcker 1989). However, isn't it a strange idea to consider living beings as being subject to development in the transitive sense: to be developed instead of developing themselves?

Within industrialized countries, the disruption of the connatural world shows that goals must be set to the economic process which cannot be reasoned economically. The economy is to serve cultural developments, including political culture in the community of humankind as well as in the community of nature; however, economic reason is not by itself devoted to culture. Within a socially and ecosocially given frame, market economies seem to be the relatively best economic organization. The exploitation of human labor as well as of the connatural world, however, shows that this frame has been neglected politically in the industrialized world. The same is seen to be true in its dealing with the countries of the Third World.

To cope with the international problem is even a much greater challenge than the national issues because there is no international sovereign. One possible policy is to shift foreign investments in Third World countries from private to public responsibility. In investments, however, governments are not necessarily better than private companies. It seems to me that political institutions would do better in confining themselves to set political limits to the economic process and not to substitute governmental for private activities. The political conditions then must be to ban exploitation of Third World countries by any activity originating in one's own country. This does not mean to interfere with other countries' sovereignties, but to refrain from any export or import exchange which does not seem to be justified with a relatively weak partner. And weak they are: three-quarters of humankind contribute less than 20% to international trade (Journalisten Handbuch 1989/90, p. 217), and yet are highly dependent on international exchange.

Next, poisonous wastes can no longer be allowed to be deposited in Third World countries; the export of environmentally or socially hazardous technologies should no longer be permitted. The import side equally deserves to be assessed. For instance, consumption patterns of the rich are indirectly co-responsible for shifting cultivation, as the apparent cause of the destruction of the tropical forests. In too many cases, commercial relations between industrialized and "developing" countries are not responsible or consistent with nondisruptive market exchange. Generally speaking, if a poor man asks a rich man to hurt him and to refund him for doing so, it is the rich man's responsibility to refuse the deal. His further-reaching responsibility is to offer a better alternative. Using this analogy, in terms of climate change, developed countries appear to not even satisfy the first condition, directly doing damage to the poor without refunding them.

PRESENT INTERGENERATIONAL EQUITIES AND INEQUITIES

Intergenerational inequity is increased by present-day humankind living at the expense of its "capital," the integrity of the connatural world, and the sociocultural integrity of the industrial as well as of the "developing" societies at the same time. Presently, no country seems to meet what might be termed an intergenerational Pareto criterion, namely to proceed at nobody's expense, not even in the future. All over the world, humankind is living (1) at the expense of the connatural world, or the community of nature, and (2) at the expense of future generations, or the community of humankind. In the Third World this is even more so because the industrialized countries live at its expense. Intergenerational equity, therefore, seems to be most endangered in the Third World.

To underestimate future goods in industrialized countries tends to be justified by "discounting" the future. This, however, is a misunderstanding of the applicability of discount rates. Future goods may be discounted to a lesser value at present when a productive investment and future activities are expected to yield profits, which in the course of time make up for the difference. With respect to present-day credits taken on future accounts, however, neither the corresponding investments are made nor the necessary activities launched.

IMPLICATIONS OF CLIMATE CHANGE FOR HUMANITY

So far the efforts to describe oncoming climate changes have been mainly directed towards climatology. On the one hand, climate modeling has reached a fairly advanced level, including atmospheric–oceanic interaction. Also regionalized scenarios may be expected in the near future. On the other hand, classical climatological reasoning continues to provide sound judgment (Flohn 1990) and should be further applied in modeling simulations. While in the 1970s climatological predictions of anthropogenic climate change were still quite varied, an impressive consensus has developed in the 1980s.

However, even if we had a complete description of climatic developments in terms of climatology (temperature, pressure, and humidity), this would still not allow a judgment with respect to the impacts of such changes and even less to their desirability or nondesirability. Therefore, political implications must be based on further reasoning. Particularly, if some climate change has been described in the language of climatology, this description must be so-to-speak translated into terms of:

- ecology, i.e., into changes in the habitat of different species, or ecosystems;
- economy, i.e., into advantages or disadvantages and their implications for particular sectors like construction, tourism, and agriculture;
- political science, i.e., into advantages or disadvantages and their implications for particular groups, or interests, and countries.

The nice thing about these different descriptions is that they all refer to the same fact, some expected climate change, like one and the same object which may be seen in different colors by means of a prism. It also turns out that there is no one person who is expert on all matters of climate change. Those, for instance, who understand climate change economically have in their terms an equally comprehensive knowledge as the climatologists in scientific terms. The concept of interdisciplinary translation is further developed in Meyer-Abich (1988).

Descriptions of climate change in terms of ecology, economy, or political science have come to be called impact studies. This wording is somewhat unfortunate because in an anthropogenically caused climate change, the impact is not directed from nonhuman natural developments to humankind but the other way around; thus humankind is making an impact on itself by means of the rest of the world. Compared to climatological studies, impact studies (in the general sense) have been enormously neglected so far. To put it in terms of research funds: while expenses for climate modeling have reached the billion dollar level, those for impact studies have been left almost two orders of magnitude behind. Even with limited funds, however, considerable progress has been made by the United Nations Environment Program (UNEP)–IIASA study (Parry et al. 1988) and the report of the Intergovernmental Panel on Climate Change (IPCC), set up in 1989 jointly by the UNEP and the World Meteorological Organization (WMO) (Tegart et al. 1990). The description of the implications of climate change for humanity in this section of my paper draws mainly on the IPCC report, particularly on the impact assessment.

The following considerations refer to a climate change scenario in which an effective doubling of CO_2 in the atmosphere between now and 2025 to 2050 implies:

- an equilibrian increase of global mean temperature in the range of 1.5°C to 4.5°C and a 1°C increase over current temperatures by the year 2030;

- an unequal global distribution of this increase with the tropical regions warming by about half, and the polar regions by about twice, the global average;
- a global mean sea-level rise of 0.3 to 0.5 m by 2050 and about 1 m by 2100.

These changes go together with a shifting of climate zones to the poles and to higher elevations, with modifications in ocean circulation (and, therefore, in carbon dioxide [CO_2]) absorption, and with changes in the variability of climate (frequency and intensity of exceptional weather). Further details are given in modeling simulations.

The following analysis refers to a number of individual predictions given in the IPCC impact assessment. Considering the state of the art in regional forecasting, it is fairly improbable that these predictions bear much certainty individually. The overall picture, however, may be taken as a good indication of how the world may be affected by climate change. While this paper is not concerned with changes in individual countries but with aggregate impacts on different parts of the world, references to individual countries may still serve (1) as illustrations for concreteness, (2) as an indication of risks (see section on CLIMATE CHANGE IN TERMS OF HUMANITY below), and (3) as a reminder that in any case global developments will ultimately spell down to individual situations and interests.

Climate Change in Terms of Ecology

The next step in making an evaluative judgment of the implications of climate change is to describe the presumed change in terms of ecology. The poleward shifting of climate zones now means that the boundaries of vegetation zones (e.g., boreal tundra, temperate forests, grasslands, etc.) may be expected to shift correspondingly several hundreds of kilometers over the next 50 years. For instance, in the northern parts of the Asiatic USSR, the vegetation zone may move 500–600 km to the north. Permafrost degradation may allow the growth of deeper rooted, broadleaved species and the establishment of denser forests of coniferous species. Again, in the European USSR the forest-steppe subzone might change while in southern portions of western Siberia the forest-steppe boundary may move up to 200 km. More to the south, in the arid zones of the Mediterranean, climate change may result in desertification of the North African and near Eastern steppes.

Considering agriculture, its climatic limits in some parts of the world are estimated to shift poleward by 200–300 km per degree of warming. Upward shifts in the mountains could be in the order of 150–200 m. Of course, the given soils as well as the terrain may not allow to take any advantage from this change. Also the climatic conditions, both at high-latitude and low-latitude boundaries of temperate and northern forests, may shift hundreds of kilometers poleward, but the forests may not be able to follow their climate because the change will occur too quickly. After all, temperatures are expected to rise between 15 and 40 times faster than in past glacial–interglacial transitions.

Generally speaking, different climate regimes may be more or less hospitable for terrestrial ecosystems so that they will bring advantages to some and disadvantages to others. To what extent plants will be able to migrate with the climate again depends on its rate of change. The rates of the movement of species are in the range of only 10–100 m/yr, depending on specific abilities to disperse and on the existence of geographical barriers. The rates of projected climate changes are likely to be faster than the ability of many species to respond. Of course, corresponding to the degree of warming, the changes in vegetation zone are projected to be greatest in polar deserts, tundras, and boreal forests.

The expected changes in climate are so large and fast that a severe disruption of ecosystems must be expected in any case. Even if some species might spread (for instance, *Melalenca quinquenervia*, a bamboo-like Australian plant) others will become less viable or become extinct. An example given by IPCC is the hard beech (*Nothofagus truncata*): a 3°C rise in temperature would increase annual respiratory carbon losses to exceed the total annual amount allocated to stem and branch growth for this species. Global biological diversity is, therefore, expected to decrease as a result of climate change. Also ecosystems will not move as units but change their structure when species distributions and abundances are altered. These changes could be enhanced by increased pest outbreaks and fire.

Most sensitive to climate change are (1) species at the edge of (or beyond) their optimal range; (2) geographically localized species (e.g., on islands, mountains, or in parks and reserves); (3) genetically impoverished species; (4) specialized organisms with specific niches; (5) poor dispersers; (6) very slowly reproducing species; and (7) localized populations of annual species. Particularly at risk are, therefore, montane and alpine, polar, island, and coastal communities as well as heritage sites and reserves.

Considering the oceans, the sea ice, for instance, is an essential element of the habitats of many marine mammals (e.g., seals, polar bears, penguins) and birds, and its reduction will also deprive land animals of migratory and hunting routes. In general, global warming will change the thermal budget of the world's oceans and shift the global ocean circulation, also affecting its capacity as a sink of atmospheric heat and CO_2.

From numerical experiments with General Circulation Models of the ocean–atmosphere system as well as palaeo-oceanographical data, IPCC suggests that global warming would lead to lower intensities of oceanic upwellings because of the decreasing meridional temperature gradient. The corresponding decrease in the productivity of marine ecosystems may be partially compensated for by some increase in the intensity of coastal upwellings as a result of an increasing temperature difference between land and water surface. Increased temperatures at high latitudes will also lead to increased productivities. As a result, a redistribution of productive zones must be expected.

Where land and water meet, too rapid a sea-level rise could reduce or eliminate many coastal ecosystems, drown coral reefs, reduce biological diversity, and disrupt the life cycles of many economically and culturally important species. Wetlands like salt, brackish and fresh marshes as well as mangrove and other swamps are vital to the ecology and economy of coastal areas. Also coral atolls are important ecological habitats with high biodiversity. If the rate of sea-level rise exceeds the maximum rate of vertical coral growth (8 mm/yr), inundation and erosive processes may begin to dominate.

The foregoing short description of climate change in terms of ecological habitat must be subjected to an evaluation as to whether and to what extent it could be justified to bring about those changes, keeping in mind that they are not occurring by themselves in the history of nature but are human-induced. We cannot, therefore, evade the question to what extent the world may be taken as human property. Does not this globe also belong to other species in their own right? I think so (cf. Meyer-Abich 1984) but will not dwell on this argument because this paper is concerned with equity within humankind.

Climate Change in Terms of Humanity

Implications of climate change for humankind concern different countries, men and women, the rich and poor, different professional groups, owners and tenants, different communities, and do not affect all of them equally. On the contrary, even the smallest changes generally go together with advantages for some and disadvantages for others. Particularly, the most basic experience in politics is that no action is equally in everybody's interest or disinterest, so that some will be in favor of it while others are against it. Considering this elementary fact of life, nobody should expect that such a far-reaching process like climate change will have equal implications for everybody all over the world. It must be assumed, therefore, that there will be advantages for some and disadvantages for others, or advantages as well as disadvantages for everybody but not to the same extent. Even if, as frequently believed, in the long run the disadvantages will outweigh the absolute advantages for everyone, it must be expected that some will have lesser disadvantages than others, and that these will still be *relative* advantages. *Absolutely or relatively, there will be winners and losers.*

With respect to climate change, it is often emphasized that we are all sitting in the same boat, but this is exactly what one must expect to hear from those who are looking forward to being the winners. The one-boat metaphor, although well-meant, should not be accepted, therefore, without reservations. On the other hand, confidence in regional estimates of critical climatic factors is low so that, for instance, precipitation and soil moisture cannot yet be reliably predicted, and data on the regional effects of climate change on crop yields, livestock productivity and production costs or other economic parameters cannot be given so far. To indicate such data would be the simplest way of assessing winners and losers. This is, however, not the only way. Instead, *risks* can be assessed; even if the economic data were available, *vulnerabilities* need also to be taken into account.

Most vulnerable to climate change are those countries which:

- heavily depend on agriculture since this is obviously the economic sector most dependent on climate;
- cannot help themselves easily when agricultural production is detrimentally affected;
- already suffer from droughts or otherwise depend on climate variability;
- will suffer from flooding when the sea level rises.

Some or all of these conditions apply to the Third World. By contrast, some industrialized countries depend on climate to some extent but can in any case help themselves. Expected flooding may even prove to be an incentive for some economic sectors. Already from this general analysis it is easily seen that the Third World once more must be expected to be on the loser's side in oncoming climate change, while the industrialized countries again tend to be the winners, at least relatively.

This is shown to be even more so when the general patterns of anthropogenic climate change are taken into account. Again, in this discussion, I refer mainly to the impact part of the IPCC report (Tegart et al. 1990). While individual forecasts in the IPCC analysis generally do not bear any certainty, the analysis may to some extent still be taken as an assessment of risks (probable losses) rather than a forecast of what will definitely happen. One is doing harm to others, however, not only by inflicting a definite (deterministic) loss upon them, but by exposing them to a (probabilistic) risk as well. Risks are facts, particular kinds of entities existing in the world. For instance, when I am at a risk of losing my life in a specific situation, I am better off as soon as the risk is over. Also the value of an economic good is reduced when a risk is burdened upon it. As soon as low-lying countries such as the Maldives, Tuvalu, and Kiribati are exposed to the risk of disappearance as sea levels rise, this risk in itself is harmful for them because property is depreciated and they will have different conditions for long-term credits. Thus apart from deterministic damages, *risks are a second category of doing harm to others.* In the following analysis I deal with both categories together, because distinctions between damages expected with certainty or with significant probability are not sufficiently warranted by current knowledge.

The IPCC impact report leaves no doubt that the implications of climate change will be regionally different and is surprisingly explicit with respect to impacts on specific regions. Distributional inequalities depend on the following factors:

1. Throughout the world "the most vulnerable populations are in developing countries, in the lower income groups, residents of coastal lowlands and islands, populations in semi-arid grasslands, and the urban poor in squatter settlements, slums and shanty towns, especially in megacities."
2. "Change in drought risk represents potentially the most serious impact of climate change on agriculture at both regional and global level." With respect to the socioeconomic implications in agriculture and forestry, "it is concluded that climate change could more likely exacerbate most current and near-term issues and tensions rather than relieve them."
3. Generally there is no doubt that changes will differ among countries and that some countries are better able than others to cope with the impacts. Particularly, "the capacity of developing nations to adapt to likely climate changes and to minimize their own contributions to it through greenhouse-gas emissions is constrained by their limited resources, by their debt problems, and by their difficulties in developing their economies on a sustainable and equitable basis." This implies a special responsibility of the industrialized countries even apart from the question of who is causing climate change. The high vulnerability of the Third World with respect to climate fluctuations has already been shown in several cases by the IIASA–UNEP study (Parry et al. 1988).
4. "The impacts of climate change on human settlement and related socioeconomic activity, including the energy transport and industry sectors, will differ regionally, depending on regional distribution of changes in temperature, precipitation, soil moisture, patterns of severe storm, and other possible manifestations of climate change." Particularly, global warming can be expected to affect the availability of water resources and biomass, both major sources of energy in many developing countries. "These effects are likely to differ between and within regions with some areas losing and others gaining water and biomass." Climate change will also affect the regional distribution of other renewable energy resources such as wind and solar power.

5. Further distributional effects will be that:
   - an increase in average temperature and any change in the distribution of seasonal wetlands will alter the temporal and spatial distribution of diseases such as malaria, filariasis, and schistosomiasis since wetlands (particularly seasonal wetlands in warmer regions) provide habitat for the breeding and growth of vectors of these diseases;
   - the predicted change of the patterns of cloud cover, stability in the lower atmosphere, circulation, and precipitation could concentrate or dilute pollutants and change their distribution patterns and transformation rates in regional or local sectors;
   - changes in coastlines will mean that some people get beach-front property while others lose it.

   These are just a few examples to which many others could be added by further studies on the relevance of climate for human habitats.
6. The effects of changes in ocean circulation, upwelling zones, etc. will vary geographically with changes in habitats, a decrease in biological diversity, and shifts in marine organisms. As a result of these changes, a redistribution of productive zones must be expected. Regional shifts in fisheries will have major socioeconomic impacts.

Beyond these general statements, the IPCC report indicates specific changes in different regions of the world. These are not generally detrimental, or even catastrophic.

Implications for the Industrialized Countries

Evidently the poleward shift of climatic zones may bring a warmer climate and, therefore, advantages to human living conditions in the higher latitudes. The course of history has been such that the industrialized countries have developed in these latitudes. They may, therefore, become beneficiaries of climate change to some extent. This has been generally confirmed by some of the IIASA–UNEP case studies (Parry et al. 1988) and is again pointed out in the IPCC analysis. Generally speaking, towards the northern edge of current core agricultural and forestry regions, warming may enhance the productive potential in climatic terms. Northern Europe and the higher latitudes of North America as well as of the USSR and Japan could considerably benefit from these increases, even if they are limited by soils and terrain.

Apart from agriculture and forestry, which are only of minor importance for the economic welfare of industrialized societies, considerable benefits may emerge through more favorable living conditions such as more sunshine and warmer winters. Together with the cultural environment, these general factors of human habitat (including recreation possibilities) are becoming more and more important for business decisions (where to settle or where to expand). Just as, for instance, a century ago the industrial economy developed where resources like coal and iron were mined, the economic shift from material and energy centeredness to communication and information technologies is accompanied by an economic trend to siting in clean and culturally, as well as climatically, attractive regions. Presently in central and northern Europe, for instance, these comparative advantages for the south have led to migration from the other parts. Climate change seems to imply that the privileged regions will be extended further to the north.

A surprising result of the IIASA-UNEP study (Parry et al. 1988) indicated that not only could less damage be expected for the industrial economies than for the Third World but that even advantages could come about with climate change. For instance, it is expected that:

- Crop yields estimated for doubled atmospheric CO_2 concentrations will be above the present-day average, a result that can be attributed to large predicted increases in precipitation, offsetting the increased water demand from crop transpiration caused by markedly higher temperatures. Parry et al. (1988) suggest that there may not be a sufficient capacity to collect and store the greater output;
- The relative profitability of different crops will change, and thus the optimum patterns according to which they are grown in different regions. With CO_2 doubling, some crops can be expected to do better than others. For instance, in the Moscow region of the USSR, crops with greater thermal requirements (such as maize) may show a greater increase in yield than cold-region crops (such as potatoes and oats);
- The enhanced growth rates of boreal (northern) forests in the long run may lead to increased supply of forest products on the world market.

Correspondingly, a similar study of the U.S. Environmental Protection Agency (EPA 1988) refers to economic modeling results, showing that the production capacity of U.S. agriculture appears to be adequate to meet domestic needs, even under the more extreme climate change scenarios. "Only small to moderate economic losses are estimated when climate change scenarios are modeled without the beneficial effects of carbon dioxide on crop yields." Of course, agricultural production will have to shift northward, for instance to Minnesota, Wisconsin, and northern Michigan. The farmers in Appalachia, the Southeast and the northern Great Plains in the U.S. will not be happy about this. However, as long as the discriminated South is part of the same country that benefits from gains in the North, this is a national problem which does not ask for too much solidarity compared to the current status of political ethics.

A clear-cut advantage is predicted by the IIASA-UNEP study for Iceland.

The model calculation for a doubled CO_2 concentration in the atmosphere predicts that the mean annual temperature and precipitation will be 4°C and 15% above the 1951–1980 baseline; the onset of the growing season is brought forward by 48 days; hay yields increase by 66% and pasture yields by 49 to 52%; the sheep-carrying capacity of improved grassland increases by 253% and of range lands by 64 to 66%; lamb carcass weight increases by 12% and sheep-feeding requirements decrease by 53%. Fairly pleasant expectations, if one accepts these preliminary calculations, can also be held for Finland and Japan. In both of these countries, again, the north appears to be comparatively better off than the south. For instance, in northern Japan the rice-growing area could be doubled.

With respect to the USSR, the study also suggests fairly favorable developments with climate change. For instance, in the central region, winter-wheat yields are expected to increase by 30% while barley yields decrease 4%. Estimates also suggest a substantial (17%) increase in spring-wheat yields for the northern part of the country. I would not be surprised, therefore, if the fairly positive assessment of the oncoming climate change, which the well-known Russian climatologist, M.I. Budyko, gave at the Hamburg Congress on Climate and Development in November, 1988, had some political relevance in the USSR (Budyko and Sedunov 1988).

Considering the oceans, an economic benefit of warming would be improved access for shipping if there is a reduction of sea ice. The IPCC report also points out that biodegradation processes will increase by up to 30–50% in higher latitudes. This would accelerate bacterial and photochemical degradation of pollutants so that their residence time in the marine environment could be reduced.

However, the IPCC message is far from M. Budyko's optimism, that a new Golden Age might emerge with climate warming. A more favorable human habitat cannot be denied as a general long-term perspective, but as soon as one looks at the changes which must be expected in the near future the probable advantages are balanced by numerous disadvantages:

- Current stands in boreal forests will mature and decline during a climate to which they are increasingly more poorly adapted. Owing to physical stress, mortality will increase, as will the susceptibility to insects, diseases, and fires.
- In the northern midlatitude regions, where summer drying may reduce productive potential (e.g., in the south and central U.S. and in southern Europe), yield potential in agriculture is estimated to fall by 10–30% under a doubled CO_2 climate by the middle of the next century.
- In western Europe, southern U.S., and western Australia, surplus production of cereals could be reduced so that exports are at risk.

- Horticultural production in midlatitude regions may be reduced owing to insufficient accumulated winter chilling.
- Temperature increases may extend the geographic range of some insect pests, diseases, and weeds, allowing their expansion to new regions as they warm and become suitable habitats.
- Vector-borne and viral diseases such as malaria, schistosomiasis, and dengue can be expected to shift to higher latitudes under warmer climatic conditions.
- General circulation model simulations for Europe show that precipitation and runoff may increase in the North (possibly causing flooding problems) while the Mediterranean countries of Europe may experience a decline in runoff, thereby increasing the already serious and frequent water supply shortages there. Agriculture will probably suffer the most adverse effects.
- Flooding problems could arise in many northern rivers of the USSR. Increased runoff would notably improve the water quality of the Caspian Sea but not suffice to save the Aral Sea.
- In the U.S. the arid and semi-arid regions, like similar areas in the world, may be most severely affected by global warming. A special study has shown that the Delaware River's runoff would decline by 9–25%, causing problems for the water supply of Philadelphia and New York.
- Further water circulation effects are expected to be prolonged periods of droughts and shorter periods of intense precipitation in Japan, an expansion of the summer-dominant rainfall area of Australia and considerably more rain in Great Britain.
- Of course, the area of snow cover (seasonally 62% of the Eurasian continent and all of North America 35°N) is expected to decrease with global warming. Reductions in both the temporal and spatial seasonal snow cover will again inadvertently affect the availability of water. Particularly sensitive may be areas such as the Alps and Carpathians, the Altai mountains of Central Asia, the Syr Dar'ya and Amu Dar'ya region of the USSR, the Rocky Mountains, and the North American Great plains, all of which are dependent on snowmelt for the majority of their spring and summer water resources. Changes in snow cover will also affect tourism and recreation-based industries and societies.
- Presently about 20–25% of the land surface of the Earth contains permafrost. Although in the permafrost region, global warming may result in expansion of human settlement poleward, thawing of the permafrost may also disrupt infrastructure and transport and adversely affect the stability of existing buildings as well as the conditions for future construction. The socioeconomic consequences of these changes should not be underestimated. It is expected that a 2°C global warming would shift the southern boundary of the climatic zone currently associated with

permafrost over most of Siberia north and northeast, by at least 500–700 km. However, the southern extent of permafrost will lag behind this, moving only 25–50 km in the next 40–50 years. During this period the depth of the active layer is expected to increase by about 1 m.
- Last but not least, as the sea level rises, much of the infrastructure in low-lying urban areas may be affected, requiring major engineering design adjustments and investments. In particular, storm water drainage and sewage systems of many cities will be affected. Coastal protection structures, highways, power plants, and bridges may require redesign and reinforcement to withstand increased flooding, erosion, storm surges, wave attack, and seawater intrusion.

All of these changes are undesirable; many of them would inflict high costs on the countries concerned. As long as the West Antarctic ice sheet does not collapse, however, none of them seems to pose a catastrophic threat. In fact, all of these problems have, in principle, been successfully handled in the past, though not as quickly as countries should respond to climate change. In some cases, losses may already have been balanced by gains elsewhere, during this transition period to the future climate, e.g., agricultural losses in the southern U.S. may have been balanced by gains in the northern states. In other cases, infrastructures in need of renewal or modernization, with respect to climate change, can be adapted slowly over decades. In any case, the actual costs will represent only a fraction of what has been spent for military armament. Therefore, if some of the industrialized countries decided that their poleward expansion was an economic incentive for approaching new frontiers by means of climate change, they would be able to pursue this goal actively since the economic resources would be available.

Climate policies of the industrialized countries then will have to balance uncertain but possible long-term advantages against fairly certain short-term disadvantages. To prevent climate change, therefore, implied short-term advantages outweigh the risk of losing long-term advantages. If governments would decide (as they usually do in environmental matters) to prefer short-term gains to reasonable long-range considerations, this kind of rationale might lead to the prevention of climate change. This is by no means a clear case, however, since a lot of money would have to be spent to prevent further climate change. By contrast, the countries of the Third World are in a completely different position.

Implications for the Third World

Even a scrupulous evaluation of the IPCC impact report produces no hint that the situation of the Third World could be generally improved by climate change in some respects. The same is true for the earlier report by Parry et al. (1988). Of course, comparative advantages of some Third World countries with respect to others may come up, e.g., in fisheries due to changes in oceanic upwelling. However, these generally do not seem to help the poor relative to the rich. There are, however, general aggravations the other way around. The Third World is not only more vulnerable than the industrialized countries, it will moreover be hit harder.

- Climatically induced changes in natural terrestrial ecosystems will have far-reaching socioeconomic implications, especially in those regions of the globe where societies and their economies directly depend on the integrity of those systems for their welfare. Changes in the availability of food, fuel, medicine, construction materials, fiber products, recreational attractiveness, and general income must be expected when those ecosystems no longer provide what they used to do.
- "One can reliably predict that certain developing countries will be extremely vulnerable to climate changes because they are already at the limits of their capacity to cope with climatic events" (Tegart et al. 1990). These include populations in low-lying coastal regions and islands, subsistence farmers, populations in semi-arid grasslands, and the urban poor. Their vulnerability to the adverse implications of climate change is high because of limited access to the necessary information, infrastructure, human expertise, and financial resources. According to IPCC, the adaptive capacity of "developing" economies is generally very low (not only in agriculture), and there is some evidence to suggest that their industries may also be particularly vulnerable to climatic change.
- In agriculture, according to the IPCC impacts report, two broad sets of regions appear to be most vulnerable to climate change: (1) some semi-arid, tropical and subtropical regions (such as western Arabia, the Maghreb, western West Africa, Horn of Africa and southern Africa, eastern Brazil) and (2) some humid tropical and equatorial regions (such as Southeast Asia and Central America). Also, the productive capacity of South America and other net exporters of cereals could decline.
- Increase in drought risk is probably the most serious impact of climate change on agriculture. Those countries suffering from drought already will suffer even more in the future. These are mainly countries of the Third World. Therefore, "shifts of moisture limits in some semi-arid and sub-humid regions could lead to significant reductions of (agricultural) potential with serious implications for regional food supplies on some developing countries." Substantial regional dislocations of access to food must be expected. Increased drought frequency also implies forest degradation and more fire risk.
- A regional case is that in the semi-arid, arid, and hyper-arid ecoclimatic

zones of the Mediterranean, greenhouse-gas-induced climate change "will reduce plant productivity and result in desertification of the North African and Near Eastern steppes owing to increased evapotranspiration. The upper limit of the deserts would migrate under the influence of climate change and most likely extend into the area that currently corresponds to the lower limits of the Semi-Arid Zone (i.e., foothills of the High, Mid and Tell Atlas and Tunisian Dorsal in Northern Africa, and of the main mountain ranges of the Near-Middle East: Taurus, Lebanon, Alaoui, Kurdistan, Zagros and Alborz)" (Tegart et al. 1990).

- With respect to water resources, the regions that appear to be at greatest risk, in terms of serious threats to sustaining the population, by climate change are:

  1. Africa: Maghreb, Sahel, the north of Africa, southern Africa;
  2. Asia: western Arabia, Southeast Asia, the Indian subcontinent;
  3. America: Mexico, Central America, southwest U.S., parts of eastern Brazil;
  4. Europe: the Mediterranean zone.

In many instances, it can be expected that changes in hydrologic extremes in response to global warming will be more significant than changes in hydrologic mean conditions. "Water shortages caused by irregular rainfall may especially effect developing countries, as seen in the case of the Zambezi river basin" (Tegart et al. 1990). Biomass (fuel, wood) provides more than 90% of the energy in most of the countries of sub-Saharan Africa. Changed moisture conditions in some areas, reducing this biomass, could pose grave problems for domestic energy needs as well as for the construction of shelter. Increased scarcity of water will also have health impacts, especially in large urban areas.

- Climate change may make some present hydroelectric power facilities obsolete. Changes in cloud cover, precipitation and wind intensity will also effect the availability of other forms of renewable energy.
- In the short term, fisheries could benefit when marshes flood, die, and decompose; however, "in the longer term, by 2050, the overall impact on fisheries and wildlife is likely to be negative." Regional shifts will also have major socioeconomic impacts.
- Climate change imperils the existence of entire countries (such as the Maldives, Tuvalu, and Kiribati) by a rise of only a few meters in sea levels. Populous river delta and coastal areas of such countries as Egypt, Bangladesh, India, China, and Indonesia are threatened by inundation from even a moderate global sea-level rise. For instance, if it rose 1 m, 12–15% of Egypt's arable land and 14% of Bangladesh's net cropped areas would be below sea-level and tens of millions of inhabitants could be displaced by inundation. Also, low-lying urban infrastructure would

be destroyed, freshwater supplies contaminated, and coastlines altered. Again, like in the industrialized countries but in a very different economic situation, coastal protection structures, highways, power plants, and bridges may require redesign and reinforcement to withstand increased flooding, erosion, storm surges, wave attack, and seawater intrusion.

Taking these implications together suggests that greenhouse-gas-induced climate change will severely aggravate the general situation of the Third World. Even if some countries may expect relative advantages with respect to others, which could be defined by further regionalizing climate modeling simulations, on the whole there is little to be seen that is positive for the Third World in these projections.

Considering the quality of regional climate projections at present, I expect that further studies will modify, as well make more specific, the changes in individual countries mentioned in the above assessment. However, there is no reason to believe that the general result will be changed if some names and data were exchanged for others. With respect to equities and inequities then, we have to face the fact that the Third World is the prospective loser. Could this picture be changed by an adequate political response? And may such a response be expected?

EQUITY IMPLICATIONS OF POLITICAL RESPONSES TO CLIMATE CHANGE

To consider political responses to inequities induced by climate change, the basic observation is that not only are the implications of such change unevenly distributed internationally, its causes are also. Countries, however, are not affected according to their share of responsibility for the climate change. Taking CO_2 as the predominant greenhouse gas, the average per capita consumption in 1985 in the industrialized countries was almost ten times as high as in the Third World, namely 3.12 and 0.36 t carbon/person. This imbalance is demonstrated in Table 22.1.

The foregoing analysis shows that those who cause about three-quarters of the climate change will be least affected by the implications or will even have absolute advantages. Those who will suffer from it most, share the responsibility only to the extent of about one-quarter. This picture remains about the same when the other greenhouse gases are also taken into account. Of course, with respect to the future, the projections are that the share of Third World countries in causing climate change would rise with economic welfare (according to Tegart et al. 1990, to 44% of 12 billion tons in 2025), but this depends on the degree of welfare to be expected.

If an international sovereign could apply the "polluter pays principle,"

Table 22.1 The emissions of carbon by sectors of the world community due to the combustion of fossil fuels (after Tegart et al. 1990)

	Gt/yr	% of Total Global
Industrialized Countries	3.83	74
Western Europe	0.85	16
North America	1.34	26
OECD Pacific	0.31	6
Centrally planned Europe	1.33	26
Third World Countries	1.33	26
Africa	0.17	3
Latin America	0.22	4
South and East Asia	0.27	5
Centrally planned Asia	0.54	10
Middle East	0.13	3

generally so well accepted within industrial market economies, the political response clearly would be:

1. that the industrialized countries, through preventive technologies as soon as possible, must reduce greenhouse-gas emissions to a level so that they would no longer live at the expense of the welfare of others and of future generations;
2. that the industrialized countries should be made liable for the harm they are doing to Third World countries as far as this has already happened or can no longer be prevented. With respect to future emissions, worldwide per capita allowances could be made tradable.

I wonder when the "developing" countries will bring this argument to bear in debt negotiations. While the "polluter pays principle" is sometimes difficult to apply because those who are liable cannot be identified, this is not the case with respect to climate change since the CO_2 emissions of different countries are known more or less.

As in national environmental liability issues, those who might be held responsible may argue that scientifically some questions remain open. This is what some national governments have been doing from conference to conference, particularly the government of the U.S. However, even if this argument is accepted, climate change should still be prevented for reasons similar to "Pascal's bet." His argument was as follows:

1. If God exists and I believe in him, it's alright.
2. If God does not exist and I do not believe in him, it's equally alright.
3. If God does not exist and I believe in him, I will refrain from certain actions but this will not hurt me too much.
4. If God exists and I do not believe in him, I will be damned, subject to an eternal catastrophe. Since no action is worth eternal damnation, it's prudent to believe in God.

Transferring Pascal's bet to the climate discussion, the issue is to believe or not to believe in the existence or nonexistence of the threat that climate will change due to greenhouse-gas emissions, as has been described in the preceding section. The four cases to be compared are:

1. Greenhouse-gas-induced climate change exists and everything is done to prevent globally harmful developments in the future. This is the ideal case, generally well above the actual performance of even the best political systems.
2. Greenhouse-gas-induced climate change does not exist and nothing is done to prevent globally harmful development in the future. This is the ideal case for the political propensity to do nothing except symbolic politics.
3. Greenhouse-gas-induced climate change does not exist and everything is done to prevent globally harmful developments in the future. This means that, for instance,
    - burning fossil fuels will be considerably reduced so that valuable resources are saved for more sophisticated uses in the future;
    - energy conservation will be enforced, which seems to be generally beneficial for technological modernization of the economy. As far as this means to increase energy prices, it should be kept in mind that empirically economic performance is fairly well correlated with energy prices (both decreasing from Japan to Europe to the U.S. to the USSR);
    - "development" policies will be cancelled and economic policies fitted into a new economic world order instead;
    - emissions of chlorofluorocarbons banned so that the ozone hole gets a chance to fill up within the next century;
    - tropical forests will be saved to some extent through the help of the industrialized countries, which have already cut their trees in the past and now legitimately compensate others for not proceeding alike.

It may be difficult to imagine any reasonable action in favor of preventing climate change which is not at the same time beneficial for others. Perhaps it has not happened so far because global interests are not easily

observed by the many national and international authorities taking care of world issues. This case then deserves to be positively denoted like the former two. Also, in politics it is fairly common for a decision to be right but not necessarily implemented for the right reasons.
4. Greenhouse-gas-induced climate change exists and nothing is done to prevent globally harmful developments in the future. This means that the loser will be the Third World. Famines of a hitherto unknown dimension may take place and tens of millions of people may be displaced from their countries. When this happens it will be too late to do something, except on the charity level. Most of those who are responsible for the catastrophe will no longer be alive.

The result of evaluating the "climate bet" seems to be even more conclusive than that of Pascal's original bet. The argument shows that if there were an international sovereign, its duty would be to prevent greenhouse-gas-induced climate change because, if it happens, present inequities would be irresponsibly increased. If this authority also considered the heritage of future generations, the argument would be enforced because, with still rising inequities, more and more human beings would be born into a world with decreasing chances of equity. Ethically, this is a clear case because the basic rule: *neminem laede*, do no harm to others, is violated by inflicting climate change risk or definite climate change or both (described in terms of human habitat) upon others. In this sense, no further philosophical discussion is required. The problem, however, is that as there is no international authority to protect the disadvantaged from the advantaged in the context of anthropogenic climate change.

Is there any chance that political perception of the consequences of climate change will lead to a response which prevents the inequity increases at least to some extent? From past experience this may not be expected. The Third World is already suffering and it is expected that it will suffer even more with climate change. There are famines as well as millions of environmental refugees crossing national borders already, and nothing appears likely to change the situation. To the contrary, wealth is transferred from the poor to the rich as the data given in the section on Present International Equities and Inequities show. Conceiving climate change in political terms which are considered realistic at present, therefore, the situation can be likened to that of someone who has to cross a river which is 1000 m wide but is strong enough to swim no more than 100 m and now additionally finds out that the river is not 1000 m but 2000 m wide. In present politics, the climate issue is reasonably perceived as (white) "chalk on the white wall" (Meyer-Abich 1980).

Years ago I argued that this lack of perception was consistent with humankind's current political organization. In the meantime, the responsibility of the industrialized countries is seen more clearly, and environmental awareness has risen internationally well above the level of present political institutions. Thus, there may be a chance that the river has narrowed somewhat and the notions of development may develop themselves so that swimming could be improved. National governments do not necessarily deny international responsibility when the public expects them to recognize it and moreover to spend tax money on it. In fact, some countries are moving now on lines beyond traditional political institutions. My country, Germany, is on the way to reduce CO_2 emissions by 25–30% in the period 1990–2005, compared to the 1987 level, so that the Toronto target (WMO 1989) may be met (Deutscher Bundestag 1990b). The parliamentary commission on protection of the atmosphere, which worked out this target (and has shown how to reach it), has also proposed internationally that the leading industrial economies should follow this example while all industrial economies in East and West should reduce their CO_2 emissions by 20% of the 1987 level on average by 2005.

It seems that some other countries will move correspondingly. Others, like the United Kingdom, are at least considering the stabilization of current emission levels by 2005. The impression, therefore, could be justified that, e.g., the European Community is not fully unaware of its political responsibility beyond legal liabilities. Considering the fact that some of the beneficial side effects in climate change abatement may also work nationally (e.g., technological modernization), this may in the long run even prove to be in the national interest.

Other countries, however, so far do not seem willing to join a cooperative action for the prevention of climate change by the industrialized countries. This is especially true for the U.S. and the USSR. Together, these countries are responsible for almost half of humankind's CO_2 emissions. At the same time their governments, so far, have not been particularly sensitive to environmental protection so that their political reasoning must be expected to correspond to the institutional framework, i.e., simply to their national interests. These are:

1. Not to compensate other countries for the losses one's national activities have inflicted upon them as long as such compensation can be avoided without losing one's own reputation.
2. To be aware of national advantages, be they absolute or relative, which might be brought about by climate change.

To defend national interest, following point (1), we should expect a government to respond to the climate issue as late as possible and then to insist on scientific uncertainties so that much more research would have to be done before any decisions could be made. To soothe public worries one should emphasize that in any case we are all sitting in one boat, so that everybody must be equally interested to prevent or mitigate climate change. The one-boat argument also helps to conceal national interest number (2) with the implication to suppress arguments on winners and losers as long as possible.[2] All of this could happen. Considering political perception, it is also easily seen that expectations (absolute or relative) of advantages through climate change can become political facts in themselves, be they scientifically justified or not.

Possibly some governments in Europe will partially act or announce their intention to act globally responsible, perhaps identifying such global reputation as a national interest and in any case pleasing environmentally minded voters. However, should one expect a global move in this direction on the basis of humankind's current political organization? To ban chlorofluorocarbons will be hard enough, and yet there are strong economic interests to do so in favor of substitutes. But again there will be advantaged and disadvantaged countries, or sectors. In the case of energy, even if there were national interests to reduce fossil-fuel consumption, the economy so far does not seem to be generally in accord with national interest. With respect to global equity, this means that the current political organization of the world from its inherent rationality is not suited to prevent rising inequities by preventing greenhouse-gas-induced climate change.

Within current political rationality, it is reasonable to shift any response to oncoming climate changes as far as possible into the future because:

- it is anyway the propensity of governments to shift costs so that others (in this case, future generations) will have to pay;
- the responsibility, or liability, issue will look differently in the future, when the debts from the past become larger than those politically in charge can be expected to cover. This is a way of "socializing losses" globally as well as intergenerationally.

Both strategies, therefore, as a rule will be favored by current political institutions. The implication for international as well as intergenerational equity is that *everything will be done to do nothing presently*. Inequities will increase the more current political institutions succeed in doing nothing. Traditional political rationality has become unreasonable at the present stage of global interrelations.

We are not, however, left completely without hope that a response to climate change might reduce the inequities which seem inevitable so far.

---

[2] The first international conference on the winner's and loser's issue took place in 1990. The report is written in a fairly restrained language which may reflect administrative expectations (Glantz 1990).

One reason is an increasing awareness that as an output of industrial economy wealth is more and more balanced by newly emerging risks. Climate change is certainly one of them, and the growing uneasiness to accept new burdens of risk may help to support policies of preventing climate change even devoid of international considerations.

Second, a lack of political rationality that usually leads to a preference of short-term goals may become helpful in climate politics. For instance, even if the possible long-term advantages of climate change for the industrialized countries would by far exceed the short-term disadvantages of the migration of some part of the population from one to another part of the country, these disruptions could be decisive for the fate of particular governments. Shortsightedness in this case, therefore, could be advantageous for the prevention of climate change. Usually in environmental policies, long-term losses are accepted for short-term gains. In the case of climate, the same bias—paradoxically enough—could help to prevent changes at the expense of the Third World.

Third, to turn off the lights of misleading optimism could help to identify emerging lights, which might permit a different view on human solidarity. With respect to the foregoing analysis, it is a surprising fact that by now about twenty countries have announced their intention to reduce or at least to stabilize their CO_2 emissions. Even if

- the two superpolluters have not yet joined them;
- announcements should not be taken as actual reductions but may fail like the Swedish nuclear energy phase-out policy;
- politics in western democracies have a growing tendency to content itself with purely symbolic actions,

this international movement could become self-enforcing, so that it gets more and more difficult to diverge from the movement without losing one's reputation. Also as soon as political reputation is won by being prominent in climate protection, a political competition to be quicker than others begins.

Finally, our political institutions were developed well before the industrial revolution took place and must not be adequate at the present stage of global interrelations. While in the 19th century social inequity was mitigated at the expense of disruption of the environment, climate inequities are brought about by disruption of the environment. This may imply a chance that the social and environmental issues are not considered separately any longer. The task then is to conceive a political philosophy of humankind in the community of nature beyond that of the 17th and 18th century, which considers peace within the nations but not peace between them and peace with nature. Such a philosophy could support the emerging forms of international solidarity when nations fall back to where they basically still are.

SUMMARY

1. Climate change will not only happen in terms of climatology but equally in terms of ecology, economics, and political science. A political judgment presupposes that these latter descriptions have been given.
2. Harm can be done to others either by inflicting a definite damage upon them or by exposing them to the risk that such damage could happen. Risks are a category of facts.
3. Climate change will not equally effect all regions and countries in the world; there will be winners and losers, be it absolutely or relatively (comparatively).
4. In economic terms, for the industrialized countries adaptation to climate change will be expensive but will not overstrain their economic power. At the same time, these investments will include modernization incentives and may in the long run be compensated by a poleward extension of climatically favorite regions.
5. The Third World is not only more vulnerable to greenhouse-gas-induced climate change, it will also be hit harder. Especially, those countries which already suffer from drought risk will do so even more in the future. Apart from comparative advantages of some Third World countries with respect to others, the Third World in general will be the loser of climate change.
6. Ethically the basic rule: *neminem laede*, do no harm to others, is clearly violated by inflicting climate change risk or definite climate change or both (described in terms of human habitat) to others.
7. If the "polluter pays principle" were accepted internationally, or if an international authority would transfer Pascal's bet to greenhouse-gas-induced climate change, the industrial economies would be held liable for the harm they are doing others.
8. While some national governments seem to recognize global responsibilities according to public environmental awareness, others define their national interests corresponding to the present level of international political organization. On this level, it must be expected that everything will be done to do nothing. The current political system is from its inherent rationality not suited to prevent rising inequities by greenhouse-gas-induced climate change.
9. There is a chance that in the industrialized countries the prevailing bias

to accept short-term gains for long-term losses may help to prevent further climate change.

REFERENCES

Budyko, M.I., and Y.S. Sedunov. 1988. Anthropogenic climate changes. Prepared for the world congress "Climate and Development," Hamburg, Nov. 7–10, 1988 (manuscript).
Enquête-Kommission. 1990a. Vorsorge zum Schutz der Erdatmosphäre zum Thema Schutz der tropischen Wälder. Bundestagsdrucksache 11/7220. Bonn, May 24, 1990.
Enquête-Kommission. 1990b. Vorsorge zum Schutz der Erdatmosphäre zum Thema Schutz der Erde. Bundestagsdrucksache 11/8030. Bonn, Oct. 2, 1990.
Flohn, H. 1990. Treibhauseffekt der Atmosphäre: Neue Fakten und Perspektiven. In: Rheinisch-Westfälische Akademie der Wissenschaften. Vorträge N 379, pp. 11–41. Opladen: Westdeutscher Verlag.
Houghton, J.T., G.J. Jenkins, and J.J. Ephraums, eds. 1990. Climate Change: The IPCC Scientific Assessment. Prepared for the Intergovernmental Panel on Climate Change by Working Group I. WMO/UNEP, 365pp. Cambridge: Cambridge Univ. Press.
IPCC (Intergovernmental Panel on Climate Change). 1990. Policymakers Summary of the Formulation of Response Strategies. Report prepared for IPCC by Working Group III. June.
Journalisten Handbuch Entwicklungspolitik 1989/90, ed. Bundesministerium für Wirtschaftliche Zusammenarbeit, Bonn.
Meyer-Abich, K.M. 1980. Chalk on the white wall? On the transformation of climatological facts into political facts. In: Climatic Constraints and Human Activities, ed. J. Ausubel and A.K. Biswas (International Institute for Applied System Analysis [IIASA] Proceedings Series Vol. 10), pp. 61–73. Oxford: Pergamon.
Meyer-Abich, K.M. 1984. Wege zum Frieden mit der natur—Praktische Naturphilosophie für die Umweltpolitik. München: Hanser.
Meyer-Abich, K.M. 1988. Wissenschaft für die Zukunft—Holistisches Denken in ökologischer und gesellschaftlicher Verantwortung. München: C.H. Beck.
Parry, M.L., T.R. Carter, and N.T. Konjin. 1988. The Impact of Climatic Variations on Agriculture. Vol. 1: Assessment in Cool, Temperate and Cold Regions. Vol. 2: Assessments in Semi-Arid Regions. Dordrecht: Kluwer (UNEP-IIASA Study).
Sachs, W. 1990. The Archaeology of the Development Idea. Six Essays. INTERculture/No. 109. XXIII. 4:2–37.
Smith, A. 1969. The Theory of Moral Sentiments (1759). Introd. by E.G. West. Indianapolis: Liberty Classics.
Tegart, W.J., G.W. Sheldon, and D.C. Griffiths, eds. 1990. Climate Change: The IPCC Impacts Assessment, (Working Group II), WMO/UNEP, 230pp. Canberra: Australian Government Pub. Service.
UNEP (United Nations Environment Programme). 1988. The potential effects of global climate change on the United States. Executive summary, ed. J.B. Smith and D.A. Tirpak; Draft October 1988.
Weizsäcker, E.U. von. 1989. Erdpolitik. Ökologische Realpolitik an der Schwelle zum Jahrhundert der Umwelt. Darmstadt: Wissenschaftliche Buchgesellschaft.

Weltentwicklungsbericht 1989. Finanzsysteme und Entwicklung. Kennzahlen der Weltentwicklung. Bonn: UNO-Verlag.

World Development Report 1988. Opportunities and Risks in Managing the World Economy. Public Finance in Development. World Development Indicators. New York: Oxford Univ. Press.

WMO (World Meterological Organization). 1989. Conference Proceedings. The Changing Atmosphere: Implications for Glocal Security. Toronto, Canada 27–30 June 1988 (WMO/OMM No. 710). Geneva: WMO.

Standing, left to right:
K.M. Meyer-Abich, T. Peter, P.H. Gleick,
I.J. Walker, H. Hilse, D.A. Tirpak
Seated, left to right:
U.E. Simonis, C.L. Caccia, U. Fritsche, M. Lönnroth

23

Group Report: What Is the Significance of the Linkage between Strategies to Reduce CO_2 Emissions and Other National and International Goals?

M. LÖNNROTH, Rapporteur
C.L. CACCIA, U. FRITSCHE, P.H. GLEICH, H. HILSE,
K.M. MEYER-ABICH, T. PETER, U.E. SIMONIS,
D.A. TIRPAK, I.J. WALKER

INTRODUCTION

This report is a summary of our group's discussions. The group's mandate was to examine the significance of the linkages between strategies to reduce CO_2 emissions and other national and international goals. Attention was given to identifying research agendas and the need for further analysis.

The discussion with the group was structured around a preliminary set of issues. The following summary is based on a framework agreed upon after the discussion.

BACKGROUND, GOALS, AND VALUES

The Intergovernmental Panel on Climate Change (IPCC) concluded that:

- there is a natural greenhouse effect;

Limiting the Greenhouse Effect: Options for Controlling Atmospheric CO_2 Accumulation
Edited by G.I. Pearman © 1992 John Wiley & Sons Ltd

- emissions resulting from human activities are increasing the atmospheric concentrations of the greenhouse gases;
- some gases are potentially more effective than others;
- atmospheric concentrations of some gases adjust only slowly;
- the longer emissions continue at the present rates, the greater reduction will have to be for concentrations to stabilize at a given level.

The IPCC also estimated, based on current model results, that net reduction of over 60% of the emissions of long-lived greenhouse gases would be necessary in order to achieve stabilization of concentration at today's level.

The group spent considerable time discussing winners and losers in relation to climate change. First, it was noted that climate change would take place against the background of existing differences in living standards. It was also argued, but not generally agreed upon, that while poverty was increasing around the world in absolute terms, it was also decreasing relatively. The last decades offered cause for pessimism as well as optimism, depending on the perspective of the observer. Second, it was argued, and generally agreed to, that the effect of climate change would be greater on poor people around the world than on the more well-to-do. This was likely to be particularly true for those living at sea levels in island states or river mouths. Third, it was argued that one has to look at the distribution of winners and losers in relation to response policies. For instance, necessary technological innovations may well bring advantages to some more than others.

Three dimensions are necessary in formulating climate-change policies: time horizons, equity, and ethics.

Time horizons enter in several ways. According to the IPCC estimates, the rate of climate change would exceed, by several orders of magnitude, the natural rate-of-change. The implications for ecosystems are impossible to foresee. Attempts have largely been based on estimating differences in temperature zones, without taking into account the impact of the rate-of-change.

On the other hand, on a political time scale, expected climate changes in response to accumulation of greenhouse gases in the atmosphere are relatively slow. The pace is similar to the rate of change of many human activities, such as infrastructure, that generate the long-lived greenhouse gases, the first factor of importance in estimating future emission rates. The same holds for the rate of change of global population, the second factor of importance in estimating future emission rates. Time horizons, time scales and rate-of-change have to be prime elements in devising response strategies.

Emission rates of CO_2 differ substantially between societies and, indeed, between groups of individuals. Americans emit some five times the global per capita average; Swedes, Japanese, and Swiss emit about twice the global average; Germans and English emit about three times the global average. Indians and Chinese emit on the average less than half of the global average. Clearly, CO_2 emissions depend strongly on gross national product (GNP) when comparing developing and developed countries, but less so when comparing different developed countries. On the other hand, different power-generation strategies and resource endowments can lead to major differences in emissions in industrialized countries (e.g., Sweden and the U.S.). At the same time, population growth is by far higher in developing countries with low per capita emission rates and, in fact, has the same doubling time as greenhouse-gas concentrations according to the IPCC "business-as-usual" scenario. Thus, equity has to be the second consideration in devising response strategies.

Ethics has to be the third consideration. Ethics concerns not only equity between humans of widely differing standards of living, or between generations, but also, and perhaps even more fundamentally, the relationship between humans and nature. Response strategies have to reflect a clear and well-argued balance between the right of humans to irreversibly change the ecosystems and the duty of humans to preserve and protect these ecosystems.

Formulating response strategies to climate change will present the world's societies with a daunting task. The responsibilities will be the heaviest for those societies that have the largest resources at their disposal, in terms of intellectual knowledge, skills, and capital. Industrialized countries will have to bear the brunt of the response.

It is here that the relationship between humans and nature becomes important, indeed critical. Certainly the historical, but also, as argued in the group, the present patterns of development in the industrialized world have been fundamentally at odds with the ethics of achieving a balance between humans and nature. Increasing production and reducing social inequities within industrialized countries has so far largely occurred at the expense of the living world. Some felt that the conflict between nature and industrialization was so fundamental that only an extremely drastic reorientation of industrial nations could even begin to redress the balance. Others thought that while the conflict historically had been strong, there was no fundamental reason why industrialization could not be pursued in a much more measured balance between humans and nature. There was general agreement on the need to establish climate-change policies on a clear ethical perspective.

It was also stated that the human perception of nature is changing towards a stronger responsibility towards the balance between humans and nature. This can be clearly seen in some countries, such as Germany and the Scandinavian countries. Several observations and possible explanations were offered, ranging from religion to generational change. Extensive value shifts appear to be possible. To detect and document such shifts in values with respect to human responsibilities towards nature is an important research task in its own right.

Studies of climate change concern not only the natural sciences, engineering, or economic sciences but also social sciences as well as the humanities. Fundamental issues of social behavior and human anthropology have to be understood. It is important that research on climate change is broadened into these fields as well. A broad, transdisciplinary scientific understanding is necessary both in order to assess proposed climate-change policies as well as the impact of climate change itself. There are also important issues for science policy in general.

LINKAGES BETWEEN DIFFERENT POLICY AREAS AND CLIMATE-CHANGE POLICIES

While time scale, equity, and ethics are three basic concerns to any climate-change policy, a fourth concern is uncertainty. While there is substantial agreement on many scientific issues with respect to the mechanisms of climate change, the uncertainty and, indeed, the scientific disagreement is still considerable. Given the scale of commitments of resources needed in order to influence decisively the future emission rates of the long-lived greenhouse gases, it is not surprising that policymakers hesitate and require more information on both costs of reducing emissions and costs of mitigating climate change. At the same time, waiting carries its own costs; required reductions will be all the larger, the longer the delays, provided the result of decreased uncertainty is an affirmation of the need for action. The precautionary principle is a more and more widely accepted principle on which to act: it is better to be prudent and thus take on the costs now. Given the scale of these costs, it is not obvious where the precautionary principle might lead.

Robust policies should both reduce the risks of climate change as well as reduce the consequences. For the foreseeable future (perhaps the 1990s), robust policies should be based on explicitly identifying other policies with other goals that also have a positive impact on perceived climate-change issues. The group spent considerable time in discussing such policy linkages. The major role of the robust policy is to buy time for reducing the uncertainties and increasing preparedness in an economically reasonable way.

There is no clear-cut taxonomy of such policy linkages. What follows is a series of areas which either are already well explored or which should be explored for their linkages to climate-change policies. They are ordered from the specific to the more general.

Ozone Depletion Policies

Several emitted gases that cause depletion of the ozone layer are also greenhouse gases. Steps to strengthen the Vienna Convention, to include the ozone-depleting gases, should also in effect buy time for climate-change policies. Such gases include CH_3CCl_3, the so-called hydrochlorofluorocarbons and hydrofluorocarbons as well as certain other gases. Phasing out chlorofluorocarbons (CFCs) earlier than the London agreement of 1990 would also buy time. While agreement states that CFCs should be phased out by the year 2000, some countries plan to have the phase-out achieved by the mid-1990s.

Reducing Acid Rain, Smog, and Other Steps to Increase Air Quality

A whole series of issues relating to emissions of SO_X, NO_X, volatile organic carbons, etc. have implications also for greenhouse-gas emissions. Switching from coal to natural gas in electricity generation occurs for a variety of reasons in the U.S. and western Europe and has implications for air quality, in general, as well as greenhouse-gas emissions. Policies to reduce volatile organic carbons from landfills in the U.S., in order to reduce smog risks, also have positive greenhouse-gas implications. Electric vehicles would, if the electricity were generated with natural gas in advanced gas turbines, reduce the CO_2 emission rate per kilometer traveled as well as improve air quality.

Phasing Out Coal in the Central European Coal Belt

Poland, Czechoslovakia, and the former Germany Democratic Republic use enormous amounts of coal in highly inefficient ways. This coal use, and its production, has disastrous environmental impacts in the areas of direct mining and use, but also far away. Three great European rivers originate in the coal-bearing mountains (the Elbe, the Oder/Neisse and the Vistula) and they all carry large amounts of pollutants to the North Sea and Baltic Sea. Modernization and economic restructuring of these old industrial areas will have major and positive impacts on human health in the region, on the environment in northern Europe as a whole as well as on the European CO_2 emission. However, modernization will also require innovative social policies for reducing the hardships of loss of employment. What may be needed is an east European equivalent of the coal and steel union of the late 1940s. The most likely alternative to coal is natural gas, with sources in north Africa, the USSR, and, in the longer term, the Middle East.

European Security

Replacing coal with natural gas has linkages to security issues as well. After the fall of the Berlin Wall and the dismantling of the Warsaw Pact, the main risks for potential conflicts are the border areas between the USSR and east and central Europe. There are lessons to be learned here from the successful defusing of tension between the almost permanent enemies of France and Germany after the second World War: economic and social integration stabilizes democracy and therefore peace. The first stage of European integration, the original six, managed to stabilize democracy within the west European heartland of former fascism. The second stage occurred when the southern belt of states with strong fascist traditions became members in the early 1980s: Spain, Portugal, and Greece. Democracy and thus peace are now also firmly rooted in these areas. A third stage will have to deal with the former East Bloc countries: Poland, Czechoslovakia, Hungary, etc. In these countries, the rate of reduction of CO_2 emission will, in all probability, be directly proportional to economic integration and thus stabilization of democracy. The same analysis holds, obviously, for the USSR, although arguably in a longer time perspective. Thus, insecurity, economic inefficiency, environmental degradation, massive coal use, and large-scale CO_2 emissions are interlinked in the so-called Second World. It should be possible to have greatly increasing standards of living in east and central Europe as well as the USSR, and at the same time have decreasing CO_2 emissions.

A somewhat similar argument can be made with respect to the other border areas of Europe, the southern part of the Mediterranean, and the Middle East. Replacing European coal with natural gas from northern Africa and the Middle East would probably both reduce CO_2 emissions and enhance economic integration and thus security.

Energy Efficiency Policies

Energy efficiency is obviously of major importance. A number of studies demonstrate that it should be possible to stabilize CO_2 emissions from industrialized countries by, say, the year 2000 through cost-effective efficiency measures alone. Given time, the potential obviously increases. Some members of the group agreed with A. Lovins and felt it was possible to reduce the emissions from industrialized countries substantially more, given another decade or so. Others were more impressed with L. Schipper's argument about the barriers to implementation. It was generally agreed that the most important research issue with respect to energy efficiency has to do with implementation. Policies to increase energy efficiency should be the linchpin of any robust strategy. Such policies include mandatory regulation, pricing and taxation, various forms of economic incentives, the roles of energy companies, and the possibility of turning them into energy service companies, etc. There is a huge and growing literature on the topic.

Policies and Cultural Patterns that Stimulate Energy Use

Policies to increase energy efficiency do not operate in a vacuum, however. Energy intensity differs substantially between countries with the same GNP per capita. These differences have both cultural and policy backgrounds. Industrial energy intensities, for example, clearly depend on the amount of international competition: according to L. Schipper, differences between the countries of the Organization for Economic Cooperation and Development (OECD) appear to be smaller now than some 20 years ago, while differences between western Europe and the USSR probably are increasing. This again illustrates the economic and efficiency cost of autarchy. The implications for the Third World will be discussed below.

There are other policies that have an impact on energy use. Subsidies for housing tend to increase housing area per capita, whether the subsidy takes place in the form of direct subsidy or through tax deductions of, e.g., mortgage rates. Some countries also have tax deductions for traveling to and from work, and some have very generous company car rules. Still in some countries, second homes are very important and thus tend to generate large amounts of traveling for leisure purposes.

The largest differences between countries in terms of energy per GNP per capita appear to take place in the sectors of the economy that are closed to international competition. Area-intensive living, whether in terms of housing only or also transportation, reflects deep-seated cultural conditions and thus takes long to change.

Energy is taxed very differently in different countries. The taxation system for capital, as well as write-off rules, can differ, often as part of deliberate policies. This is true for the industrialized world as well as for the USSR and the Third World. Many western European countries tax fuels and electricity directly, which is one reason why gasoline prices differ so much from the U.S. However, some countries also subsidize national fuels, e.g., coal in Germany. India subsidizes rural electricity. The USSR subsidizes energy in general. As climate-change policies have to evolve, international agreements and guidelines appear to be necessary, at least for those countries that demand increased assistance for coping with CO_2 reduction policies. This topic could be brought up in the context of the International Monetary Fund and the World Bank as well as within the OECD. A major research effort is necessary in order to understand the role of taxation and subsidization of CO_2-intensive activities in different countries. International agreements may be necessary.

Finally, it was mentioned that there does not appear to be any stiff link between energy price levels and international competitiveness. Both Japan and the F.R. Germany have high energy prices and high competitiveness. In fact, at least among the industrialized countries, it could be argued that low energy prices carry with them the risks of "locking in" inefficient technologies and patterns of living.

Development Policies

There are obvious, but difficult to unravel, linkages between climate-change policies and development policies. Start with population. Simple arithmetic demonstrates that climate policy geared to stabilize climate change requires major changes in present development trajectories. Taken together with the need to reduce CO_2 emissions from developed countries, the challenges are, to put it mildly, daunting. The concept of a robust strategy, aiming at buying time, offers some order to the thoughts.

First, there are complex lessons relating environment, and thus climate change, to underdevelopment and poverty, insecurity, intolerance, threats to peace and stability, and the preconditions for development. Also, the impact of climate change differs considerably between different regions. An extremely general statement might be that while poverty and destitution threaten the environment and cause CO_2 emissions in large parts of Africa, economic development and increasing standards of living threaten the environment and cause CO_2 emissions in large parts of southeast Asia. The mechanisms differ, however.

Some countries are developing more rapidly than others. Compared to what was generally believed some thirty years ago, large parts of Asia have developed more rapidly than thought, while Africa has had a more troublesome period. Latin America is more heterogeneous but falls somewhere between Asia and Africa. Various forecasts for the next decade would indicate that the linkages between development policies and climate-change policies look roughly like the following:

In certain parts of the world, notably Africa, the main linkage is to assist in reducing poverty and poverty-induced forest management, including large-scale deforestation and slash-and-burn agriculture. Deforestation is an important source of CO_2 in Africa and, at the same time, severely threatens the biological diversity as well as the ecosystem upon which the population has to live.

In Asia, the problem is rather different, and again differs between regions. Poverty is declining in several countries as economic development occurs and the integration with the industrialized world proceeds. This has two important effects: (1) it promotes deforestation due to logging and timber exports to Japan and, (2) encourages increased use of fuel-intensive commodities such as cars, electric and other appliances, electricity in general, etc. An increasing standard of living could clearly lead to massive environmental damage in general, along the industrialization experiences of the U.S. and Europe, but also to increases in CO_2 emissions in particular. The robust strategy must include higher efficiency in energy use as well as less CO_2-intensive electricity generation. Here again, natural gas may be an alternative in several areas, compared to coal and oil. Also, solar electric generation should be developed further. It is quite likely that the major breakthrough of photovoltaics will occur in southeast Asia, with its frequently complicated geography. Mountains, rivers, and waters are barriers to centrally generated electricity which should make decentralized photovoltaics more interesting. It is also important, as was stressed in the group, that electric appliances are as efficient as possible. Here again suppliers, frequently Japanese, should take notice.

This points to the conclusion that the southeast Asian electric utility sector is a key element in linking climate-change policies with development. Also, the logging industry as well as forestry policies should be considered.

It was also stressed that certain habitats face virtual extinction due to the possible rise of sea level due to climate change. Some types of societies are very much more threatened than others.

The role of international aid institutions is crucial. They will be further discussed in the next section.

There were strong feelings within the group about the debt issue. This was seen by many as a major hindrance to development, and thus reinforcing major ecological disasters, such as further deforestation in rain forests. A deeper analysis would probably show the relationship to differ significantly from country to country, but there is hardly any doubt that the extremely high debt service burdens of many countries are adding to the hardships that sustain poverty, degradation, and thus adverse environmental effects.

Finally, the group spent considerable time discussing the linkage between security issues in the Third World and climate change. It was generally seen that continued poverty and degradation would lead to extended social instability and security risks. After all, peace is strongly correlated with democracy and democracy with industrialization, acceptable standards of living and reduced social inequities. Climate change was seen as a possible factor in environmental degradation that would lead to continued poverty and thus constitute a security risk. It was added that environmental degradation would lead to increasing numbers of refugees, so-called "environment refugees." There was no agreement on the extent to which this was a major addition compared to already existing causes of poverty.

INTERNATIONAL INSTITUTIONS, ORGANIZATIONS, AND REGIMES

The group discussed three different sets of issues: the existing international organizations, possible CO_2-reduction agreements, and verification issues.

Generally it was felt that the existing organizations, particularly when seen as a network, are by and large sufficient for the foreseeable future. The Montreal Protocol, with its related funding mechanism, was seen as a demonstration of adaptability.

The European Community and European Free Trade Area cover all different aspects of CO_2-related issues on the western European scene and have already agreed to develop a common climate-change policy. The framework is there. The western European integration is also the natural starting point for more extensive agreements on aid and development and joint industrial policies with respect to the energy sectors between western and eastern Europe as well as the USSR.

The OECD is the meeting place for several crucial issues, including agreements between industrialized countries as well as coordinated policies with respect to development and aid to the Third World. A series of issues have to be agreed upon between the industrialized countries.

The Economic Commission for Europe (ECE) consists of Europe plus the U.S. and Canada. It is a meeting place for agreements on, e.g., coal use and policies.

The new European Bank for Reconstruction and Development has an important role to play in the economic modernization of the eastern European countries and the USSR. Here again the energy sector and, in particular, coal and natural gas utilization become crucial.

The International Monetary Fund and the World Bank have crucial roles to play with respect to development and climate change. The industrialized countries should integrate climate-change issues into their development policies, as reflected in the governance of the institutions.

Several United Nations agencies (such as the Food and Agriculture Organization, United Nations Environmental Project, United Nations Development Project, and World Meterological Organization) have special roles to play. It might be preferable to have more clear-cut division of responsibilities between these and other agencies. Also, cooperation with the multilateral financial institutions should be strengthened. The Global Environment Facility is a case in point.

General Agreement of Trade and Tariffs (GATT) is a special case that warrants close examination. There has been considerable discussion of whether certain aspects of the Montreal Protocol are consistent with GATT rules. Trade in certain goods and services may well become contentious also from a climate-change perspective.

These international institutions are frequently large organizations. It was strongly felt within the group that there is a need for transparency and surveillance of their work. The many nongovernmental organizations that operate on a global scale, or out of Washington, D.C., are important in this. They raise awareness, they act as watchdogs, they alert their respective national member organizations to issues for which they should contact their respective national governments. Examples are the World Wildlife Fund, Greenpeace, Environment Defense Fund, Friends of the Earth, etc.

There are also other forms of international action. One of the more interesting ones concerns the way certain cities have taken to looking directly for inspiration and cooperation with like-minded cities in other countries. This ranges from twinning between eastern and western Europe, which probably will have considerable importance as channels of communications and knowledge, to cooperation between the larger western European cities on common issues such as transportation, etc. If cities act together, they can act more decisively when it comes to procurement and technical development. The World Health Organization project "Healthy Cities" was also mentioned. Finally, the case of the Danish/Schleswig-Holstein Brundtland cities were mentioned.

One important suggestion would be to attempt to extend or duplicate the CGIAR (the international centers for agricultural research) into similar energy centers with the particular mandate to work with renewable energy sources and other advanced technologies that might provide CO_2 low energy or energy services.

A possible CO_2 reduction regime requires considerable thought on goal formulation and agreements on joint reduction strategies. It was seen to be very important to look into different suggestions here, as well as into the lessons that, e.g., the SO_X agreement of the ECE or the large-scale combustion plant directive within the European Community could bring. Economic incentives such as tradeable permits or common carbon taxes should also be looked into. Flat rate reduction of, e.g., 25% across the board, are hardly reasonable. Countries with low levels of emissions and still developing are obviously in a different position than say fully developed countries with high levels of emissions. Countries with net immigration, such as Australia, would be at a disadvantage. Also, the special case of the inefficient coal producers in eastern Europe must be taken into account. It is very likely that the European Community will be the forward mover in Europe; it is the only organization which has managed to negotiate an agreement on differentiated emission control measures.

There was no strong support for a Climate Fund, modeled upon the chlorofluorocarbon fund of the Montreal Protocol. The existing development funding mechanisms were, for the time being, seen as sufficient, although the levels of funding were not.

The third issue that was discussed was goal setting and verification. Here it was felt that much work needs to be done, and urgently. Some countries, notably the F.R. Germany and Holland, have come the furthest. Precise agreements were seen as desirable. Transparency of emission data is important, and it may be possible to borrow from the experiences of the disarmament negotiations in Europe. CO_2 verification issues can hardly be more complex than that. It was stated, however, that there are serious problems with the monitoring of the existing agreements on SO_X reductions in the ECE which should be looked into so as to be avoided in the next stage.

METHODOLOGIES

The group spent considerable time discussing methodologies in analyzing CO_2 reduction strategies and robust climate-change policies. This was seen to be an important issue since there are strong requests from some groups that climate-change strategies have to be "properly costed" before nations can commit themselves to far-reaching international agreements. Considerable skepticism was voiced over the traditional methods of cost/benefit analysis.

First, it was strongly felt that attention should be given to both value conflicts and action conflicts. This was partially related to the classic discussion within cost/benefit analysis of options versus existence values. Cost/benefit analysis was not seen to be relevant for analyzing the "costs" or the "benefits" of decisions with irreversible consequences, such as loss of a species or the survival of the small island states.

Second, there were serious problems with time scales. The long time periods involved made discounted values meaningless. The sheer size of the uncertainty, sometimes to the point of direct ignorance, creates problems. Agriculture in fifty years' time was mentioned as one example. The problems of estimating costs and benefits were practically insurmountable in several cases.

Third, cost/benefit analysis was seen to be a possible tool for certain, more limited studies, but needed to be complemented with other methods and conceptual approaches. Experiences from risk analysis have shown that the most important differences between, say, different energy systems did not lend themselves to quantification. Comparing large-scale, longer-term use of coal with nuclear energy used on a similar scale would require weighing very different types of risks. Similarly, the experiences with Environment Impact Statements show that such statements differ considerably between different decision-making contexts. The Canadians have had considerable experiences in relation to large-scale development proposals, such as the Mackenzie River Delta. It was also felt that analyses needed to include distributional aspects.

Nevertheless, the need for more analysis of the possible response to climate change was noted. It was argued that this might be more feasible if done on a regional basis. For several reasons, the IPCC might be such a forum. One is that the impact studies so far are very rudimentary. They should be seen as illustrations of what might happen rather than predictions. Some conclusions from the climate models are relatively robust, but regional weather patterns are not. Another reason is that the whole question of winners and losers are much more visible on the well-defined regional level. Implications for sea-level habitats are a case in point.

A rather special but important reason for regional and collaborative research is the need to anchor firmly conclusions in several countries. Even scientists, not to speak of policymakers, have a certain cultural bias. The U.K. appears to have had considerable problems accepting the Scandinavian claim of acid rain and policymakers did not appear to accept opinions from non-British scientists. The continental Europeans had trouble accepting U.S. ozone depletion studies in the 1970s, etc. Third World countries frequently have problems accepting results about deforestation, etc. Few developing countries have the research base to develop their own views independently. Brazil and India are notable exceptions.

Author Index

Ahuja, D.R.   535–559
Arrhenius, E.   513–534
Ashton, W.B.   303–326, 463–474
Ausubel, J.   513–534
Bashmakov, I.A.   59–82, 463–474
Benedick, R.E.   513–534
Boyle, S.T.   229–260
Büttner, S.   463–474
Caccia, C.L.   603–615
Cantor, R.A.   165–188
Chandler, W.U.   13–58, 463–474
Davidson, O.   513–534
Delene, J.G.   165–188
Edmonds, J.A.   13–58
Fritsche, U.   603–615
Fulkerson, W.   165–188, 229–260
Gleick, P.H.   561–572, 603–615
Hilse, H.   603–615
Jain, A.K.   513–534
Jenny, L.L.   283–302, 463–474
Jochem, E.K.   327–349, 463–474
Jones, J.E.   165–188
Kane, R.L.   189–227
Kim, J.-H.   513–534
Klingholz, R.   229–260
Lönnroth, M.   603–615
Lovins, A.B.   351–442, 463–474
Lovins, L.H.   351–442
McDonald, S.C.   303–326
Meyer-Abich, K.M.   573–601, 603–615
Mintzer, I.M.   83–109, 229–260
Munasinghe, M.   487–511
Nicholls, A.K.   303–326
Pearman, G.I.   1–12, 229–260
Perry, A.M.   165–188
Peter, T.   603–615
Pinchera, G.   229–260
Rayner, S.   513–534
Reilly, J.   111–123, 229–260
Richels, R.G.   463–474
Schipper, L.   443–460, 463–474
Schlomann, B.   513–534
Simonis, U.E.   603–615
South, D.W.   189–227
Staiß, F.   229–260
Swart, R.J.   229–260
Tirpak, D.A.   535–559, 603–615
Ueberhorst, R.   513–534
von Moltke, K.   475–485, 513–534
Walker, I.J.   603–615
Winter, C.-J.   135–163, 229–260
Wuebbles, D.   13–58
Yamaji, K.   261–281, 463–474

Subject Index

100–200 MW, Program 156
1,000 Rooftops Program 156

acid rain reduction 607
Advanced Liquid Metal Reactor (ALMR) 174, 175, 176–7, 179
advanced nuclear reactors 174–9
advanced storage batteries 269
aerosols 37
African Academy of Sciences 526, 529
aggressive response scenario 83, 85, 86–8, 101–9, 300
agrichemicals 405–6
agriculture
    benefits of climatic change 586–7
    burning of residues 553
    cattle, methane emissions and 30
    climatic stablization and 376–8, 405–8
    crop residues, carbon dioxide emission of 27
    disadvantages of climatic change 587–8, 590
    livestock
        feed additives 551
        feedlot waste management 551
        biogas plants 551–2
        management 551–2
        production, energy-efficient 378–9
    low-input sustainable 379–82
    organic 380–1
    tools 417
    vegetarianism 552
agroforestry 382–5
    see also agriculture: forestry
air conditioning 274
air pollution 86
    emitted by transporation 288
    role in MWC scenario 99–100
    smog reduction 607
airborne fraction model (AFM) 42
aircraft energy efficiency 363–4, 449–50
Antarctic Treaty System 477
atmosphere
    changing composition of 3
    processes 46–9
atmospheric decay model (ADM) 42, 45
atmospheric fluidized-bed combustion (AFBC) 205, 214–16
automobile transport
    atmospheric contaminants from 288–9
    car ownership 447
    combustion engines 296–7
    driving restrictions 549
    effect on global warming 290
    efficiency 359–65, 388, 448, 547
    electric vehicles 293, 299–300
    emissions 287–9
    emission controls 291, 294, 547
    energy use 287, 336
    evolution of 283–6
    fuels 237–8, 295–6, 548
        alternative energy sources 292–3
        conservation 292, 294
        hydrogen 293, 297–9
    implications for future action 289–90
    long-term outlook 300–1
    near-term actions 290–6
    numbers of vehicles 86
    pattern of energy use
        in aggressive response scenario 88
        in business-as-usual scenario 85–6
        role in MWC scenario 104

Subject Index

automobile transport (*cont.*)
  technological improvements 291–3
  transportation system
    characteristics 300

Balbina dam, Brazil 576
barriers
  to implementation of energy-
    efficient technologies/practices
    in buildings 319–20
  policies to overcome barriers
    320–4
    economic incentives 320–1
    education and information
      322–3
    new markets or market
      instruments 321–2
    regulation 322
    research, development, and
      demonstration 323
    social engineering policies
      323–4
  to reducing carbon dioxide
    emissions—513–33
    diffusion of technology 527–30
    economic distortions 522–4
    lifestyle 530–2
    organizational design 525–7
    plural rationalities 514–17
    processes of consensus
      formation 517–20
    technological leapfrogging
      527–30
    time horizons 520–2
  to reducing greenhouse gas
    emissions in developing
    countries 487–510
batteries
  advanced storage 269
  for electric engines 299–300
"best buys first", need for 393–4
biomass
  carbon dioxide emission of 27
  combustion, methane emissions
    and 30, 32
  global consumption 69
  in industrial energy budget 72
  residential and commerical energy
    consumption 73
  terrestrial, carbon emissions 27–8,
    167

transport energy consumption 72–3
  in WRC scenario 106
biomass-burning cookstoves 553
biomass converters, unit output of
  148
box-diffusion models 43–4
Brayton cycle engines 297
building materials 387–8
buildings
  energy efficiency 335, 343–4,
    452–6
  energy savings in 356, 357, 365–6,
    391
  greenhouse-gas emissions from
    303–24
    global energy use in, and
      304–11
    complementary policies, need
      for 324
    consumption by end-use
      309–11
    determination of 306–9
      building operations 308
      characteristics of the building
        stock 307
      demand for energy services
        306–7
      operating practices 307
      principal drivers 308
      wasteful practices 308
    policies to reduce emissions
      316–24
      barriers to implementation of
        energy-efficient
        technologies/practices 319–20
      policies to overcome barriers
        320–4
        economic incentives 320–1
        education and information
          322–3
        new markets or market
          instruments 321–2
        regulation 322
        research, development, and
          demonstration 323
        social engineering policies
          323–4
    technical options to reduce
      emissions 311–16
      building design, siting and
        construction 313–14

Subject Index 621

equipment selection, operation and maintenance 314
fuel substitution 314–15
identifying high-priority technologies and practices 315–16
restructure aggregate building energy-service demand 312–13
heating 315, 452–3
business-as-usual scenario 83, 84, 85–6, 95–101, 123

Carajas mining project 576
carbon
  atmospheric fluxes 40–1
  emissions
    anthropogenic net, and atmospheric carbon dioxide accumulation 40–1
    from combustion of fossil fuels 593
    instruments for control 121–2
    in MWC scenario 107
    targets for 121–2
    from terrestrial ecosystems 37–8
  missing 40
  ocean uptake 53
  sinks for 40
carbon cycle 37–46, 53
  background on 39–40
  global 39
  models 114
carbon dioxide 3
  atmospheric chemistry as source of 28
  atmospheric concentrations of 24–8
  atmospheric fluxes 40–1
  and climate change 112
  contribution to greenhouse effect 191
  in developing countries 193
  fossil-fuel combustion and 193–5
  global emission, growth of 78, 79, 80
  historical emissions 26
  infrared radiation absorption of 16, 17
  lifetime of, in atmosphere 44–6

  models of atmospheric retention of 41–4
  removal, recovery and disposal of 220–5
  role in MWC scenarios 93–4, 98
  role of nuclear power in controlling emissions 165–86
  sources of 25
  strategies for controlling 166–7
  targets 2
  see also carbon dioxide emission reduction
carbon dioxide emission reduction 177–85
  barriers to 513–33
  choosing R&D strategies for 252–4
  costs of 111, 245–8, 432–3
  feasibility evaluations 248–9
  fossil-energy options 203–25
  hardware approaches to 233
  implementation mechanisms 244–56
  international policy instruments and institutions 254–6
  options for reducing 7–8, 10
  role in MWC scenario 98
  strategies 603–15
    acid rain and smog reduction 607
    coal, phasing out, in Central European coal belt 607
    development policies 610–11
    energy efficiency policies 608–9
    European security 608
    international institutions, organizations and regimens 612–14
    methodologies 614–15
    ozone depletion policies 607
    policies and cultural patterns that stimulate energy use 609–10
    technical feasibility, definition 231
    technical limits, definition 231
    technology transfer in developing countries 250–2
    in transport 237
carbon dioxide fertilization effect 28, 44, 117
  extent of 129
carbon dioxide scrubbing 220–2, 225

carbon monoxide 37, 53
  emissions, historical 38
  role in MWC scenario 99
carbon tax 113, 255, 415–16, 422
carbon tetrachloride 35
cement manufacture, carbon dioxide
    emissions and 25
CFCs 3, 35–7, 549
  applications 36–7
  infrared radiation absorption of 17
  phase-out, US 556
  photodissociation of 46
  reduction, costs 128
  role in MWC scenarios 94, 98
  substitution 385–6, 408–19
CFC-11 35
  global production 36
  infrared radiation absorption of 17
  role in MWC scenario 98, 100,
    108
CFC-12
  global production 36
  infrared radiation absorption of 17
  radiative forcing of 35
  role in MWC scenario 99, 100,
    108
CFC-13 35
CFC-22 35
CFC-113 35
Chernobyl nuclear power disaster
    78, 156, 167, 170
chlorofluorocarbon gases see CFCs
Clean Air Act (US) 555, 556
clean coal technologies (CCTs) 190,
    205–6, 225
climate change
  advantages of 586–7
  "business as usual" societal changes
    and 4, 6
  and carbon dioxide 112
  climate forcing 3–4
  disadvantages of 587–9, 590–2
  global, principal uncertainties
    191–3
  impacts of 565
  implications of, for humanity
    578–92
    ecology 580–2
    industrialized countries 585–9
    Third World 589–92
  and international peace 561–71

  predictions 4–6
  regional 5
  and trace gas emissions 112
  see also climatic stabilization, least-
    cost
Climate Change Convention 256
Climate Protection Fund 423, 613
climatic stabilization, least-cost
    351–435
  benefits 427–31
  implementation techniques
    394–427
    CFC substitution 408–19
    in developing countries 417–19
    energy 396–405
    farming and forestry 405–8
    in Formerly Centrally Planned
      Economies (FCPES) 408–17
    institutional parallels 421
    lending institutions 421–4
    marketing for diversity 419–21
    OECD countries 394–5
    technology transfer 424–7
  major abatement terms 352–4
  money savings of 351–2
  myths about abating global
    warming 432–5
  technological and economic
    options 354–94
    "best buys first", need for
      393–4
    CFC substitution 385–6
    conversion and distribution
      efficiency 372–3
    electricity 355–9
    energy efficiency 354–5
    farming and forestry 376–8
    livestock 378–9
    low-input sustainable
      agriculture 379–82
    multigas abatements 386–8
    oil 359–70
    renewable energy sources 373–6
    supply curves for abatement
      388–93
    sustainable forestry 382–94
closed-loop energy systems 138–9
cloud feedback 22, 53
coal
  advanced combustion and flue-gas
    desulfurization 213–14

carbon content per unit energy 27, 207
consumption
  China 418
  global 69
  conventional combustion and wet limestone flue-gas desulfurization 212–13
  fuel switching coal to gas 540–1
  fuel switching coal to oil 541
  global reserves of 198–203
  in industrial energy budget 72
  methane emissions and 30, 32
  in MWC scenarios 97
  phasing out, in Central European coal belt 607
  projected demand in developing countries 202, 204
  residential and commerical energy consumption 73
  substitution with natural gas or oil 167, 207–9
  trade 75
  transport energy consumption 72–3
  in WRC scenario 106
coal-carbon content per unit of energy 27, 207
coal-fired power plants
  costs 151
  yield factor 142
combined cycle gas co-generation plants 232
combustion engines 296–7
compact fluorescent lamps (CFL) 541
compressed air energy storage (CAES) 269
compressed natural gas (CNG) 548
conservation, energy see energy conservation
Convention on Long-Range Transboundary Air Pollution, Geneva 477, 484
Council of Academies of Engineering and Technological Sciences 526
coupled atmosphere, ocean–ice–land, climatic system 23
crop residues, carbon dioxide emission of 27
cultural theory 515

Dahlem Conference, aim of 7–10

deforestation
  benefits of halting 242
  carbon dioxide emissions and 25, 28
  estimates 242
  in MWC scenarios 97
  need for policies to limit 243
  reducing 552
  as source of carbon emissions 38
diesel engines 548
dioxin 550
Doppler broadening 15
Drive+ feebate proposal 399–400
dung, carbon dioxide emission of 27
dynamic general equilibrium models of the economy 114

ECE 612
Edmonds–Reilly model 84, 89, 91–2, 178, 201, 248
EFTA 612
electric advanced heat pumps 276
electric appliances
  home, efficiency 273–6, 454–6
  standards, US 555
electric end-use technologies 270–1
electric furnaces, industrial 278
electric lighting efficiency 271–3
electric motor vehicles 293, 299–300
electric motors 276–8
Electric Power Research Institute 352
electric utilities
  climatic stabilization and 396–9
  hydrocarbon fuels 368
electricity
  and climatic stabilization 355–9
  co-generation systems 270
  efficiency in use 270–8, 355–9
  generation
    carbon dioxide emissions in 151, 154, 195, 234–6
    demand for 178
    efficiency in 261–80
    losses in transmission and distribution systems 268–9
    global primary energy consumption 69
    possible contributions of nuclear power to 178–85

electricity (cont.)
　　prospects for realizing the
　　　　technical potentials 279–80
　　role of 262–4
　　technical potential for efficiency
　　　　improvements 264–70
　　impact of price structures on energy
　　　　conservation 341
　　in industrial energy budget 72
　　in MWC scenarios 96
　　needs of developing countries
　　　　490–3
　　residential and commercial energy
　　　　consumption 73
　　pattern of energy use
　　　　in aggressive response scenario 88
　　　　in business-as-usual scenario
　　　　　　85–6
　　savings 388, 391
　　transport energy consumption 72–3
　　in WRC scenario 104–5
electricity storage systems 269
electro-arc induction 235
end-energy economy 145–8
energy
　　challenge of the future 79–81
　　and the environment 77–9
　　pricing 76–7, 466–7
　　　　in MWC scenarios 96
　　　　in WRC scenario 106
　　production
　　　　developments in 136–8
　　　　methane emissions and 30, 32
　　sources 373–8
　　see also renewable energies
energy balance, global 15, 16
energy budget
　　of residential and commerical
　　　　sector 75
　　world, structure of 68–76
energy conservation 27, 541
　　banks 471
　　definition 464
　　law of 68
　　policy 338–48
　　potentials 65–6
energy consumption
　　commercial 61, 68
　　in developing countries 193–5
　　and economic growth 63–8
　　global distribution 64, 69

　　historical analysis 61–3
　　industrial 71–2
　　mechanism of variations 61
　　non-energy uses 75, 76
　　primary
　　　　history and forecasts 60–1
　　　　per capita 64–5
　　　　residential and commercial
　　　　　　sector 72–3, 75
　　　　transport 72, 73
　　　　world (1860–1990) 63, 68–9
energy conversion chains 139, 140
　　central primary/secondary 144–5
　　local (decentral)
　　　　primary/secondary 144
energy-development forecasts,
　　historical 59–63
energy efficiency
　　characteristics 464
　　and climatic stabilization 354–5
　　costs 464–5, 495
　　policies for reducing carbon dioxide
　　　　emissions 608–9
　　regulatory standards 468
　　in the Soviet Union 412–13
　　see also energy use efficiency;
　　　　energy use reduction
energy efficiency technologies 354–5
energy–environmental issues,
　　economics of 493–8
　　energy efficiency 495
　　nonquantifiable national
　　　　environmental costs 495–6
　　quantifiable national environmental
　　　　costs 495–6
　　policy measures for the LDCs
　　　　497–8
　　transnational environmental costs
　　　　496–7
energy gain factor 143
energy/GDP ratio, 61, 62, 463–73
energy intensity, rates of economic
　　growth and 67–8
energy payback time 142, 143
energy use efficiency 327–48
　　capital replacement 467
　　conservation policy 338–48
　　　　international policy aspects
　　　　　　345–8
　　　　obstacles and market
　　　　　　imperfections 339–42

disparity in rate-of-return
  expectations 340–1
impact of electricity and gas
  price structures 341
improved energy efficiency,
  strategic policies 342–5
lack of access to capital and
  investment patterns 339–40
lack of knowledge, know-how
  and technical skills 339
legal and administrative
  obstacles 341–2
separation of expenditure and
  benefit 340
energy savings
  conversion and distribution
    efficiency 372–3
  by improved energy
    efficiency 334–5
  proved technical and economic
    potentials 331–4
  by reducing energy services
    337–8
  by reducing specific useful
    energy demand 335–7
in the US and wealthy countries
  443–58
initiatives, US 555
international improvements 458
measuring 444
sectoral comparisons 446–57
  manufacturing 450–2
  residential sector 452–6
  services sector 456–7
  transporation 447–50
energy use reduction
  options available for 463–73
  communicating opportunities for
    energy efficiency 472–3
  energy efficiency, potential
    464–5
  modeling the potential 469–70
  overcoming barriers to more-
    efficient energy use 465–8
  transferring technology for energy
    efficiency 470–2
Enquête-Kommission 338, 519, 526
Environment Defense Fund 613
EPA energy model 248
equities and inequities, implications
  429–30

political responses to climate
  change 592–8
climate change
  for humanity 582–5
  for industralized countries 585–9
  for the Third World 589–92
  on ecology 580–2
present intergenerational 578
present international 573–8
  population growth 575–6
  indebtedness 576–8
ethanol 548
Euratom 160
European Bank for Reconstruction
  and Development 612
European Commission 255, 612

farming *see* agriculture
fee-bates 322, 397, 399–401
fertilizers, reductions in 382, 552
firewood, carbon dioxide emission
  of 27
flue gas desulfurization systems 546
  and advanced coal combustion
    213–14
  wet limestone, and conventional
    coal combustion 212–13
food, impact of climatic change on
  565–6
Food and Agriculture Organization
  612
forestry 376–8, 552–3
  climatic stabilization and 405–8
  conservation, potential to reduce
    carbon dioxide accumulation
    241–4
  cost-effective management
    practices 243
  national forest plans 243–4
  sustainable 382–94
  urban 384, 552–3
Formally Centrally Planned
  Economies (FCPES) 408–17
fossil containment 139
fossil fuels
  advanced fossil-energy
    technologies 212–20
  carbon dioxide emissions 193–5
  costs of reducing 124–7
  and use 25–7
  efficiency improvement 166–7

fossil fuels (*cont.*)
   global consumption   69
   methane emissions   30, 32
   options to reduce carbon dioxide
      emissions   203–25
   resource base   195–203
   as source of nitrous oxide
      emissions   32–4
   substitution between   207–12
   see also biomass; coal; natural gas
Friends of the Earth   613
fuel cell
   with coal gasification technology
      205–6
   with natural gas technology   205
fuel-cycle costing   249

gas *see* natural gas
gasifiers   241
gasoline   548
   prices   76, 78
General Agreement on Trade and
   Tariffs (GATT)   252, 255, 612
general circulation models (GCM)
   22, 114, 192
General Electric PRISM   175
global change policy   120–3
   bargaining, buyouts and unilateral
      action   122–3
   instruments   121–2
   no regrets policies   120–1
   scale of impact   121
   scale of action   121
   targets   121–2
global warming
   international response to   6–7
   mechanism   14–15
   myths about abating   432–5
   prediction   84
   scientific basis   2–6
   see also climate change; climatic
      stabilization, least-cost
global warming potential (GWP)   5,
   49–53, 117
   definition   4
greenhouse effect   14–24
   climate response   22–4
   direct radiative influence   15–19
   role of fossil-fuel emissions   190–1
greenhouse gases   19–21
   atmospheric composition of   46–9

contribution to global climate
   change   191
emission profiles   8
global warming potentials   5
heat-trapping mechanism of   14
historical and current atmospheric
   concentration of   4
infrared radiation absorption of
   15–18
primary anthropogenic sources of
   24
relationship between solar radiation
   and   14–15
relative importance of   3
see also greenhouse gas control
greenhouse gas control
   comprehensive assessment   537–40
   costs   493–8
   in developing countries   487–510
   economic incentives   503–6
   environmental costs   507–8
   evaluating   554–5
   examples of control   540–54
      energy recovery   550–1
      utilities sectors   540–6
   global environmental issues   508–10
   initiatives in the US   555–6
   multigas abatement   386–8
   national energy-environmental
      policy analysis and options
      498–508
   national level
      organizations/institutional
      options   506–7
   policy tools and constraints   500–1
   regimes for   475–84, 535–57
      current status   476–7
      international regime   479–80
      negative outcomes   570
      positive outcomes   569–70
      precedents   477–9
      regionalization   480–4
   supply curves   388–93
   technological options   501–3
Greenpeace   613

halons   35
heat pump systems
   electric advanced   276
   unit output of   148
heating

Subject Index

energy-efficient, in buildings, 315, 452–3
  in industrial energy budget 72
  residential and commercial energy consumption 73
high-temperature heat 366–8
Houston Economic Declaration 8
hydrocarbon economy 147
hydrochlorofluorocarbons 607
  role in MWC scenario 94, 98
  HCFC-22 35
hydroenergy 263
hydrofluorocarbons 607
hydrogen
  production 269
  as transport fuel 293, 297–9
hydropower 27
  effect of global warming on 591
  global consumption 69
hydroxyl radical in climate change 48–9
  production 49
  sinks for 49

ice-albedo feedbacks 22
inequities *see* equities and inequities
Institute of Nuclear Power Operations (INPO) 170
insulation, thermal 335, 386
Integrated gasification combined cycle (IGCC) 167, 205, 217
integrated national energy planning (INEP) 499–500, 501
Integrated Resource Planning 321
Intergovernmental Panel on Climate Change (IPCC) 2–3, 4, 189, 579
International Council of Scientific Unions 526
International Energy Efficiency 255
International Geosphere-Biosphere Program 526
International Law Commission 571
International Monetary Fund 255, 609, 612

Kohlepfennig subsidy 416

labor productivity, effect on GNP 90, 103

lamps 354, 420–1, 423, 541
land-use changes, carbon dioxide emissions and 25, 26, 27–8
landfill 387, 550–1
  methane emissions and 30, 32
  US 556
landfill gas plants 232
law of energy conservation 68
Least Cost Utility Planning 321
light water reactor (LWR) 174, 179
lighting
  capacity of devices 541, 546
  energy-efficient electric 271–3, 315
  lamps 354, 420–1, 420–1, 423, 541
liquefied natural gas (LNG) 265
livestock
  feed additives 551
  feedlot waste management 551
  biogas plants 551–2
  management 551–2
  methane emissions and 30
  production, energy-efficient 378–9
London Ozone Convention 529
London Protocol 476, 477, 556
Lorentz line wings 17
low-temperature heat 365–6
LUZ-solar electricity generating systems (SEGS) 150

Mackenzie River delta 614
magnetohydrodynamic (MHD) generation 205, 206, 219–20, 265
manufacturing, energy efficiency in 450–1
methane 3, 29–32
  atmospheric concentrations 29
  emissions
    from coal 540–1
    global, growth of 79, 80
    historical 31
    sources 29–30
    from landfill 550
  infrared radiation absorption of 17, 49
  radiative forcing of 18
  recent trends 30
  role in MWC scenarios 94, 99, 100, 108
methane cycles 381
methanol 237–8, 548–9
methyl chloroform 35

mineral resources, impact of climatic change on 567
missing carbon 40
Model of Warming Commitment (MWC) 84
  input assumptions for 88–95
    human factors 89–91
    energy technology 91–3
    other greenhouse gases 93–5
  model results 95–109
    aggressive response scenario 101–9
    business-as-usual scenario 95–101
Modular High-temperature Gas-Cooled reactor (MHTGR) 174–5, 177, 179
molten carbonate fuel cell (MCFC) 265
Montanunion 160
Montreal protocol 94, 98, 108, 255, 476, 477, 482, 509, 556, 612, 613
Moonlight Project 276
motor vehicles see automobile transportation sector
multi-fired boilers 207
multi-lateral development bank (MDBs) 255
multigas abatements 386–8

national forest plans 243–4
natural gas
  carbon content per unit energy 27
  compressed (CNG) 548
  global consumption 69
  global reserves of 195, 197–8, 199
  impact of price structures on energy conservation 341
  in industrial energy budget 72
  methane emissions and 30, 32
  residential and commercial energy consumption 73
  substitution of coal by 167
  trade 75
  transport energy consumption 72–3
  in WRC scenario 105–6
natural-gas liquids (NGL) 370
nature myths 515
nitrogen dioxide
  biotic emissions in MWC scenario 107

  cycles 381
  infrared radiation absorption of 17
  photodissociation of 46
  in WRI scenario 98, 99
nitrous oxide 3, 32–5
  concentrations 32, 33
  emissions
    historical 34
    human activities associated with 32–3
    sources 32
no regrets actions 129, 432
no regrets policies 120–1, 253, 520, 536
nonfossil fuels, limitations of 166
nongovernmental organizations (NGOs) 255
nuclear containment 139, 144
nuclear energy 27, 136, 263, 301
  advanced reactors 174–9
  global consumption 69
  in MWC scenarios 97
  potential role in controlling carbon dioxide emissions 165–86
  probability of nuclear accident 171
  risk of proliferation for weapons 171–2
  scenarios 179–80
  total social costs of 168–77
    indirect 168–72
    total direct 172–7
  uranium resources and requirements 172, 183–4
  worldwide enterprise for 185
  yield factors 142–4
nuclear reactors, advanced 174–9
Nuclear Regulatory Commission 170
nuclear waste disposal 185
  entombment 175–6
  monitored retrievable storage 176
  reprocessing costs 176

Oak Ridge Industrial Model 248
ocean–atmosphere interactions 22–3
ocean carbon processes, models of 43–4
ocean thermal energy conversion 157
OECD 255, 612
  climatic stabilization and 394–5
oil
  carbon content per unit energy 27

climatic stabilization and 359–70
efficiency 399–402
electric-utility hydrocarbon fuels 368
global consumption 69
global reserves remaining 195, 197, 198, 199
high-temperature heat 366–8
in industrial energy budget 72
low-temperature heat 365–6
miscellaneous uses 368–9
potential for saving 359–70
production in MWC scenarios 96
prices 76, 431
residential and commercial energy consumption 73
as security 429
trade 75
transport energy consumption 72–3, 359–65
in WRC scenario 104
open-loop energy systems 138–9
Operational Biomass Burning Systems 243
ozone 3
depletion policies 607
infrared radiation absorption of 17
radiative forcing of 18
Ozone Defense fund 509

paraboloid/Stirling units, unit output of 148
photovoltaic generators 374
cost trends for 148, 149
production 136
polluter pays principle 422, 592, 593, 599
pollution
air 86, 99–100, 288, 607
effect on energy utilization 77
population
distribution 85, 87
effect on GNP 90
growth, assumed
for aggressive response scenario 89, 103
for business-as-usual scenario 89, 103
world, 59
pressure broadening 15

pressurized fluidized-bed combustion (PFBC) 205, 216–17
pressurized water reactor (PWR) 172
probabilistic risk assessments (PRA) 171
procyclical energy demand policy 329
pulverized coal/flue-gas desulfurization 205

radiative forcing 17–19, 111, 113
definition 18
railway equipment efficiency 364–5
rankine cycle engines 297
rational energy use *see* energy use, rational
recycling 336, 387
reforestation
US 556
in WRC scenario 107
see also forests
refrigerators 423
Regional Seas Program 255, 477
renewable energies 135–61, 373–8, 546
costs 151
definition 135
domestic and imported 144–5
and economics 148–53
future role of 153–9
15 years 153–7
50 years 157–8
100 years 158–9
unlocking energy 139–44
unsolved problems 160–1
yield factor 142
Response Strategies Working Group 554
rice production
irrigated, promotion of 554
methane emissions and 30

scale of action 121
scale of impact 121
sea-level rise 582
impact of climatic change on 428–9, 567

security
  climatic changes and 563–7
  international, environment and 561–3
  reducing risk of conflict over climatic change 570–1
  winners and losers 567–9, 573–99
services sector, energy efficiency in 456–7
ship efficiency 365
Sizewell pressurized water reactor (PWR) 172
soil erosion 377
solar central receiver technology 148, 151, 152
solar containment 144
solar energy 14–15, 27, 136
  yield factor 142
Solar Energy Agency 255
solar power plants 139, 141, 142
SolarEurop 160
solid oxide fuel cell (SOFC) 265
Southern California Air Quality Management Plan 238
steel production, energy efficiency 450–1
Stirling engines 297
subsidies 609
  coal, W. Germany 415–16
  energy 410
  role of 248
sulfur dioxide emissions 493
superconductive magnetic field energy storage (SMES) 269
Swedish Process Inherent Ultimate Safety LWR 174, 175, 179, 183

taxes
  carbon 113, 255, 415–16, 422
  energy 609
  incentives for energy efficiency 467–8
  World Resource 255
technology transfer 88, 103, 482–4
  benefits and disadvantages 424–7
  in developing countries 250–2
  for energy efficiency 470–2
Third World Academy of Sciences 526
Three Mile Island accident 170–1
trace gas indices

  comparison of 50, 52, 119
  damage function parameters 118
  economic interpretation 112–15
  estimated 115–20
trace gases
  characteristics of 116
  concentrations, costs of limiting 123–8, 129
  emissions
    and climate change 112
    unilateral action for reducing 122–3
transport
  alternatives to 363
  energy use efficiency 447–50
  global energy consumption 72–3
  oil and 359–65
  potential role of electricity in 236
  reducing carbon dioxide emissions in 237
  reducing greenhouse-gas emissions 283–301
  systems and structures 238–41
  see also aircraft; automobile transport; railway
Treaty of Rome 255
Tropical Forest Action Planning 244

United Nations Center for Trade and Development (UNCTD) 252
United Nations Conference on Environment and Development 7
United Nations Conference on the Law of the Sea, Third 478
United Nations Environment Program (UNEP)-IIASA study 579, 585–6
United Nations Intergovernmental Panel on Climate Change (IPCC) 2–3, 4, 189, 579

vegetarianism 552
Villach report 2, 4

waste disposal
  nuclear 175–6, 185
  in Third World 577–8

water
  greenhouse feedback 22
  impact of climatic change on 566–7
  infrared radiation absorption of 16, 17
  water-use efficiency effect 129
  wetlands 553–4
  whole tree burner concept 241
  wind converters, unit output of 148
  wind energy 136
  wind parks 232
  window region 17

World Association of Nuclear Operators (WANO) 170
World Bank 255, 471, 490, 609, 612
World Climate Conference, Second (Geneva) 3, 9
World Climate Program 525
world energy policy 159–60
World Health Authority (WHO) 255, 613
World Meteorological Organization (WMO) 579, 612
World Resource Tax 255

Index compiled by Annette J. Musker